Recent Developments in Petroleum Science

Recent Developments in Petroleum Science

Edited by Eduardo Hanks

SYRAWOOD
PUBLISHING HOUSE

New York

Published by Syrawood Publishing House,
750 Third Avenue, 9th Floor,
New York, NY 10017, USA
www.syrawoodpublishinghouse.com

Recent Developments in Petroleum Science
Edited by Eduardo Hanks

International Standard Book Number: 978-1-68286-743-3 (Hardback)

Cataloging-in-Publication Data

Recent developments in petroleum science / edited by Eduardo Hanks.
     p. cm.
Includes bibliographical references and index.
ISBN 978-1-68286-743-3
1. Petroleum. 2. Petroleum engineering. 3. Petroleum--Geology.
4. Petroleum--Prospecting. I. Hanks, Eduardo.
TN870 .R43 2019
665.5--dc23

TABLE OF CONTENTS

PREFACE

Petroleum science is a broad term that integrates the principles of petroleum engineering, petroleum geology and exploration geophysics. Petroleum geology studies the origin, movement and exploration of hydrocarbon fuels. It evaluates different elements of sedimentary basins such as reservoir, seal, trap, timing, maturation and migration. Geophysical survey data is applied in exploration geophysics to analyze potential reservoirs of petroleum. Methods used for finding such deposits include seismic reflection, well logging, remote sensing techniques, etc. Petroleum engineering is a field of engineering that deals with the production of crude oil or natural gas. It focuses on the optimal recovery of hydrocarbons from reservoirs. This book is a valuable compilation of topics, ranging from the basic to the most complex advancements in the field of petroleum science. It presents this complex subject in the most comprehensible and easy to understand language. With state-of-the-art inputs by acclaimed experts of this field, this book targets students, researchers, experts and professionals.

All of the data presented henceforth, was collaborated in the wake of recent advancements in the field. The aim of this book is to present the diversified developments from across the globe in a comprehensible manner. The opinions expressed in each chapter belong solely to the contributing authors. Their interpretations of the topics are the integral part of this book, which I have carefully compiled for a better understanding of the readers.

At the end, I would like to thank all those who dedicated their time and efforts for the successful completion of this book. I also wish to convey my gratitude towards my friends and family who supported me at every step.

Editor

Forecasting of China's natural gas production and its policy implications

Shi-Qun Li[1] · Bao-Sheng Zhang[1] · Xu Tang[1]

Abstract With the vigorous promotion of energy conservation and implementation of clean energy strategies, China's natural gas industry has entered a rapid development phase, and natural gas is playing an increasingly important role in China's energy structure. This paper uses a Generalized Weng model to forecast Chinese regional natural gas production, where accuracy and reasonableness compared with other predictions are enhanced by taking remaining estimated recoverable resources as a criterion. The forecast shows that China's natural gas production will maintain a rapid growth with peak gas of 323 billion cubic meters a year coming in 2036; in 2020, natural gas production will surpass that of oil to become a more important source of energy. Natural gas will play an important role in optimizing China's energy consumption structure and will be a strategic replacement of oil. This will require that exploration and development of conventional natural gas is highly valued and its industrial development to be reasonably planned. As well, full use should be made of domestic and international markets. Initiative should also be taken in the exploration and development of unconventional and deepwater gas, which shall form a complement to the development of China's conventional natural gas industry.

Keywords Natural gas · Production forecast · Generalized Weng model · Energy structure · Policy implication

✉ Bao-Sheng Zhang
  bshshysh@cup.edu.cn

[1] School of Business Administration, China University of Petroleum, Beijing 102249, China

Edited by Xiu-Qin Zhu

1 Introduction

Since 2000, China's natural gas production and consumption have both maintained rapid growth, with an average annual growth rate of 11.9 % and 15.6 %, respectively. In 2014, China's natural gas production reached 134.5 billion cubic meters, consumption totaled 185.5 billion cubic meters (BP 2015), which shows that China's natural gas industry has entered a rapid development stage. All these are largely attributed to China's vigorous promotion of energy conservation and implementation of a clean energy strategy, as well as the clean and efficient characteristics of natural gas itself. China's natural gas industry has entered a "golden period" (Qiu 2008), and natural gas will play an increasingly important role in China's energy structure. Therefore a reasonable and accurate forecast of natural gas production will greatly assist the healthy development of the natural gas industry.

Many scholars have conducted research into energy forecast models. Jebaraj and Iniyan included in detail all available models (Jebaraj and Iniyan 2006) and divided them into major categories such as energy planning, energy supply and demand, energy forecast, and optimization, whereby a full picture of developments in energy demand and supply models was fairly presented. This sets the basis for this paper's overview of energy forecasting models. With regard to natural gas forecasting models, scholars have used methods like Artificial Neural Network (ANN) and scenario analysis to forecast natural gas consumption in areas such as Europe, Asia–Pacific, and East Africa (Szoplik 2015; Aguilera et al. 2014; Demierre et al. 2015) and countries like Italy, Croatia, and Turkey (Taşpınar et al. 2013; Bianco et al. 2014; Primoz et al. 2014). While the most famous forecasting model abroad for natural gas is still the Hubbert model. The United States renowned geologist Hubbert in 1956 put forward that the rate of mineral resources production over time would resemble a bell curve (Hubbert 1956), based on which the Hubbert model was proposed to forecast oil and gas production (Hubbert 1959). The model accurately predicted the oil and gas production in the Lower 48 US states and has become the main model for oil and gas production forecasts in foreign countries. All the models afterward were actually updates and improvements of the Hubbert model, among which the Multicyclic Hubbert model put forward by Al-Fattah and Startzman was the most well known (Al-Fattah and Startzman 1999), and was utilized in natural gas production forecast globally with great effect. In China, the famous geophysicist and academician Weng Wenbo, based on the theory that everything in the world follows the natural process of "rise-grow-mature-decay" (Tayfun 2007), proposed in 1984 the Weng model for oil and gas production forecast (Weng 1984), and for the first time predicted China's natural gas production. Since then, with the development of the geological theory inside China, oil and gas production forecasting models began to emerge continuously. Among them, oilfield statistics-based HCZ model (Hu et al. 1995) and the Generalized Weng model (Chen 1996), gained the most attention, and are currently used in a wide range in China's oil and gas production forecast; besides, there have also been studies (Feng et al. 2010; Boqiang and Ting 2012) in China on Multicyclic Hubbert models based on the research results of Al-Fattah and Startzman. It is to be noted that some studies on China's natural gas production were based on combined forecasting (Yuan et al. 2007; Li et al. 2009), where predictions from different models were combined through optimization algorithms (Granger and Ramanathan 1984). From a mathematical point of view, combined forecasting could indeed promote forecasting accuracy (Wu 2007); while practically due to the uncertainty of every model's forecasting result, combing the forecasts is no doubt a superposition of all the uncertainties, leading to discrepancies with the actual production in oil and gas fields, thus indicating a limitation therein. Taking all the models above, this paper takes the Generalized Weng model to forecast the growth trend of China's natural gas production, based on which countermeasures and strategies for the natural gas industrial development are analyzed.

2 Methodology and data

2.1 Methodology

Table 1 shows the mathematical expressions of Hubbert, HCZ, and Generalized Weng models.

Table 1 Typical forecasting models for oil & gas production

Forecasting models	Basic equation	$N_{p\max}$ & N_R
Hubbert model	$Q = \dfrac{abN_R e^{-bt}}{(1+ae^{-bt})^2}$	$N_{p\max} = \frac{1}{2}N_R$
HCZ model	$Q = aN_R e^{\left[-(a/b)e^{-bt} - bt\right]}$	$N_{p\max} = \frac{1}{e}N_R$
Generalized Weng model	$Q = at^b e^{-t/c}$	NA

Q oil & gas annual output, N_R recoverable oil & gas resources, a, b, c constants, q_{max} production peak, t_{max} time of production peak, N_{pmax} cumulative production till peak production, $N_{p\max} = \int_t^0 \max Q dt$

As can be seen from the relationship between $N_{p\max}$ and N_R, production forecast by the Hubbert model assumes a strictly symmetrical production peak. The HCZ model expects the production peak will occur when cumulative production reaches 36.8 % (1/e) of estimated total recoverable resources; while for the Generalized Weng model, production peak and recoverable resources are not necessarily correlated. Therefore, for oil & gas fields that have entered the declining stage after 50 % of reserves is recovered, a good result can be achieved when using the Hubbert model to make production forecasts; for fields that decline occurs after 36.8 % of reserves is recovered, the HCZ model would bring about a relatively accountable result; while the Generalized Weng model demonstrates a strong flexibility in forecasting with no specific requirements on the timing of the oil & gas fields' decline, thus being more widely used in China. That is why this paper also adopts this model to make forecasts for natural gas production.

2.2 Interpretation of Generalized Weng model

(1) *Traditional interpretation* Suppose a sequence Q_t for historical natural gas production of a given oil & gas field:

$$Q_t = (Q_1, Q_2, \ldots, Q_n), \quad t = 1, 2, \ldots, n. \tag{1}$$

The cumulative production as of year m amounts to N_{pm}:

$$N_{pm} = \sum_{t=1}^{m} Q_t, \quad m \in [1, n]. \tag{2}$$

Then the Generalized Weng model can be applied in the natural gas production forecast and ultimate recoverable resources as follows:

$$Q = at^b e^{-t/c} \tag{3}$$

$$N_R = ac^{b+1}\Gamma(b+1). \tag{4}$$

Through a logarithm process of formula (3), Professor Chen (1996) proposed the traditional interpretation of the Generalized Wend model, where semi-logarithmic relations between production (Q) and time (t) are

$$\ln\frac{Q}{t^b} = \ln a - \frac{1}{c}t. \tag{5}$$

Suppose the value of b can be inferred empirically, and an optimal segment selected from the time sequence generated through formula (5), then the values for both a and c can be determined using linear regression, whereby forecast production Q_t^f can be obtained as

$$Q_t^f = (Q_1^f, Q_2^f, \ldots, Q_n^f), \quad t = 1, 2, \ldots, n. \tag{6}$$

The correlation coefficient R regarding sequences of actual historical production and forecast production goes like

$$R = \frac{\sum\limits_{t=1}^{n}\left[\left(Q_t - \overline{Q_t}\right)\left(Q_t^f - \overline{Q_t^f}\right)\right]}{\sqrt{\sum\limits_{t=1}^{n}\left(Q_t - \overline{Q_t}\right)^2}\sqrt{\sum\limits_{t=1}^{n}\left(Q_t^f - \overline{Q_t^f}\right)^2}}. \tag{7}$$

Value b may not be exactly right due to the fact that it is given empirically beforehand. Chen (1996) then proposed that the value may only be optimal when on a linear trial and error basis, the highest degree of fitting (R) is reached. Linear trial and error process can be conducted via the spreadsheet program Excel's Goal Seek tool. With the values for a, b, and c, respectively, being determined, this field's future production can be forecast by trend extrapolation.

(2) *The second interpretation* In order to determine value b more simply, other scholars (Zhao et al. 2009) have proposed the second interpretation of the Generalized Weng model, where formula (5) is translated as

$$\ln Q = \ln a + b \ln t - \frac{1}{c}t. \tag{8}$$

Make $T_1 = \ln t$, $T_2 = \ln - t$, formula (5) and (8) can then be translated into a linear equation with two unknowns:

$$\ln Q = \ln a + bT_1 + \frac{1}{c}T_2. \tag{9}$$

Through some mathematical tools, the semi-log relationship between historical production (Q) and T_1/T_2 can be drawn into a three-dimensional space, by selecting data segments with better plane relations and conducting binary regressions thereto, the values for a, b, and c can then be determined.

(3) *New interpretation* There are several issues regarding the interpretation that Chen proposed (Chen 1996).

First is that estimated recoverable resources (NR) are not taken into account in the calculation process, since resources determine production and will undoubtedly constrain the growth trend thereof. Also, relationship between historical cumulative production and estimated residual recoverable resources, that constitute the whole recoverable resources, is not fully considered. For natural gas production forecast of any oil & gas field, it should always be based on estimated residual recoverable resources, since the historical cumulative production is just "as is" and has no impact on future production. Therefore, of all the estimated recoverable resources, it is the residual recoverable resources that determine the future production growth trend. The traditional Generalized Weng model is a full-cycle one, where the values of parameters are determined from regression to certain data segments, then comes the forecast based on all the estimated recoverable resource; since historical cumulative production is not excluded, a high forecast maybe suggested. Furthermore, after value b is given then regression being made, randomness and subjectivity are implied when optimal segment selected from the time sequence generated on semi-logarithmic relations between historical production (Q) and time (t), which may in turn, affect the forecast result. Lastly, determining value b from higher coefficient R may not be solidly based since the latter shall not be the sole element to decide the accuracy of the forecast. Even though the second interpretation does describe and improve the process for determining value b, the above-mentioned issues still apply. Also, data segments with better plane relations are located from a three-dimensional space, where difficulty and subjectivity is implied, making no guarantee to the accuracy of the forecast. The dependence on some advanced mathematical tools would further restrict the wider application of this interpretation.

In order to more accurately forecast China's natural gas production, based on the above interpretations and China's current conditions, this paper proposes a new interpretation of the Generalized Weng model.

(1) For formula (5), value b is still given empirically to determine the semi-logarithmic relations between historical production (Q) and time (t).

(2) The regression data segment (S) for a given oil & gas block's natural gas production is implied when $t = m$:

$$S = \left(\ln\frac{Q_m}{t^b}, \ln\frac{Q_{m+1}}{t^b}, \ldots, \ln\frac{Q_n}{t^b}\right), \quad t < m. \tag{10}$$

(3) Linear trial and error is still needed in getting the right value for b, while the coefficient R, which is more mathematical oriented, is ignored in this paper, and such index more reflective of oil & gas field's actual conditions as estimated residual recoverable resource is adopted. Also, as is well known, that with the new round of evaluation for China's oil and gas resources having been successfully implemented, a clearer picture is presented regarding the oil and gas resources for different regions of China. For a given oil & gas field, when reaching the stage of rapid growth, its cumulative natural gas production goes like

$$N_{pm} = \sum_{t=1}^{m} Q_t, \quad m \in [1, n). \tag{11}$$

The residual natural gas estimated recoverable resource N_{Rr} at this time point can be inferred as

$$N_{Rr} = N_R - N_{pm} = N_R - \sum_{t=1}^{m} Q_t, \quad m \in [1, n). \tag{12}$$

When the values for a, b, and c are determined through regression, the estimated recoverable natural gas resources N_R^f of China can be deduced from formula (4):

$$N_R^f = ac^{b+1} \Gamma(b + 1). \tag{13}$$

Before the time point when $t = m$, forecast cumulative natural gas production N_{pm}^f would be

$$N_{pm}^f = \sum_{t=1}^{m} Q_t^f, \quad m \in [1, n). \tag{14}$$

As it is mentioned above, it is the estimated residual natural gas resources that determine future production trends. Residual recoverable resources at time point m can be inferred based on formula (13) and (14):

$$N_{Rr}^f = N_R^f - N_{pm}^f = ac^{b+1} \Gamma(b+1) - \sum_{t=1}^{m} Q_t^f, \ m \in [1, n). \tag{15}$$

In getting the right b value via Goal Seek, forecast residual recoverable resources after the time point of $t = m$ must be equal to that of the actual, whereby the conclusiveness of residual estimated recoverable resources on future production trend can be reflected. The corresponding mathematical expressions are as follows:

$$\begin{cases} N_{Rr} = N_R - N_{pm} = N_R - \sum_{t=1}^{m} Q_t \\ N_{Rr}^f = N_R^f - N_{pm}^f = ac^{b+1} \Gamma(b+1) - \sum_{t=1}^{m} Q_t^f \\ Opt.b, s.t. N_{Rr} = N_{Rr}^f \\ m \in [1, n) \end{cases} \tag{16}$$

As can be inferred from above, that model-based forecast of recoverable resources N_R^f may not be equal to actual recoverable resources N_R, which is mainly due to the inequality of forecast and actual cumulative production before forecasting point ($t = m$); whereas after the forecasting point, Goal Seek of b ensures the estimated residual recoverable resource to be an actual one, thus guaranteeing the accuracy of the forecasting base.

(4) Finally, the values for a, b, and c can be decided through the above processes, then future natural gas production for a given region can be forecast by trend extrapolation.

By proposing a new interpretation of the Generalized Weng model, this paper aims at promoting the reasonableness of forecasts. Although many studies do suggest a Multicyclic Generalized Weng Model, we are of the opinion that it is the residual natural gas resources that determine future production trends. Hence, the forecasting base should be the future production trend instead of the full-cycle trend, given that in different periods, the production growth may vary greatly, thus it might be only reasonable to make forecasts based on production growth in recent years; also, while fitting the whole historical production enhances accuracy, it does not bring strong reasonableness. Therefore, this paper adopts the traditional Generalized Weng model in making the forecast.

In order to better forecast and analyze China's natural gas production development on a regional basis, this paper divides China into 9 geological regions as follows:

- Northeast Region: Heilongjiang, Jilin, Liaoning;
- Bohai Region: Beijing, Tianjin, Hebei, Shandong;
- Yangtze River Delta: Shanghai, Jiangsu, Zhejiang;
- Southeast Region: Fujian, Guangdong, Guangxi, Hainan;
- Central Region: Inner Mongolia, Shaanxi, Shanxi, Ningxia;
- Mid-South Region: Henan, Hubei, Hunan, Anhui, Jiangxi;
- Western Region: Xinjiang, Qinghai, Gansu;
- Southwest Region: Sichuan, Chongqing, Tibet, Yunnan, Guizhou;

Table 2 Regional distribution of China's recoverable natural gas resources

Regions	Estimated proved recoverable reserves, *tcm*	Estimated recoverable resources, *tcm*	Proved rate
The Northeast	0.05	0.87	5.7 %
Bohai Rim	0.15	0.43	34.5 %
Yangtze River Delta	0	0.09	0
Southeast	0.0004	0.0037	10.8 %
Central	0.71	3.07	23.1 %
Mid-South	0	0.14	0
Western	0.70	7.45	9.4 %
Southwest	0.51	4.73	10.8 %
Offshore	0.33	5.25	6.3 %
Total	2.45	22.03	11.1 %

Data source A new round of China's Oil and Gas Resources Evaluation in 2014 from Chinese Ministry of Land and Resources (2014)

- Offshore: including offshore part of Bohai Bay, East China Sea and Yellow Sea, Yingge Sea, Beibu Gulf, while the South China Sea is not included.

2.3 Data

In the forecast, statistics for China's estimated recoverable natural gas resources come from the evaluation results made by Ministry of Land and Resources, while data for oilfields are based on company data.

Resources are the prerequisite for production forecasts. The national oil and gas resource evaluation shows that China's estimated recoverable natural gas resources are about 22 trillion cubic meters with the proved rate being 11.1 percent. Estimated recoverable resources and proved

recoverable reserves for the above 9 regions are shown in Table 2.

By attributing gas production in each oilfield to its corresponding geological region, this paper comes up with gas production in the 9 regions of China shown in Fig. 1. Since the Southeast Region did not have gas production until 2002, it is difficult to make a reasonable forecast based on historical production statistics; also, given the fact that the estimated recoverable natural gas resources of this region only totals 3.7 billion cubic meters and has little impact on the national output in the future, production in this region is therefore not included in the whole forecasting picture. Yangtze River Delta will not be shown in the figures because of the low percentage, but will be presented in the legend as it is included in calculation of national production.

3 Results

Based on the estimates of recoverable natural gas resources, estimated remaining recoverable resources and historical production data of each region, forecasting of their gas production was made with the Generalized Weng model by linear trial and error, with results shown in Fig. 2. Accuracy of the forecasting is secured by basing it on the actual estimated remaining recoverable resources.

Mid-South Region is an exception, the production there has gradually been declining in recent years; while the estimated residual recoverable resources still make up 73.7 % of total recoverable resources, whereby a potential for production increase can still be inferred. Therefore, it is forecast that a new growth trend will emerge soon in this area soon.

Results are obtained as follows based on Fig. 2:

(1) *Rising output of natural gas in all regions* From the perspective of the estimated resource base shown in

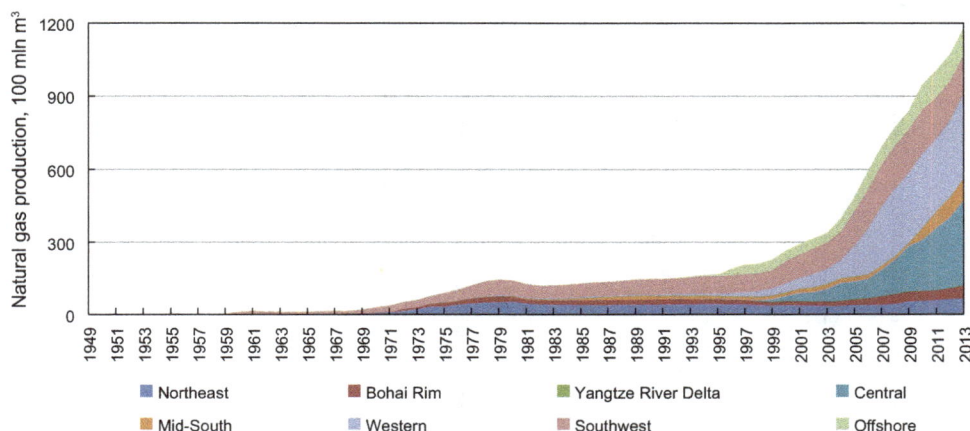

Fig. 1 China's natural gas output by regions

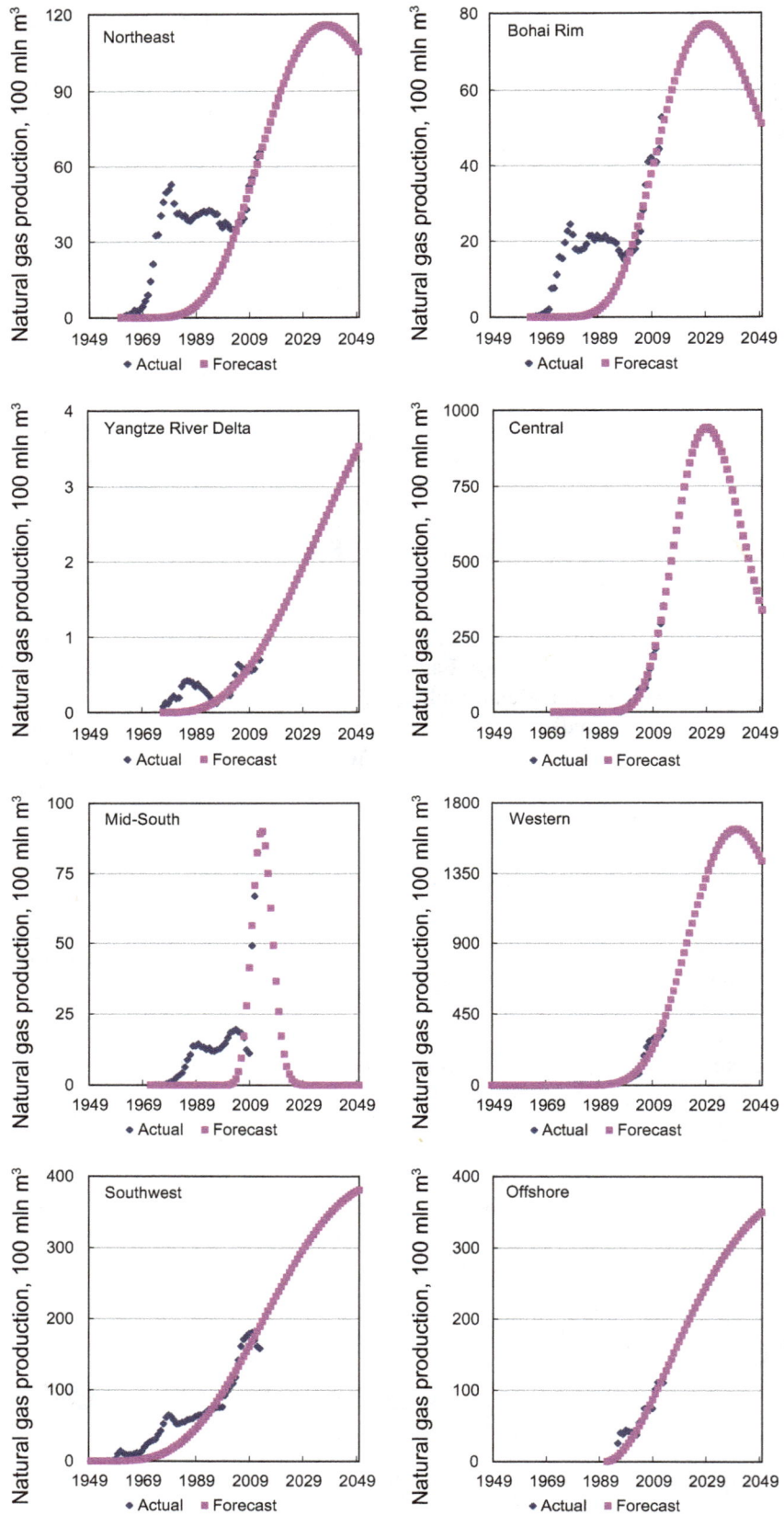

Fig. 2 Forecast of China's natural gas production in regions

Table 3 Forecast results of China's regional natural gas output

Regions	Estimated remaining recoverable resources , tcm	Remaining recoverable Ratio, %	Time of peak gas	Peak annual production, 100 mln cubic meters	Current output, 100 mln cubic meters pa
Northeast	0.69	79.6	2038	115.68	65.8
Bohai Rim	0.34	78.0	2030	77.00	52.8
Yangtze River Delta	0.085	99.4	2098	5.28	0.7
Central	2.88	93.6	2029	941.71	351.5
Mid-South	0.07	51.6	2014	89.87	89.1
Western	7.16	96.1	2040	1630.51	348.5
Southwest	4.35	91.9	2060	390.84	158.4
Offshore	5.14	97.8	2069	376.41	111.0

Table 3, apart from the Mid-South Region of China, all the other regions demonstrate a remaining estimated recoverable to original gas in place ratio of over 75 %, which on one hand means that abundant resources are available for future production, but on the other also implies that China's natural gas industry is still in its early stage of development. Furthermore, in light of output, all the regional natural gas production will continue to rise, among which the Yangtze River Delta has the longest rising period with only small production scales, imposing no substantial effect on the overall growing trend of national production. The Mid-South Region, though in recent years having achieved rapid growth in natural gas production, will have a short period of high production due to its weak resource base, which means that the current production scale will be difficult to maintain for a long time and a rapid downward trend is soon to come unless additional reserves are discovered. The Western, Central, Southwest, and Offshore Regions will be the main battlefields for the development of China's natural gas industries in the future. Among them, the Western and Central regions will both turn into "one hundred billion cubic meter p.a. natural gas zones"; and even though the current natural gas production in the Western Region is lower than that of the Central Region, however the huge resource potential in the Western Region will bring a faster production growth than the latter, making it the main natural gas producing area of China with a peak annual production of over 160 billion cubic meters.

(2) *Rapid upward trend of national natural gas production* As of 2014, China's cumulative natural gas production totals 1.5 trillion cubic meters, accounting for only 5 % of estimated recoverable resources, which indicates that a great growth potential still lies

ahead. A superposition of production in all regions could come up with the growth trend of national natural gas production as shown in Fig. 3. As can be seen therein, China's natural gas production will continue to maintain a rapid growth, and output breakthroughs of 200/300/323 billion cubic meters will occur in year 2020/2030/2036, respectively, followed by a gradual decline. Also, there shall be room for China's natural gas production that is three times as high as in 2014; and China's annual natural gas production is estimated to remain above 300 billion cubic meters p.a. for 15 consecutive years from 2030 to 2044, making natural gas an important energy source for China's economic and social development. With the same approach used in natural gas forecasting, China's oil production is also predicted with results shown in Fig. 3. Projections indicate that China's oil production will continue to grow, reaching a peak production of 214 million tonnes p.a. or 237 billion cubic meters of gas equivalent[1] in 2024. However, China's oil production growth potential is relatively small with growth being moderate. With the rapid rise in gas production, China will move into a stage where oil and natural gas will both dominate the energy structure around year 2020. Natural gas will play an increasingly important role in China's energy industry, and oil and natural gas together will provide more abundant energy sources for China's economic and social development.

(3) *Uneven regional distribution of China's natural gas production to remain* Results also show that uneven regional distribution will continue for China's natural gas production, as shown in Fig. 4. In 2014, the four regions of top natural gas production were

---

[1] Per BP Statistical Review of World Energy, 1 ton of oil = 1110 cubic meters natural gas equivalent.

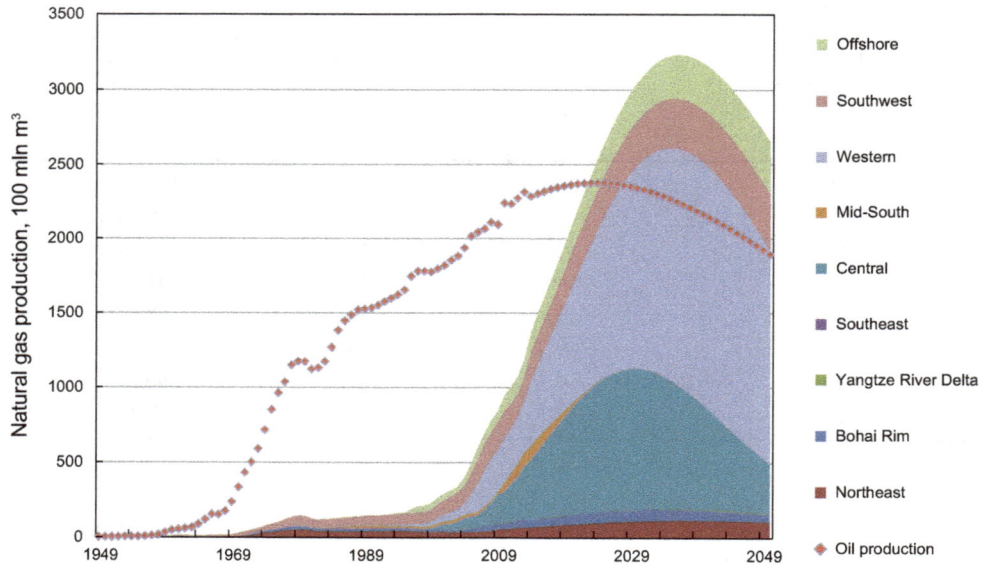

Fig. 3 Past and forecast annual natural gas production in China and comparison with that of oil

Western, Central, Southwest, and Offshore, which together accounted for 82.2 % of China's total gas production. In the future, these four regions will still be the most important sources of China's natural gas output, but with a different scenario. The Western Region's natural gas output will make up about 50 % of China's overall natural gas production, up from the 32.0 % in 2014. The Central region will have an early gas production peak and fast growth, and this will also be later followed by a fast declining rate of gas production, signifying a downward trend of its gas output ratio in China's total gas production from 2020; Natural gas production in the Southwest and Offshore Regions will increase moderately with a long growth period, and their ratios of gas output in China's total gas production will show a change from decline to increase. Overall, the Central and Western Regions will be the main sources of China's natural gas production with the Offshore Region being the new growth point. The Eastern Region, although the production of the old gas fields there still grows, it will gradually account for a lower proportion of China's nationwide output, leading to a decline of its status in China's natural gas industry. Therefore, based on Central and Western Regions, exploring the Offshore Region and stabilizing the Eastern Region will be the reasonable choice for the current development of China's natural gas industry.

(4) *Comparison of forecast results* In IEA's *World Energy Outlook 2015* and BP's *Energy Outlook*

2035, China's natural gas production has been forecast as in Table 4. These two forecasts, together with the one made in this paper, are all optimistic in general, with an increase of average annual production of 4.1 %, 5.1 %, and 4.4 % respectively, a conclusion drawn from the estimated resources base and government policy support. Also the minor annual increase difference among these three also justifies the forecast in this paper. For the year of 2030, IEA's forecast result shows a slightly lower estimate, while BP's results show almost no differences, in comparison with the forecast result in this work; while from 2030, BP foresees the fastest growth due to the perception that nonconventional natural gas will grow dramatically then in China.

4 Discussions and policy implications

Based on the forecast results, the following views are drawn:

- *Natural gas will change China's energy consumption structure* In 2014, natural gas accounted for only 6.5 % of China's total energy consumption, a ratio that is far less than the developed world average. The proportion of natural gas in a country's energy consumption structure is an important criterion in deciding a scientific and reasonable energy consumption structure of that country. For a long time, China has been a coal consuming country and has not paid enough attention to

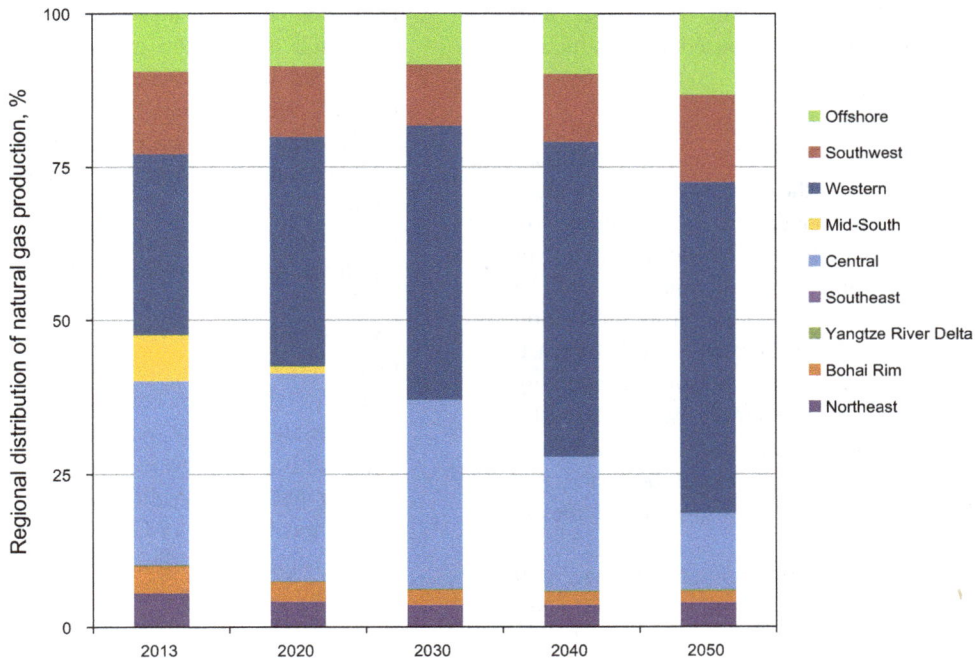

Fig. 4 Regional distribution of natural gas production in China

Table 4 Comparison of natural gas production forecast

	2020	2025	2030	2035
IEA	1720	2120	2600	3090
BP	1812	2324	2981	3822
Forecast in this work	2037	2587	2987	3159

natural gas development, which resulted in a relatively low natural gas yield. If China wants to change the long-standing coal-dominated energy consumption structure to achieve a cleaner and more efficient development of the national economy, increasing natural gas consumption is the most realistic and scientific choice (Ma and Li 2010). With the rapid increase in natural gas production, as well as the establishment of a large number of natural gas pipelines and other facilities, China's natural gas consumption is sure to rise dramatically.

- *Natural gas will promote the new development of the oil and gas industry* After a long period of growth, our country's oil production will inevitably begin to decline, which requires new resources to be in place in order to meet continuous economic development. In 2020, China's natural gas production will exceed that of the oil, and peak gas will come later with a much higher yield, providing a richer source of energy than oil. Therefore, under the current circumstances, natural gas

is the best strategic resource to replace the petroleum resources. Oilfields discovered in last century like Daqing and the Bohai Bay Basin have contributed to the substantial increase of our country's oil production and reserves, prompting the first rapid development of China's oil and gas industry; whereas now, when oil production lacks steam in growth and may soon begin to fall, natural gas, being an important substitute resource, will promote the new development of China's oil and gas industry through its rapidly growing production and ever rising status.

- *Unconventional gas resources will form an important supplement to China's natural gas industry* In recent years, with the evidence from the US shale gas revolution, unconventional gas resources have attracted universal attention. China has abundant unconventional gas resources (Jia et al. 2012), which will gradually become economically available as technological breakthroughs occur. Nevertheless, these resources are not included into the forecast in this paper due to the bloom prospect and the very fact that current low international oil prices have affected high-cost projects, particularly the development of these unconventional gas resources. However, once it is economically viable to exploit these resources, they shall form a great supplement to the development of China's natural gas industry.

Rapid growth in natural gas production will promote fast development of China's natural gas industry. To promote scientific and reasonable development, this paper also proposes the following suggestions based on production forecast results:

(1) *Strengthening planning to increase natural gas production* Early in 2005, Qiu and Fang (2005) suggested to strengthen the planning of oil production and postponing peak oil by lowering oil production to a proper level. The same will go for our country's natural gas production (Liu and Li 2005; Zhang and Li 2006). Although its production peak is a long way off according to the forecast, limited total resources require reasonable planning and economical use; therefore, with domestic resources and market taken into basic account, foreign natural gas resources might be made full use of (Hu 2014) to meet fast-growing domestic consumer demands.

Increase in natural gas imports from abroad will bring greater dependence. Oil and gas security has always been a hot topic in our country. But a country's energy security cannot simply be measured by its dependence on foreign oil and gas, obtaining a stable supply of oil and gas at a reasonable price is the real point, as it is for Japan. China imported about 58 billion cubic meters of natural gas in 2014, a 32.4 % dependence on foreign gas which implied gas security in general. The current international oil prices have brought forward a favorable opportunity which we should seize. The import of foreign natural gas resources could be strengthened. Also a moderate slowdown in domestic natural gas production to conserve resources or acceleration in the discovery of additional domestic gas reserves will both help make full use of domestic and foreign markets to meet the domestic gas demand in economic and social development.

(2) *Enhancing natural gas exploration and development* While taking advantage of current favorable circumstances to fully utilize foreign natural gas resources is an important aspect in promoting China's natural gas industrial development, the domestic market would remain its cornerstone. More efforts should be put into the development of domestic natural gas resources. Exploration should be strengthened to find more natural gas resources (Xu et al. 2009) and development should be reinforced to enhance oil recovery, so that more reserves could be turned into production. All our previous Gas Resources Assessments have shown that China's natural gas estimated resources have entered a period of rapid growth. Natural gas reservoir genesis needs to be further studied and considered with theoretical and technological innovations being strengthened, based on which exploration and development should be reinforced in order to achieve new breakthroughs. Central, Western, and Offshore Regions host relatively more natural gas resources, thus should be the focus of future exploration and development areas in China.

(3) *Implementing inter-regional natural gas transport pipelines* As outlined in the forecasts western and offshore areas are to be important sources of our country's natural gas production, while the output of mature fields in the eastern region will gradually decline. Natural gas, being a source of cleaner and more efficient energy, will have a great development potential in the future in China, where the consumption for natural gas will increase dramatically. The natural gas industry in China should be domestically based, supplying sufficient energy for the national economic development. Therefore, we should put great efforts into developing our country's natural gas industry while implementing inter-regional natural gas transport pipelines in the meantime, as the main production areas shift from the east to the west and offshore. China needs to push for sustainable and orderly development of the natural gas industry.

(4) *Speeding up technological innovation* Technological innovation can help promote development breakthroughs and lower cost. Our country has complex geological conditions, especially in mid-western and offshore areas, thus only advanced technology can help achieve breakthroughs. For those mature oilfields in the eastern region, it is also necessary to make use of technological innovation to lower costs and extend the life of oilfields. In general, our country's technology level still falls behind some foreign countries, only through more focus and innovation input into technology can our oil and gas industry develop sustainably.

(5) *Concentrating more on unconventional and deep-water gas resources* China is rich in unconventional oil and gas, among which the coal bed methane recoverable resource is estimated to exceed 11 trillion cubic meters and oil sands a total of 2.3 billion tonnes. Under certain technical and economic conditions, these resources can gradually be exploited and utilized. Therefore, the focus on conventional natural gas resources should also be accompanied by the strengthening of study of unconventional oil and gas resources to promote their exploration and development. Recently, CNPC and Sinopec have achieved breakthroughs in shale gas exploiting, with which Sinopec has planned to

build 5 billion cubic meters and 10 billion cubic meters per annum of shale gas production capacity, respectively, in 2015 and 2017. This is also pointed out in the *Chinese Energy Development Strategy Action Plan 2014–2020*, China's shale gas and CBM production shall both reach 30 billion cubic meters p.a.by 2020, thus forming an important complement to conventional natural gas production in China. In addition, deepwater natural gas exploration and development is another important direction of China's natural gas industry, in particular in the resource-rich South China Sea deepwater areas, where China National Offshore Oil Corp. (CNOOC) has commenced the commercial exploitation in the northern part. The exploration and development of natural gas resources lags in its southern part of the South China Sea, which should be an important area of our country's deepwater gas development.

5 Conclusions

This paper uses the Generalized Weng model to forecast Chinese regional natural gas production, This shows that China's natural gas production has great growth potential, that production will keep growing for a long time in most regions; also that the rapid growth of natural gas production will help China enter a stage where oil and natural gas will together dominate the energy structure, providing abundant sources for China's energy consumption. Furthermore, the uneven regional distribution of output makes "based in the Central and Western Regions and exploring the Offshore Region" a reasonable choice for the development of China's natural gas industry.

The forecast also shows that China's natural gas production will maintain a rapid growth with peak gas of 323 billion cubic meters coming in 2036; in 2020, natural gas production will surpass that of oil.

Large-scale production has yet to come for unconventional natural gas resources in China, thus they not included into the full forecast in this paper. With future technological breakthroughs, these resources will probably enter a gradual exploitation phase, forming a critical supplement to our country's existing natural gas production; therefore continuous attention needs to be paid to them.

Acknowledgments The authors would like to thank the National Social Science Funds of China (13&ZD159) and the, National Natural Science Foundation of China (71303258, 71373285), MOE (Ministry of Education in China) Project of Humanities and Social Sciences (13YJC630148), and Science Foundation of China University of Petroleum, Beijing (ZX20150130) for sponsoring this joint research.

References

Aguilera RF, Inchauspe J, Ripple RD. The Asia Pacific natural gas market: large enough for all? Energy Policy. 2014;65:1–6.

Al-Fattah SM and Startzman RA. Analysis of worldwide natural gas production. Charleston: SPE Eastern Regional Meeting held in Charleston; 1999.

Bianco V, Scarpa F, Tagliafico LA. Scenario analysis of nonresidential natural gas consumption in Italy. Appl Energy. 2014;113:392–403.

BP Statistical Review of World Energy June 2015. London: BP. 2015. From http://www.bp.com/ reports and publications.

Chen YQ. Derivation and application of the generalized Weng model. Nat Gas Ind. 1996;3:22–6 (**in Chinese**).

Chinese Ministry of Land and Resources. A new round of China's oil and gas resources evaluation in 2014. Beijing: Geological Publishing House; 2015. p. 1–54.

Demierre J, Bazilian M, Carbajal J, et al. Potential for regional use of East Africa's natural gas. Appl Energy. 2015;143:414–36.

Feng LY, Wang JL, Zhao L. Construction and application of multicyclic models for natural gas production forecasting. Nat Gas Ind. 2010;30(7):110–2 (**in Chinese**).

Granger C, Ramanathan R. Improved methods of combining forecasting. J Forecast. 1984;2:197–204.

Hu B. Oil and gas cooperation between China and Central Asia in an environment of political and resource competition. Pet Sci. 2014;11:596–605.

Hu JG, Chen YQ, Zhang SZ. A new model for predicting production and reserves of oil and gas fields. Acta Pet Sin. 1995;1:79–86 (**in Chinese**).

Hubbert MK. Nuclear energy and the fossil fuels. Meeting of the Southern District, Division of Production, American Petroleum Institute. San Antonio: Shell Development Company; 1956.

Hubbert MK. Techniques of prediction with application to the petroleum industry. 44th Annual Meeting of the American Association of Petroleum Geologists. Dallas: Shell Development Company; 1959. p. 43.

Jebaraj S, Iniyan S. A review of energy models. Renew Sustain Energy Rev. 2006;10:281–311.

Jia CZ, Zheng M, Zhang YF. China's unconventional hydrocarbon resources and their exploration and development prospects. Pet Explor Dev. 2012;39(2):129–36 (**in Chinese**).

Lin Boqiang, Wang Ting. Forecasting natural gas supply in China: production peak and import trends. Energy Policy. 2012;49:225–33.

Li JC, Dong XC, Gao J. Optimized combination forecasting and analysis of our country's natural gas production. Futur Dev. 2009;7:65–9 (**in Chinese**).

Liu CX, Li HX. Problems and strategies of China's natural gas industry. Pet Sci. 2005;9:80–92.

Ma YF, Li YL. Analysis of the supply-demand status of China's natural gas to 2020. Pet Sci. 2010;7:132–5.

Primoz P, Bozidar S, Goran Š, et al. Comparison of static and adaptive models for short-term residential natural gas forecasting in Croatia. Appl Energy. 2014;129:94–103.

Qiu ZJ, Fang H. Some opinions about sustainable development of petroleum resources in China. Acta Pet Sin. 2005;26(2):1–2 (**in Chinese**).

Qiu ZJ. China's natural gas industry to usher in golden development [EB/OL]. Xinjiang Information Network. 2008. From http://www.xj.cei.gov.cn/e/DoPrint/?classid=39&id=32037, (in Chinese).

Szoplik J. Forecasting of natural gas consumption with artificial neural networks. Energy. 2015;85:1–13.

Taşpınar F, Çelebi N, Tutkun N. Forecasting of daily natural gas consumption on regional basis in Turkey using various computational methods. Energy Build. 2013;56:23–31.

Tayfun B. Development of mature oil fields—a review. J Pet Sci Eng. 2007;57:221–46.

Weng WB. The foundation of the forecasting theory. Beijing: Petroleum Industry Press; 1984 **(in Chinese)**.

Wu YL. Study of a concise algorithm for approximately optimal combining forecasting with non-negative weights. Colleague Math. 2007;1:132–3 **(in Chinese)**.

Xu ZY, Yue DL, Wu SH, et al. An analysis of the types and distribution characteristics of natural gas reservoirs in China. Pet Sci. 2009;6:38–42.

Yuan AW, Sun GS, Yang XY, et al. A new model for forecasting natural gas production. Nat Gas Ind. 2007;2:84–6 **(in Chinese)**.

Zhang BS, Li J. Oil and gas supply and demand in China and its development strategy. Pet Sci. 2006;3(1):92–6.

Zhao L, Feng LY, Lu XA, et al. Comparison of two methods for generalized Weng model Xinjiang. Pet Geol. 2009;3:658–60.

2

Interactive plant simulation modeling for developing an operator training system in a natural gas pressure-regulating station

Yongseok Lee[1] · Changjun Ko[1] · Hodong Lee[1] · Kyeongwoo Jeon[1] ·
Seolin Shin[1] · Chonghun Han[1]

Handling editor: Jian Shuai

Abstract This study proposes a method of interactive plant simulation modeling which delivers the online simulated results to the field operators and induces them to take proper actions in the case of pre-identified accident scenarios in a chemical plant. The developed model integrates the real-time process dynamic simulation with 3D-CFD accident simulation in a designed interface using object linking and embedding technology so that it can convey to trainees the online information of the accident which is not available in existing operator training systems. The model encompasses the whole process of data transfer till the end of the training at which a trainee operates an emergency shutdown system in a programmed model. In this work, an overall scenario is simulated which is from an abnormal increase in the main valve discharge (second) pressure due to valve malfunction to accidental gas release through the crack of a pressure recorder, and the magnitude of the accident with respect to the lead time of each trainee's emergency response is analyzed. The model can improve the effectiveness of the operator training system through interactively linking the trainee actions with the simulation model resulting in different accident scenarios with respect to each trainee's competence when facing an accident.

✉ Chonghun Han
chhan@snu.ac.kr

[1] School of Chemical and Biological Engineering, Seoul National University, Seoul, South Korea

Edited by Yan-Hua Sun

Keywords Operator training system · Dynamic process simulation · Accident simulation

1 Introduction

In 2011, the total revenue of the whole chemical industry came to about 100 billion euros (CEFIC 2011). While the development of modern chemical plants has created high economic profits, the issues of reduced operability and increased risk are inevitably brought about due to complicated processes and large quantity and variety of treating chemicals. Chemical accidents result in productivity loss, equipment and environment damage, and fatalities which we can observe in several cases from the Bhopal toxic gas release accident in 1984 to the Texas BP refinery explosion in 2005. According to several studies dealing with main causes of the accidents, maloperation of plant equipment by human error is one of the most frequent causes (Antonovsky et al. 2014), and accidents occur mainly due to an inefficient structure of information sharing between each operator and insufficient education about past accident cases (Kletz 1998). Particularly, fast and accurate communication between physically separated operators in the large sites of chemical plants requires high competence.

Established operator training systems based on dynamic process simulation like UniSim® OTS (operator training simulator) by Honeywell (2005) and Aspen® OTS by AspenTech have played great roles in training of proficiency in operation procedures and control of risk factors in the chemical process for control room operators (CROPs). However, they have difficulty in enhancing training efficacy due to limited information delivery for field operators (FOPs). Generally, they are not equipped with systematic knowledge of the process simulation. As field operations are performed manually based on communication with the control room, FOPs are highly reliant on the control room orders and heuristics based on personal experience in the case of an accident. However, existing education for FOPs is limited to study of operation and emergency text manuals or the handling method of each item of equipment. Therefore, it is necessary to build a systematic FOP training system which can induce the series of processes from correct assessment of accidental situations to active management.

Cha et al. (2012) developed a fire suppression training program which generates a fire scenario in virtual reality, calculates the fire effect using 3D-computational fluid dynamics (CFD), and delivers the situation information to trainees so that they can actively suppress the fire using an avatar (Cha et al. 2012). Schneider Electric (2014) performed and evaluated this operator training with 30 operation scenarios (15 scenarios each for CROPs and FOPs, respectively) via the EYESIM® immersive training package in a virtual reality simulation of the plant. Even though the above two strands of research try to develop FOP customized training solutions by integration of accident simulation or process simulation with immersive virtual reality systems, respectively, they are not able to train the whole process of accident initiated from process upsets and terminated at emergency response as they do not link the process simulation to accident simulation directly. To solve this limitation, Manca et al. (2013) interlinked the process simulation with a self-developing accident simulator, AXIM by object linking and embedding (OLE) technology, and implemented this module into the immersive virtual reality. Through this combined model, they realized the pool formation and pool fire scenario by liquid release, and let the fire results affect the process simulation model so that trainees could experience the fault propagation realistically (Manca et al. 2013). Nazir et al. (2015) evaluated and validated the training efficacy by applying this model to FOPs directly. The AXIM simulator is based on parametric calculation with simple heat and material balances only for a liquid phase, so the accuracy of this model is not sufficient for vapor phase or two-phase jet release, dispersion, and fire and explosion calculation.

This study aims to develop an interactive plant simulation model in which the dynamic process simulator and self-developing discharge model are directly linked via Visual Basic, and especially for a gas dispersion scenario, pre-calculated offline 3D-CFD data are processed real time with respect to trainees' emergency actions. Simulated results are delivered to trainees so that they can correctly understand the abnormal situation based on the information from the model and actively take proper actions in the programmed interface. Then, the actions affect the process and accident simulation simultaneously. As our model utilizes a commercial 3D-CFD simulator to calculate the effect of an accident, given that the proper modeling is assured, the model guarantees sufficient accuracy for a vapor phase jet release scenario and following gas dispersion calculation. A case study deals with a natural gas pressure-regulating station in South Korea and evaluates the applicability of our model to practical operator training by generating process upsets and accident scenarios, constructing dynamic process and accident simulation models, and developing a demonstration program.

2 Interactive plant simulation modeling

2.1 Model structure

The interactive plant simulation model interlinks process and accident simulation models in an overall training scenario from process upsets to accident occurrence and propagation (Fig. 1). In this model, three simulations are linked based on a certain sequence of accident scenarios: dynamic process simulation, discharge calculation, and pre-calculated 3D-CFD simulation. As the scenarios are initiated with certain process upsets like equipment failure, dynamic process simulation firstly calculates the effect of the failure to the process each time. The real-time results are automatically conveyed to an integration domain like Microsoft Excel via the export port of the simulator. When the accidental release conditions are met at certain times as an error is accumulated, the discharge model built in the domain is activated to calculate the discharge at the leakage point in the equipment. The calculated discharge flowrate is transmitted back to the dynamic process simulator via the export port of the domain and affects the simulator to realize the leak through generating an additional stream. At the same time, pre-calculated 3D-CFD simulation results for dispersion and explosion of discharged fluid are selectively sent to the domain for each time via the export port of a CFD database according to the leakage conditions like pressure, temperature, and hole size. This whole sequence partly by OLE technology is visualized in the training system so that trainees can see the results and take actions in it.

Through this simulation linking structure, the model leads trainees to actively analyze process variable trends based on the simulation results and take proper actions with their own decisions to stabilize the variables or minimize operational losses. When stabilization fails and an accident occurs, associated results like gas cloud concentration and explosion overpressure at each time and position are additionally provided to the trainees. Their actions like emergency shutdown can be inputted by clicking a mouse or a joystick control.

Training interface

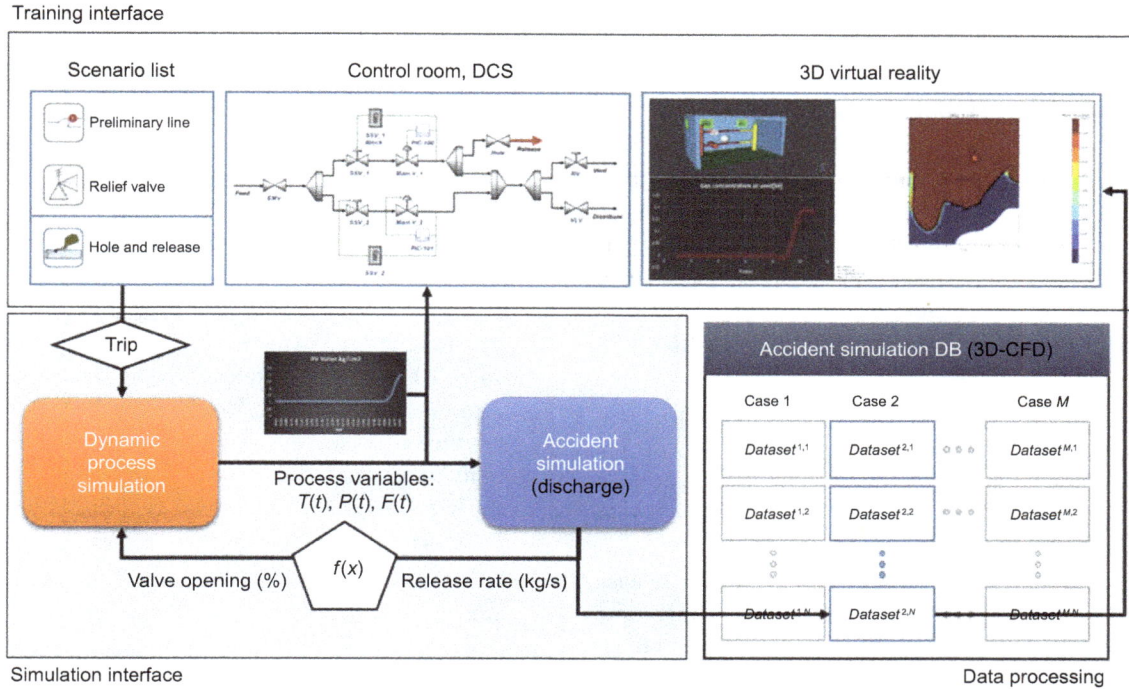

Fig. 1 Schematic design of the interactive simulation model

2.2 Dynamic process simulation engine

Dynamic process simulation gives trainees almost the same trend of process variables as that of a real plant. As a scenario is initiated, physical and thermodynamic calculations are conducted online and variable trends deviating from set points or being stabilized to those points can be analyzed. As errors are accumulated and the variables reach the pre-defined conditions of an accidental release scenario, values of the variables at that time are automatically inserted into the accident simulation model. When trainees' actions like emergency shutdown by clicking a manual valve in the training environment are taken, the associated signal is transferred to the process simulation model so that the actions are reflected in the model.

2.3 Real-time accident simulation module

In order to separate the linking point with the process simulation model, the accident simulation model is divided into two sub-models: One is the 'discharge model' calculating the release conditions of a fluid from inside the equipment to the outside through an orifice. And the other is '3D-CFD model' calculating indoor or outdoor dispersion and fire and explosion effects after the discharge.

The discharge model calculates the release mass flow-rate Q [kg/h] from given process simulation results at the

time when a fluid starts to release [Eq. (1)] and transfers the results to the 3D-CFD model. As this release should be simultaneously reflected in the process simulation model, we generate an additional stream and a valve at the release position right behind the main valve in the process model and automatically adjust the valve openings (Eq. (2): $f(x)$ in Fig. 1) so that the fluid is to be released with the quantity calculated from the discharge model.

$$Q_1 = C_{\text{disc}} A \sqrt{\gamma \cdot \rho_i \cdot P_i \cdot (2/\gamma + 1)^{(\gamma+1)/(\gamma-1)}} \qquad (1)$$

$$Q_2 = k\sqrt{V_{\text{open}} \cdot \mathrm{d}P \cdot \rho} \qquad (2)$$

where C_{disc} is the discharge coefficient; A is the hole area, m^2; γ is the specific heat ratio, $\gamma = C_p/C_v$; ρ_i is the inlet fluid density, kg/m^3; P_i is the inlet pressure, kPa; k is the conductance, kg/h/(kPa kg/m^3)$^{0.5}$; V_{open} is the valve opening, %; $\mathrm{d}P$ is the friction delta pressure, kPa.

The 3D-CFD model utilizes commercial software (FLACS$^{®}$ by Gexcon) to guarantee the accuracy of the dispersion calculation. As the CFD calculation requires a heavy computational load unlike the discharge model, this study develops a method of real-time processing of offline CFD data for applying the CFD model to our training system in which the real-time data transfer between the simulation model and a training environment is essential. For this purpose, we construct a big database to save the CFD results with respect to each scenario, and provide them to trainees selectively as they take a certain action in the training interface.

2.4 Real-time 3D-CFD data processing method

Training with pre-defined operating scenarios and pre-calculated CFD data holds a low degree of freedom in which the trainees cannot do anything but certain actions designated by the system in advance. In order to overcome this limitation, this study suggests a real-time CFD data processing method (Fig. 2) and increases the training effectiveness of our model.

1. Trainee action list—generate trainee action list in a certain scenario and process. For the case of the pressure-regulating station, 'manually close the emergency shutdown valve inside the station' is a representative action in the case of a gas release.
2. Release duration—determine the range of release duration based on a field operator's average site arriving time, 15 min for pressure-regulating station, and the mission fails if the training time exceeds the maximum time without a series of proper actions.
3. 3D-CFD database—divide the range of release duration (15–30 min) into 1-min intervals, and save total 16 simulation results, labeling each gas concentration dataset $C^i(x, y, z, t)$ as $Dataset^i (i = 15, 16, \ldots, 30)$.

4. Data processing with respect to trainee action—as the release starts ($t = t_{rel}$), first the dataset of the maximum release duration $Dataset^{30}$ is transferred to a trainee in the training environment in real time. When the trainee receives the message to move, an avatar in the program heads for the site instead by trainee's manipulation. When the avatar closes the shutdown valve at certain time ($t = t_{act}$), the CFD data after that time are replaced by those in $Dataset^{(t_{act}-t_{rel})}$ not in $Dataset^{30}$ (Eq. (3), Fig. 3). For instance, if a gas release occurs 10 min after the training starts ($t_{rel} = 10$) and a trainee closes the valve 20 min after the release ($t_{act} = 30$), concentration data of $Dataset^{30}(= C^{30}(x, y, z, t), 0 \leq t \leq 20)$ during the time between release and action ($10 \leq t \leq 30$) are transferred in real time, and after that time ($30 < t \leq t_{max}$) the data are replaced by those of $Dataset^{20}(= C^{20}(x, y, z, t), 20 \leq t \leq (t_{max} - 10))$

$$C(x, y, z, t) = \begin{cases} 0 & 0 \leq t < t_{rel} \\ C^{30}C^{(t_{act}-t_{rel})}(x, y, z, (t - t_{rel})) & t_{rel} \leq t < t_{act} \\ C^{(t_{act}-t_{rel})}(x, y, z, (t - t_{rel})) & t_{act} \leq t \leq t_{max} \end{cases}$$

(3)

3 Case study—pressure-regulating station

3.1 Pressure-regulating station

Natural gas in South Korea is supplied from the LNG receiving terminal to residences or offices through KOGAS (Korea Gas Corporation) supply management stations at 6.86 MPa and then two pressure-regulating stations operating at 0.8 or 0.6 MPa, respectively. Pressure-regulating stations reduce pressure of the high-pressure supplied gas toward the proper level of 2 kPag for safe distribution (Lee et al. 2010).

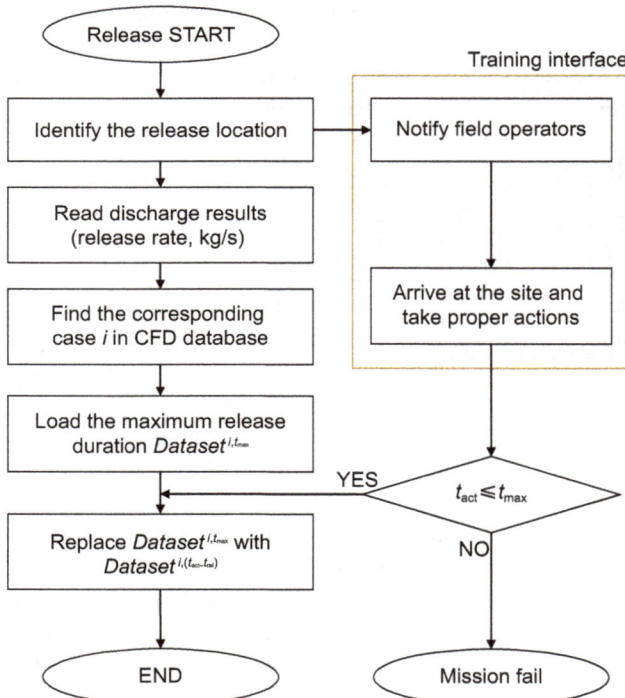

Fig. 2 Real-time CFD data processing method

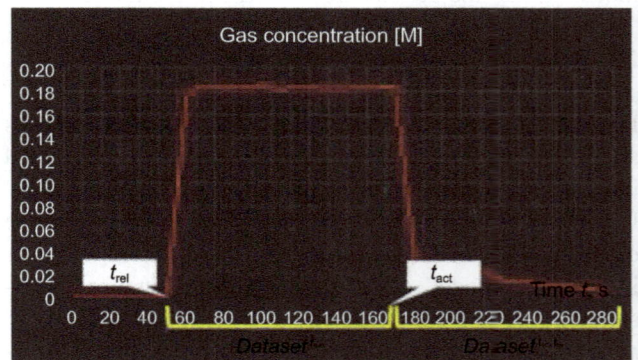

Fig. 3 Real-time CFD data of the gas concentration

In this study, the pressure-regulating station near the residential area is chosen as the target process for implementing our model as it has a high risk of fire and explosion accident (Fig. 4). It consists of main (upper) and preliminary (lower) lines including main valves for reducing and controlling the gas pressure, gas heater for compensating lowered temperature due to abrupt expansion, gas filter for preventing inflow of other substance, slam shutoff valve (SSV) to automatically block the flow and relief valves in case of an emergency.

3.2 Process simulation modeling

A process model can be constructed (Fig. 5) based on controller set pressures of the pressure-regulating station (Table 1). The yellow region of the figure is one additional stream and valve in order to simulate the gas release right behind the main valve. When a gas releases, the opening percentage of the valve is set by discharge calculation in the discharge model; otherwise, it is set to be zero at the normal operation. Dynamic simulations of normal controller operations are tested in Fig. 6. As shown in the figure, the main valve controller tracks the setpoint change well and the SSV controller blocks the gas stream at the set pressure.

The model uses AspenHysys v.8.4 as a process simulator and PR-LK EOS as a thermodynamic model for simulating the natural gas (C_1:C_2:C_3:n-C_4:i-C_4 = 0.90:0.05:0.03:0.01:0.01). The main valve type is 1098-EGR, and the pressure–flow correlation at the choked flow is as the following equation (Emerson Process Management 2016).

$$Q_{main} = P_i \cdot C_g \cdot 1.29 \qquad (4)$$

where Q_{main} is the gas flowrate through the main valve, SCFH, and C_g is the regulator or wide-open gas sizing coefficient.

3.3 Scenario generation

Scenarios are generated based on the historical data of process upsets or accidents. As these cases are documented with real process data and event sequence, a scenario generation process is initiated at the case-based analysis of historical data. In this case study, data from a pressure regulator in South Korea are classified into three representative scenarios listed in Table 2.

In the mild case, when the second pressure reaches the SSV set pressure, the SSV is closed immediately and the main valve in the preliminary line opens to stabilize the gas flowrate and the second pressure. In the relief case, as the SSV fails to block the supply and the gas pressure reaches the relief valve set pressure, the valve vents pressurized gas outside the station as much as the quantity its size is capable of. For the worst case, gas is released at the high pressure due to a series of malfunction of all safety devices. In this study, the worst-case scenario is performed among three main scenarios in order to evaluate our model linking the process and accident simulation for operator training system in the pressure-regulating station.

3.4 Accident simulation modeling

After the onset of training, the instructor starts the second pressure rise scenario. The stepwise course of training is as follows. First, the value of the second pressure from the

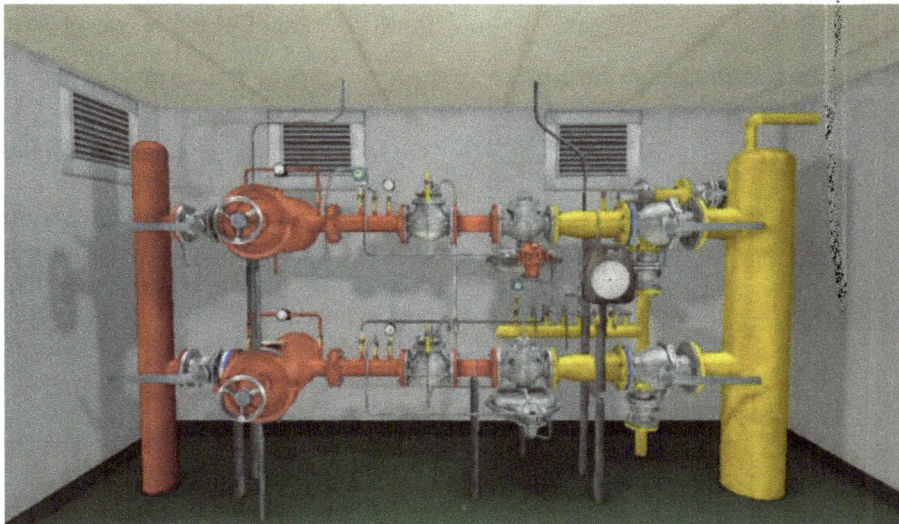

Fig. 4 3D image of a pressure-regulating station

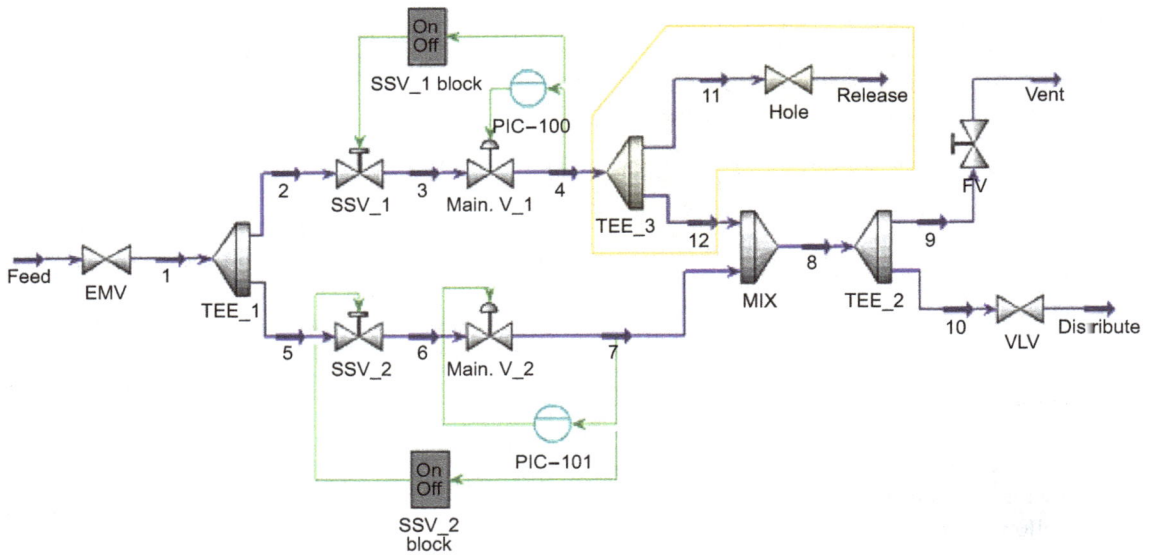

Fig. 5 Process model of the pressure-regulating station

Table 1 Controller set pressure

Controller		Set pressure, kPa
Failure alarm	Lower limit	1.2
	Upper limit	3.2
SSV-1 (main line)		3.6
Relief valve		4.0
SSV-2 (preliminary line)		4.4

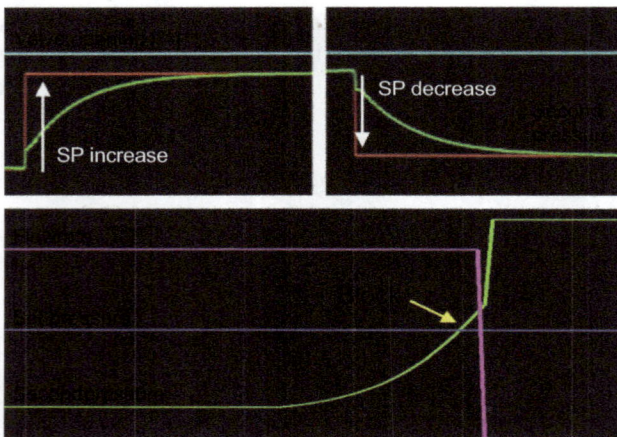

Fig. 6 Controller operation (*top* main valve and *bottom* SSV)

dynamic process simulation model is transmitted to the trainee in real time. Next, the second pressure reaches the release pressure which was set by the instructor (202.6 kPa—changeable depending on the scenario), which leads to the automatic discharge calculation based on the values of process variables like temperature and pressure at the leakage spot. At the same time, the corresponding 3D-CFD dispersion simulation results in the database whose input is from the discharge calculation are provided to the trainee. Finally, based on this process variables and accident data the trainee is induced to take appropriate actions.

Figure 7 indicates the accident simulation model of the pressure-regulating station in FLACS, and Fig. 8 indicates the gas concentration (red: 1.0, green: 0.5, blue: 0.0 m^3/m^3) from 10 to 600 s at the height of ventilation when a gas releases for 5 min near the pressure recorder due to the second pressure rise (202.6 kPa).

3.5 Interactive plant simulation modeling

The demonstration version of the interactive simulation program in a natural gas governor station was designed as shown in Fig. 9. The training interface is as follows. In the center is a process flow diagram (PFD) of the pressure-regulating station, on the left are process upsets and accident scenarios, at the top are controllers and process variable information, on the right is

Table 2 Three representative scenarios

Case	Scenario
Mild	Second pressure increase → SSV_1 block → Preliminary line operation
Relief	Second pressure increase → SSV_1 block fail → Relief valve operation → Supply block
Worst	Second pressure increase → SSV_1 block fail → Relief valve fail → Gas release

a single variable chart, and at the bottom is the gas concentration of dispersed gas. The trainee can shut off gas supply by clicking the red circle above the emergency shutdown valve.

The message above the PFD indicates the status of controller alarms such as high or low when the second

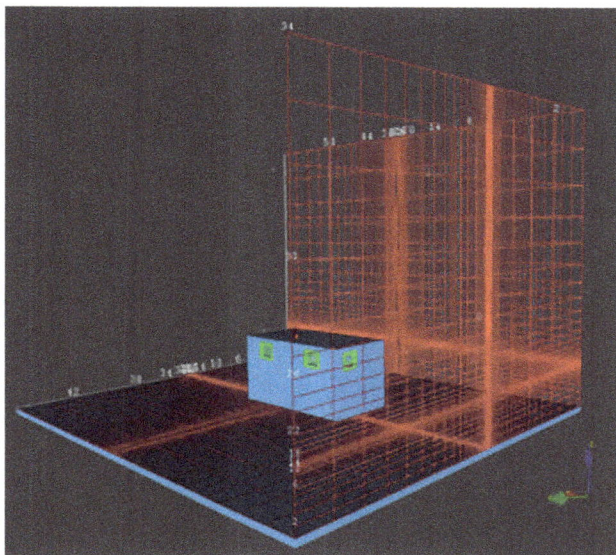

Fig. 7 3D-CFD accident simulation model using FLACS

pressure goes beyond or below each limit. The process upsets and accidentt scenarios on the left list were set up to initiate the desired scenario by clicking the button. Each scenario is identical to the mild case, relief case, and the worst case in Sect. 3.3. On the top are several tables of dynamic integrator, regulator operation, discharge model, and main and preliminary controller. In the dynamic integrator table, the trainer can specify the simulation speed and display interval, and in the regulator operation table the opening % of supply and safety valves is displayed. In the discharge model table, the discharge calculation is done in real time, and in the main and preliminary table, the status of controllers (PV, OP, SP) is displayed. By clicking the right end cell of each table, trainee can monitor the trend of the univariate chart on the right corner. Dispersed gas concentration results at the bottom are only activated in the worst case among three scenarios, which shows the 3D geometry of the pressure-regulating station, gas concentration at the height of ventilation in the form of the univariate chart, and 2D and 3D image. Figure 9 displays the process of the scenario.

By clicking the play button on the left top corner, training starts from the initial state. When the trainer clicks the 'Hole and release' scenario button on the left, the second pressure starts to rise and the second pressure keeps

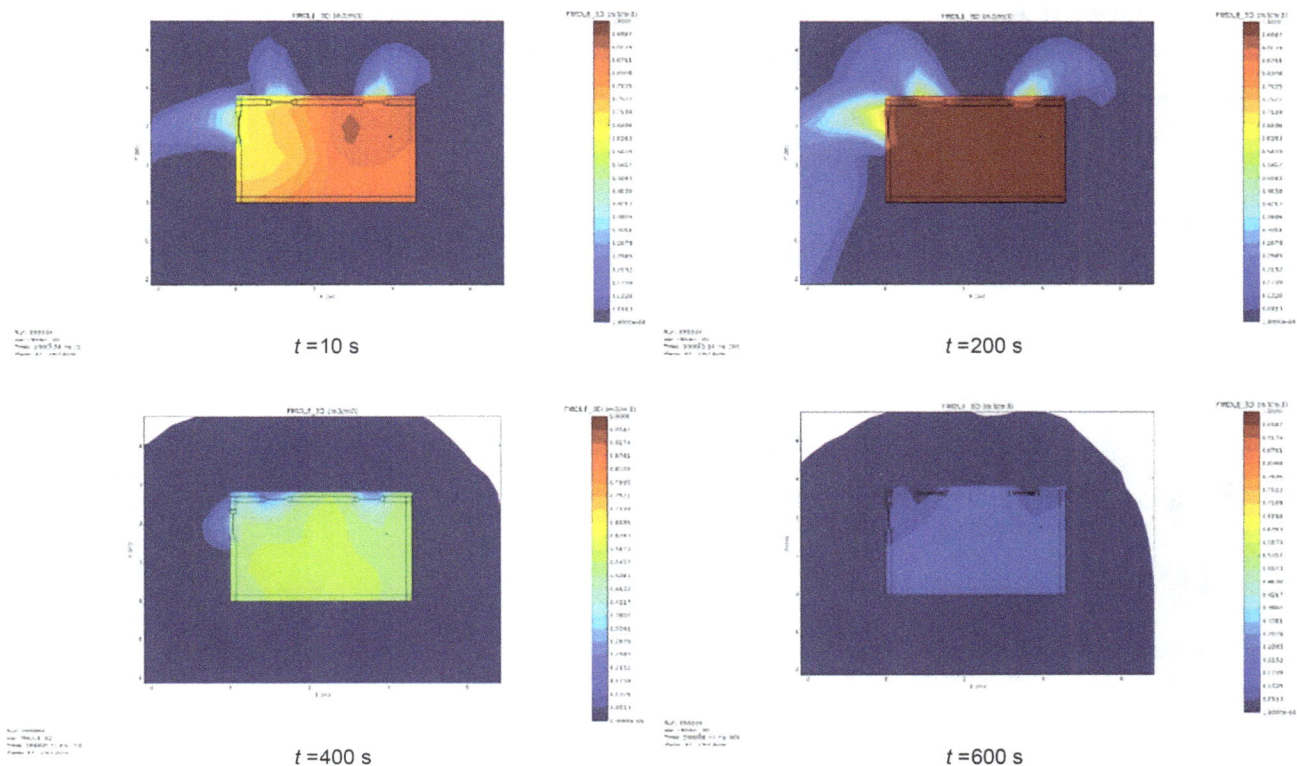

$t = 10$ s

$t = 200$ s

$t = 400$ s

$t = 600$ s

Fig. 8 Gas concentration results at the height of ventilating hole in the station

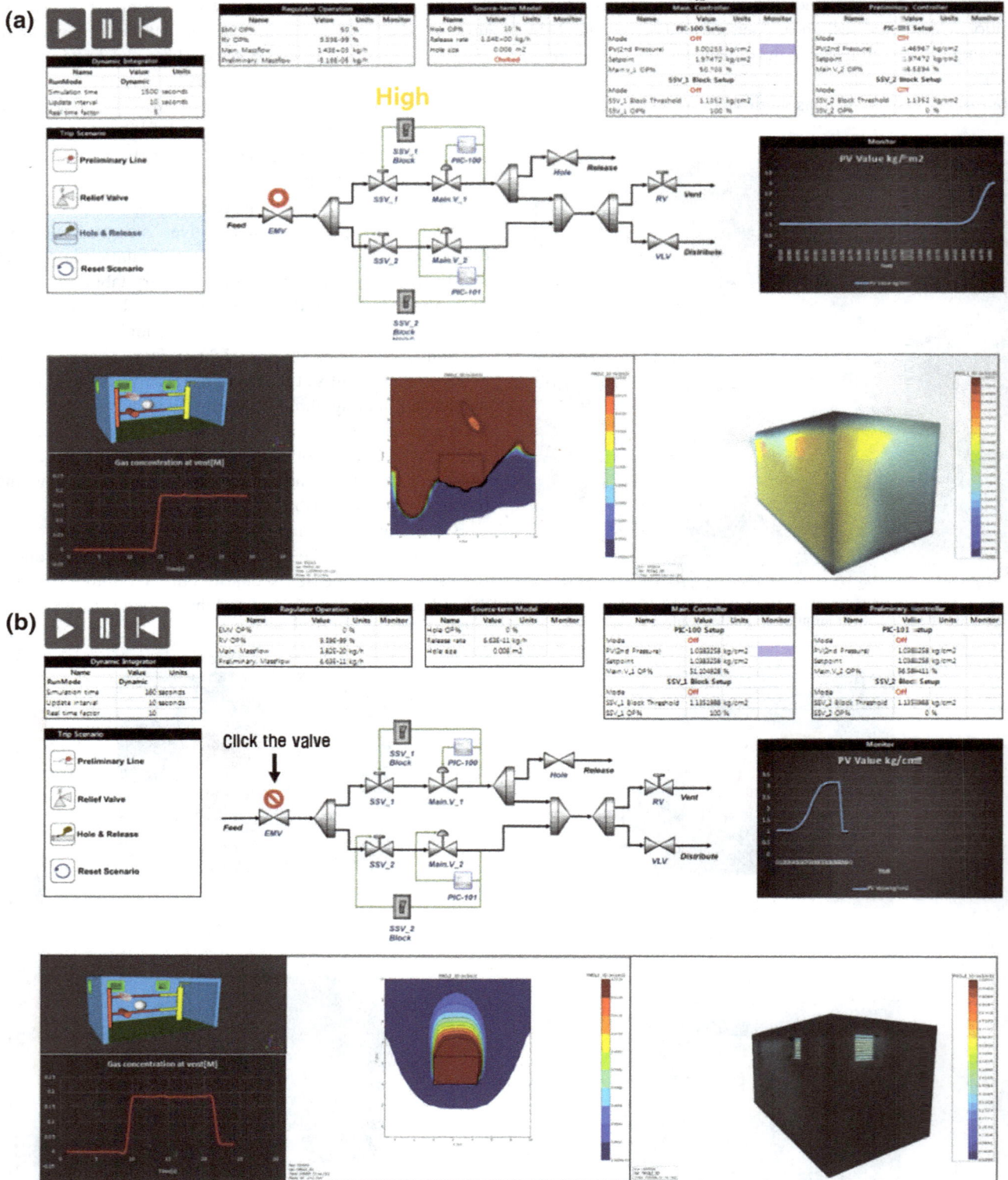

Fig. 9 Prototype of the model. **a** Release starts. **b** Emergency shutdown

rising beyond the high alarm limit. When it reaches the release pressure which was set up in advance, gas release begins through the valve named 'Hole'. If release occurs,

the release rate is calculated in the discharge model and the opening % of 'Hole' is adjusted automatically to meet the calculated release rate. In addition, the corresponding dispersion simulation result which accepts the release rate as an input is displayed at the bottom from the data bank (Fig. 9a). If the trainee clicks the EMV, the inlet supply is shutoff and only the gas released until that time would disperse outside (Fig. 9b).

3.6 Training evaluation

By using the data processing method in Sect. 2.4, a rating system was designed to evaluate the performance of the trainee. For instance, two dispersion cases with lead time of 20 and 30 min to take measures are simulated and the resulting dispersion outcome to surrounding area is compared as shown in Fig. 10. As it appears in the figure, a stark difference can be seen in the dispersion results. When this technique is applied to the operator training system, it could help trainees to make a correct and prompt decision and accordingly to minimize accident damage.

4 Future works

The developed interactive simulation model in this study could provide a training interface between operators and a training instructor and guarantee the reality of the upset situation by process and accident simulations. For more

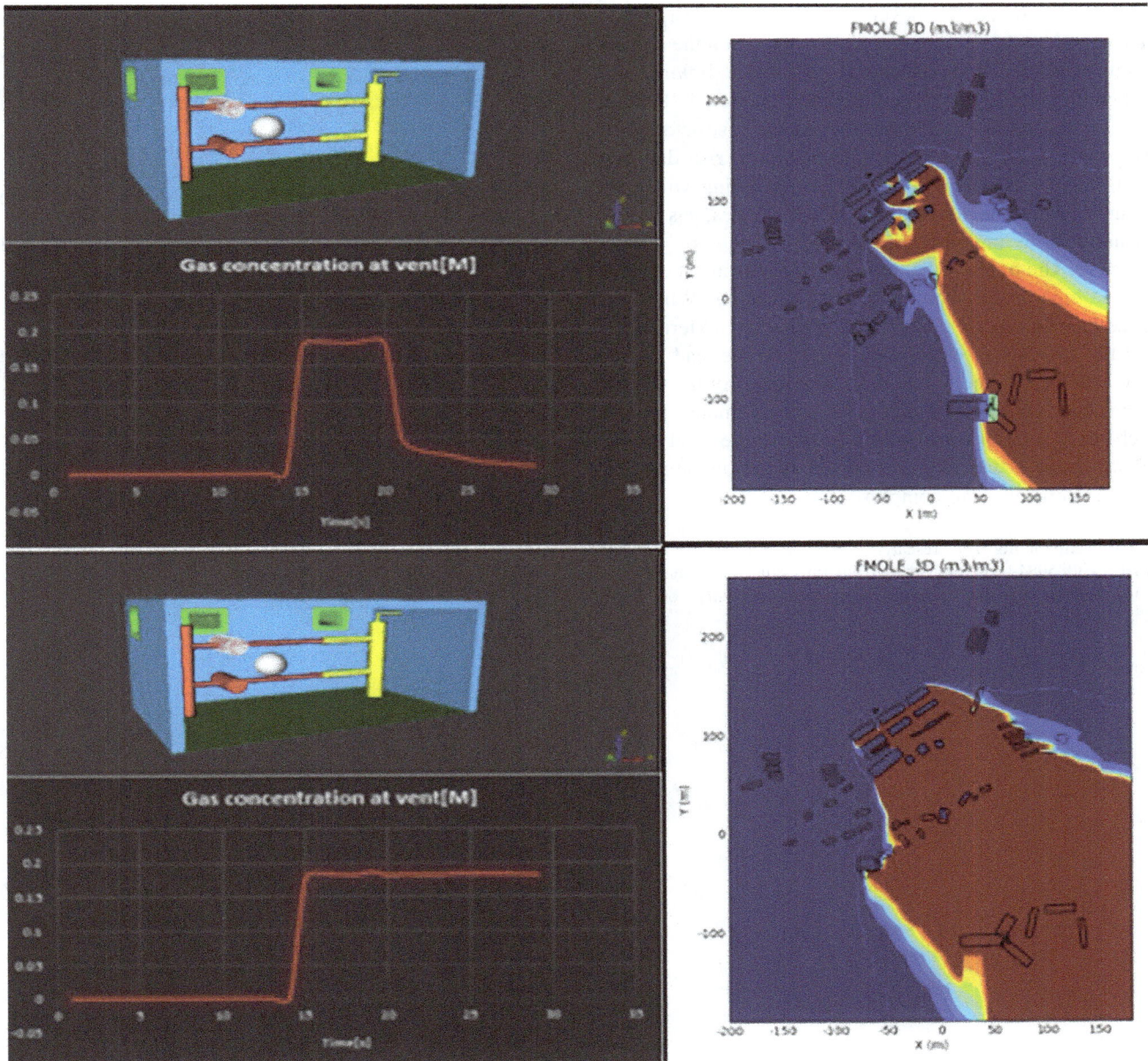

Fig. 10 Magnitude of accident with respect to the lead time of trainees' action (*top* 20 min and *bottom* 30 min)

detailed training evaluation, a quantitative risk assessment based on the calculated gas amounts from the model or additional calculations for fire and explosion effects is now in progress. Then, the training results of each trainee can be compared and the associated analysis would affect the emergency response manual in detail and the controller design in an emergency shutdown logic.

5 Conclusions

In this research, an interactive plant simulation model was developed and the performance was rated. It will be used as an internal engine of operator training targeting at pressure-regulating stations. The developed model was designed to take correct and prompt measures depending on the process upsets and accident scenarios via object linking and embedding (OLE). A representative scenario of 'Hole & Release' is studied as follows. When the instructor starts the scenario, the second pressure begins to rise due to a main valve malfunction. When it keeps rising with total failure of safety devices and reaches 202.6 kPa, gas release occurs near a pressure recorder via a hole of 10 mm diameter. The trainee who noticed this accident should shut off the emergency shutdown valve manually. When this procedure is properly done, the training scenario terminates and the simulation results with respect to the lead time to take actions are used to rate the performance of the trainee. The model could be applied to a more complex process such as a petrochemical plant in the future, and higher effectiveness of training is expected as the operating procedure becomes more complex.

Acknowledgements This research was supported by a Grant No. (14IFIP-B085984-03) from Smart Civil Infrastructure Research Program funded by the Korea Government Ministry of Land, Infrastructure and Transport (MOLIT) and The Korea Agency for Infrastructure Technology Advancement(KAIA), and by Korea Ministry of Environment (MOE) as 'the Chemical Accident Prevention Technology Development Project' (No. 201500_950003).

References

Antonovsky A, Pollock C, Straker L. Identification of the human factors contributing to maintenance failures in a petroleum operation. Hum Factors. 2014;56(2):306–2 . doi:10.1177/0018720813491424.

Cha MH, Han SH, Lee JK, Choi BG. A virtual reality based fire training simulator integrated with fire dynamics data. Fire Saf J. 2012;50:12–24. doi:10.1016/j.firesaf.2012.01.004.

Emerson Process Management. Types 1098-EGR and 1098H-EGR pressure reducing regulators. Bulletin. 2016;71(2):1098–EGR.

European Chemical Industry Council (CEFIC). The chemical industry in Europe: towards sustainability. Report 2011/2012, 2012.

Honeywell. Guide to the naphtha hydro desulphurization standard model. 2nd ed. Honeywell Process Solutions. 2005.

Kletz T. Process plant: A handbook for inherently safer design. Park Drive: Taylor Francis; 1998.

Lee SR, Sung JG, Kwon JR, Lee YS. Study of the safety for small-scale governor station. J Korean Inst Gas. 2010 1:15–146.

Manca D, Brambilla S, Colombo S. Bridging between virtual reality and accident simulation for training of process-industry operators. Adv Eng Softw. 2013;55:1–9. doi:10.1016/j.advengsoft.2012.09.002.

Nazir S, Sorensen LJ, Overgard KI, Manca D. Impact of training methods on distributed situation awareness of industrial operators. Saf Sci. 2015;73:136–45. doi:10.1016/j.ssc.2014.11.015.

Schneider Electric. Trainee guidebook: EYESIM immersive training system generic virtual crude unit. Engineering Development Research Center (EDRC) of Seoul National University. 2014.

3

Green and efficient dry gel conversion synthesis of SAPO-34 catalyst with plate-like morphology

Chun-Yu Di[1] · Xiao-Feng Li[1] · Ping Wang[1] ·
Zhi-Hong Li[1] · Bin-Bin Fan[1] · Tao Dou[2]

Abstract SAPO-34 catalyst with plate-like morphology was designed and synthesized for the first time, by the dry gel conversion method using cheap triethylamine as a structure-directing agent assisted with seed suspension containing nanosheet-like SAPO-34 seed. The latter played an important role in formation of SAPO-34 (CHA-type) with plate-like morphology. In addition, the yield of the product in the synthesis system containing seed suspension reached 97%, 15% higher than that obtained in the corresponding synthesis system without the seed suspension. Meanwhile, the plate-like SAPO-34 catalysts synthesized by this method exhibited higher selectivity to light olefins and longer lifetime in methanol-to-olefins (MTO) reaction than the traditional cubic SAPO-34 catalyst. This work provides a new technical route for green and efficient synthesis of SAPO-34 catalysts with improved MTO performance.

Keywords SAPO-34 · Dry gel conversion · Green chemistry · Seed suspension · Plate-like · MTO

✉ Bin-Bin Fan
  fanbinbin2002@yahoo.com

✉ Tao Dou
  dtao1@163.com

[1] College of Chemistry and Chemical Engineering, Taiyuan University of Technology, Taiyuan 030024, Shanxi, China

[2] CNPC Key Laboratory of Catalysis, College of Chemical Engineering, China University of Petroleum, Beijing 102249, China

Edited by Xiu-Qin Zhu

1 Introduction

Ethylene and propylene are the most widely used basic organic chemical materials, and they play an important role in the petrochemical industry. At present, light olefins are mainly produced by cracking light hydrocarbon (naphtha and light diesel oil). However, with the shortage of oil resources, developing alternative routes for ethylene and propylene production has attracted intense attraction, in which the conversion of natural gas or coal to light olefins via methanol is the most promising route (Chen et al. 2005). The conversion process of coal or natural gas to light olefins via methanol is an effective way to solve such problems such as the limited oil resource and increasing olefins demands (Qi et al. 2005), such as UOP/Hydro's methanol-to-olefins (USA) (Chen et al. 2005), syngas via dimethylether to olefins of Dalian Institute of Chemical Physics (China) (Tian et al. 2015) and Lurgi's methanol to propylene (Rothaemel and Holtmann 2002).

Silicoaluminophosphate (SAPO) molecular sieves have been widely studied because of their many technological applications. Among the SAPOs, SAPO-34 with a chabazite-related structure has exhibited excellent catalytic performance in the methanol-to-olefin (MTO) conversion due to its relatively small pore diameter (Pastore et al. 2005), medium acid strength and high hydrothermal stability (Marchi and Froment 1991; Wei et al. 2012; Wilson and Barger 1999). However, SAPO-34 is easily deactivated by coke, which can heavily block the internal channels of the SAPO-34 crystals and decrease both activity and selectivity, resulting in a short catalyst lifetime (Qi et al. 2007; Lee et al. 2007). During the MTO process, coke formation is related to many factors, such as Si/Al ratio (Xu et al. 2008), acidity (Ye et al. 2011), crystal morphology, as well as crystal size (Chen et al. 1999; Álvaro-Muñoz et al. 2012), in which the catalyst acidity and size are two important factors. Studies on the crystal size and morphology have shown that SAPO-34 catalysts with small crystal size or with nanosheet-like morphology generally exhibit better catalytic activity, selectivity and longer lifetime due to enhancing the accessibility of methanol into its cages and promoting the diffusion of the products. However, work on synthesis of SAPO-34 with nanosheet-like or plate-like morphology is very limited. Recently, Sun et al. (2014) prepared nanosheet-like SAPO-34 and SAPO-18 molecular sieves with different silicon contents under hydrothermal conditions by using tetraethylammonium hydroxide (TEAOH) as the template, and the catalysts showed high catalytic activity in the MTO reaction. However, the employed hydrothermal synthesis method has the following problems: (1) producing large amounts of waste and harmful gas; (2) using expensive TEAOH; and (3) difficulties in solid–liquid separation (Roth et al. 2014; Choi et al. 2009). In order to overcome these problems, a new alternative synthesis method, i.e., zeolites-dry gel conversion (DGC), used in the synthesis of SAPO molecular sieves has recently shown good potential. This method involves treating pre-dried gel powder at elevated temperatures and pressures to form crystalline molecular sieves (Xu et al. 1990; Rao et al. 1998). Compared with the traditional hydrothermal method, DGC does not produce mother liquor (Yang et al. 2012), can avoid complicated separation processes (Yang et al. 2010) and can give high product yield (Matsukata et al. 1999; Cundy and Cox 2003). In addition, DGC uses a lower amount of organic template and the organic template is easy to recycle and reuse. Therefore, DGC is more environmentally friendly and economical. Many kinds of aluminophosphates (AlPO) and silicoaluminophosphates (SAPOs) have been synthesized by dry gel conversion (Askari et al. 2014). For example, Hirota et al. (2010) synthesized SAPO-34 with an average crystal size of 75 nm by DGC, using TEAOH as the structure-directing agent. However, to our best knowledge, nanosized crystalline catalysts often suffer from some problems, such as low product yield, low hydrothermal stability and crystalline defects due to the intergrowth of crystals. Hence, a novel synthesis strategy is desirable for design and synthesis of highly hydrothermal-stable and plate-like SAPO-34 molecular sieve with a high yield by DGC. In addition, SAPO-34 molecular sieves with low silicon content generally have longer lifetime in the MTO reaction due to their low acid strength (Dahl et al. 1999; Izadbakhsh et al. 2009a, b). But reports on synthesis of SAPO-34 molecular sieves with low silicon content by DGC are scarce.

In this work, we developed a seed suspension-assisted method (containing nanosheet-like SAPO-34 seed) for preparation of SAPO-34 molecular sieves with plate-like morphology by DGC (Fig. 1) using cheap triethylamine (TEA) as the structure-directing agent (SDA). The results showed that this novel strategy could synthesize highly hydrothermal-stable and plate-like SAPO-34 molecular sieve product with a low silicon content, and the product yield was up to 97%. Compared with the traditional cubic SAPO-34 molecular sieve, the plate-like SAPO-34 catalysts synthesized by this method exhibited high selectivity to light olefins (ethylene + propylene) and long lifetime in methanol-to-olefins (MTO) reaction. The selectivity to light olefins increased from 81% to 87%, and the catalyst lifetime was more than doubled.

2 Experimental

2.1 Synthesis of SAPO-34

The SAPO-34 seed suspension was prepared by hydrothermal crystallization from a gel with a molar composition of 1 Al_2O_3: 0.9 P_2O_5: 0.3 SiO_2: 1.6 TEAOH: 60 H_2O. The gel was put into a stainless steel autoclave lined with Teflon, and then it was heated at 150 °C for 6 h. The obtained slurry mixture was the seed suspension (denoted as SS), and the seed (denoted as SD) of SAPO-34 molecular sieve can be obtained by separating and drying the solid product in the crystal seed suspension.

The SAPO-34 catalysts were synthesized by DGC (Fig. 1) with triethylamine (TEA) as the SDA. The molar composition of the initial gel was 1.0 Al_2O_3: 2.5 TEA: 0.9 P_2O_5: 0.3 SiO_2: 60 H_2O. Different mass percentages of SS were added into the initial gel. The detailed synthesis procedures were as follows. At first, pseudoboehmite (73 wt% Al_2O_3), TEA (Alfa Aesar), silica sol (40 wt% SiO_2) and distilled water were mixed and stirred at room

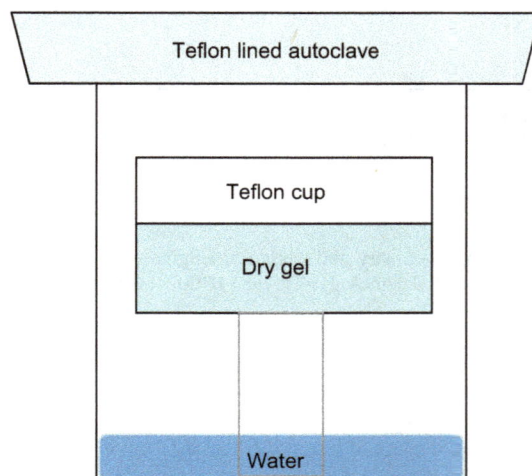

Fig. 1 Diagram of the reaction vessel used in the DGC method

temperature. Then, SS was added to the mixture and stirred. Finally, phosphoric acid (85 wt% H_3PO_4) was added dropwise to the mixture under stirring. The synthesis mixture was stirred and then dried at 110 °C to obtain a dry gel. The dry gel with different amounts of SS was placed in an autoclave, and crystallization was performed at 170 °C for 48 h.

The solid product was washed only once by centrifuging with distilled water and then dried at 100 °C over night. The as-synthesized products were calcined at 550 °C for 5 h to remove the template. The synthesis conditions of different samples are given in Table 1.

2.2 Characterization

The powder X-ray diffraction (XRD) patterns of as-synthesized samples were obtained on a Rigaku D/max-3C X-ray diffractometer (Rigaku Co., Japan) with Cu Kα radiation at 36 kV and 40 mA. The patterns were recorded from 5° to 35° with a step size of 0.02°. N_2 adsorption–desorption was performed at −196 °C using a Quantachrome AUTOSORB-1C instrument (Quantachrome Co., USA). Before the measurement, the samples were vacuum degassed at 300 °C for at least 10 h. The specific surface area (S_{BET}) and the micropore volume were calculated according to Brunauer–Emmett–Teller (BET) equation and t-plot method, respectively. Scanning electron microscope (SEM) images were recorded on a HITACHI S-570 scanning electron microscope. Each sample had been placed onto a carbon membrane, and an Au sputter coating was applied to reduce charging effects. The temperature-programmed desorption of ammonia (NH_3-TPD) was performed on a Micromeritics ASAP 2020C instrument. The sample (0.1 g) was pretreated at 450 °C for 2 h in an Ar flow of 20 mL min^{-1}. After cooling to 100 °C, the sample was saturated with 10 vol% NH_3/Ar, and then the sample was purged with Ar for 1 h to eliminate physically absorbed NH_3. Desorption of NH_3 was carried out from 100 to

600 °C at a heating rate of 5 °C min^{-1}. ^{29}Si MAS NMR spectra were recorded in 7 mm ZrO_2 rotors at 79.5 MHz on a Varian Infinity-plus 400 WB spectrometer, fitted with a BBO (broadband observe) probe. The spinning rate of the samples at the magic angle was 4 kHz. The internal standard for chemical shifts was 2,2-dimethyl-2-silapentane-5-sulfonate sodium salt (DSS).

2.3 Catalytic performance

Catalytic activity measurements were carried out in a quartz tubular fixed-bed reactor. First, the catalysts were pressed, crushed and sieved to obtain particle sizes between 250 and 500 μm. Second, 1.0 g of the shaped catalyst was placed into a quartz tube (inner diameter 10 mm) between two quartz-wool plugs. Prior to reaction, the catalyst was activated at 550 °C in air (30 mL min^{-1}) for 2 h. Aqueous methanol solution (95 wt%) was fed into the reactor under atmospheric pressure. The volume hourly space velocity (volume of methanol aqueous solution flowing through a unit volume of catalyst in a unit time) was 3 h^{-1}, and the reaction temperature was 450 °C. The analysis of the reaction products was performed using an on-line gas chromatograph Agilent GC (6890 N), equipped with a flame ionization detector (FID) and Plot-Q column. The conversion and selectivity were calculated on CH_2 basis, and dimethyl ether (DME) was considered to be a reactant for the calculation.

3 Results and discussion

3.1 Crystalline structure and morphological features of SAPO-34 seed

Figure 2a shows the XRD patterns of the as-synthesized SAPO-34 seed (SD) contained in the seed suspension (SS). From Fig. 2a, it can be seen that the representative diffraction peaks at 9.5°, 13.0°, 16.2°, 20.7°, 26.0° and 31.0° were observed in the XRD pattern of the SD, corresponding to pure SAPO-34 with a CHA-structure (Rao and Matsukata 1996). Furthermore, the peaks of the SD are wider than the other as-synthesized samples, indicating that the crystallite size of SD is smaller (Li et al. 2014). This can be further confirmed by the SEM images. As shown in Fig. 2b, the SD has nanosheet-like morphology with an average particle below 200 nm, indicating that the precursor gel at low temperature and short crystallization time can transform to nanosheet-like crystals. This is in agreement with the results reported in the literature (Lei et al. 2013).

Table 1 Synthesis conditions, yield and phase of the synthesized SAPO crystals

Samples	SS^a, wt%	$SiO_2/Al_2O_3^b$	Yield, $\%^c$	Phase identity
S-1	0	0.3	82	CHA-type
S-2	10	0.3	88	CHA-type
S-3	20	0.3	97	CHA-type
S-4	30	0.3	91	CHA-type

^{a} SS content in the synthesis suspension

^{b} SiO_2/Al_2O_3 in dry gel

^{c} Product yield calculated by the mass ratio of calcined solid product to SiO_2, Al_2O_3 and P_2O_5 in the synthesis gel

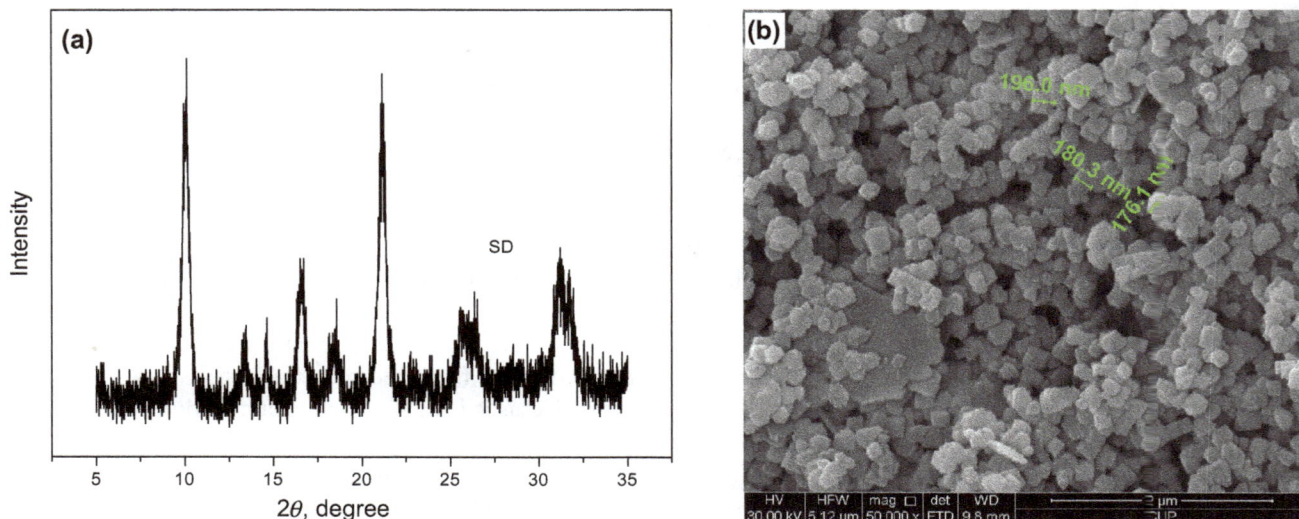

Fig. 2 XRD pattern (**a**) and SEM image (**b**) of the SAPO-34 seed (SD)

3.2 Physicochemical properties of SAPO-34

Figure 3 shows the XRD patterns of the SAPO-34 samples prepared by dry gel conversion with different amounts of SAPO-34 SS. It can be seen that all the synthesized samples showed the typical diffraction patterns of CHA-structure SAPO-34 without the presence of other impurity phases (Lee et al. 2007). However, the SS added into the initial gel influenced the relative intensity of the different diffraction peaks. For S-2, S-3 and S-4 samples, their peak intensities at $2\theta = 20.7°$ were much higher than that of S-1, while their peak intensities at $2\theta = 13.0°$ were much lower than that of S-1. These phenomena indicated that the SS had great influence on the growth of different crystal faces. In addition, as can be seen from Table 1, the SS can significantly improve product yield and promote

conversion of more amorphous materials into SAPO-34 molecular sieve. The yield of the sample S-3 was 97%, 15% higher than that of the sample S-1.

Figure 4 shows the SEM images of the different samples. The S-1 sample had a larger cubic crystal structure of about 4 μm, whereas the other synthesized samples exhibited a more plate (or sheet)-like morphology. The reason maybe that seed solution not only provided more crystal nucleus (crystal growth site) but also hindered the growth rate of crystals in one dimension in the dry gel conversion system. Furthermore, with the increase in the amount of SS in the dry gel, the framework of SAPO-34 particles transformed to a plate-like structure with a decrease in thickness, suggesting that the particles mainly grew on the periphery of the nanosheet-like structure of the SD. The thickness of particle S-2 was about 1 μm, while the thickness of particle S-3 was about 500 nm. But when the amount of SS was further increased to 30%, the particles of the obtained S-4 sample became smaller and thinner. Moreover, their plate-like morphologies became very regular. This can be attributed to the addition of the SS to the dry gel. The added SS provided a large number of nucleation sites and the formed silicoaluminophosphate species aggregated into a plate structure.

Based on the XRD and SEM results we can propose a scheme that qualitatively describes the formation of plate-like SAPO-34 with the assist of seed suspension under the DGC conditions (Fig. 5). The seed suspension contained a large number of very thin SAPO-34 molecular sieve particles. At the same time, there were a large number of SAPO-34 molecular sieve secondary structure units in the seed suspension (Sun et al. 2016). In the dry gel conversion system, the microcrystalline structure guided the surrounding silicon aluminum phosphate species to continue

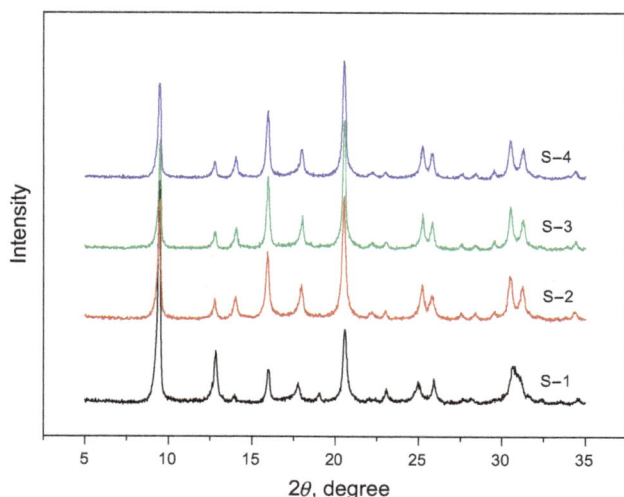

Fig. 3 XRD patterns of the synthesized SAPO-34 samples

Fig. 4 SEM images of the synthesized SAPO-34 samples

to grow in situ along its edges and eventually form a plate (or sheet-like) SAPO-34 molecular sieve. In addition, Zhang et al. (2011) studied of the crystallization process of the synthesis of SAPO-34 in the dry gel conversion system and found that in the early stage of the crystallization, the gel samples generated a semi-crystalline layered phase. So, they speculate that the layered phase is rich in the double six ring structure, which is very important for the synthesis of SAPO-34 molecular sieve by the dry gel conversion method.

The NH$_3$-TPD plots of the four catalyst samples are shown in Fig. 6. Four samples gave peaks at approximately 210 and 380 °C, which correspond to the weak and strong acid sites, respectively. The desorption peak at low temperature was attributed to the hydroxyl groups (–OH) bounded to the defect sites, i.e., POH, SiOH and AlOH (Campelo et al. 2000; Dumitriu et al. 1997). As shown in Fig. 6, the four samples had similar acid strength and weak

acid amounts, whereas their strong acid amounts slightly decreased with the increase in the added SS.

The nitrogen adsorption–desorption isotherms of four catalyst samples are presented in Fig. 7 with corresponding textural data listed in Table 2. All of the samples had similar micropore volumes in the range of 0.23–0.27 cm^3 g^{-1}, and the samples synthesized using seed suspension exhibited higher BET surface areas.

In the MTO reaction, the catalyst is always used in a harsh high temperature hydrothermal environment. Therefore, it is very necessary to study the hydrothermal stability of different types of catalysts. In this study, four catalyst samples with different morphologies were treated by high temperature hydrothermal aging, and the specific surface area (S_{BET}) was used to reflect their hydrothermal stability. As shown in Fig. 8, the specific surface area (S_{BET}) of the sample S-4 with nanosheet-like structure decreased from 608 to 360 m^2/g after 80 h of hydrothermal aging. In

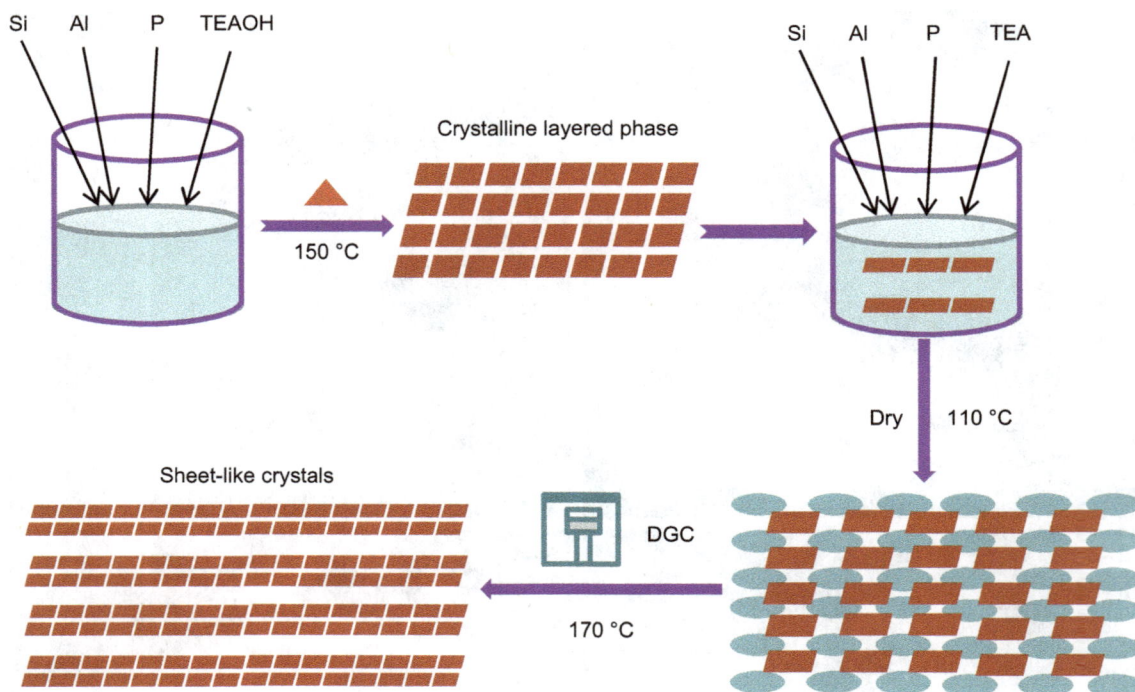

Fig. 5 Illustration of the formation of plate-like SAPO-34 under DGC conditions

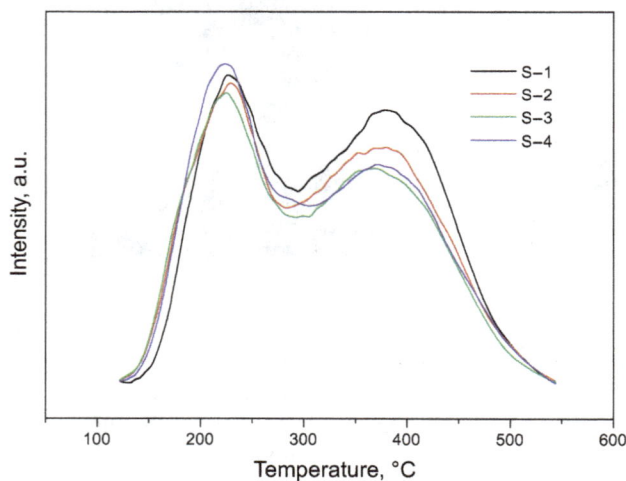

Fig. 6 NH₃-TPD profiles of the synthesized SAPO-34 samples

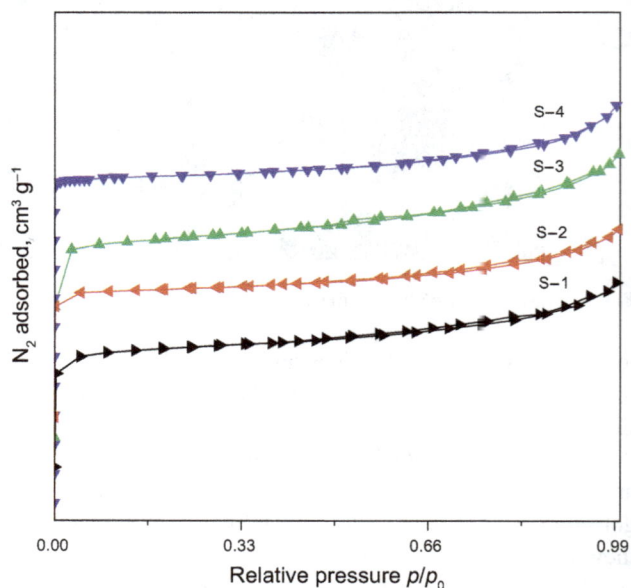

Fig. 7 Nitrogen adsorption–desorption isotherms of the synthesized SAPO-34 samples

contrast, the specific surface areas of cubic or plate-like SAPO-34 molecular sieve samples decreased slightly after the same treatment. These results demonstrated that cubic (S-1) or plate-like SAPO-34 samples (S-2 and S-3) had better hydrothermal stability than the nanosheet-like S-4.

3.3 Catalytic performance

As shown in Fig. 9, SAPO-34 samples synthesized at different conditions showed different methanol conversions and lifetimes. It can be seen that methanol was completely converted over all the catalysts with ethylene and propylene as the main products. Herein, catalyst lifetime is defined as the time when the methanol conversion reaches 98%, and the highest selectivity to light olefins (ethylene and propylene) in the catalyst lifetime is used to represent

Table 2 Physical properties of the synthesized SAPO-34 samples

Samples	Surface area, m²/g		Micro pore volume, cm³/g	Particle size, μm
	S_{BET}	$S_{External}$		
S-1	530	8	0.23	4.0
S-2	586	14	0.27	3.0
S-3	612	20	0.28	1.5
S-4	608	22	0.25	1.5

Table 3 MTO reaction results over SAPO-34 samples

Sample	Lifetime, min	Selectivity to product, %								Selectivity to $C_2^=+C_3^=$, %
		CH_4	C_2H_4	C_2H_6	C_3H_6	C_3H_8	C_4	C_5	C_6+	
S-1	80	1.6	40.9	0.5	40.1	1.9	12.4	2.3	0.3	81.0
S-2	120	1.2	44.0	0.3	39.8	0.8	11.5	2.0	0.4	83.8
S-3	180	0.8	45.6	0.3	41.7	0.3	9.7	1.5	0.2	87.2
S-4	150	1.0	45.1	0.3	41.4	0.6	9.1	1.9	0.6	86.6

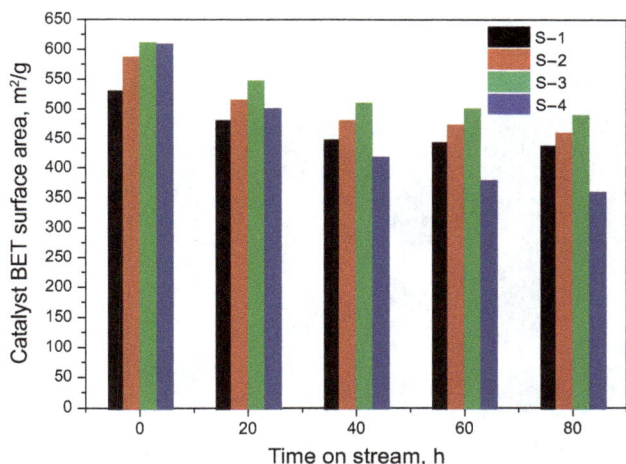

Fig. 8 Duration of hydrothermal aging at 800 °C and 100% steam

the catalytic activity according to the literature (Zhu et al. 2010). Figure 9 and Table 3 show that under the same reaction conditions, the lifetime of sample S-1 was about 80 min with 81% selectivity to ethylene and propylene, whereas the lifetime of S-3 sample with the thinner layers and the largest surface area could reach 180 min with 87% selectivity to ethylene and propylene. The significant improvement of catalyst lifetime and activity attributed to the morphology of the SAPO-34 samples. In MTO reaction, the successive polymerization, which would result in coke formation, may be partly avoided over plate-like SAPO-34 catalysts due to their short diffusion length. The fast deactivation of the S-1 catalyst (conventional large cubic morphology) can be attributed to the coke formation that occurs near the external surface of the catalyst

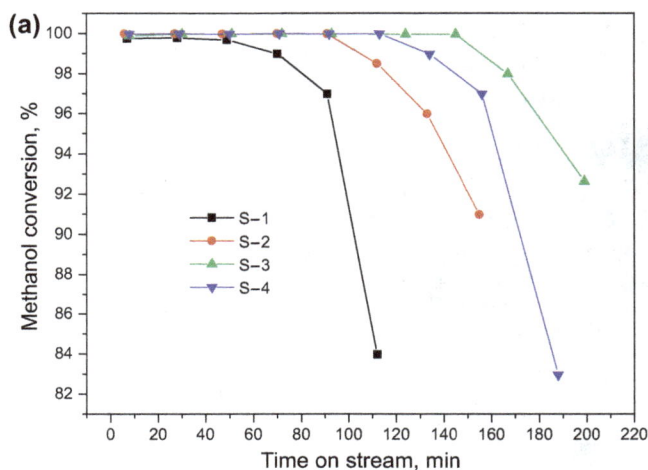

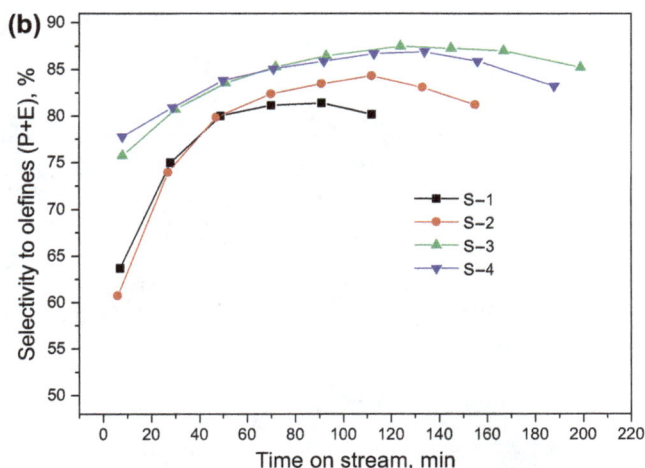

Fig. 9 Catalytic performance of SAPO-34 samples in MTO reaction at 450 °C and 3.0 h⁻¹

Table 4 Synthesis conditions, yield and phase of the synthesized SAPO samples

Samples	SS, wt%	SiO_2/Al_2O_3	Yield, %	Phase identity
S-3	20	0.3	97	CHA-type
S-5	20	0.2	95	CHA-type
S-6	20	0.1	94	CHA-type
S-7	20	0.05	85	CHA-type/AEI-type
S-8	0	0.1	76	CHA-type/AEI-type

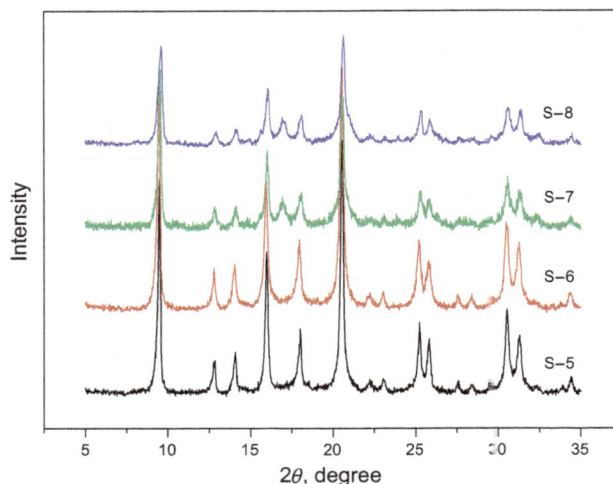

Fig. 10 XRD patterns of the synthesized SAPO-34 samples

particles, gradually blocking the diffusion path of oxygenates to the inner core of catalysts. However, for S-4 catalyst with the thinnest sheet, its deactivation can partially attribute to its poor hydrothermal stability, which could cause its crystal structure destruction under the MTO reaction conditions.

Fig. 11 SEM images of the synthesized SAPO-34 samples

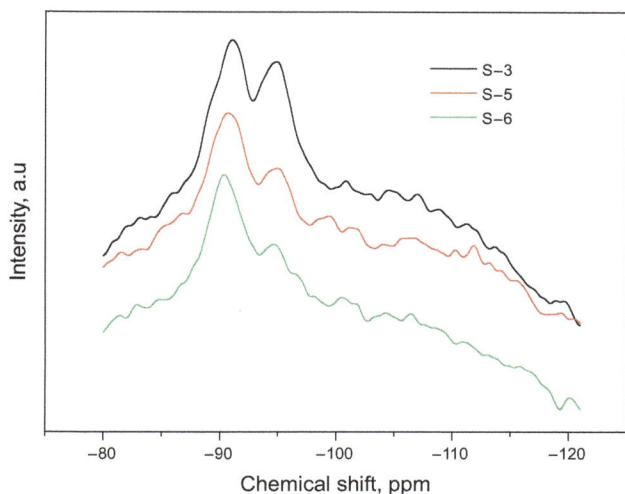

Fig. 12 ^{29}Si MAS NMR spectra of prepared SAPO-34 samples

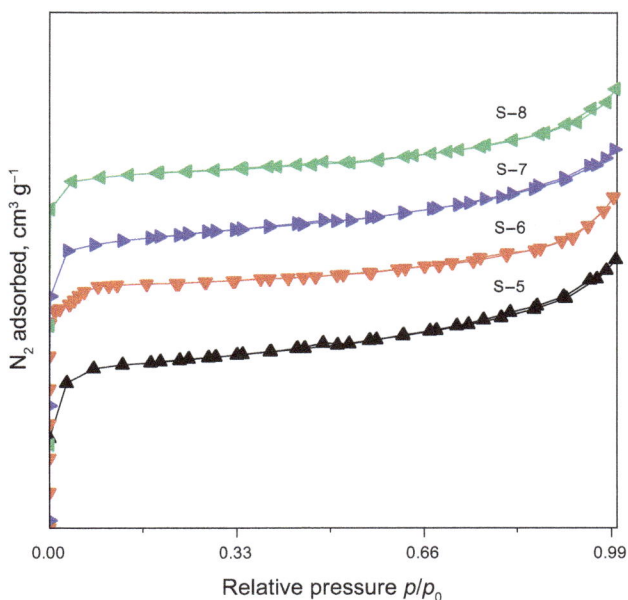

Fig. 13 Nitrogen adsorption–desorption isotherms of the synthesized SAPO-34 samples

3.4 Physicochemical properties of SAPO-34 with low silicon content

Apart from the morphology of SAPO-34 catalyst, the SiO$_2$/Al$_2$O$_3$ ratio of it is also an important factor for the catalytic lifetime (Sastre et al. 1997; Izadbakhsh et al. 2009a, b; Tan et al. 2002). In order to further improve the catalytic performance of SAPO-34 catalyst in MTO reaction, a series of plate-like SAPO-34 molecular sieve with low SiO$_2$/Al$_2$O$_3$ ratio were synthesized by DGC (Table 4).

Samples S-5, S-6 and S-7 were synthesized by DGC with the same compositions as S-3 except for silicon content. The XRD patterns (Fig. 10) of S-5 and S-6 were in agreement with that simulated from the CHA framework type (pure SAPO-34), whereas a small amount of the framework type of AEI was observed in the XRD patterns of S-7 and S-8 (Simmen et al. 1991), indicating that addition of seed suspension (SS) is beneficial to the synthesis of SAPO-34 with low Si contents by DGC.

The SEM images (Fig. 11) showed that S-5, S-6 and S-7 samples exhibited uniform plate-like morphology with the thickness of each plate in the range of 400-500 nm, whereas S-8 synthesized in the absence of the SS showed conventional cube-like morphology with a particle size of about 4 μm.

As shown in Fig. 12, all of the samples showed two major resonances at around −91 and −95 ppm, which were assigned to Si (4Al) and Si (3Al), respectively. The peaks at around −100, −105 and −109 ppm correspond to the signal of Si (2Al), Si (1Al) and Si (0Al), respectively (Shen et al. 2012). Based on the integrated areas of the resonance at −95 ppm, the concentration of Si (4Al) species in the three samples had the following order: S-6 > S-5 > S-3. It seems that under the DGC conditions, the Si species existing in the samples were incorporated into the frameworks via both SM2 (one Si substitution for one P, which forms Si (4Al) species) and SM3 [double Si substitution for pairs of Al and P, which forms Si (nAl) (n = 0–3) species] substitution mechanisms (Tian et al. 2013). Decreasing the Si content in the dry gel can efficiently promote the formation of the Si (4Al) unit. The

Samples	Surface area, m^2/g		Micro pore volume, cm^3/g	Particle size, μm
	S_{BET}	$S_{External}$		
S-5	610	19	0.26	1.7
S-6	636	23	0.28	1.2
S-7	536	20	0.23	1.4
S-8	508	9	0.21	3.5

Table 5 Physical properties of the synthesized SAPO-34 samples

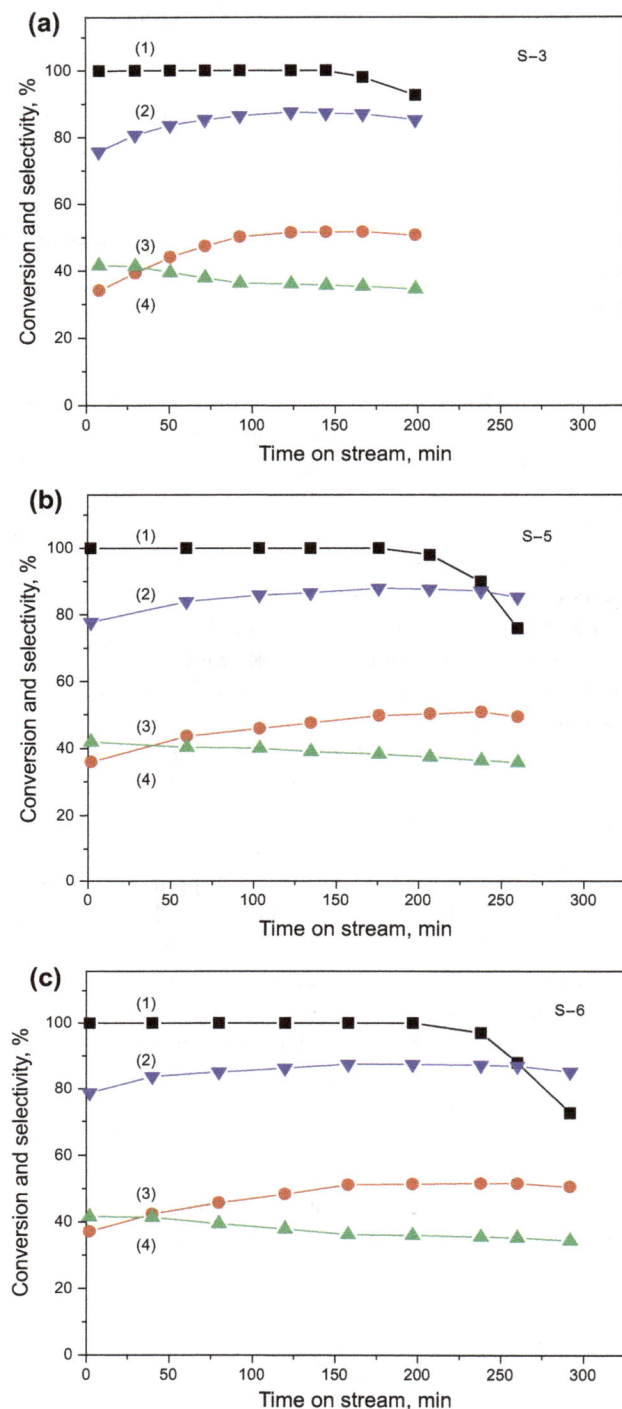

Fig. 14 Performance of SAPO-34 samples in MTO reaction at 450 °C and 3.0 h^{-1} (*1*) Methanol conversion (*black filled squares*), (*2*) selectivity to C$_2$H$_4$ and C$_3$H$_6$ (*blue filled triangles*), (*3*) selectivity to C$_2$H$_4$ (*green filled triangles*), (*4*) selectivity to C$_3$H$_6$ (*red filled circles*)

displayed the characteristic type I isotherms, confirming the microporosity of the samples. The BET surface area of S-5 and S-6 was 610 and 636 m^2/g, respectively.

3.5 Catalytic performance of plate-like SAPO-34 with low silicon contents

As shown in Fig. 14, all the catalysts exhibited a long catalyst lifetime and high selectivity to ethylene and propylene because the plate-like crystals can greatly enhance the mass transfer of reactant and generated products during MTO process. Significantly, the S-6 sample synthesized with SiO$_2$/Al$_2$O$_3$ of 0.1 and the highest Si (4Al) content (reducing the acidity of SAPO-34 catalysts) can retain more coke species than the others without fast deactivation (Izadbakhsh et al. 2009a, b). At the same time, for the S-6 sample, the selectivity to ethylene and propylene increased slightly faster than others, indicating that its acidity is more suitable for MTO reaction than others.

4 Conclusion

This work provides a new technical route of green and efficient synthetic strategies to create SAPO-34 molecular sieve with plate-like morphology. SAPO-34 was synthesized by the dry gel conversion method using cheap triethylamine (TEA) as structure-directing agent with the assistance of seed suspension containing nanosheet-like SAPO-34 seed. In addition, the yield of the product in the synthesis system containing seed suspension reached 97%, 15% higher than that obtained in the corresponding synthesis system without the seed suspension. Compared with the traditional cubic SAPO-34 molecular sieve, the selectivity for olefins (ethylene + propylene) for the plate-like SAPO-34 reached 87%, and the catalyst lifetime was more than doubled.

References

Álvaro-Muñoz T, Márquez-Álvarez C, Sastre E. Use of different templates on SAPO-34 synthesis: effect on the acidity and catalytic activity in the MTO reaction. Catal Today. 2012;179(1):27–34. doi:10.1016/j.cattod.2011.07.038.

Askari S, Sedighi Z, Halladj R. Rapid synthesis of SAPO-34 nanocatalyst by dry gel conversion method templated with morphline: investigating the effects of experimental parameters. Microporous Mesoporous Mater. 2014;197(10) 229–36. doi:10.1016/j.micromeso.2014.06.028.

Campelo JM, Lafont F, Marinas JM, et al. Studies of catalyst deactivation in methanol conversion with high, medium and small pore silicoaluminophosphates. Appl Catal A Gen. 2000;192(1):85–96. doi:10.1016/S0926-860X(99)00329-4.

Chen JQ, Bozzano A, Glover B, et al. Recent advancements in ethylene and propylene production using the UOP/Hydro MTO process. Catal Today. 2005;106(1–4):103–7. doi:10.1016/j.cattod.2005.07.178.

textual properties of the SAPO-34 samples determined with N$_2$ adsorption–desorption measurement are shown in Fig. 13 and summarized in Table 5. All the isotherms

Chen D, Moljord K, Fuglerud T, et al. The effect of crystal size of SAPO-34 on the selectivity and deactivation of the MTO reaction. Microporous Mesoporous Mater. 1999;29(1–2): 191–203. doi:10.1016/S1387-1811(98)00331-X.

Choi M, Na K, Kim J, et al. ChemInform abstract: stable single-unit-cell nanosheets of zeolite MFI as active and long-lived catalysts. Nature. 2009;461(7261):246–9. doi:10.1038/nature08288.

Cundy CS, Cox PA. The hydrothermal synthesis of zeolites: history and development from the earliest days to the present time. Chem Rev. 2003;103(3):663–702. doi:10.1002/chin.200319217.

Dahl IM, Mostad H, Akporiaye D, et al. Structural and chemical influences on the MTO reaction: a comparison of chabazite and SAPO-34 as MTO catalysts. Microporous Mesoporous Mater. 1999;29(1):185–90. doi:10.1016/S1387-1811(98)00330-8.

Dumitriu E, Azzouz A, Hulea V, et al. Synthesis, characterization and catalytic activity of SAPO-34 obtained with piperidine as templating agent. Microporous Mater. 1997;10(1–3):1–12. doi:10.1016/S0927-6513(96)00107-1.

Hirota Y, Murata K, Tanaka S, et al. Dry gel conversion synthesis of SAPO-34 nanocrystals. Mater Chem Phys. 2010;123(2–3): 507–9. doi:10.1016/j.matchemphys.2010.05.005.

Izadbakhsh A, Farhadi F, Khorasheh F, et al. Effect of SAPO-34's composition on its physico-chemical properties and deactivation in MTO process. Appl Catal A Gen. 2009a;364(1):48–56. doi:10.1016/j.apcata.2009.05.022.

Izadbakhsh A, Farhadi F, Khorasheh F, et al. Key parameters in hydrothermal synthesis and characterization of low silicon content SAPO-34 molecular sieve. Microporous Mesoporous Mater. 2009b;126(1–2):1–7. doi:10.1016/j.micromeso.2008.12.009.

Lee YJ, Baek SC, Jun KW. Methanol conversion on SAPO-34 catalysts prepared by mixed template method. Appl Catal A Gen. 2007;329(10):130–6. doi:10.1016/j.apcata.2007.06.034.

Lei W, Liu Z, Lin X, et al. Effect of SAPO-34 molecular sieve morphology on methanol to olefins performance. Chin J Catal. 2013;34(7):1348–56. doi:10.1016/S1872-2067(12)60575-0.

Li J, Li Z, Han D, et al. Facile synthesis of SAPO-34 with small crystal size for conversion of methanol to olefins. Powder Technol. 2014;262:177–82. doi:10.1016/j.powtec.2014.04.0820032-5910.

Marchi AJ, Froment GF. Catalytic conversion of methanol to light alkenes on SAPO molecular sieves. Appl Catal. 1991;71(1): 139–52. doi:10.1016/j.apcata.2007.06.034.

Matsukata M, Ogura M, Osaki T, et al. Conversion of dry gel to microporous crystals in gas phase. Top Catal. 1999;9(1):77–92. doi:10.1023/A:1019106421183.

Pastore HO, Coluccia S, Marchese L. Porous aluminophosphates: from molecular sieves to designed acids catalysts. Cheminformatics. 2005;35(44):351–95. doi:10.1002/chin.200544244.

Qi G, Xie Z, Yang W, et al. Behaviors of coke deposition on SAPO-34 catalyst during methanol conversion to light olefins. Fuel Process Technol. 2007;88(5):437–41. doi:10.1016/j.fuproc.2006.11.008.

Qi G, Xie Z, Zhong S, et al. Advances in process research on coal or natural gas to light olefins via methanol. Mod Chem Ind. 2005;. doi:10.16606/j.cnki.issn0253-4320.2005.02.003 (in Chinese).

Rao PRHP, Leon CALY, Ueyama K, et al. Synthesis of BEA by dry gel conversion and its characterization. Microporous Mesoporous Mater. 1998;21(4–6):305–13. doi:10.1016/S1387-1811(98)00033-X.

Rao PRHP, Matsukata M. Dry-gel conversion technique for synthesis of zeolite BEA. Chem Commun. 1996;12(12):1441–2. doi:10.1039/CC9960001441.

Roth WJ, Nachtigall P, Morris RE, et al. Two-dimensional zeolites: current status and perspectives. Chem Rev. 2014;114(9): 4807–37. doi:10.1021/cr400600f.

Rothaemel M, Holtmann HD. Methanol to propylene MTP—Lurgi's way. Oil Gas. 2002;28(1):27–30. doi:10.1016/S0167-2991(07)80 142-X.

Sastre G, Lewis DW, Catlow CRA. Mechanisms of silicon incorporation in aluminophosphate molecular sieves. J Mol Catal A Chem. 1997;119(1–3):349–56. doi:10.1016/S1381-1169(96)00498-0.

Shen W, Li X, Wei Y, et al. A study of the acidity of SAPO-34 by solid-state NMR spectroscopy. Microporous Mesoporous Mater. 2012;158(8):19–25. doi:10.1016/j.micromeso.2012.03.013.

Simmen A, McCusker LB, Baerlocher C, et al. The structure determination and rietveld refinement of the aluminophosphate AIPO4-18. Zeolites. 1991;11(7):654–61. doi:10.1016/S0144-2449(05)80167-8.

Sun Q, Ma Y, Wang N, et al. High performance nanosheet-like silicoaluminophosphate molecular sieves: synthesis, 3D EDT structural analysis and MTO catalytic studies. J Mater Chem A. 2014;2(42):17828–39. doi:10.1039/C4TA03419H.

Sun Q, Wang N, Bai R, et al. Seeding induced nano-sized hierarchical SAPO-34 zeolites: cost-effective synthesis and superior MTO performance. J Mater Chem A. 2016;4:14978–82. doi:10.1039/c6ta06613e.

Tan J, Liu Z, Bao X, et al. Crystallization and Si incorporation mechanisms of SAPO-34. Microporous Mesoporous Mater. 2002;53(1–3):97–108. doi:10.1016/S1387-1811(02)00329-3.

Tian P, Li B, Xu S, et al. Investigation of the crystallization process of SAPO-35 and Si distribution in the crystals. J Phys Chem C. 2013;117(8):24–9. doi:10.1021/jp311334q.

Tian P, Wei Y, Ye M, et al. Methanol to olefins (MTO): from fundamentals to commercialization. ACS Catal. 2015;5(3): 1922–38. doi:10.1021/acscatal.5b00007.

Wei Y, Li J, Yuan C, et al. Generation of diamondoid hydrocarbons as confined compounds in SAPO-34 catalyst in the conversion of methanol. Chem Commun. 2012;48(25):3082–4. doi:10.1039/c2cc17676a.

Wilson S, Barger P. The characteristics of SAPO-34 which influence the conversion of methanol to light olefins. Microporous Mesoporous Mater. 1999;29(1–2):117–26. doi:10.1016/S1387-1811(98)00325-4.

Xu W, Dong J, Li J, et al. A novel method for the preparation of zeolite ZSM-5. J Chem Soc Chem Commun. 1990;10(10):755–6. doi:10.1039/C39900000755.

Xu L, Du A, Wei Y, et al. Synthesis of SAPO-34 with only Si (4Al) species: effect of Si contents on Si incorporation mechanism and Si coordination environment of SAPO-34. Microporous Mesoporous Mater. 2008;115(3):332–7. doi:10.1016/j.micromeso. 2008.02.001.

Yang H, Liu Z, Gao H, et al. Synthesis and catalytic performances of hierarchical SAPO-34 monolith. J Mater Chem. 2010;20(16): 3227–31. doi:10.1039/B924736J.

Yang N, Yue M, Wang Y. Synthesis of zeolites by dry gel conversion. Progress Chem. 2012;24(2):253–61. doi:10.16085/j.issn.1000-6613.2012.3.0253-09 (in Chinese).

Ye L, Cao F, Ying W, et al. Effect of different TEAOH/DEA combinations on SAPO-34's synthesis and catalytic performance. J Porous Mater. 2011;18(2):225–32. doi:10.1007/s10934-010-9374-4.

Zhang L, Bates J, Chen D, et al. Investigations of formation of molecular sieve SAPO-34. J Phys Chem C. 2011;115(45): 22309–19. doi:10.1021/jp208560t.

Zhu J, Cui Y, Zeeshan N. In situ synthesis of SAPO-34 zeolites in kaolin microspheres for a fluidized methanol or dimethyl ether to olefins process. Chin J Chem Eng. 2010;18(6):979–87. doi:10.1016/j.cjche.2010.18.6.979-987.

4

Extractive oxidative desulfurization of model oil/crude oil using KSF montmorillonite-supported 12-tungstophosphoric acid

Ezzat Rafiee[1,2] · Sadegh Sahraei[3] · Gholam Reza Moradi[3]

Abstract 12-Tungstophosphoric acid (PW) supported on KSF montmorillonite, PW/KSF, was used as catalyst for deep oxidative desulfurization (ODS) of mixed thiophenic compounds in model oil and crude oil under mild conditions using hydrogen peroxide (H_2O_2) as an oxidizing agent. A one-factor-at-a-time method was applied for optimizing the parameters such as temperature, reaction time, amount of catalyst, type of extractant and oxidant-to-sulfur compounds (S-compounds) molar ratio. The corresponding products can be easily removed from the model oil by using ethanol as the best extractant. The results showed high catalytic activity of PW/KSF in the oxidative removal of dibenzothiophene (DBT) and mixed thiophenic model oil under atmospheric pressure at 75 °C in a biphasic system. To investigate the oxidation and adsorption effects of crude oil composition on ODS, the effects of cyclohexene, 1,7-octadiene and o-xylene with different concentrations were studied.

Keywords Tungstophosphoric acid · Montmorillonite · Catalyst · Oxidative desulfurization · Clean fuel

✉ Ezzat Rafiee
 e.rafiei@razi.ac.ir; ezzat_rafiee@yahoo.com

[1] Faculty of Chemistry, Razi University, Kermanshah 67149, Iran

[2] Institute of Nano Science and Nano Technology, Razi University, Kermanshah 67149, Iran

[3] Department of Chemical Engineering, Faculty of Engineering, Razi University, Kermanshah 67149, Iran

Edited by Xiu-Qin Zhu

1 Introduction

A major cause of air pollution is the burning of fossil fuels and producing compounds such as NO_x and SO_x, which is released from factories and vehicle exhausts. Sulfur content in crude oil is highly significant due to the high toxicity, corrosivity and air pollution that results from the burning of sulfur-containing fuel products (Collins et al. 1997). In accordance with the increasingly stringent environmental regulations on sulfur concentration in transportation fuels and the demand for diesel fuel with low sulfur content, removal of sulfur-containing compounds is an important issue for the petroleum refining industry (Mei et al. 2003).

To achieve this goal, most sulfur-containing compounds are easily desulfurized by hydrodesulfurization (HDS), in which H_2 gas is used to remove the S-compounds through the formation of hydrocarbons and H_2S. It is difficult to decrease the sulfur content from several hundred ppm to a few ppm by the HDS method which is extremely efficient in removing thiols, sulfides and disulfides, but less effective for thiophenic compounds (García-Gutiérrez et al. 2008). Due to the high capital and operating costs of the HDS process due to factors such as reduced catalyst lifetime, higher hydrogen consumption, high temperature and pressure, oxidative desulfurization (ODS) is proposed as an appropriate substitute for the traditional HDS process and complementary method with ambient pressure relatively low temperature, high selectivity and no consumption of hydrogen (Khenkin and Neumann 2011; Caero et al. 2005). The integration of an ODS unit with a conventional hydrotreating unit can improve the economics of diesel desulfurization process in comparison with the current HDS technology (Stanislaus et al. 2010).

Generally, an ODS process consists of two steps: The first step is the oxidation of sulfur; various oxidants such as hydrogen peroxide (H_2O_2) (Rafiee and Eavani 2013; Gao et al. 2010a, b), formic acid (Shojaei et al. 2014), acetic acid (Abdalla et al. 2009a, b), O_2 (Dooley et al. 2013) and t-BuOOH (Hui-Peng et al. 2007; Ishihara et al. 2005) are used in ODS process. Among these oxidants, H_2O_2 is mostly chosen as an oxidant due to its innocuous side product (H_2O). Thus, some studies of H_2O_2 have been conducted for investigation of its use in ODS reactions (Rafiee and Mirnezami 2014; Lei et al. 2013; Da Silva and Dos Santos 2013). The second step is the removal of products (sulfoxide or sulfone) by solvent extraction. These compounds are removed by polar solvents such as methanol (Abdalla and Li 2012), N,N-dimethylformamide (DMF) (Bakar et al. 2012), acetonitrile (MeCN) (Li et al. 2013) and DMSO (Hassan et al. 2013). In recent years, there has been an increasing interest in polyoxometalates (POMs) as catalysts due to their composition, size, shape, photochemical response, ionic charge, acidic properties and tunable redox properties (Zhang et al. 2013). Heteropoly acids (HPAs) and supported HPAs are widely used as catalysts in various acid-catalyzed reactions and partial oxidation reactions because of their strong acidity and oxidation activity (Chamack et al. 2014; Zhang et al. 2011a, b; Te et al. 2001). Among the different types of HPAs, Keggin-type HPAs and their salts have been widely used for acid–base and oxidation reactions (Zhang et al. 2011a, b). In general, $H_3PMo_{12}O_{40}$ (PMo), $H_3PW_{12}O_{40}$ (PW), $H_4SiMo_{12}O_{40}$ (SiMo) and $H_4SiW_{12}O_{40}$ (SiW) are employed in the first instance for many applications (Katsoulis 1998). The results show that Keggin-type HPA catalysts with H_2O_2 may be promising for ODS of model oils (Kozhevnikov 1998). Previous reports have demonstrated the effective catalytic performances of supported HPAs for the oxidation of thiophene (T) and its derivatives with different oxidants (Hui-Peng et al. 2007; Kukovecz et al. 2002; Yan et al. 2009b). However, most of the research up to now has investigated homogenous or amphiphilic catalysts and oxidants (Gao et al. 2006; Zhang et al. 2012a, b; Maity et al. 2009; Qiu et al. 2009; Xue et al. 2012; Bhutto et al. 2016). In these systems, the catalysts are usually dissolved in the continuous phase or inside the emulsion droplets; thus, it is difficult to separate the catalysts and recycle them from the emulsion systems.

The current study aims to investigate a kind of catalyst which is not soluble in the oil phase or extractant solvent. PW supported on KSF (PW/KSF) was prepared by impregnation and used as catalyst for the ODS of model oil. Thiophene (T), benzothiophene (BT) and dibenzothiophene (DBT) in n-hexane were the simulated model oils. Different extractants were EtOH, MeCN and DMF. H_2O_2 was chosen as an oxidant. The effect of various parameters such as temperature, oxidant-to-S-compounds molar ratio (O/S), type of extractant, reaction time and amount of the catalyst was investigated and optimal conditions of the process were evaluated for real oil. The catalytic activity of PW/KSF was investigated for crude oil before and after the HDS process. Additionally, for assessing the influence of crude oil composition, 5 vol%, 15 vol% and 25 vol% of o-xylene, cyclohexene and 1,7-octadiene were, respectively, added to 1000 ppm mixed model oil. For testing real catalytic activity of PW/KSF, ODS of a crude oil and HDS-treated crude oil was conducted under optimal conditions.

2 Experimental

2.1 Materials and methods

$H_3PW_{12}O_{40}$ (PW) hydrate was purchased from Aldrich, and KSF montmorillonite was obtained from the Fluka Company. n-Hexane (98 %), ethanol (99 %), acetonitrile (98 %), N,N-dimethylformamide (DMF, 99.5 %), T (98 %), BT (98 %), DBT (98 %), H_2O_2 (30 %), cyclohexene (99 %), o-xylene (99 %) and 1,7-octadiene (97 %) were purchased from the Merck Company and used without further purification.

The ODS experiments were carried out at atmospheric pressure in a 50-mL glass batch reactor, equipped with a temperature controller, a condenser and a mechanical stirrer. Fourier transform infrared spectroscopy (FTIR) spectra were recorded with KBr pellets using a Rayleigh WQF-510 FTIR. The total sulfur content was analyzed with a Multi EA 3100 Element Analyzer (Analytik Jena AG Company). The Multi EA 3100 model can achieve detection limits of 10 ppb for sulfur determination by UV fluorescence and 50 ppb for nitrogen determination by chemiluminescence. All experiments were conducted and repeated twice more to ensure reproducibility of the results.

2.2 Catalyst preparation

The PW/KSF was prepared by the impregnation method, which was developed by our research group (Rafiee et al. 2009). For preparation of the PW/KSF, KSF montmorillonite was oven-dried at 120 °C for 2 h prior to its use as support. An appropriate amount of PW (to produce 40 wt% of PW to support) was dissolved in 5 mL of dry methanol and added dropwise to pre-dried KSF while stirring with a glass rod. Initially with addition of PW solution, the PW/KSF clay was in a powdery form, but with further addition of PW solution, the clay turned to a paste form. The paste changed to free flowing powder with further stirring as the methanol evaporated. Characterization of the catalyst was reported previously (Rafiee et al. 2009, 2011).

2.3 Activity test

2.3.1 Oxidation of DBT

In a typical experiment, the water bath was first heated to reach and maintain the reaction temperature (55, 65 and 75 °C), and then model oil was prepared by dissolving DBT in n-hexane/EtOH (50/50 vol. ratio) to get 1000 ppm, 500 ppm and 100 ppm solution of DBT. In addition, required amounts of 30 vol% aq. H_2O_2 solution (O/S molar ratio 8:1, 10:1, 12:1 and 14:1) and certain amounts of the catalyst (0.02 g (0.0028 mmol), 0.03 g (0.004 mmol) and 0.04 g (0.0056 mmol)) were added to the mixture and stirred vigorously while still in the water bath to start the reaction.

With the aid of thin-layer chromatography (TLC) analysis (with n-hexane/ethyl acetate, 2:1 volume ratio as eluent), the oxidized DBT (sulfones and sulfoxides) spots could be seen by illuminating the TLC plate with ultraviolet (UV) light. After completion of the reaction (80 min), the mixture was left to settle for 30 min at room temperature and the products were extracted by extractant solvent. The heterogeneous catalyst was separated by centrifuging at 1500 rpm. For quantification of ODS products, upper liquid layer (n-hexane) was withdrawn and analyzed for S by the Analytik Jena AG—Multi EA 3100 Element Analyzer.

2.3.2 Oxidation of mixed thiophenic model oil

A typical reaction procedure for ODS of mixed model oil was as follows: Mixed model oil was prepared by dissolving required amounts of DBT, BT and T (DBT/BT/T, 500:250:250 ppm) in 5 mL n-hexane to get mixed model oil with sulfur concentration of 1000 ppm. PW/KSF (0.03 g, 0.004 mmol), 5 mL extraction phase and required amount of 30 % aq. H_2O_2 solution (O/S molar ratio 10:1) were added to the model oil, and the resulting mixture was stirred at atmospheric pressure at 75 °C for 80 min. After completion of the reaction, the mixture was left to settle for 30 min at room temperature and the corresponding products were extracted by extractant solvent. TLC was performed with n-hexane/ethyl acetate, in 3:1 volume ratio, as eluent. The ODS of 500 ppm and 100 ppm (sulfur concentration) mixed model oil was performed in the same way for ODS of 1000 ppm (sulfur concentration) mixed model oil, with O/S molar ratio fixed at 10:1 in the presence of PW/KSF (0.03 g) at 75 °C.

In order to test the catalytic recyclability of PW/KSF in the ODS of 1000 ppm (sulfur concentration) mixed model oil, the heterogeneous catalyst was separated by centrifuge with 1500 rpm, washed three times with 2 mL MeCN and then dried at room temperature for 24 h. The recovered catalyst was charged into the next run with fresh reactants.

2.3.3 Oxidation of crude oil

In the same manner as the oxidation of the mixed thiophenic model oil, PW/KSF (0.03 g) catalyst, 50 µL 30 % aq. H_2O_2 and 5 mL extractant (EtOH, DMF or MeCN) were added to 5 mL crude oil (with sulfur concentration of 1000 ppm, API 41.67). The mixture was stirred vigorously at atmospheric pressure at 75 °C. After completion of the reaction, the polar phase was analyzed for S. For crude oil after HDS, 5 mL of HDS-treated crude oil (with sulfur concentration of 300 ppm), PW/KSF (0.03 g) catalyst and 15 µL 30 % aq. H_2O_2 (under the same conditions as above mentioned) were added and the mixture stirred at 75 °C for 80 min. The upper phase was withdrawn and analyzed by the total sulfur analyzer.

3 Results and discussion

3.1 DBT oxidative desulfurization (ODS)

In order to investigate the activity of PW/KSF and optimize the reaction conditions, deep desulfurization of 1000 ppm DBT in n-hexane was carried out, with EtOH as extractant, at different temperatures, O/S molar ratios and amounts of the catalyst (Table 1). As the temperature increased from 55 °C to 75 °C, the conversion of DBT increased greatly from 31 % to 99 % at an O/S molar ratio of 10:1. The corresponding $TON = N_{product}/N_{PW}$ was calculated (where TON is the mole number of converted DBT per mole of PW supported on KSF). The TON value increased with an increase in reaction time at each reaction temperature (Table 1, Nos. 5, 6 and 8). With an increase in temperature, the TON value for ODS increased at each reaction time. Above an O/S molar ratio 10:1, TON value increased very little.

Table 1 shows that at 75 °C, the conversion of DBT and corresponding reaction time changed very little when the amount of the catalyst increased from 0.03 g to 0.04 g. In addition, the ODS activity of PW/KSF increased only slightly when the O/S molar ratio was higher than 10:1. Based on the above experimental results, it was found that optimum conditions for ODS were O/S (10:1, temperature (75 °C) and amount of the catalyst (0.03 g. These optimum conditions were used in the next experiments. Prior to investigating the effect of the catalyst (PW/KSF) loading on ODS, two more experiments were conducted for comparison: one with O/S molar ratio 10:1 in the presence of 0.03 g of KSF and another without using any catalyst at optimum reaction conditions. The results showed that the conversion of DBT was 46 % and 28 %, respectively, which did not meet the requirement of deep desulfurization. It was due to the lack of a suitable reaction medium,

Table 1 Optimization of reaction conditions for oxidation of DBT (1000 ppm) in n-hexane, using EtOH as extractant and H_2O_2 as oxidant

No.	Catalyst, g:mmol	O/S molar ratio	Temperature, °C	Time, min	Conversion of DBT, %	TON value
1	0.03:0.004	10	55	80	31	2.1
2	0.03:0.004	10	65	80	68	4.62
3	0.03:0.004	10	75	80	99	6.73
4	0.03:0.004	8	75	80	92	6.26
5	0.03:0.004	12	75	60 (80)	94 (99)	6.39 (6.73)
6	0.03:0.004	14	75	60 (80)	95 (99)	6.46 (6.73)
7	0.02:0.0028	10	75	80	87	8.45
8	0.04:0.0056	10	75	60 (80)	96 (99)	4.66 (4.81)
9	–	10	75	80	28	–
10	0.03 g KSF	10	75	80	46	–

and DBT could not effectively contact the H_2O_2 (Zhang et al. 2013). In both cases, noticeable transformation of DBT into sulfone was not observed on the TLC plate.

During the reaction, it was found that the mixture (EtOH, n-hexane and DBT) tended to form two phases. The active polyoxoperoxo (peroxo-metallate complexes) species converted DBT to $DBTO_2$ and the oxidized DBT moved into the EtOH phase. The catalyst was a separate phase during the entire reaction process (Scheme 1).

The experiment was also conducted with 500 ppm and 100 ppm DBT in n-hexane with EtOH as extractant. Figure 1 shows the catalytic performance of PW/KSF for 1000, 500 and 100 ppm of DBT in n-hexane at optimum reaction conditions. The results showed that the catalyst was active for all the concentrations of DBT in n-hexane with EtOH as extractant. It can be seen that DBT was almost fully converted by PW/KSF. Table 2 shows the comparison of catalytic activity of PW/KSF with that of other reported catalysts. The results indicated that PW/KSF was quite reliable for ODS in comparison with other

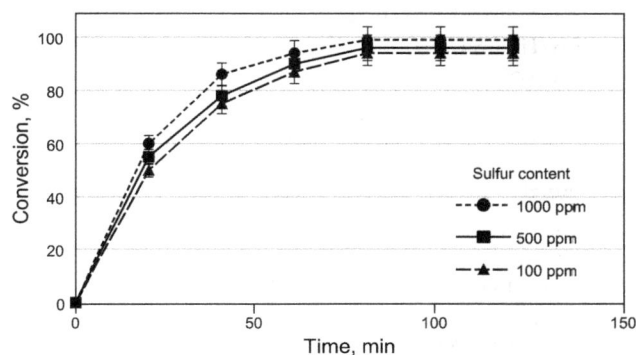

Fig. 1 Deep desulfurization of DBT in n-hexane and ethanol (O/S = 10, T = 75 °C, 0.03 g of PW/KSF)

catalysts. There are some disadvantages of these reported systems (compared with our system) such as long reaction time, high reaction temperature or large amount of catalyst and using ionic liquids as additives.

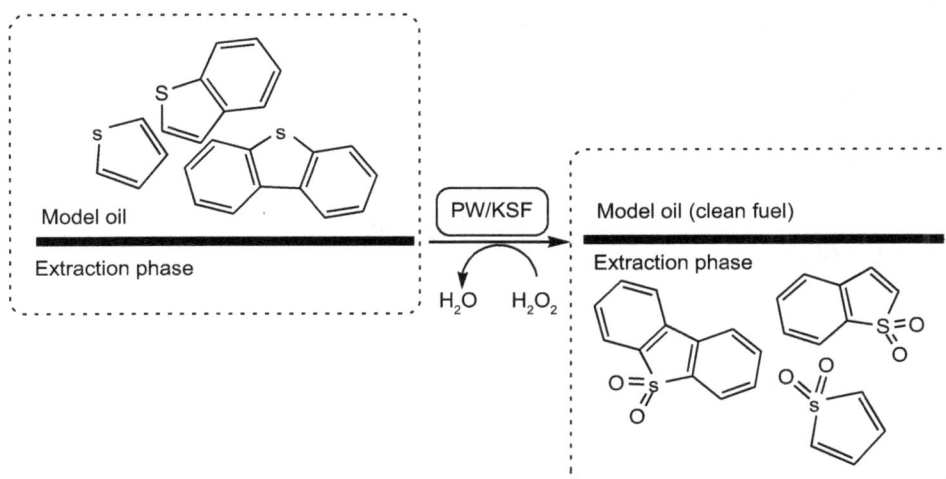

Scheme 1 Desulfurization process

Table 2 Comparison of the reaction data in this work with that using other catalyst

No.	Catalyst	S-compounds	Concentration, ppm	Reaction conditions (O/S; Temp, °C; time, min; additives)	Conversion, %	Reference
1	PW/KSF	DBT	1000	(10, 75, 80)	99	–
2	V_2O_5/TiO_2	DBT	445	(4, 70, 20)	>80	Caero et al. (2005)
3	[PyPs]PW [PhPyPs]PW	Mixed thiophenic model	1000	(1.6, 60, 120)	95 98	Rafiee and Eavani (2013)
4	$PW/SiO_2–Al_2O_3$	DBT	200	(3, 70, 120)	79.9	Hui-Feng et al. (2007)
5	MoO_3/SiO_2	DBT	2000	(4, 50, 120)	90	Ishihara et al. (2005)
6	[PhPyBs]PW	Sulfide	1000	(1, 70, 50)	98	Rafiee and Minezami (2014)
7	PW/SiO_2	BT, DBT, 4,6-DMDBT	500	(12, 60, 120)	94.8, 99.6, 97.6	Lei et al. (2013)
8	PW/CeO_2	DBT	1000	(6, 30, 30, [C_8mim]BF_4)	99.4	Zhang et al. (2013)
9	Ag-modified PW/SiO_2	DBT	800	(12, 70, 240)	89.8	Yan et al. (2009b)
10	$PW/TiO_2–SiO_2$	DBT	1000	(12, 70, 120)	96	Yan et al. (2013)
11	PW/AC	T	489	(4 mL/20 mL model oil, 90, 120)	90	Xiao et al. (2014)
12	PW/TiO_2	BT, DBT, 4,6- DMDBT	500	(12, 80, 120)	94.6, 100, 97.8	Yan et al. (2009a)
13	PW/SPC	DBT	500	(3, 60, 120)	98.6	Li et al. (2011)
14	$Mn–Co/Al_2O_3$	4,6- DMDBT	450	(Air, 150, 25)	65	Sampanthar et al. (2006)
15	WO_3/Al_2O_3	Diesel fuel	320	(11, 60, 60)	70	García-Gutiérrez et al. (2014)
16	$[(C_4H_9)_4N]\{PO_4[MoO_2)_2]_4\}$	DBT	758	(2, 70, 180, [C_8mim]PF_6)	94	Lo et al. (2003)
17	[C_4mim]HSO_4	DBT	1000	(5, Room Temp., 120)	85	Gao et al. (2010a, b)
18	[Bmim]HSO_4	Sulfide	10 mmol/4 mL solvent	(2, 25, 240)	83	Zhang et al. (2012a, b)

The mechanism of the catalytic desulfurization process (Scheme 2) is assumed to include the following steps: (1) transformation of $[PW_{12}O_{40}]^{3-}$ (denoted as PW) into its peroxide form $[PO_4\{WO(O_2)_2\}_4]^{3-}$ (denoted as PW_4) in the presence of H_2O_2; (2) oxidation of the extracted DBT to DBTO (corresponding sulfoxide) and $DBTO_2$ (corresponding sulfone) by PW_4, and reduction of PW_4 to $[PO_4\{WO(O_2)\}_4]^{3-}$; (3) regeneration of PW_4 via oxidizing its reduced form $[PO_4\{WO(O_2)\}_4]^{3-}$ by H_2O_2; (4) transformation of PW_4 into PW_{12} species with free tungsten species, after the desulfurization (Zhang et al. 2012a, b).

3.2 Desulfurization of mixed thiophenic model oil

Another model oil was prepared by dissolving required amounts of DBT, BT and T in n-hexane to get mixed model oil with sulfur concentration of 1000, 500 and 100 ppm, respectively. The experiments were performed under the conditions of O/S molar ratio 10:1, 0.03 g of PW/KSF, 75 °C and reaction time of 80 min (Figure 2). The result showed that the catalyst was effective for mixed

thiophenic compounds models. It was clear that the conversion rate of 1000, 500 and 100 ppm mixed model could reach 97 %, 95 % and 93 %, respectively. As shown in Figs. 1 and 2, the initial S-compounds concentration did not have much more effect on ODS after 80 min.

It is instructive to compare the catalytic activity of PW/KSF with different extractants (EtOH, DMF and MeCN) in the oxidation process (Fig. 3). In general, the performance of the extractant solvent depends considerably on the solubility of the sulfur oxidation products in the reaction mixture. During ODS process, the oxidized S-compounds are transferred to the extractant solvent, and the solvent can influence the mass transport and subsequently have diffusional problems, especially with porous catalysts (Caero et al. 2005). As shown in Fig. 3, EtOH was the best extractant for removing oxidized thiophenic compounds in the model oil. It can be found that the catalytic activity of PW/KSF depended on the kind of the extractant (Fig. 3). To study the recyclability of the catalyst, 1000 ppm mixed model oil was used successively seven times and the catalytic activity decreased a little (Fig. 4). The recycled

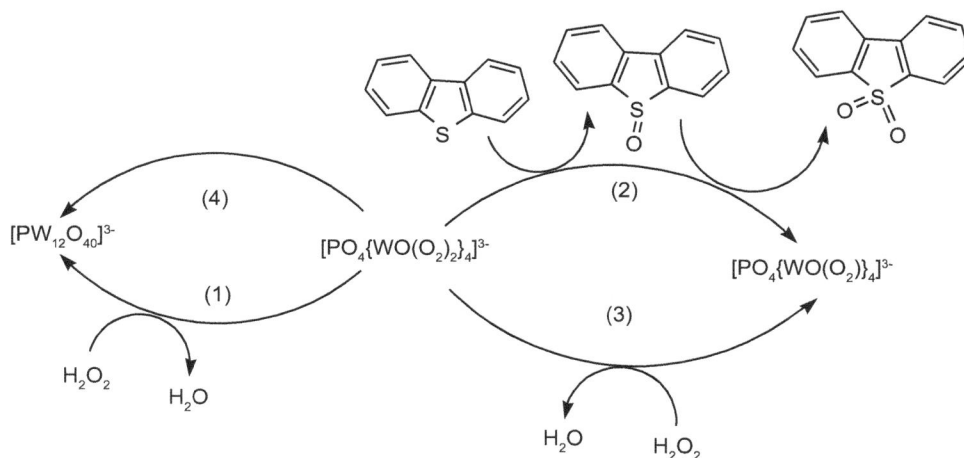

Scheme 2 Probable mechanism for catalytic desulfurization process using KSF/PW and H_2O_2

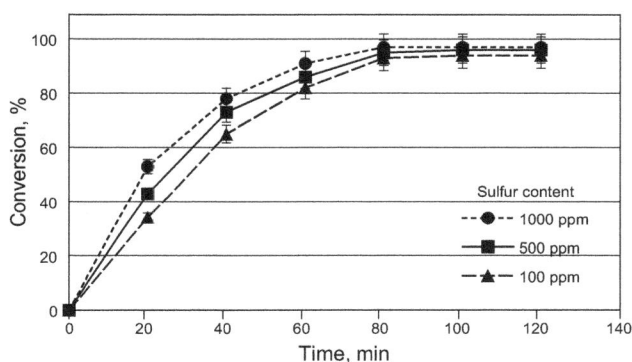

Fig. 2 ODS deep desulfurization of mixed model oil (DBT/BT/T, 2:1:1 in mass ratio) in n-hexane and EtOH (O/S = 10, T = 75 °C, 0.03 g of PW/KSF)

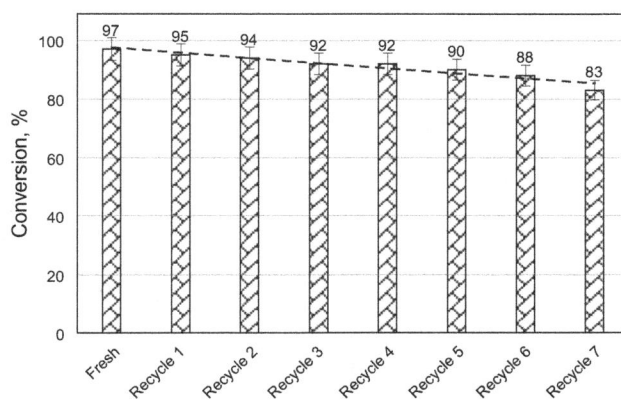

Fig. 4 Influence of the recycle times on ODS deep desulfurization mixed model oil (DBT/BT/T, 500:250:250 ppm) containing 1000 ppm of sulfur in n-hexane and EtOH (O/S = 10, T = 75 °C, 0.03 g of PW/KSF, reaction time = 80 min.)

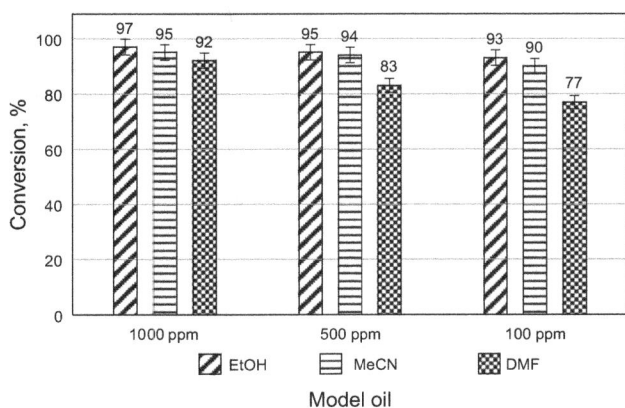

Fig. 3 ODS desulfurization of mixed model oil (DBT/BT/T, 2:1:1 in mass ratio) containing 1000, 500 or 100 ppm of elemental sulfur in n-hexane with different extractants (EtOH, DMF and MeCN), reaction time = 80 min, model oil/solvent ratio = 1:1 (O/S = 10, T = 75 °C, 0.03 g of PW/KSF)

catalyst showed 83 % ODS removal after seventh recycling. The average recovery weight was ∼93 % and ∼92 % for PW/KSF and model oil, respectively. The FTIR spectrum of

the recovered PW/KSF after fourth run confirmed no significant change in the Keggin structure of the catalyst (Fig. 5). The $PW_{12}O_{40}^{-3}$ Keggin ion structure consists of a PO_4 tetrahedron surround by four W_3O_{13} groups formed by edge-sharing octahedral. These groups are connected with each other by corner-sharing oxygens. This structure consists of four types of oxygens, being responsible for the fingerprint bands of the Keggin ion between 700 cm^{-1} and 1200 cm^{-1}. Bulk PW shows the typical bands for absorptions of P–O (1080 cm^{-1}), W = O$_t$ (985 cm^{-1}), W–O$_c$–W (890 cm^{-1}) and W–O$_e$–W (814 cm^{-1}). For PW supported on KSF, the bands at 1078, 986, 899 and 802 cm^{-1} are attributed to stretching vibrations of P–O, W=O$_t$, W–O$_c$–W corner-shared bonds and W–O$_e$–W edge-shared bonds, respectively, which indicates the encapsulation of PW in the KSF frameworks, respectively (Rafiee et al. 2009; Zhang et al. 2011a, b). After catalytic experiments on 1000 ppm mixed model oil using ethanol as extractant solvent, the PW/KSF exhibits a TON = 8.69 at optimum conditions.

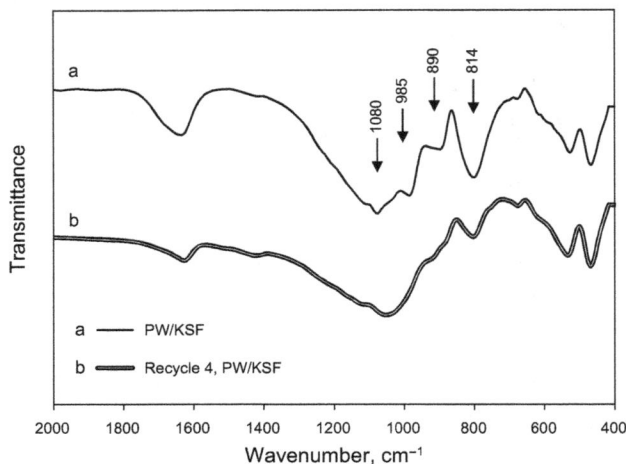

Fig. 5 FTIR spectrum of PW/KSF and fourth recovered PW/KSF, mixed model oil (DBT/BT/T, 500:250:250 ppm) containing 1000 ppm of sulfur in *n*-hexane and EtOH (O/S = 10, T = 75 °C, 0.03 g of PW/KSF, reaction time = 80 min.)

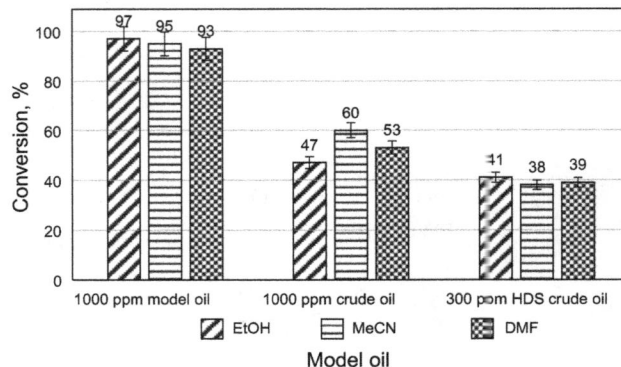

Fig. 6 ODS deep desulfurization of crude oil, HDS-treated crude oil and mixed model oil (DBT/BT/T, 500:250:250 ppm) in *n*-hexane and different extractants, reaction time = 80 min, model oil/solvent ratio = 1:1 (O/S = 10, T = 75 °C, 0.03 g of PW/KSF)

3.3 ODS desulfurization of crude oil

For investigating the industrial performance of PW/KSF, the ODS was tested using 1000 ppm crude oil as a real model oil with different extractants (MeCN, DMF and EtOH). The results showed that the extractability of S-compounds decreased in the order of MeCN > DMF > EtOH (Fig. 6). The conversion of S-compounds of 1000 ppm crude oil only reached 60 %, 53 % and 47 % by employing MeCN, DMF and EtOH, respectively, as extractant.

Desulfurization of crude oil has been a challenge for a long time, and ODS can be a promising complementary method for hydro-treated crude oil. So, the HDS-treated crude oil was selected for testing performance of the catalyst. By using EtOH, MeCN and DMF as extractants at optimum conditions, the conversion of S-compounds of 300 ppm HDS crude oil only reached 41 %, 38 % and 39 %, respectively (Fig. 6). Indeed, when crude oil was first desulfurized with HDS process, thiols, sulfides and disulfides were removed but aromatic S-compounds (thiophenic compounds) were remained in crude oil. It should be noted that EtOH is the most effective extractant for crude oil after HDS process but it is not as efficient for crude oil without HDS.

3.4 Effect of composition of mixed model oil on the ODS conversion

With respect to the considerable decrease in ODS conversion of crude oil in comparison with simulated model oil (see Fig. 6), the effect of crude oil composition on ODS conversion was investigated. Crude oil is the most

important source of aromatics and olefins, and they may have a significant effect on ODS (Abdalla et al. 2009a, b; Yan et al. 2013; Xiao et al. 2014). In these systems (PW catalysts), the active sites of the catalyst with H_2O_2 exhibited considerable activity in epoxidation of alkenes (Wenjia et al. 2013; Aoto et al. 2014). For investigation of the effects of composition of crude oil and epoxidation of alkenes, 5 vol%, 15 vol% and 25 vol% (according to crude oil properties of west Iranian oil wells) of *o*-xylene, cyclohexene and 1,7-octadiene were, respectively, added to the 1000 ppm model oil (DBT/BT/T, 500:250:250 ppm) to examine their effects on ODS using the PW/KSF catalyst with H_2O_2 at optimum reaction conditions. Figure 7 shows the ODS conversion of the model oil with addition of *o*-xylene at various concentrations. The ODS conversion (97 %) of model oil decreased by 10 %, 24 % and 28 %, respectively, after 80 min of reaction time when 5 vol%, 15 vol% and 25 vol% of *o*-xylene were added to the model oil.

Figure 8 shows that when 5 vol%, 15 vol% and 25 vol% of cyclohexene were added to the model oil, the ODS conversion decreased to 63 %, 52 % and 47 %, respectively, compared to that of model oil. Under the same reaction conditions, the ODS conversion of model oil decreased to 84 %, 72 % and 66 % by the addition of 5 vol%, 15 vol% and 25 vol% of 1,7-octadiene (Fig. 9). The results of olefins and aromatics addition showed strong negative effects on ODS. Therefore, the presence of aromatics and olefins in crude oil will decrease ODS conversion.

The high electron-donating ability of the olefins and aromatics double bonds is considered to be the problem factor in ODS of crude oil (Te et al. 2001; Xiao et al. 2014). Figures 7, 8 and 9 show that the inhibiting effect on ODS conversion increased in the order of *o*-xylene < 1,7-

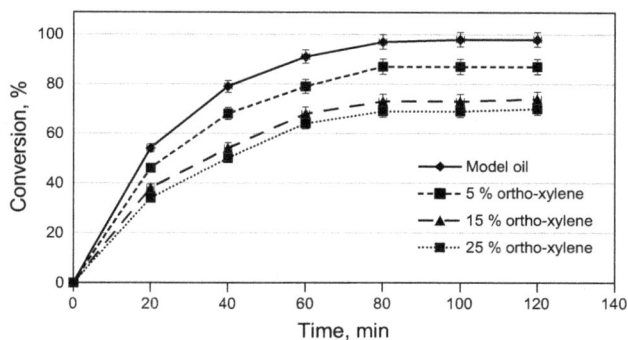

Fig. 7 Effect of *o*-xylene on ODS conversion of mixed model oil (DBT/BT/T, 500:250:250 ppm) in *n*-hexane and EtOH (O/S = 10, $T = 75\ °C$, 0.03 g of PW/KSF)

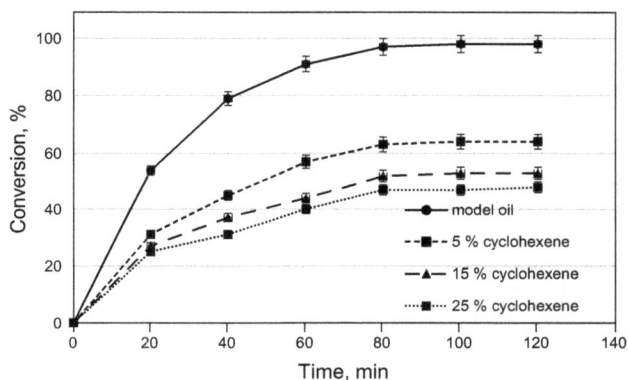

Fig. 8 Effect of cyclohexene on ODS conversion of mixed model oil (DBT/BT/T, 500:250:250 ppm) in *n*-hexane and EtOH (O/S = 10, $T = 75\ °C$, 0.03 g of PW/KSF)

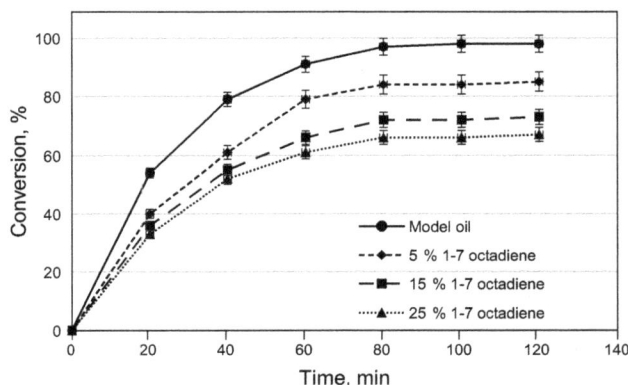

Fig. 9 Effect of 1,7-octadiene on ODS conversion of mixed model oil (DBT/BT/T, 500:250:250 ppm) in *n*-hexane and EtOH (O/S = 10, $T = 75\ °C$, 0.03 g of PW/KSF)

octadiene < cyclohexene. To explain this trend, the electronic and steric effects should be taken into consideration. The partial electron charge on the alkenes and aromatics plays a detrimental role for oxidation reactivity of the catalyst (Xiao et al. 2014; Yan et al. 2009a).

4 Conclusions

Catalytic activity of PW/KSF was investigated for ODS of DBT and mixed thiophenic model oils and crude oil. First, effects of temperature, O/S molar ratio, amount of the catalyst and type of the extractant solvent on ODS of DBT were investigated. The results of the experiments show that DBT conversion in model oil was excellent for 100 ppm, 500 ppm and even 1000 ppm and conversion for 1000, 500 and 100 ppm mixed thiophenic compounds was 97 %, 95 % and 93 %, respectively. For 1000 ppm crude oil and 300 ppm HDS-treated crude oil, catalytic activities were 47 % and 41 % by using EtOH as extractant. It was discovered that EtOH is the best extractant solvent for removing oxidized S-compounds of mixed model oil. In the second part of the study, effects of cyclohexene, 1,7-octadiene and *o*-xylene were investigated for mixed model oil and the ODS decreases in order of: *o*-xylene < 1,7-octadiene < cyclohexene.

Acknowledgments The authors thank the Razi University Research Council for support of this work.

References

Abdalla ZEA, Li B, Tufail A. Preparation of phosphate promoted Na_2WO_4/Al_2O_3 catalyst and its application for oxidative desulfurization. J Ind Eng Chem. 2009a;15:780–3.

Abdalla ZEA, Li B, Tufail A. Direct synthesis of mesoporous $(C_{19}H_{42}N)_4H_3(PW_{11}O_{39})/SiO_2$ and its catalytic performance in oxidative desulfurization. Coll Surf A. 2009b;341:86–92.

Abdalla ZEA, Li B. Preparation of MCM-41 supported $(Bu_4N)_4H_3(PW_{11}O_{39})$ catalyst and its performance in oxidative desulfurization. Chem Eng J. 2012;200–202:113–21.

Aoto H, Matsui K, Sakai Y, et al. Zirconium(IV)- and hafnium(IV)-containing polyoxometalates as oxidation precatalysts: homogeneous catalytic epoxidation of cyclooctene by hydrogen peroxide. J Mol Catal A: Chem. 2014;394:224–31.

Bakar WAWA, Ali R, Kadir AAA, Mokhtar WNAW. Effect of transition metal oxides catalysts on oxidative desulfurization of model diesel. Fuel Process Technol. 2012;101:78–84.

Bhutto AW, Abro R, Gao S, et al. Oxidative desulfurization of fuel oils using ionic liquids: a review. J Taiwan Inst Chem Eng. 2016;62:84–97.

Caero LC, Hernández E, Pedraza F, Murrieta F. Oxidative desulfurization of synthetic diesel using supported catalysts: part I.

Study of the operation conditions with a vanadium oxide based catalyst. Catal Today. 2005;107:564–9.

Chamack M, Mahjoub AR, Aghayan H. Cesium salts of tungsten-substituted molybdophosphoric acid immobilized onto platelet mesoporous silica: efficient catalysts for oxidative desulfurization of dibenzothiophene. Chem Eng J. 2014;255:686–94.

Collins FM, Lucy AR, Sharp C. Oxidative desulphurisation of oils via hydrogen peroxide and heteropolyanion catalysis. J Mol Catal A: Chem. 1997;117:397–403.

Da Silva MJ, Dos Santos LF. Novel oxidative desulfurization of a model fuel with H_2O_2 catalyzed by $AlPMo_{12}O_{40}$ under phase transfer catalyst-free conditions. J Appl Chem. 2013;2013:1–8.

Dooley KM, Liu D, Madrid AM, Knopf FC. Oxidative desulfurization of diesel with oxygen: reaction pathways on supported metal and metal oxide catalysts. Appl Catal A: Gen. 2013;468:143–9.

Gao J, Wang S, Jiang Z, et al. Deep desulfurization from fuel oil via selective oxidation using an amphiphilic peroxotungsten catalyst assembled in emulsion droplets. J Mol Catal A: Chem. 2006;258:261–6.

Gao G, Cheng S, An Y, et al. Oxidative desulfurization of aromatic sulfur compounds over titanosilicates. ChemCatChem. 2010a;2:459–66.

Gao H, Guo C, Xing J, et al. Extraction and oxidative desulfurization of diesel fuel catalyzed by a brønsted acidic ionic liquid at room temperature. Green Chem. 2010b;12:1220–4.

García-Gutiérrez JL, Fuentes GA, Hernández-Terán ME, García P, Murrieta-Guevara F, Jiménez-Cruz F. Ultra-deep oxidative desulfurization of diesel fuel by the Mo/Al_2O_3–H_2O_2 system: the effect of system parameters on catalytic activity. Appl Catal A: Gen. 2008;334:366–73.

García-Gutiérrez JL, Laredo GC, García-Gutiérrez P, et al. Oxidative desulfurization of diesel using promising heterogeneous tungsten catalysts and hydrogen peroxide. Fuel. 2014;138:118–25.

Hassan SI, Sif El-Din OI, Tawfik SM, Abd El-Aty DM. Solvent extraction of oxidized diesel fuel: phase equilibrium. Fuel Process Technol. 2013;106:127–32.

Hui-Peng L, Jian S, Hua Z. Desulfurization using an oxidation/catalysis/adsorption scheme over $H_3PW_{12}O_{40}/SiO_2$–Al_2O_3. Pet Chem. 2007;47:452–6.

Ishihara A, Wang D, Dumeignil F, Amano H, Qian EW, Kabe T. Oxidative desulfurization and denitrogenation of a light gas oil using an oxidation/adsorption continuous flow process. Appl Catal A: Gen. 2005;279:279–87.

Katsoulis DE. A survey of applications of polyoxometalates. Chem Rev. 1998;98:359–88.

Khenkin AM, Neumann R. Desulfurization of hydrocarbons by electron transfer oxidative polymerization of heteroaromatic sulfides catalyzed by $H_5PV_2Mo_{10}O_{40}$ polyoxometalate. ChemSusChem. 2011;4:346–8.

Kozhevnikov IV. Catalysis by heteropoly acids and multicomponent polyoxometalates in liquid-phase reactions. Chem Rev. 1998;98:171–98.

Kukovecz Á, Balogi Z, Kónya Z, et al. Synthesis, characterisation and catalytic applications of sol–gel derived silica–phosphotungstic acid composites. Appl Catal A. Gen. 2002;228:83–94.

Lei J, Chen L, Yang P, Du X, Yan X. Oxidative desulfurization of diesel fuel by mesoporous phosphotungstic acid/SiO_2: the effect of preparation methods on catalytic performance. J Porous Mater. 2013;20:1379–85.

Li B, Liu Z, Liu J, et al. Preparation, characterization and application in deep catalytic ODS of the mesoporous silica pillared clay incorporated with phosphotungstic acid. J Coll Interface Sci. 2011;362:450–6.

Li J, Hu B, Tan J, Zhuang J. Deep oxidative desulfurization of fuels catalyzed by molybdovanadophosphoric acid on amino-functionalized SBA-15 using hydrogen peroxide as oxidant. Transit Metal Chem. 2013;38:495–501.

Lo W-H, Yang H-Y, Wei G-T. One-pot desulfurization of light oils by chemical oxidation and solvent extraction with room temperature ionic liquids. Green Chem. 2003;5:639–42.

Maity P, Mukesh D, Bhaduri S, et al. A water soluble heteropoly-oxotungstate as a selective, efficient and environment friendly oxidation catalyst. J Chem Sci. 2009;121:377–85.

Mei H, Mei BW, Yen TF. A new method for obtaining ultra-low sulfur diesel fuel via ultrasound assisted oxidative desulfurization. Fuel. 2003;82:405–14.

Qiu J, Wang G, Zeng D, et al. Oxidative desulfurization of diesel fuel using amphiphilic quaternary ammonium phosphomolybdate catalysts. Fuel Process Technol. 2009;90:1538–42.

Rafiee E, Eavani S. Organic–inorganic polyoxometalate based salts as thermoregulated phase-separable catalysts for selective oxidation of thioethers and thiophenes and deep desulfurization of model fuels. J Mol Catal A: Chem. 2013;380:18–27.

Rafiee E, Eavani S, Rashidzadeh S, et al. Silica supported 12-tungstophosphoric acid catalysts for synthesis of 1, 4-dihydropyridines under solvent-free conditions. Inorg Chim Acta. 2009;362:3555–62.

Rafiee E, Mahdavi H, Joshaghani M. Supported heteropoly acids offering strong option for efficient and cleaner processing for the synthesis of imidazole derivatives under solvent-free condition. Mol Divers. 2011;15:125–34.

Rafiee E, Mirnezami F. Keggin-structured polyoxometalate-based ionic liquid salts: thermoregulated catalysts for rapid oxidation of sulfur-based compounds using H_2O_2 and extractive oxidation desulfurization of sulfur-containing model oil. J Mol Liq. 2014;199:156–61.

Sampanthar JT, Xiao H, Dou J, et al. A novel oxidative desulfurization process to remove refractory sulfur compounds from diesel fuel. Appl Catal B Environ. 2006;63:85–93.

Shojaei AF, Rezvani MA, Loghmani MH. Comparative study on oxidation desulphurization of actual gas oil and model sulfur compounds with hydrogen peroxide promoted by formic acid: synthesis and characterization of vanadium containing polyoxometalate supported on anatase crushed nanoleaf. Fuel Process Technol. 2014;118:1–6.

Stanislaus A, Marafi A, Rana MS. Recent advances in the science and technology of ultra low sulfur diesel (ULSD) production. Catal Today. 2010;153:1–68.

Te M, Fairbridge C, Ring Z. Oxidation reactivities of dibenzothiophenes in polyoxometalate/H_2O_2 and formic acid/H_2O_2 systems. Appl Catal A: Gen. 2001;219:267–80.

Wenjia C, Yan Z, Renlie B, et al. Catalytic epoxidation of cyclohexene over mesoporous-silica immobilized Keggin-type tungstophosphoric acid. Chin J Chem. 2013;34:193–9.

Xiao J, Wu L, Wu Y, et al. Effect of gasoline composition on oxidative desulfurization using a phosphotungstic acid/activated carbon catalyst with hydrogen peroxide. Appl Energy. 2014;113:78–85.

Xue X, Zhao W, Ma B, et al. Efficient oxidation of sulfides catalyzed by a temperature-responsive phase transfer catalyst $[(C_{18}H_{37})_2(CH_3)_2N]_7PW_{11}O_{39}$ with hydrogen peroxide. Catal Commun. 2012;29:73–6.

Yan X-M, Mei P, Lei J, et al. Synthesis and characterization of mesoporous phosphotungstic acid/TiO_2 nanocomposite as a novel oxidative desulfurization catalyst. J Mol Catal A: Chem. 2009a;304:52–7.

Yan X-M, Su G-S, Xiong L. Oxidative desulfurization of diesel oil over Ag-modified mesoporous HPW/SiO_2 catalyst. J Fuel Chem Technol. 2009b;37:318–23.

Yan X-M, Mei P, Xiong L, et al. Mesoporous titania–silica–polyoxometalate nanocomposite materials for catalytic oxidation desulfurization of fuel oil. Catal Sci Technol. 2013;3:1985–92.

Zhang B, Zhou M-D, Cokoja M, et al. Oxidation of sulfides to sulfoxides mediated by ionic liquids. RSC Adv. 2012a;2:8416–20.

Zhang H, Gao J, Meng H, et al. Removal of thiophenic sulfurs using an extractive oxidative desulfurization process with three new phosphotungstate catalysts. Ind Eng Chem Res. 2012b;51: 6658–65.

Zhang J, Wang A, Li X, et al. Oxidative desulfurization of dibenzothiophene and diesel over $[Bmim]_3PMo_{12}O_{40}$. J Catal. 2011a;279:269–75.

Zhang M, Zhu W, Xun S, Li H, Gu Q, Zhao Z, Wang Q. Deep oxidative desulfurization of dibenzothiophene with POM-based hybrid materials in ionic liquids. Chem Eng J. 2013;220:328–36.

Zhang Z, Zhang F, Zhu Q, et al. Magnetically separable polyoxometalate catalyst for the oxidation of dibenzothiophene with H_2O_2. J Coll Interface Sci. 2011b;360:189–94.

5

Effects of U-ore on the chemical and isotopic composition of products of hydrous pyrolysis of organic matter

Yu-Wen Cai[1,2,3] · Shui-Chang Zhang[2,3] · Kun He[2,3] · Jing-Kui Mi[2,3] ·
Wen-Long Zhang[2,3] · Xiao-Mei Wang[2,3] · Hua-Jian Wang[2,3] · Chao-Dong Wu[1]

Abstract In order to investigate the impact of U-ore on organic matter maturation and isotopic fractionation, we designed hydrous pyrolysis experiments on Type-II kerogen samples, supposing that the water and water–mineral interaction play a role. U-ore was set as the variable for comparison. Meanwhile, anhydrous pyrolysis under the same conditions was carried out as the control experiments. The determination of liquid products indicates that the presence of water and minerals obviously enhanced the yields of C_{15+} and the amounts of hydrocarbon and non-hydrocarbon gases. Such results may be attributed to water-organic matter reaction in the high-temperature system, which can provide additional hydrogen and oxygen for the generation of gas and liquid products from organic matter. It is found that δD values of hydrocarbon gases generated in both hydrous pyrolysis experiments are much lower than those in anhydrous pyrolysis. What is more, δD values are lower in the hydrous pyrolysis with uranium ore. Therefore, we can infer that water-derived hydrogen played a significant role during the kerogen thermal evolution and the hydrocarbon generation in our experiments. Isotopic exchange was facilitated by the reversible equilibration between reaction intermediaries with hydrogen under hydrothermal conditions with uranium ore. Carbon isotopic fractionations of hydrocarbon gases were somehow affected by the presence of water and the uranium ore. The increased level of $i\text{-}C_4/n\text{-}C_4$ ratios for gas products in hydrous pyrolysis implied the carbocation mechanism for water-kerogen reactions.

Keywords Organic–inorganic interaction · Hydrous pyrolysis · Stable isotopes · U-ore · Carbocation mechanism

1 Introduction

Organic and inorganic compounds coexist in most geological environments. Experimental and field studies have demonstrated that the presence of water, minerals and catalytically active transition metal elements in sedimentary basins can influence petroleum generation and accumulation (Goldstein 1983; Huizinga et al. 1987a, b; Pan et al. 2009, 2010; Tannenbaum and Kaplan 1985a, b; Mango 1992; Mango and Joe 1997). The classical hydrous pyrolysis system described by Lewan et al. had verified that many organic reactions cannot occur without water (Mayer 1994), which also plays an essential role in the process of petroleum generation, such as influencing the stability of crude oil and the secondary cracking of long-chain hydrocarbons (Lewan 1993, 1997). In addition, with increasing temperature, more chemical types of hydrogen in the organic matter become exchangeable with those in the water and result in isotopic shifts among kerogen, bitumen and oil, but only small changes in organic $^{13}C/^{12}C$ (Schimmelmann et al. 1999). Mineral catalysis has been suggested as an important factor in petroleum generation based on the catalytic activity of acidic clays during the refining process. Experiments using minerals ubiquitous in

✉ Shui-Chang Zhang
sczhang@petrochina.com.cn

1   School of Earth and Space Sciences, Peking University,
    Beijing 100087, China

2   Research Institute of Petroleum Exploration and
    Development (RIPED), Beijing 100083, China

3   State Key Laboratory of Enhanced Oil Recovery,
    Beijing 100083, China

Edited by Jie Hao

sedimentary basins as substrates, such as iron-bearing minerals, aluminum-containing minerals and silicates, suggested that these compounds might oxidize petroleum and generate organic acids and CO_2 (Seewald 2001, 2003; Seewald et al. 1998; Surdam et al. 1985, 1993). The minerals with catalytic activity, such as montmorillonite, that catalyze reactions under dry conditions, may also be inhibited by water (Tannenbaum and Kaplan 1985a).

Transition metal elements in source rocks have been pinpointed to be catalytic in the conversion of hydrogen and n-alkenes into natural gas. However, specific mechanisms remain equivocal (Mango 1992; Mango and Joe 1997; Lewan et al. 2008). Uranium, as a special transition metal element, has long been known to be associated with organic matter. Under reducing conditions, U(VI) is reduced to the immobile tetravalent state, U(IV), which mainly occurs in the organic-rich shales (Partin et al. 2013; Lev and Filer 2004; Lev et al. 2008; Fisher and Wignall 2001; Galindo et al. 2007). A lot of research has been carried out to investigate the role of organic components in metal mineralization, while the catalytic effects of uranium on oil cracking have rarely been studied or reported. Previous studies showed that U-bearing ore in source rocks acts as a catalyst, determining the degree, composition and timing of natural gas generation (Liu et al. 2006, 2013; Mao et al. 2012a, b, 2014; Li et al. 2008; Lu et al. 2007). This hypothesis may have important implications concerning gas generation in unconventional shale-gas accumulations. But since U-ore has a degree of radioactivity and serves as a complex catalyst under geological conditions, it is difficult to make breakthroughs in this field. Therefore, additional experiments are needed to better evaluate the significance of uranium in hydrocarbon generation from petroleum source rocks.

Based on oil cracking experiments in a confined pyrolysis system (gold tubes), we investigated the catalytic effects of U-ore on oil cracking, identified the chemical and isotopic features of gas hydrocarbons, and evaluated possible mechanisms in the system. Type-II kerogen samples with low maturity from the Xiamaling Formation were selected for isothermal anhydrous pyrolysis and hydrous pyrolysis under two conditions, with and without U-bearing ore.

2 Materials and methods

2.1 Samples

The kerogen samples used in our pyrolysis experiments were separated from the organic-rich immature source rocks from Xiamaling Formation, Xiahuayuan Area, Hebei Province, North China, with an age of 1384.4 ± 1.4 Ma (Zhang et al. 2015). The uranium ore was collected from a depth of 170 m below the surface of the Zoujiashan Deposit, Xiangshan Uranium Field, South China. Before pyrolysis, source rock samples were firstly crushed into powder (80–100 mesh), and then treated with solvents to obtain purified kerogen samples. The kerogen was isolated from the rocks by rinsing with HF/HCl, Soxhlet extraction for 72 h, washed with distilled water and dried at a low temperature (40 °C).

The geochemical features of the source rock, kerogen and minerals are listed in Table 1. The source rock has a TOC of 3.82%, hydrogen index (HI) of 369.09 mg/g TOC and T_{max} of 433 °C. The kerogen has an H/C ratio of 1.19 and an O/C ratio of 0.02, which is classified as Type-II kerogen. The U-bearing ore we used in the experiment is composed of 50% clay, 17.4% quartz, 26.1% fluorite and 6.6% feldspar, among which uranium mainly exists in the clay minerals, with the elemental content of 0.85% (Table 1).

2.2 Gold-tube pyrolysis

All pyrolysis experiments were performed in a gold-tube pyrolysis system at constant temperatures of 300, 330, 350, 360, 370, 390, 400 and 420 °C for time intervals between 24 and 360 h at a constant pressure of 50 MPa. These were carried out at the Petroleum Geology Research and Laboratory Center (PGRLC) of the Research Institute of Petroleum Exploration and Development (Zhang et al. 2013). The error for the temperature was less than ±1 °C. The tubes used in our experiments were 50 mm in length with an inner diameter (ID) of 4.0 mm and a wall thickness of 0.5 mm. The reactants were accurately measured and loaded into the tube. A series of anhydrous samples and two series of hydrous samples were loaded into the gold tubes. In anhydrous experiments, only 50 mg of kerogen samples were loaded, while in hydrous experiments samples were separated into two sets, one with kerogen plus 50 μg deionized water ($\delta D_{H2O} = -55‰$), and the other with the same amount of water and additional 50 mg of U-ore in order to magnify the impact of water and differentiate the effects of added minerals on the hydrocarbon generation. Air in the tube was removed by argon flow, and the other end of the tube was crimped and sealed by an argon arc welder with most of the sealed end submerged in liquid nitrogen. When reaction ended, the pressure was relieved, and the gold tubes were withdrawn. Before analysis, the tubes were weighed and compared with the preheated weight to determine whether leakage had occurred during pyrolysis.

After pyrolysis, the volatile components in the tubes were collected with a custom-made gas inlet device with a known volume. Prior to piercing a tube, the device was first evacuated to a pressure of less than 0.1 MPa, then gas was allowed to escape slowly into the evacuated inlet device, and

Table 1 Geochemical characteristics of the original samples and U-ore

Kerogen analyses		Whole rock analyses		Mineral and elemental analyses	
TOC wt%[a]	57.03	TOC wt%[a]	3.82	%Clay[c]	49.9
C%[b]	55.6	T_{max} (°C)[a]	433	%Fluorite[c]	26.1
Atomic H/C ratio[b]	1.19	S_1 (mg/g rock)[a]	2.12	%Quartz[c]	17.4
Atomic O/C ratio[b]	0.02	S_2 (mg/g rock)[a]	369.09	%Feldspar[c]	6.6
$\delta^{13}C$ (per mil, V PDB)	−33.2	S_3 (mg/g rock)[a]	1.11	%U[d]	0.85
δD (per mil, V SMOW)	−110.4	HI (mg S_2/g TOC)[a]	420	%Fe[d]	1.2
R_b%	0.5			%Al[d]	1.2

[a] Rock–Eval pyrolysis
[b] Elemental analyses
[c] X–ray diffraction
[d] ICP–MS analyses

the final pressure was recorded. The device was directly connected to a gas chromatograph (GC; 7890N, Agilent) to analyze the organic and inorganic gas compositions.

2.3 Product analyses

Qualitative and quantitative analysis of the individual hydrocarbon and non-hydrocarbon gas components were conducted by a two-channel Agilent 7890 Series Gas Chromatograph (GC) integrated with an auxiliary oven, which was custom configured by Wasson-ECE instrumentation (Fort Collins, CO). The instrument was fitted with two capillary and six packed analytical columns, a flame ionization detector (FID) and two thermal conductivity detectors (TCDs). The carrier gases for FID and TCD were high-purity N_2 and He, respectively. The temperature program of the GC oven was: heating from initial 68 °C (held for 7 min) to 90 °C (held for 1.5 min) at a rate of 10 °C/min, then to 175 °C (held for 5 min) at a rate of 15 °C/min. The analytical precision was ≤1 mol‰ (Zhang et al. 2007; He et al. 2011a).

Stable carbon isotopes ($\delta^{13}C$) were measured on the generated methane, ethane and propane gas with a GC/combustion/isotope ratio mass spectrometer (IRMS) (Finnigan MAT Delta, Thermo Fisher Scientific). Original stable carbon isotopes in the kerogen were determined by IRMS (Delta Plus V Advantage). Individual hydrocarbon components (C_{1-5}) and CO_2 were initially separated by using a chromatographic column. The chromatographic column temperature was increased from 35 to 80 °C at a rate of 8 °C/min, then to 260 °C at a rate of 5 °C/min, and the final oven temperature was held for 10 min. Each gas sample was analyzed twice, with an analytical precision of ±0.5‰ compared with the Vienna Pee Dee Belemnite (VPDB) standard.

The hydrogen isotopic (δD) analysis of the gases was performed by GC/thermal conversion/IRMS (Finnigan MAT 253 Delta). Gas components were separated by a chromatographic column, and helium was used as the carrier gas. Methane was injected in split mode (split ratio 1:7) with a constant temperature of 40 °C. Ethane was injected in splitless mode and was kept at an initial temperature of 40 °C for 4 min, heated to 80 °C at a rate of 10 °C/min, heated to 140 °C at a rate of 5 °C/min, and finally heated to 260 °C at a rate of 30 °C/min. The analytical precision was less than ±5‰.

Generated liquid products (C_{15+} compounds) were collected by dichloromethane (DCM) extraction of solid residues after pyrolysis. In a typical procedure, the gold tubes were cut in a glass bottle, weighed as G_0, with dichloromethane (DCM). The bottle was ultrasonicated twice for 3 min until the solvent and residues were totally separated. Then, we carefully filtered the solvent into another glass bottle, which was weighed as G_1, through the chromatographic membrane several times until the liquid was clean. After the volatilization of DCM in both two bottles, we weighed each glass bottle again, recorded as G_2 and G_3. The differences between G_2 and G_0 as well as G_3 and G_1 were due to C_{15+} compounds and solid residues after the pyrolysis, respectively (He et al. 2011a).

Total organic carbon (TOC) and rock evaluation analyses were conducted on the kerogen with a Rock–Eval 6 carbon analyzer. Elemental analyses (C, H, N and O) were determined with an elemental analyzer (vario MICRO cube, Elementar) using acetanilide as a standard for carbon, hydrogen and oxygen. Stable carbon isotope ratios of the residues were measured online with a combustion system (Flash EA–ConFlo IV, Thermo Fisher Scientific) connected to an isotope ratio mass spectrometer (IRMS; Delta Plus V Advantage, Thermo Fisher Scientific). $\delta^{13}C$ and δD values of kerogen were measured with a precision of ±0.5‰ and ±5‰, respectively. All $\delta^{13}C$ and δD values are reported in the permillage (‰) relative to Vienna Pee Dee Belemnite (VPDB) and Vienna Standard Mean Ocean Water (VSMOW), respectively. All the analyses were performed at the Petroleum Geology Research and Laboratory Center (PGRLC) of the Research Institute of Petroleum Exploration and Development.

Table 2 Detailed experimental conditions, yields, and carbon and hydrogen isotope ratios during pyrolysis of Type-II kerogen samples from the Xiamaling Formation at 50 MPa

Temp., °C	Time, h	EasyR_o (%)	Total C_{15+} oils, mg/g TOC	Gas Yields (mL/g TOC)											C_1/C_{1-5}	iC_4/nC_4	Carbon isotope (‰, vs. VPDB)			Hydrogen isotope (‰, vs. VSMOW)		
				C_1	C_2	C_3	iC_4	nC_4	iC_5	nC_5	CO_2	H_2	H_2S	C_{2-5}			$\delta^{13}C_1$	$\delta^{13}C_2$	$\delta^{13}C_3$	δD_1	δD_2	δD_3
Kerogen																						
300	24	0.6	30.5	2.2	1	0.4	0	0.1	0	0	4.5	–	1.1	1.5	0.59	0.35	-51.9	-39.2	-39.7	-225	-146	–
	24	0.79	80.5	6.5	5.4	2.6	0.2	1.0	0.1	0.4	3.9	–	4.2	9.6	0.4	0.21	-51.5	-40.1	-38	–	–	–
330	72	0.89	136.8	12	10	6.1	0.6	3.1	0.3	1.2	4.3	–	6.2	21.3	0.37	0.2	-52.3	-39.7	-36.5	-236	-152	-156
	120	0.95	286.9	13.2	10.7	6.6	0.7	3.3	0.3	1.3	3.8	–	8.2	22.9	0.36	0.2	-52.3	-39.5	-36.7	-251	-168	-150
	240	1.05	523.2	19.4	14.1	8.3	0.8	4	0.3	1.5	3.5	–	10	29.1	0.4	0.21	-51.5	-38.3	-36.6	-252	-172	-142
350	120	1.17	403.7	26.2	15.8	9.2	1.0	4.3	0.4	1.5	4.4	–	12.4	32.1	0.45	0.23	-49.5	-37.6	-36	-254	-164	-121
	240	1.29	331.2	33	17.5	10.4	1.1	4.7	0.4	1.5	5.1	–	11.2	35.7	0.48	0.24	-48	-37.7	-36.6	-254	-162	-106
360	168	1.35	295.3	37.2	19.5	11.4	1.2	4.7	0.4	1.3	5.1	–	11.1	38.5	0.49	0.26	-48.3	-37.6	-36.1	-260	-161	-104
	240	1.42	225.1	41.3	19.8	11.7	1.4	5.1	0.5	1.6	5.7	–	11.7	40.1	0.51	0.27	-47.9	-37.5	-36.1	-254	-161	-102
370	168	1.49	179.3	46.6	21.6	13.3	1.6	5.8	0.5	1.8	5.7	–	11.3	44.7	0.51	0.28	-47	-36.7	-36	-255	-146	-95
	240	1.59	131.9	51.3	22.8	14.2	1.7	6	0.6	1.9	6.3	–	11.5	47.2	0.52	0.29	-47	-36.2	-36	-248	-138	-102
	360	1.66	92.2	68.5	16.5	16.4	1.7	5.3	0.6	1.7	8.1	–	10.3	42.2	0.62	0.32	-46.8	-37.5	-35.8	-238	–	–
390	120	1.71	76.6	66.7	25.1	14.7	1.6	4.8	0.5	1.5	9.3	–	9.6	48.2	0.58	0.33	-46	-37.1	-34	-232	–	–
	168	1.8	69.9	71.5	29.4	18.2	2.2	6.3	0.7	1.9	10.8	0.1	10.9	58.8	0.55	0.35	-45.7	-36.7	-34.5	-230	-108	-95
	240	1.89	33.1	79	32.9	19.7	2.3	5.7	0.6	1.5	10.9	0.2	11.2	62.7	0.56	0.4	-45.4	-37	-34.5	-224	-96	-86
400	360	2.0	–	100.9	39.6	23.8	3.3	7.7	0.9	2.2	12.6	0.4	9.7	77.6	0.57	0.43	-45.3	-36.8	-34.1	-223	-83	-86
	240	2.07	–	97.7	39.6	24.7	3.4	7.9	0.9	2	12	0.6	8.7	78.4	0.55	0.43	-44.9	-36.2	-33.6	-212	-90	-81
	360	2.19	–	110.7	43.5	26.6	4.2	8.7	0.8	1.7	12.9	0.8	10.3	85.5	0.56	0.49	-44.6	-36.1	-32.4	-185	-77	-56
420	168	2.35	–	134	47	24.1	3.3	5	0.3	0.6	13.3	1.4	10.4	80.4	0.62	0.65	-43.9	-35.4	-32.4	-180	-57	-42
	264	2.5	–	140.8	48	23.5	3.3	4.8	0.3	0.6	14.2	1.6	10.8	80.4	0.64	0.68	-42.3	-34.5	-29.3	–	-29	–
Kerogen plus deionized water and U-ore																						
300	24	0.6	67.8	5.8	2.2	1.2	0.2	0.3	0	0	1.6	–	0.1	3.9	0.6	0.92	-45.3	-35.6	-24.4	–	–	–
	24	0.79	137.2	5.7	2.2	0.9	0.1	0.2	0	0	5.1	–	0.1	3.4	0.63	0.65	–	-40.4	–	–	–	–
330	72	0.89	281.7	9.8	4.8	2.3	0.3	0.7	0	0	6.4	–	1	7.9	0.55	0.41	-46.6	-40	-36.2	-282	-233	-203
	120	0.95	354.1	13.3	6.6	3.2	0.4	1	0.1	0	7.2	–	0.5	11.2	0.54	0.43	-46.3	-39.7	-35.8	-286	-232	-204
	240	1.05	679.3	19.9	9.3	4.8	0.6	1.5	0.2	0	7.9	–	0.4	16.5	0.55	0.43	-46.8	-39.5	-35.7	-286	-217	-193
350	120	1.17	425.3	27.9	13.6	8.5	1.5	3.6	0.9	0.8	8.2	–	0.6	28.8	0.5	0.4	-48.4	-38.7	-34.9	-284	-208	-190
	240	1.29	304.2	38	16.9	10.3	1.8	4.4	1.1	1	8.1	–	0.6	35.5	0.52	0.42	-48.2	-37.7	-34.9	-282	-204	-188
360	168	1.35	221.9	41.5	18.2	11.6	2.1	5.1	1.2	1.1	8.7	–	0.5	39.2	0.51	0.41	-47.5	-37.5	-34.6	-280	-199	-180
	240	1.42	168.1	51.5	21.4	14	2.8	6.1	1.6	1.3	9.3	–	0.8	47.2	0.52	0.46	-47.1	-37.3	-34.3	-278	-200	-176
370	168	1.49	104.6	60.6	23.4	14.1	2.3	4.6	1.1	0.9	9.8	–	1.8	46.4	0.57	0.5	-46.8	-37.1	-34.1	-276	-181	-172
	240	1.59	102.9	70.9	26.7	16.9	3.2	6.3	0.1	1.5	11.5	0.4	2.5	54.7	0.56	0.51	-46.7	-37	-34	-272	-177	-166
	360	1.66	98.7	82.1	29.8	18.8	3.5	6.4	0.1	1.4	12.9	0.7	3.2	59.9	0.58	0.54	-45.3	-37.2	-34	–	–	–

Table 2 continued

Temp., °C	Time, h	$EasyR_o$ (%)	Total C_{15+} oils, mg/g TOC	C_1	C_2	C_3	iC_4	nC_4	iC_5	nC_5	CO_2	H_2	H_2S	C_{2-5}	C_1/C_{1-5}	iC_4/nC_4	$\delta^{13}C_1$	$\delta^{13}C_2$	$\delta^{13}C_3$	δD_1	δD_2	δD_3
						Gas Yields (mL/g TOC)											Carbon isotope (‰, vs. VPDB)			Hydrogen isotope (‰, vs. VSMOW)		
390	120	1.71	95.4	91	33	22	4.3	7.5	0.1	1.5	13.2	0.7	5.2	68.4	0.57	0.57	−45.2	−36.7	−33.6	−274	−168	−142
	168	1.8	85	103.2	36.2	24.3	5.1	8.7	0.1	1.7	14.2	1.1	5.3	76	0.58	0.59	−44.7	−36.4	−33.1	−272	−151	−146
	240	1.89	49.9	104.2	36.7	24.2	4.5	8.1	0.1	1.6	14.4	1.4	5.4	75.1	0.58	0.56	−44.9	−36.1	−33.7	−274	−137	−137
	360	2.0	–	130.4	44.5	31.6	8.1	12.1	0.1	1.9	15.3	1.6	8.5	82.3	0.57	0.67	−44	−36.1	−31.3	−268	−134	−126
400	240	2.07	–	140.3	42.7	29	6.3	9.4	0.1	1.5	19	2.2	10.6	89	0.61	0.67	−43.4	−35.7	−30.5	−264	−127	−131
	360	2.19	–	162.1	52.7	32	7.1	9.6	0.1	1.8	21.2	–	–	103.3	0.62	0.73	−42.8	−34.3	–	−251	–	–
420	168	2.35	–	185.3	53.5	33.1	7.9	10.2	1	0	–	3.5	–	104.9	0.61	0.87	−42.4	−33.7	−28	−235	−122	−128
	264	2.5	–	198.5	58.5	34.5	6.6	6.8	0.7	0.4	22.5	3.9	11.0	107.4	0.65	0.97	−40.9	−33.1	−27	−224	−112	−123
Kerogen plus deionized water																						
300	24	0.6	87.8	–	–	–	–	–	–	–	–	–	–	–	–	–	−41.4	–	–	–	–	–
	24	0.79	197.2	3.5	1.8	0.9	0.2	0.2	0	0.4	2.2	–	0.3	3.4	0.51	0.94	–	–	–	–	–	–
330	72	0.89	355	–	–	–	–	–	–	–	–	–	–	–	–	–	−45.6	−39.1	−37	−288	−219	−204
	120	0.95	510.2	8	3.7	1.7	0.3	0.4	0.1	0.1	4.5	–	0.7	6.2	0.56	0.78	−45.5	−39.8	−37.2	−286	−208	−196
	240	1.05	648.3	21.1	10.6	6.1	1.2	2.1	0.6	0.4	9.1	–	1.9	21.2	0.5	0.55	−45.4	−40	−38.2	−291	−209	−200
350	120	1.17	480.1	27.6	12.9	7.7	1.7	3	1.1	1	10.5	–	2.9	27.4	0.5	0.56	−45.8	−39.5	−37.8	−291	−202	−196
	240	1.29	332.9	41.6	17.8	11.2	2.6	4.5	1.7	1.4	11.3	–	4.6	39.3	0.51	0.58	−45.3	−39.2	−37.6	−284	−204	−179
360	168	1.35	256.2	47.2	19.9	13	3.2	5.2	2.1	1.5	11.6	–	5.7	45	0.51	0.62	−46	−38.4	−37.4	−274	−193	−160
	240	1.42	203	52.6	21.3	14	3.3	5.5	2.1	1.5	12.2	–	5.3	47.6	0.52	0.6	−46	−38.6	−37.6	−276	−188	−164
370	168	1.49	133.6	61.7	24.3	16.3	4.1	6.3	2.6	1.8	13.5	0.5	6.8	55.4	0.53	0.64	−45.7	−38.2	−37.2	−280	−176	−157
	240	1.59	91.2	78	29	20.6	5.7	8.3	3.9	2.5	14.3	0.7	8.9	70	0.53	0.68	−44.6	−37.4	−36.9	−274	−169	−145
	360	1.66	42.4	85.8	31	22	6.3	8.2	0.1	2.1	15.4	0.9	13.8	69.8	0.55	0.77	−44.2	−37.9	−36.9	−277	−147	−128
390	120	1.71	–	87.4	32	23.6	6.7	9.5	4.5	2.6	16.9	1.4	16.5	78.9	0.53	0.7	−43.9	−37.6	−36.7	−261	−148	−130
	168	1.8	–	94.9	34.5	25.5	7.2	9.5	4.2	2.2	18.4	1.4	17.5	83	0.53	0.76	−44	−37.4	−36	−257	−126	−121
	240	1.89	–	103.2	34.9	24.7	6.8	9.3	4	2.4	20.1	1.5	19.7	82.2	0.56	0.73	−43.6	−37.4	−36	−250	−113	−116
	360	2.0	–	124.3	40.8	29.2	8.7	9.8	3.6	1.8	22.8	2	20.4	93.9	0.57	0.89	−42.7	−37.2	−35.7	−240	−106	−103
400	240	2.07	–	132.8	43	30.5	8.6	9.9	0.1	1.8	27.1	2.8	23.6	93.9	0.59	0.87	−43.1	−37	−34.4	−235	−103	−101
	360	2.19	–	133.8	42.6	30.7	8.9	10	3.2	1.7	27.6	3.2	25.4	97.2	0.58	0.89	−43.2	−36.8	−33.3	−226	−99	−100
420	168	2.35	–	148.5	45.1	30.8	7.7	8.8	1.9	1.1	27.9	3.6	27.3	95.5	0.61	0.88	−42.5	−36.4	−34	−221	−95	−104
	264	2.5	–	159.3	45.7	31.4	8.3	8.6	1.9	1	28.2	3.6	26.4	96.8	0.62	0.96	−42.4	−36	−33.1	−211	−95	−98

$EasyR_o$% values were calculated in Matlab according to Sweeney and Burnham (1990))

The amounts of kerogen, uranium ore and deionized water were 50 mg, 50 mg and 50 µL, respectively

– not detected

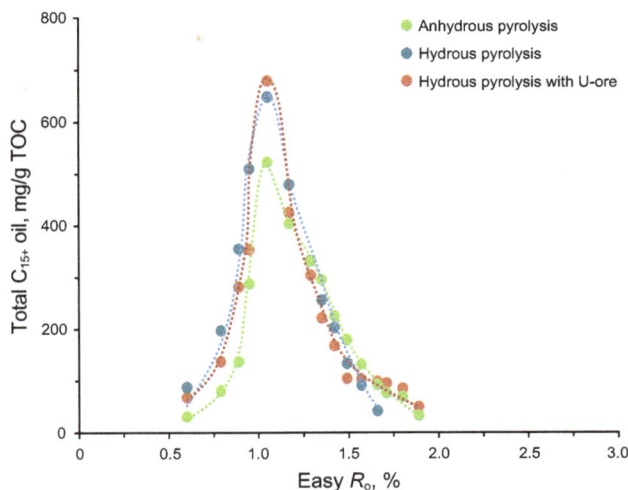

Fig. 1 Yields of total C_{15+} oils versus EasyR_o% of kerogen generated under conditions of no added water (*green points*), added water (*blue points*), and added water plus U-ore (*red points*)

3 Results

3.1 Yields of C_{15+} oils

After heating for 1, 3, 5, 7, 10 and 15 days at different temperatures, equivalent R_o values (EasyR_o% using the Sweeney and Burnham (Sweeney and Burnham 1990) kinetic model to estimate maturity) were calculated, ranging from 0.6% to 2.5% (Table 2).

Total C_{15+} oils from anhydrous and hydrous experiments were given relative to TOC (mg/g) and EasyR_o% (normalized percentage) (Table 2; Fig. 1). As demonstrated in Fig. 1, all three experiments show a similar trend concerning the relationship between temperature and EasyR_o%. In anhydrous and hydrous experiments with and without uranium ore, yields of C_{15+} oils at first increased rapidly from the initial values, 30.5, 67.8 and 87.8 mg/g TOC, to the maximum values, 523.2, 679.3 and 648.3 mg/g TOC with an EasyR_o% of 1.05%, and then decreased with further increased heating time. Experiments with water produced more oil than those without water, and the hydrous experiments with added U-ore produced more oil as well. Obviously, the presence of water and U-ore apparently enhanced oil yields from Type-II kerogen on different levels. Nevertheless, oil generation peaks were not affected in terms of the same EasyR_o% (Fig. 1).

3.2 Yields of gas products

The amounts and ratios of gas compositions produced during experiments are shown in Table 2 and Fig. 2. The amounts of generated methane decrease in the order of

hydrous experiments with added uranium ore (5.8–198.5 mL/g TOC), those without uranium ore (3.5–159.3 mL/g TOC), and those without water and U-ore (2.2–140.8 mL/g TOC) (Fig. 2a).

In contrast to the continuously increasing yields of methane, the yields of wet gases (C_{2-5}) first increased to maximum values and then barely changed (Fig. 2b). Maximum values are 107.4 mL/g TOC, 96.8 mL/g TOC and 80.4 mL/g TOC for the kerogen pyrolysis under conditions with water plus uranium ore, with water, and without water, respectively.

Among non-hydrocarbon gases, CO_2 is the most abundant, followed by H_2S (Table 2). The yields of CO_2 generated from pyrolysis with added water generally exceeded those without water. As shown in Fig. 2c, hydrous experiments with U-ore produced less CO_2 than those without U-ore, ranging from 1.6 to 22.5 mL/g TOC, and 2.2 to 28.2 mL/g TOC, respectively.

The yields of H_2S show different trends compared with those of CO_2. For anhydrous experiments, the yields of H_2S increased from 1.1 to 10.8 mL/g TOC before EasyR_o% reached 1%, and then the yields remained unchanged with increasing temperature and EasyR_o%. On the contrary, in both hydrous experiments, the yields of H_2S were substantially lower when EasyR_o% was less than 1.5%, especially in the experiments with uranium ore. When EasyR_o% is larger than 1.5%, yields of hydrous experiments without uranium ore (8.9–26.4 mL/g TOC) exceeded those of anhydrous experiments. Meanwhile, the yields of experiments with U-ore gradually increased, but were still lower than those of the anhydrous pyrolysis. The phenomenon may be due to H_2S being removed by dissolution in water, and rapid reaction with the iron or uranium contained in the mineral assemblage to form pyrrhotite ($Fe_{1-x}S$) (Lewan 1993). Both hydrous pyrolysis tests displayed a very similar evolution pattern in terms of the yields of H_2 (Table 2; Fig. 2d). The values increased from the initial value of 0.4 mL/g TOC (EasyR_o% = 1.59%) to 3.9 mL/g TOC (EasyR_o% = 2.5%). They are substantially higher than those in anhydrous experiments, with the values from 0.1 mL/g TOC (EasyR_o% = 1.8%) to 1.6 mL/g TOC (EasyR_o% = 2.5%).

For the three experiments, the C_1/C_{1-5} ratios first decreased to minimum values in the EasyR_o% range of 0.6%–1.2%, and then increased consistently with temperature and EasyR_o%. Under conditions with lower EasyR_o%, hydrocarbons with longer alkyl chains including C_{15+} oils and C_{2-5} were dominant products. As temperature increased, the early generated oils and the methyl bonded in kerogen structures were cracked to generate more methane. The minimum values are 0.36 (R_o = 0.95%), 0.50 (R_o = 1.05%) and 0.50 (R_o = 1.17%) in the experiments under conditions with exclusive kerogen, kerogen

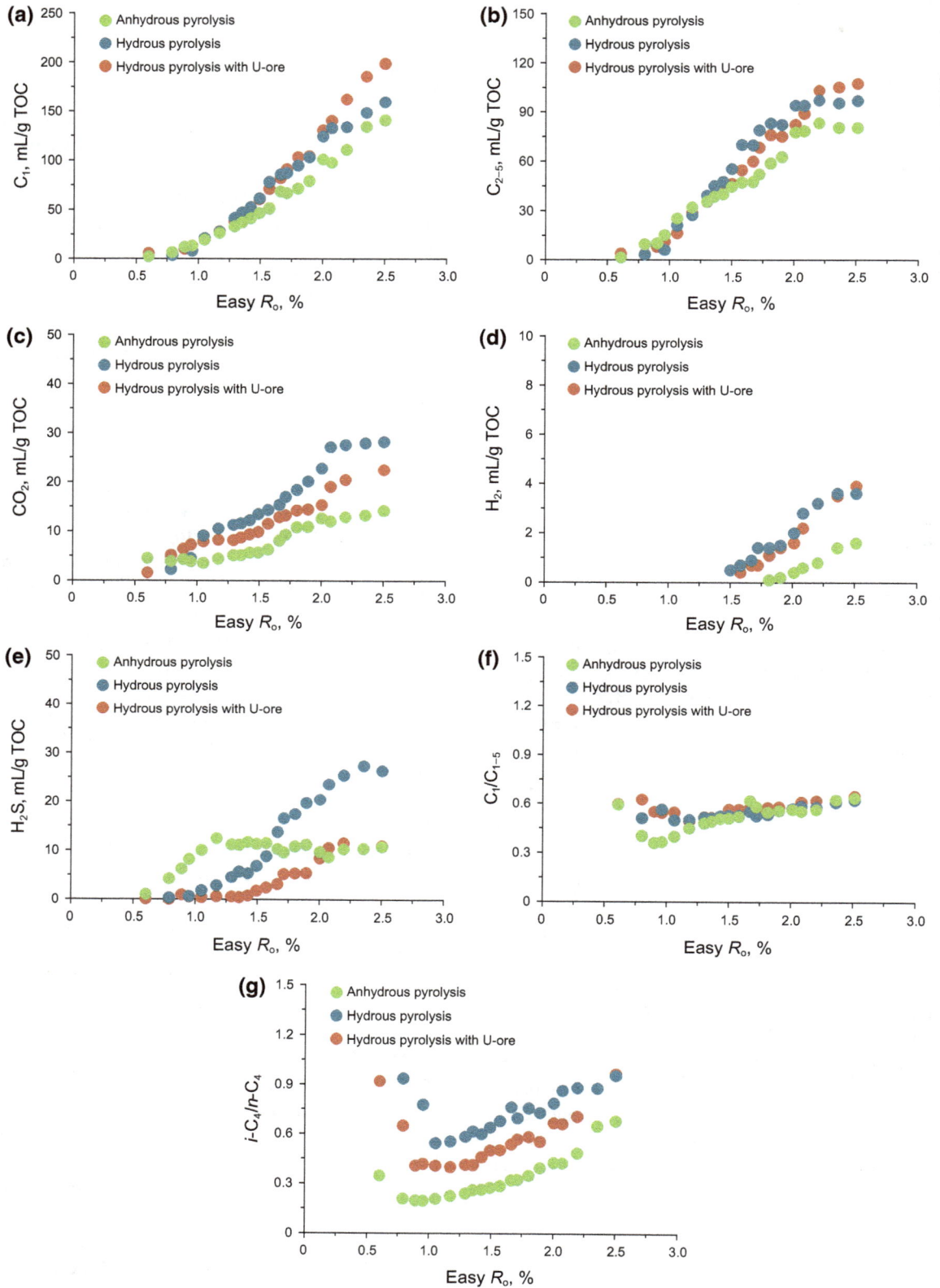

Fig. 2 Yields of **a** methane, **b** wet gases (C_{2-5}), **c** CO_2, **d** H_2, **e** H_2S, **f** C_1/C_{1-5} and **g** i-C_4/n-C_4 of kerogen generated under conditions with no added water (*green points*), added water (*blue points*), and added water plus U-ore (*red points*)

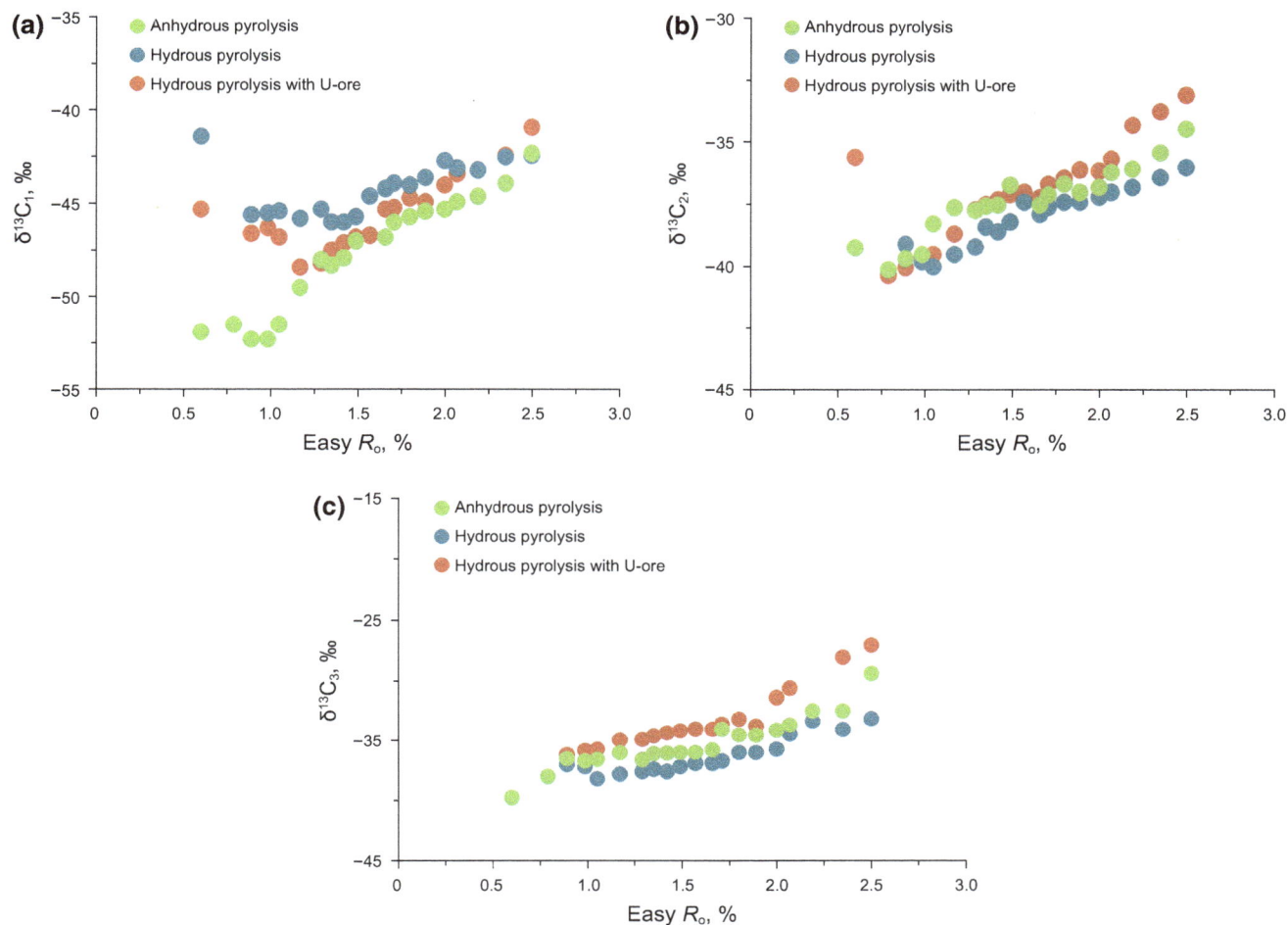

Fig. 3 Carbon isotopic compositions of **a** methane, **b** ethane, and **c** propane of kerogen generated under conditions with no added water (*green points*), added water (*blue points*), and added water plus U-ore (*red points*)

plus water, kerogen plus water and uranium ore, which corresponded closely to the maximum values of total C_{15+} oils (Table 2). This was also observed in previous experimental studies on Type-I and Type-II kerogens (Lorant et al. 1998; Pan et al. 2009).

The trends of the i-C_4/n-C_4 ratios for kerogen without water are relatively lower than those with water. For the ratios in anhydrous and hydrous pyrolysis with and without uranium ore, the values first decreased rapidly from high values of 0.35, 0.92 and 0.94 to the lowest values, 0.2, 0.4 and 0.58, and then increased again to 0.68, 0.97 and 0.96.

3.3 Carbon and hydrogen isotopes of hydrocarbon gases

3.3.1 Carbon isotopes

Stable carbon and hydrogen isotope ratios of produced hydrocarbon gases during thermal cracking of kerogen under controlled hydrous laboratory conditions are useful

to correlate characteristic isotopic responses with differences in source materials and experimental conditions. The carbon isotope compositions of methane, ethane and propane are shown in Table 2 and Fig. 3. The $\delta^{13}C$ values of methane ($\delta^{13}C_1$) became more negative with the EasyR_o% of about 0.95% in all the experiments. The lowest values are –52.3, –48.4 and –46.0‰ for the pyrolysis of exclusive kerogen, kerogen plus water and kerogen plus water and U-ore (Table 2; Fig. 3a). After the minimum values were reached, all the experiments showed consistent monotonic $\delta^{13}C_1$ enrichment with increasing maturity, ranging from –52.3 to –41.5‰ for anhydrous experiments, and from –46 to –40.9‰, from –48.4 to –42.4‰ for hydrous experiments with and without uranium ore, respectively. The $\delta^{13}C$ values of ethane ($\delta^{13}C_2$) showed the same trends with $\delta^{13}C_1$. The minimum values are –39.7, –38.2 and –36.2‰ in the EasyR_o% interval of 0.95–1.14% for the anhydrous experiments and hydrous experiments without and with uranium ore, and all the values increased to about– 30.0 ± 3‰ (Fig. 3b). The trends of $\delta^{13}C$ value of propane

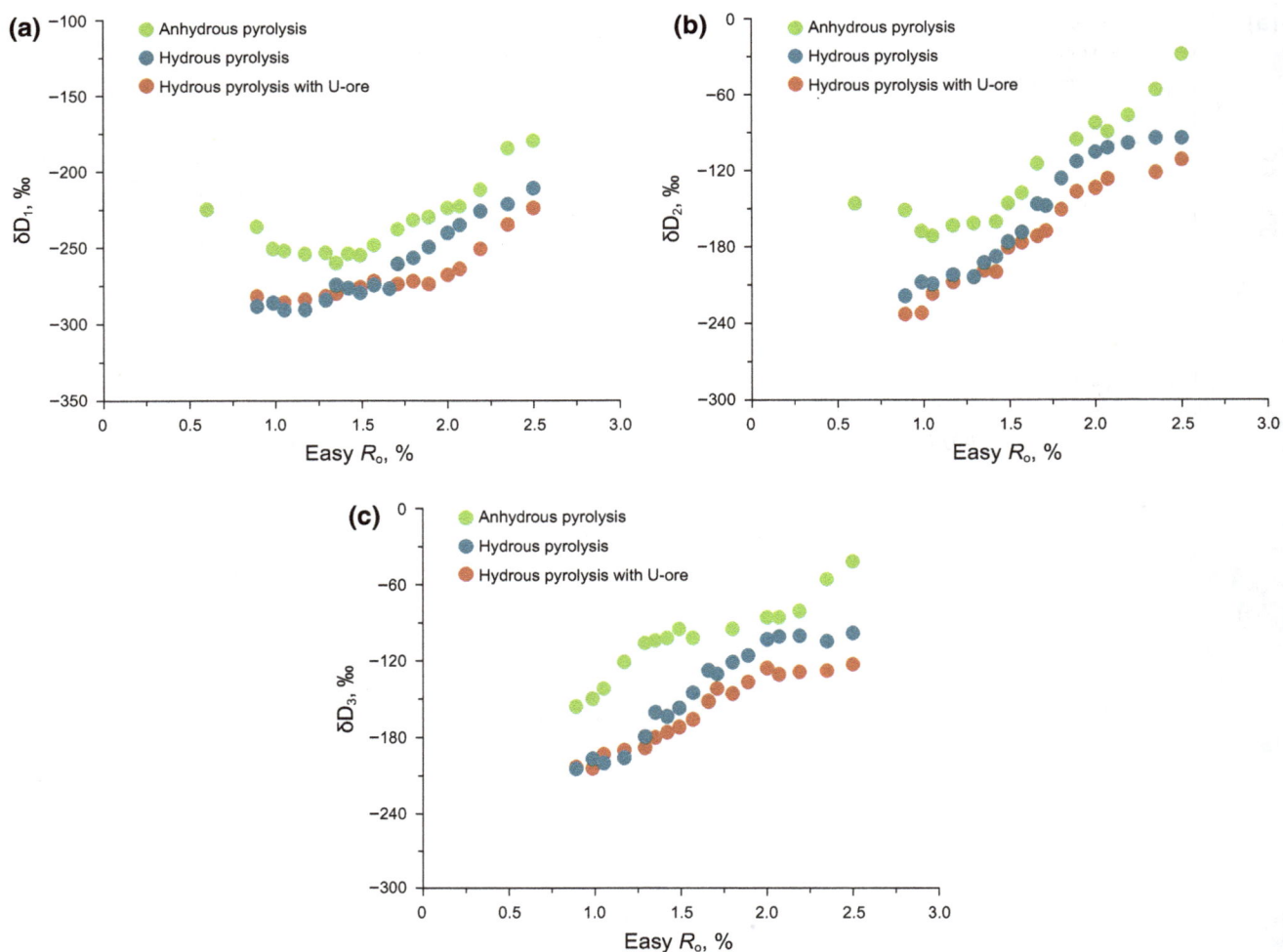

Fig. 4 Hydrogen isotopic compositions of **a** methane, **b** ethane, and **c** propane from kerogen generated under conditions with no added water (*green points*), added water (*blue points*), and added water plus U-ore (*red points*)

($\delta^{13}C_3$) are similar at the same temperature among all the sequences, with isotopic differences ranging between about −39.7 and −28.4‰ (Fig. 3c). And within all the experiments, $\delta^{13}C_1$ values are always more negative than the corresponding $\delta^{13}C$ values of ethane ($\delta^{13}C_2$) and propane ($\delta^{13}C_3$).

3.3.2 Hydrogen isotopes

The δD trends in the three experiments are shown in Table 2 and Fig. 4. In contrast to the carbon isotope compositions, the δD values for the hydrous experiments are less negative than those for the anhydrous experiments. In the experiments without water, δD values of methane (δD_1) range from −260 to −180‰. In the experiments with water and water plus uranium ore, δD values of methane (δD_1) range from −291 to −211‰ and from −286 to −224‰, respectively, with increasing temperature and

EasyR_o. δD values of ethane (δD_2) and propane (δD_3) also show consistent enrichment throughout pyrolysis, for hydrous experiments without uranium ore, and the values are between −219 and −95‰, and −204 and −98‰, respectively. For hydrous experiments with uranium ore, the value ranges are −233 to −112‰ and −203 to −123‰. And in anhydrous pyrolysis, the value ranges are −172 to −29‰, and −156 to −42‰ (Table 2).

3.4 Isotopic compositions and elemental ratios of residues

3.4.1 Elemental ratios of residues

The isotopic compositions of residues and elemental ratios obtained from anhydrous and hydrous pyrolysis experiments are presented in Table 3 and Fig. 5. The effect of increasing thermal maturation and consequent conversion of kerogen into oil and gas was indicated by the decrease in

Table 3 Isotopic and elemental ratios of the residues during anhydrous and hydrous pyrolysis with and without U-ore

Temp., °C	Time, h	EasyR$_o$ (%)	Carbon isotopes (‰, vs. VPDB)	Hydrogen isotopes (‰, vs. VSMOW)	Elemental ratios	
					H/C	O/C
Anhydrous condition						
300	24	0.6	−33.4	−110.9	1.06	0.01
	24	0.79	−33.9	−104.1	0.90	0.02
330	72	0.89	−33.3	−93.0	0.76	0.03
	120	0.95	−33.4	−89.5	0.72	0.04
	240	1.05	−33.3	−89.8	0.61	0.02
350	120	1.17	−33.4	−87.8	0.58	0.02
	240	1.29	−33.2	−85.4	0.55	0.02
360	168	1.35	−33.2	−87.6	0.59	0.02
	240	1.42	−33.3	−88.6	0.55	0.02
370	168	1.49	−33.0	−88.4	0.52	0.02
	240	1.59	−33.0	−88.9	0.51	0.02
	360	1.66	−32.9	−89.4	0.51	0.02
390	120	1.71	−32.9	−87.8	0.50	0.02
	168	1.8	−33.0	−87.5	0.51	0.02
	240	1.89	−33.1	−87.2	0.50	0.01
	360	2.0	−32.9	−85.9	0.47	0.01
400	240	2.07	−33.1	−86.0	0.48	0.02
	360	2.19	−33.0	−83.7	0.46	0.02
420	168	2.35	−33.0	−81.9	0.46	0.02
	264	2.5	−33.0	−77.6	0.44	0.02
Kerogen plus deionized water and U−ore						
300	24	0.6	−33.4	−108.1	1.23	0.08
	24	0.79	−32.8	−109.3	1.13	0.11
330	72	0.89	−33.3	−111.7	1.04	0.11
	120	0.95	−33.3	−112.2	1.02	0.12
	240	1.05	−32.9	−104.6	0.82	0.10
350	120	1.17	−33.2	−104.7	0.87	0.12
	240	1.29	−33.4	−107.2	0.79	0.12
360	168	1.35	−33.3	−103.2	0.84	0.09
	240	1.42	−33.4	−104.4	0.76	0.12
370	168	1.49	−33.3	−106.2	0.68	0.10
	240	1.59	−33.3	−105.1	0.64	0.08
	360	1.66	−33.3	−107.4	0.64	0.08
390	120	1.71	−33.3	−106.3	0.58	0.06
	168	1.8	−33.2	−103.4	0.61	0.09
	240	1.89	−33.2	−101.5	0.66	0.09
	360	2.0	−33.1	−102.3	0.58	0.05
400	240	2.07	−33.2	−108.0	0.56	0.07
	360	2.19	−33.2	−106.3	0.56	0.05
420	168	2.35	−33.1	−106.5	0.55	0.05
	264	2.5	−33.1	−104.8	0.53	0.06
Kerogen plus deionized water						
300	24	0.6	−33.5	−122.4	1.14	0.02
	24	0.79	−33.4	−129.7	1.04	0.02
330	72	0.89	−33.3	−130.1	0.95	0.02

Table 3 continued

Temp., °C	Time, h	EasyR_o (%)	Carbon isotopes (‰, vs. VPDB)	Hydrogen isotopes (‰, vs. VSMOW)	Elemental ratios	
					H/C	O/C
	120	0.95	−33.2	−128.5	0.39	0.03
	240	1.05	−33.3	−127.4	0.33	0.04
350	120	1.17	−33.3	−126.0	0.57	0.05
	240	1.29	−33.2	−127.7	0.52	0.05
360	168	1.35	−33.2	−120.6	0.50	0.05
	240	1.42	−33.1	−126.4	0.59	0.04
370	168	1.49	−33.3	−129.3	0.55	0.04
	240	1.59	−33.4	−135.0	0.53	0.04
	360	1.66	−33.4	−129.5	0.53	0.04
390	120	1.71	−33.4	−132.8	0.51	0.05
	168	1.8	−33.4	−135.8	0.55	0.04
	240	1.89	−33.4	−137.2	0.53	0.04
	360	2.0	−33.3	−134.3	0.47	0.05
400	240	2.07	−33.4	−131.4	0.47	0.06
	360	2.19	−33.5	−132.5	0.49	0.05
420	168	2.35	−33.4	−134.1	0.45	0.05
	264	2.5	−33.5	−144.1	0.47	0.05

atomic H/C and O/C ratios in both experiments (Table 3; Fig. 5a, b). These geochemical parameters decreased with increasing maturity, as a result of losing hydrogen and carbon during the oil and gas generation and expulsion. Compared with the anhydrous experiments, the hydrous experiments showed a considerable increase in elemental atomic ratios (H/C and O/C), and the ratios of hydrous pyrolysis generally exceeded those from the experiments with no water added. In the hydrous experiments, the ratios of H/C ranged from 1.23 to 0.53 for experiments with added U-ore and from 1.14 to 0.47 for experiments without added U-ore. The ratios of O/C were 0.08 to 0.06 and 0.02 to 0.05 for experiments with and without U-ore. In the anhydrous experiments, the H/C ratios were 1.06 to 0.44, while O/C ratios were 0.01 to 0.02 (Fig. 5a). This result implies that both hydrogen and oxygen were incorporated during pyrolysis, which is consistent with the higher amount of CO_2 produced in the hydrous experiments (Table 2).

3.4.2 Isotopic compositions of residues

Changes in the carbon isotope compositions of the residues ($\delta^{13}C_{residues}$) show similar trends with those of the gases in Sect. 3.3 (Fig. 3), generally ranging from −33.4 to −33‰ for the three experiments (Table 3; Fig. 5c). All hydrogen isotope ratios of the residues ($\delta D_{residues}$) from the pyrolysis experiments with added water were significantly more negative than those without water (Table 3; Fig. 5d), which

are consistent with the notably lower values of δD in the gases produced in the hydrous experiments (Fig. 4). However, δD values of hydrous experiments with added uranium ore range from −108.1 to −104.8‰, which are higher than those without uranium ore, with the values between −130.1 and −144.1‰.

4 Discussion

The apparent increase in liquid and gas products, as well as the depletion of deuterium in hydrocarbon gases in hydrous pyrolysis, demonstrated that water and uranium ore could participate in the thermal evolution of organic matter and provide hydrogen for hydrocarbon generation. Indeed, previous experimental work by Lewan (1997) found that water plays a positive role in the generation of oil from organic matter. Liquid water can act as an exogenous hydrogen source that inhibits cross-link reactions during anhydrous pyrolysis and promotes hydropyrolysis by free-radical reactions or carbocation reactions (Lewan 1997; He et al. 2011b). Moreover, the increase in H_2, the slowly decreasing H/C ratios, and the lower δD values of solid residues or kerogen during hydrous pyrolysis also demonstrated that water and mineral-derived hydrogen have been introduced, which reacted with kerogen (Hoering 1984; Leif and Simoneit 2000; Lewan 1997; Schimmelmann et al. 1999). Hence, water and uranium ore were critical in the generation of oil and

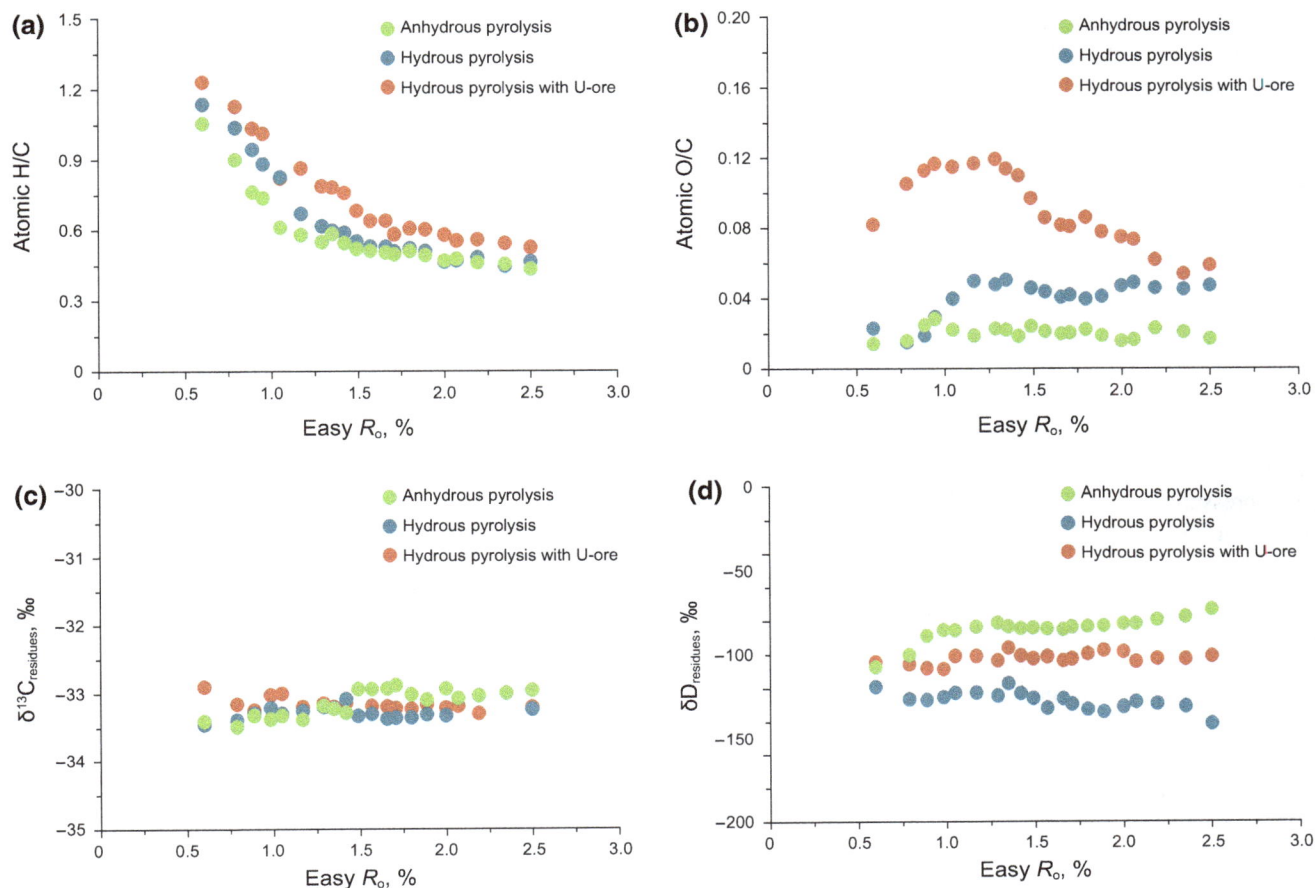

Fig. 5 Isotopic and elemental ratios of **a** H/C, **b** O/C, **c** $\delta^{13}C_{residues}$, and **d** $\delta D_{residues}$ of kerogen generated under conditions with no added water (*green points*), added water (*blue points*), and added water plus U-ore (*red points*)

gas as well as the thermal evolution of kerogen matter in our pyrolysis. Meanwhile, the mechanisms in the two tests of hydrous pyrolysis may differ a lot compared with that for kerogen cracking under anhydrous conditions. As is known, the generation of normal alkanes and isomeric alkanes during the cracking of organic compounds usually reflects free-radical reactions and carbocation reactions, respectively (Johns 2003; Kissin 1987). As shown in Table 2 and Fig. 2, the i-C_4/n-C_4 ratios for gas products were evidently higher in the hydrous experiments than those in anhydrous experiments. Such results demonstrated that the carboncation mechanism which was induced by the H^+ dissociated from the water should be presented (Siskin and Katritzky 1991; Leif and Simoneit 2000; Seewald 2001; Kissin 1987). What is more, the i-C_4/n-C_4 ratios in the hydrous experiments for kerogen plus U-ore were consistently lower compared with those without added U-ore, which is somehow opposite to the increased ratios in the previous study on kerogen with added clay mineral and water (Sweeney and Burnham 1990; Pan et al. 2010). The results indicated that the presence of U-ore inhibited the carbocation mechanism.

However, the increased yields of oil and gas implied there should be another reaction mechanism dominating hydrogenation during the hydrous pyrolysis with carbocations.

Additional insights into the difference among the three experiments were obtained from the isotopic data. The presence of added water with an initial $\delta D = -55‰$ caused about 20‰ deuterium-depletion in methane, which indicated that the isotopic transfer of water hydrogen into methane was affected by a kinetic fractionation favoring [1]H (Schimmelmann et al. 1999). Hydrogen atoms from organic compounds and other available hydrogen atoms, including those from water, might participate in thermal maturation. The transfer between water-derived hydrogen and organic compound-derived hydrogen had been observed in several studies using D_2O (Hoering 1984; Leif and Simoneit 2000; Lewan 1997) and deuterium-enriched water (Schimmelmann et al. 1999) under artificial thermal maturation conditions. The hydrogen transfer mechanisms that rely on the activation energy of various organic moieties vary. For example, in organic compounds, hydrogen bound to O, N or S can easily react with water vapor at low temperature. Hydrogens connected to C=O and COOH, or in condensed aromatic systems, are exchangeable at elevated temperatures and suitable pH via enolization. Alkyl hydrogens are non-exchangeable owing to the strong, nonpolar covalent bond to carbon (Sessions et al. 2004). Similar trends in isotopic exchange were also observed in $\delta^{13}C_1$. In the pyrolysis experiments, $\delta^{13}C_1$ initially decreased from a high value to a minimum value, and then increased with temperature and $EasyR_o$. This reversal implies that carbon cracking changed from labile bonds (between carbon and heteroatoms) to C–C bonds (Cramer 2004), and this observation is consistent with many previous studies (Liu et al. 2004; Lorant et al. 1998; Tang et al. 2000; Tian et al. 2012; Xiao et al. 2005; Xiong et al. 2004). All pyrolysis experiments were conducted under the same pressure, excluding the possibility of varying isotopic fractionation during gas generation and expulsion under different pressures.

As shown in Fig. 4 and Table 2, a higher depletion of hydrogen isotopic compositions of the gas products was observed in the hydrous pyrolysis with added U-ore than those without uranium ore, and the isotopic change accelerated with increasing temperature. This difference might be most readily interpreted by the different reaction pathways between organic matter and water. In hydrous pyrolysis without the uranium ore, hydrogen from water simply exchanged with organic matter through a carbocation mechanism or by direct addition to the double bonds in the pyrolysis (Lewan 1997; Schimmelmann et al. 1999). However, in the experiments with uranium ore, more complex mechanisms might be involved, and water could at first react with minerals during the pyrolysis to produce hydrogen gas (Seewald 2001, 2003), and the latter would continually react with kerogen to produce hydrocarbon. The secondary-fractionation of hydrogen caused the values of δD_{H2} to be relatively lower than the values of water or organic matter, and further affected the hydrogen isotope ratio of products (Schimmelmann et al. 1999; Seewald 2001; Leif and Simoneit 2000). Furthermore, as proposed by Seewald (2001, 2003), in the conditions with water and U-ore the aqueous n-alkanes decomposed through a series of oxidation and hydration reactions, which produced reaction intermediaries, including alkenes, alcohols, ketones and organic acids. The chemical changes during the pyrolysis would affect the exchangeability of hydrogen atoms. For example, a hydrogen atom adjacent to electron-withdrawing groups, such as carbonyl, carboxyl, alcohol or amino groups, can exchange with ambient water over relatively short geologic timescales (Seewald 2001; Larcher et al. 1986). Thus, the attainment of this rapidly reversible metastable thermodynamic equilibrium provided a pathway for hydrogen exchange and enhanced the exchange rates (Reeves et al. 2012).

5 Conclusions

In this study, we compared the chemical and isotopic compositions of thermogenic gas and oil generated during pyrolysis of Type-II kerogen with variation of $EasyR_o\%$, an index reflecting maturity, from 0.6% to 2.5%. Pyrolysis experiments were carried out in the presence and absence of water and U-ore in sealed gold tubes under different conditions with eight temperatures ranging from 300 to 420 °C. The conclusions are as follows:

1. In the hydrous experiments with and without U-ore, the yields of C_{15+} oils increased from 67.8 mg/g TOC and 87.8 mg/g TOC to 679.3 mg/g TOC and 648.3 mg/g TOC, respectively, which are 29% higher than those of anhydrous pyrolysis. Amounts of gas products, including methane, ethane, CO_2 and H_2 also increased to different degrees. The results demonstrated that water and U-ore provided sources of hydrogen and oxygen in the hydrocarbon generation.

2. The gas products and residues of hydrous pyrolysis were deuterium-depleted, and the residues had a higher elemental atomic ratio than the anhydrous pyrolysis counterparts, which suggested that water-derived hydrogen and oxygen, as well as uranium ore derived oxygen were incorporated into the gas products of hydrous pyrolysis, and were exchanged with the kerogen during the process. The δD values of the gas products were much lower in hydrous pyrolysis with uranium ore, indicating that the generation of H_2 occurred in hydrous pyrolysis with U-ore, which accounts for the higher depletion of deuterium in the gas products.

3. The main reaction in the pyrolysis of kerogen with water is the carbocation mechanism, reflected by a higher ratio of i-C_4/n-C_4 in the hydrous pyrolysis than that in the anhydrous pyrolysis. The lower i-C_4/n-C_4 ratios of the hydrous experiments plus U-ore indicated that the carbocation mechanism was inhibited to some extent.

References

Cramer B. Methane generation from coal during open system pyrolysis investigated by isotope specific, Gaussian distributed reaction kinetics. Org Geochem. 2004;35(4) 379–92. doi:10.1016/j.orggeochem.2004.01.004.

Fisher QJ, Wignall PB. Palaeoenvironmental controls on the uranium distribution in an Upper Carboniferous black shale (Gastrioceras listeri, Marine Band) and associated strata England. Chem Geol. 2001;175(3):605–21. doi:10.1016/S0009-2541(00)00376-4.

Galindo C, Mougin L, Fakhi S, et al. Distribution of naturally occurring radionuclides (U, Th) in Timahdit black shale (Morocco). J Environ Radioact. 2007;92(1):41–54. doi:10.1016/j.jenvrad.2006.09.005.

Goldstein TP. Geocatalytic reactions in formation and maturation of petroleum. AAPG Bull. 1983;67(1):52–159. doi:10.1306/03b5acd7-16d1-11d7-8645000102c1865d.

He K, Zhang SC, Mi JK. Research on the kinetics and controlling factors for oil cracking. Nat Gas Geosci. 2011a;22(02):211–8. doi:10.11764/j.issn.1672-1926.2011.02.211 (in Chinese).

He K, Zhang S, Mi J, et al. Mechanism of catalytic hydropyrolysis of sedimentary organic matter with MoS₂. Pet Sci. 2011b;08(2):134–42. doi:10.1007/s12182-011-0126-0.

Hoering TC. Thermal reactions of kerogen with added water, heavy water and pure organic substances. Org Geochem. 1984;5(4):267–78. doi:10.1016/0146-6380(84)90014-7.

Huizinga BJ, Tannenbaum E, Kaplan IR. The role of minerals in the thermal alteration of organic matter-III. Generation of bitumen in laboratory experiments. Org Geochem. 1987a;11(6):591–604. doi:10.1016/0016-7037(85)90128-0.

Huizinga BJ, Tannenbaum E, Kaplan IR. The role of minerals in the thermal alteration of organic matter-IV. Generation of n-alkanes, acyclic isoprenoids, and alkenes in laboratory experiments. Geochim Cosmochim Acta. 1987b;1(5):1083–97. doi:10.1016/0016-7037(85)90128-0.

Johns WD. Clay mineral catalysis and petroleum generation. Annu Rev Earth Planet Sci. 2003;7(1):183–98. doi:10.1146/annurev.ea.07.050179.001151.

Kissin YV. Catagenesis and composition of petroleum: origin of n-alkanes and isoalkanes in petroleum crudes. Geochim Cosmochim Acta. 1987;51(9):2445–57. doi:10.1016/0016-7037(87)90296-1.

Larcher AV, Alexander R, Rowland SJ, et al. Acid catalysis of alkyl hydrogen exchange and configurational isomerisation reactions: acyclic isoprenoid acids. Org Geochem. 1986;10(4–6):1015–21. doi:10.1016/S0146-6380(86)80040-7.

Leif RN, Simoneit BRT. The role of alkenes produced during hydrous pyrolysis of a shale. Org Geochem. 2000;31(11):1189–208. doi:10.1016/S0146-6380(00)00113-3.

Lev SM, Filer JK. Assessing the impact of black shale processes on REE and the U–Pb isotope system in the southern Appalachian Basin. Chem Geol. 2004;206(3):393–406. doi:10.1016/j.chemgeo.2003.12.012.

Lev SM, Filer JK, Tomascak P. Orogenesis vs. diagenesis: Can we use organic-rich shales to interpret the tectonic evolution of a depositional basin? Earth Sci Rev. 2008;86(1):1–14. doi:10.1016/j.earscirev.2007.07.001.

Lewan MD. Laboratory simulation of petroleum formation. Org Geochem. 1993;. doi:10.1007/978-1-4615-2890-6_18.

Lewan MD. Experiments on the role of water in petroleum formation. Geochim Cosmochim Acta. 1997;61(17):3691–723. doi:10.1016/S0016-7037(97)00176-2.

Lewan MD, Kotarba MJ, Więcław D, et al. Evaluating transition-metal catalysis in gas generation from the Permian Kupfer-schiefer by hydrous pyrolysis. Geochim Cosmochim Acta. 2008;72(16):4069–93. doi:10.1016/j.gca.2008.06.003.

Li B, Meng ZF, Xia B, et al. Hydrocarbon-generating thermal simulation of uranium-bearing minerals. Acta Pet Mineral. 2008;27(01):52–8. doi:10.3969/j.issn.1000-6524.2008.01.007 (in Chinese).

Liu DY, Liu JZ, Peng PA, et al. Carbon isotope kinetics of gaseous hydrocarbons generated from different kinds of vitrinites. Chin Sci Bull. 2004;49(1):72–8. doi:10.1007/BF02890456 (in Chinese).

Liu CY, Zhao HG, Tan CQ, et al. Occurrences of multiple energy mineral deposits and mineralization/reservoiring system in the basin. Oil Gas Geol. 2006;27(2):131–42. doi:10.3321/j.issn:0253-9985.2006.02.001 (in Chinese).

Liu CY, Mao GZ, Qiu XW, et al. Organic-inorganic energy minerals interactions and the accumulation and mineralization in the same sedimentary basins. Chin J Nat. 2013;35(1):47–55. doi:10.3969/j.issn.0253-9608.2013.01.006 (in Chinese).

Lorant F, Prinzhofer A, Behar F, et al. Carbon isotopic and molecular constraints on the formation and the expulsion of thermogenic hydrocarbon gases. Chem Geol. 1998;147(3):249–64. doi:10.1016/S0009-2541(98)00017-5.

Lu HX, Meng ZF, Li B, et al. Effect of uranium substance on hydrocarbon generation from liginite by hydrous pyrolysis. Xinjiang Pet Geol. 2007;28(06):718–20. doi:10.3969/j.issn.1001-3873.2007.06.016 (in Chinese).

Mango FD. Transition metal catalysis in the generation of petroleum: a genetic anomaly in Ordovician oils. Geochim Cosmochim Acta. 1992;56(10):3851–4. doi:10.1016/0016-7037(92)90153-A.

Mango FD, Joe Hightower. The catalytic decomposition of petroleum into natural gas. Geochim Cosmochim Acta. 1997;61(24):5347–50. doi:10.1016/S0016-7037(97)00310-4.

Mao GZ, Liu CY, Liu BQ, et al. Effects of uranium on hydrocarbon generation of low-mature hydrocarbon source rocks containing kerogen type I. J China Univer Pet. 2012a;36(2):172–81. doi:10.3969/j.issn.1673-5005.2012.02.030 (in Chinese).

Mao GZ, Liu CY, Zhang DD, et al. Effects of uranium (type II) on evolution of hydrocarbon generation of source rocks. Acta Geol Sin. 2012b;86(11):1833–40. doi:10.3969/j.issn.0001-5717.2012.11.013 (in Chinese).

Mao GZ, Liu CY, Zhang DD, et al. Effects of uranium on hydrocarbon generation of hydrocarbon source rocks with type-III kerogen. Sci China Earth Sci. 2014;57:1168–79. doi:10.1007/s11430-013-4723-1 (in Chinese).

Mayer LM. Relationships between mineral surfaces and organic carbon concentrations in soils and sediments. Chem Geol. 1994;114(3–4):347–63. doi:10.1016/0009-2541(94)90063-9.

Pan CC, Geng AS, Zhong NN, et al. Kerogen pyrolysis in the presence and absence of water and minerals: amounts and compositions of bitumen and liquid hydrocarbons. Fuel. 2009;88(5):909–19. doi:10.1016/j.fuel.2008.11.024.

Pan CC, Geng AS, Zhong NN, et al. Kerogen pyrolysis in the presence and absence of water and minerals: steranes and triterpenoids. Fuel. 2010;89(2):336–45. doi:10.1016/j.fuel.2009.06.032.

Partin C, Bekker A, Planansky N, et al. Large-scale fluctuations in PreCambrian atmospheric and oceanic oxygen levels from the record of U in shales. Earth Planet Sci Lett. 2013;369–370(3):284–93. doi:10.1016/j.epsl.2013.03.031.

Reeves EP, Seewald JS, Sylva SP. Hydrogen isotope exchange between n-alkanes and water under hydrothermal conditions. Geochim Cosmochim Acta. 2012;77:582–99. doi:10.1016/j.gca.2011.10.008.

Schimmelmann A, Lewan MD, Wintsch RP. D/H isotope ratios of kerogen, bitumen, oil, and water in hydrous pyrolysis of source rocks containing kerogen types I, II, IIS, and III. Geochim Cosmochim Acta. 1999;63(22):3751–66. doi:10.1016/S0016-7037(99)00221-5.

Seewald JS. Aqueous geochemistry of low molecular weight hydro-carbons at elevated temperatures and pressures: constraints from mineral buffered laboratory experiments. Geochi Cosmochim Acta. 2001;65(10):1641–64. doi:10.1016/S0016-7037(01)00544-0.

Seewald JS. Organic-inorganic interactions in petroleum-producing sedimentary basins. Nature. 2003;426(6964):327–33. doi:10.1038/nature02132.

Seewald JS, Benitez-Nelson BC, Whelan Jk. Laboratory and theoretical constraints on the generation and composition of natural gas. Geochim Cosmochim Acta. 1998;62(9):1599–617. doi:10.1016/S0016-7037(98)00000-3.

Sessions AL, Sylva SP, Summons RE, et al. Isotopic exchange of carbon-bound hydrogen over geologic timescales 1. Geochim Cosmochim Acta. 2004;68(7):1545–59. doi:10.1016/j.gca.2003.06.004.

Siskin M, Katritzky AR. Reactivity of organic compounds in hot water: geochemical and technological implications. Science. 1991;254:231–7. doi:10.1126/science.254.5029.231.

Surdam RC, Crossey LJ, Eglinton G, et al. Organic-inorganic reactions during progressive burial: key to porosity and permeability enhancement and preservation [and discussion]. Philos Trans R Soc Lond A Math Phys Eng Sci. 1985;315(315):135–56. doi:10.1098/rsta.1985.0034.

Surdam RC, Jiao ZS, MacGowan DB. Redox reactions involving hydrocarbons and mineral oxidants: a mechanism for significant porosity enhancement in sandstones. AAPG Bull. 1993;77(9):1509–1518.

Sweeney JJ, Burnham AK. Evaluation of a simple model of vitrinite reflectance based on chemical kinetics. AAPG Bull. 1990;74(10):1559–70.

Tang Y, Perry JK, Jenden PD, et al. Mathematical modeling of stable carbon isotope ratios in natural gases. Geochim Cosmochim Acta. 2000;64(15):2673–87. doi:10.1016/S0016-7037(00)00377-X.

Tannenbaum E, Kaplan IR. Low-Mr hydrocarbons generated during hydrous and dry pyrolysis of kerogen. Nature. 1985a;317(6039):708–9. doi:10.1038/317708a0.

Tannenbaum E, Kaplan IR. Role of minerals in the thermal alteration of organic matter - I: Generation of gases and condensates under dry condition. Geochim Cosmochim Acta. 1985b;49(12):2589–604. doi:10.1016/0016-7037(85)90128-0.

Tian H, Xiao XM, Wilkins RW, et al. An experimental comparison of gas generation from three oil fractions: implications for the chemical and stable carbon isotopic signatures of oil cracking gas. Org Geochem. 2012;46:96–112. doi:10.1016/j.orggeochem.2012.01.013.

Xiao XM, Zeng QH, Tian H, et al. Origin and accumulation model of the AK-1 natural gas pool from the Tarim Basin China. Org Geochem. 2005;36(9):1285–98. doi:10.1016/j.orggeochem.2005.04.001.

Xiong YQ, Geng AS, Liu JZ. Kinetic-simulating experiment combined with GC-IRMS analysis: application to identification and assessment of coal-derived methane from Zhongba Gas Field (Sichuan Basin, China). Chem Geol. 2004;213(4):325–38. doi:10.1016/j.chemgeo.2004.07.007.

Zhang T, Ellis GS, Wang KS, et al. Effect of hydrocarbon type on thermochemical sulfate reduction. Org Geochem. 2007;38(6):897–910. doi:10.1016/j.orggeochem.2007.02.004.

Zhang SC, Mi JK, He K. Synthesis of hydrocarbon gases from four different carbon sources and hydrogen gas using a gold-tube system by Fischer–Tropsch method. Chem Geol. 2013;349–350(4):27–35. doi:10.1016/j.chemgeo.2013.03.016.

Zhang SC, Wang XM, Hammarlund EU, et al. Orbital forcing of climate 1.4 billion years ago. Proc Natl Acad Sci U S A. 2015;112(12):1406–13. doi:10.1073/pnas.1502239112.

6

Catalytic aquathermolysis of Shengli heavy crude oil with an amphiphilic cobalt catalyst

Yan-Bin Cao[1,2] · Long-Li Zhang[1] · Dao-Hong Xia[1]

Abstract An interfacially active cobalt complex, cobalt dodecylbenzenesulfonate, was synthesized. Elemental analysis, atomic absorption spectroscopy, Fourier transform infrared spectroscopy (FT-IR), thermogravimetric analysis, and surface/interfacial tension determination were performed to investigate the properties of the catalyst. Results showed that the synthesized catalyst showed active interfacial behavior, decreasing the surface tension and interfacial tension between heavy oil and liquid phase to below 30 and 1.5 mN/m, respectively. The catalyst was not thermally degraded at a temperature of 400 °C, indicating its high thermal stability. Catalytic performance of the catalyst was evaluated by carrying out aquathermolysis. The viscosity determination showed that the viscosity of the heavy oil decreased by 38 %. The average molecular weight, group compositions, and average molecular structure of various samples were analyzed using elemental analysis, FT-IR, electrospray ionization Fourier transform ion cyclotron resonance (ESI FT-ICR MS), and ^1H nuclear magnetic resonance. Results indicated that the catalyst could attack the sulfur- and O_2-type heteroatomic compounds in asphaltene and resin, especially the compounds with aromatic structure, leading to a decrease in the molecular weight and then the reduction in the viscosity of heavy oil. Therefore, the synthesized catalyst might find an application in catalytic aquathermolysis of heavy oil, especially for the high-aromaticity heavy oil with high oxygen content.

Keywords Interfacially active CoDBS · Catalytic aquathermolysis · Oxygen-contained groups · Heavy oil

1 Introduction

Although heavy oil is an important supplement to traditional crude oil, especially in Canada, Venezuela, and China, it is difficult to recover and transport due to its highly viscous character (Shah et al. 2010; Hart 2014; Santos et al. 2014; Zhao et al. 2014). Many techniques have been proposed to reduce its viscosity during production, transportation, and processing, such as oil/water soluble viscosity reducing agents, thermal recovery, thermolysis, aquathermolysis, and visbreaking. Viscosity reducing agents are usually polymers or surfactants. The viscosity of heavy oil will be reduced notably if it is heated to a relative high temperature of 150–200 °C. The high viscosity will return when it cools down because no chemical reaction occurred during the thermal recovery. Thermolysis, or thermal cracking, can rupture heavy oil molecules, such as asphaltenes and resins, resulting in an irreversible viscosity reduction. However, the reaction temperature for heavy oil thermal cracking is usually above 300 °C (Hyne et al. 1982; Clark et al. 1983, 1984, 1987; Clark and Hyne 1984), which is very difficult to reach under reservoir conditions.

Catalysts and water have been adopted in thermal cracking of heavy oil to lower the reaction temperature after Hyne's pioneering work and the process was named catalytic aquathermolysis (Hyne et al. 1982). After more than

✉ Dao-Hong Xia
xiadh@upc.edu.cn

Long-Li Zhang
llzhang@upc.edu.cn

[1] State Key Laboratory of Heavy Oil Processing, China University of Petroleum, Qingdao 266580, Shandong, China

[2] Research Institute of Petroleum Engineering and Technology, Shengli Oilfield Company, Sinopec, Dongying 257000, Shandong, China

Edited by Yan-Hua Sun

30 years of research, many types of catalysts have been used. These catalysts can be classified into three groups: (1) inorganic solid particles, such as mineral particles, metal oxides, metal sulfides (Clark et al. 1988; Kök 2009); (2) water soluble metal salts, including nickel, vanadium, iron salts (Chen et al. 2008, 2009; Yi et al. 2009; Luo et al. 2011); and (3) oil soluble metal compounds, such as molybdenum oleate (Zhao et al. 2008; Yi et al. 2009; Jeon et al. 2011; Chao et al. 2012). More recently, catalysts with surface and interfacial activity have attracted researchers' attention. They are usually metal complex compounds with a catalytically active metal ion and amphiphilic ligands, such as aromatic sulfonic iron, molybdenum, and nickel (Chen et al. 2008, 2010; Wang et al. 2010; Chao et al. 2012). Because most of the cracking reactions take place at the water/oil interface, amphiphilic catalysts are expected to show better performance than other type of catalysts. Amphiphilic catalysts are apt to adsorb at the interface, thus decreasing the interfacial tension and increasing the degree of mixing of water and heavy oil. More importantly, active metal ions are also concentrated at the interface, which improves the catalytic performance.

In this paper, an amphiphilic cobalt complex catalyst was synthesized and applied to reduce the viscosity of Shengli heavy oil. The ligand is dodecylbenzenesulfonic acid (DBSA), which provides interfacial activity to the compound, and the active metal Co^{2+}. The catalyst was cobalt dodecylbenzenesulfonate, CoDBS. The catalytic aquathermolysis of heavy oil was carried out in a high pressure autoclave at 250 °C. The viscosities of the heavy crude oil and the cracked heavy oil samples were determined at both 80 and 50 °C after dehydration strictly to avoid the influence of wax, and/or the effect of dilution and emulsification of water. The changes of heavy oil compositions, SARA (saturate, aromatic, resin, and asphaltene) compositions, and average molecular structures were studied to investigate the mechanism of heavy oil viscosity reduction during catalytic aquathermolysis.

2 Experimental

2.1 Experimental material

In experiments, the heavy oil used was from the Shengli Oilfield, China, with a viscosity of 8960 mPa s at 80 °C and 167,400 mPa s at 50 °C.

Almost all of the chemicals used in the study were of analytical grade and purchased from the Sinopharm Chemical Reagent Co. Ltd. Dodecylbenzenesulfonic acid (DBSA) was provided by the Tokyo Chemical Industry Co. Ltd and its purity was above 90 %. All reagents were used as-received without further purification.

2.2 Synthesis of the active cobalt catalyst

About 120 mL of 0.1 mol/L cobalt nitrate ($Ni(NO_3)_2$) solution was accurately measured and introduced into a flask and incubated at 60 °C for 15 min in an oil bath. A 1 mol/L NaOH solution was slowly added drop-by-drop to keep the pH value of this solution at 11. After stirring for 20 min, the solution became opaque and there was vacuum-filtered with a Whatman membrane. The precipitate was washed three times with deionized water and then added to another flask, to which 10 g of DBSA was added. The flask was heated in an oil bath at 100 °C for 4 h, while the solution was stirred at 800 rpm. After that, the product was extracted with chloroform, and the by-products or unreacted solids were removed via centrifugation. The product was separated from chloroform by vacuum distillation and then dried under vacuum at 80 °C for 24 h. The collected product was a light blue powder, which was the desired catalyst.

2.3 Catalyst characterization

The elemental contents of C, H, S, and Co of the synthesized catalyst were determined with an elemental analyzer (Vario EL III, Elementar), combined with an atomic absorption spectrometer (ContrAA 700, Analytik Jena), and the content of O was calculated by subtracting the content of all other elements. The molecular structure was also characterized qualitatively with a Fourier transform infrared spectrometer (Nicolet 6700, Thermo) after being mixed with KBr.

The surface activity and interfacial activity between heavy oil and the catalyst solution were measured with a surface tensiometer (EasyDyne, Krüss) and a spinning drop interfacial tensiometer (TX 500C, CNG).

The catalyst stability was evaluated with a thermogravimetric analyzer (WCT-1D, Beijing Optical Instruments) coupled with a differential thermal analyzer (DTA) under N_2 atmosphere. The temperature was increased to 600 °C at 20 °C/min.

2.4 Catalytic aquathermolysis of Shengli heavy oil

A total of 140 g of heavy oil, 60 g of water, and 0.28 g of catalyst (0.2 wt%, mass ratio of the catalyst to heavy oil) was added to an autoclave. The temperature was raised to 250 °C in less than 1 h and was kept at 250 °C for 24 h. Then the autoclave was cooled down to room temperature with cooling water in less than 10 min. After that, the mixture was taken out, and dehydrated by distillation below 120 °C. Over 95 % water was removed. Then the viscosity of the heavy oil after aquathermolysis was measured with a Brookfield DV-III viscometer at both 80 and

50 °C. In order to obtain accurate data, the torque was kept at 50 % during measurements. The aquathermolysis experiments of heavy oil were repeated at least 3 times, and the data were averaged.

2.5 Analysis of heavy oil samples

Fourier transform ion cyclotron resonance mass spectrometry (FT-ICR MS) is an effective technique for analyzing the molecular structure of complex hydrocarbon systems, such as heavy oil (Qian and Robbins 2001; Hughey et al. 2002). Combined with soft ionization methods, e.g., electrospray ionization (ESI), FT-ICR MS was applied to investigate catalytic aquathermolysis of heavy oil (Wu et al. 2005; Smith et al. 2008; Colati et al. 2013; Wang et al. 2013). 10 mg oil sample was firstly dissolved in 1 mL toluene, and then diluted with toluene/methanol (1:3, v/v) to yield a 0.2-mg/mL solution. After that, 15 μL ammonium hydroxide (28 %) solution was added to the 1-mL oil sample solution. The solution was vortex-mixed before analysis. The MS analysis was carried out on a Bruker Apex-ultra FT-ICR mass spectrometer with a 9.4 T superconducting magnet. The diluted sample solution was injected at 180 μL/h. The atmospheric pressure photo ionization source was purchased from Bruker Daltonics. Nitrogen was used as the drying and nebulizing gas. The nebulizing gas temperature was 200 °C. The nebulizing gas flow rate was 1.2 L/min, and the drying gas flow rate was 5.0 L/min. For ESI, the main operating parameters were emitter voltage, 4.0 kV; capillary column introduction voltage, 4.5 kV; and capillary column end voltage, −320 V. Ions were accumulated for 0.01 s in a hexapole with a 2.4-V direct current (DC) voltage. The optimized mass for Q1 was 250 Da. The extraction period for ions from the hexapole to the ICR cell was set to 1.2 ms. The RF excitation was attenuated at 12.5 dB and used to excite ions over the range of 115–800 Da. Spectra comprising 4 M data points were collected. The signal-to-noise ratio was improved by summing 128 scan FT-ICR datasets.

Mass spectra were internally calibrated using an extended homologous alkylation series (molecular ions of aromatic hydrocarbons and thiophenes) of high relative abundance in a mixed heavy oil within the mass range of 300–1000 Da. Data analysis was performed using custom software and the procedure has been described in detail elsewhere (Shi et al. 2009). The compounds were characterized by class (numbers of N, O, and S heteroatoms), type [rings plus double bond equivalence (DBE)], and carbon number. Species and their isotopes with different DBE values and carbon number were searched within a set ±0.001 Kendrick mass defect (KMD) tolerance (Hughey et al. 2001).

The saturate, aromatic, resin, and asphaltene (SARA) fractions of heavy oil samples before and after catalytic aquathermolysis were separated and determined using alumina column chromatography. The average molecular weights of oil samples, resins, and asphaltenes were determined by vapor pressure osmometry (VPO) in a toluene solution. The C, H, S, and N contents of resins and asphaltenes were measured with an elemental analyzer (Elementar Vario EL III). FT-IR spectra of the resins and asphaltenes were collected on a Nicolet 6700 Fourier transform infrared spectrometer. The samples were daubed on a KBr disk to obtain a thin film. ^{1}H NMR spectra of resins and asphaltenes were recorded on a Bruker ARX400 spectrometer. CDCl$_3$ was used as the solvent and TMS as the internal chemical shift standard.

2.6 Analysis of gas products

Gas products of heavy oil after catalytic aquathermolysis were analyzed by an Agilent 7890A gas chromatograph with an Abel column AB-5MS (60 m × 0.25 mm × 0.25 μm) and a flame ionization detector. High purity nitrogen was used as carrier gas.

3 Results and discussion

3.1 Characterization of the synthesized catalyst

The FT-IR spectra of DBSA and the synthesized cobalt catalyst are shown in Fig. 1. The typical peaks of the benzenesulfonic acid group are the stretching vibration peak of O–H (907 cm^{-1}) and the stretching vibration peaks of sulfate S=O (1363, 1177, and 1129 cm^{-1}) in the DBSA spectrum. In the spectrum of the catalyst, the stretching vibration peak of O–H at 907 cm^{-1} vanished, and the stretching vibration peaks of sulfate S=O also shifted to higher wave number due to the inductive effect of the cobalt atom. Therefore, the synthesized compound was cobalt dodecylbenzenesulfonate (CoSDB). The element analysis results proved the molecular structure and composition of the synthesized catalyst (Table 1). The elemental composition is consistent with the stoichiometric values. The slightly higher H and O contents of the experimental results compared to those expected theoretically might be due to water absorption during the sample preparation.

The surface tension of the CoDBS aqueous solution decreased to approximately 27 mN/m when the CoDBS concentration was 0.02 wt% (Fig. 2). The interfacial tension between the CoDBS aqueous solution and heavy oil was also measured by a spinning drop method. At a

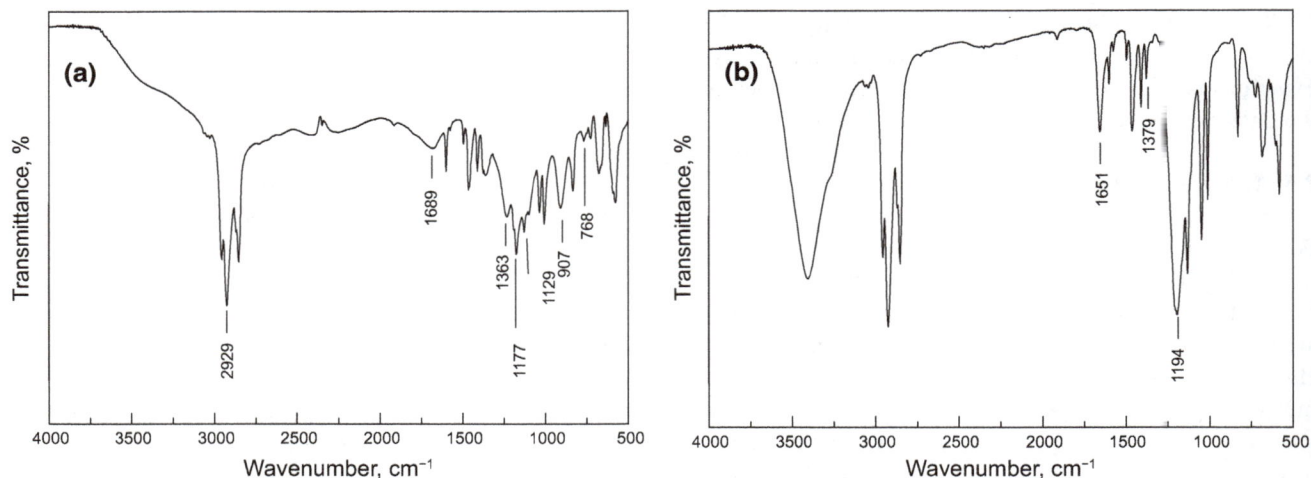

Fig. 1 FT-IR spectra of DBSA (**a**) and CoDBS (**b**)

Table 1 Elemental composition of the synthesized active cobalt catalyst

Elemental content	C, %	H, %	S, %	O, %	Co, %
Measured	57.37 ± 0.59	8.94 ± 0.17	8.59 ± 0.15	16.67 ± 0.18	8.43 ± 0.12
Theoretical	60.94 ± 0.44	8.18 ± 0.14	9.03 ± 0.13	13.54 ± 0.21	8.31 ± 0.13

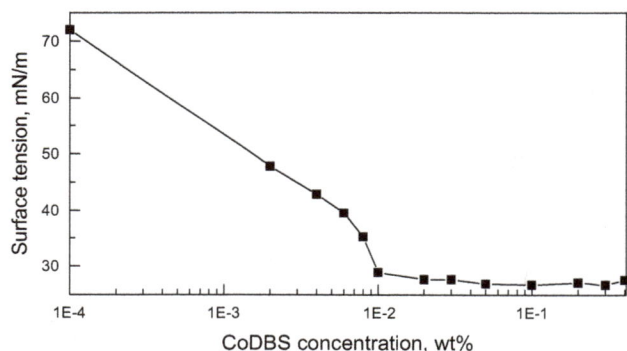

Fig. 2 Surface tension of the synthesized CoDBS catalyst

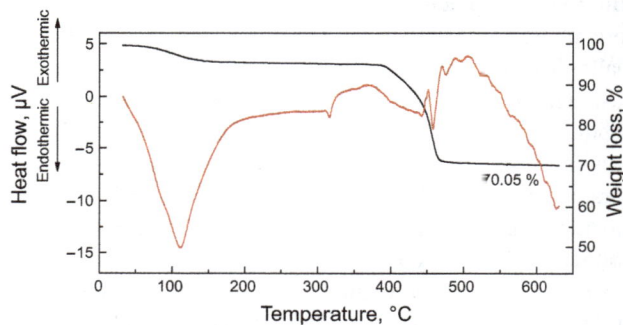

Fig. 3 TGA (*black line*) and DTA (*red line*) curves of the CoDBS catalyst

concentration of 0.2 wt%, CoDBS could reduce the interfacial tension from about 20 mM/m to 0.7 mM/m. The interfacial activity of the CoDBS catalyst will benefit the emulsification of water and heavy oil. The better mixing and increased interfacial area between water and heavy oil might improve the cracking of asphaltene and resin in heavy oil, as asphaltene is partially surface active and apt to adsorb at the interface during aquathermolysis.

TGA-DTA (thermogravimetric analysis and differential thermal analysis) data of the synthesized catalyst are shown in Fig. 3. There is a significant weight loss and an exothermic peak at about 400 °C. This indicated that the catalyst does not decompose below 400 °C. That is to say, the CoDBS catalyst is thermally stable during the aquathermolysis process, usually below 300 °C.

3.2 Viscosity reduction and properties of the heavy oil after aquathermolysis

The viscosity of heavy oil was measured before and after catalytic aquathermolysis. The SARA compositions and average molecular weights of SARA components were also compared, as shown in Table 2. The viscosities of the heavy oil samples measured at 80 and 50 °C were reduced by 37.5 % and 57.5 %, respectively, after catalytic aquathermolysis. Because of the effect of microcrystalline wax, the reduction in oil viscosity measured at 80 °C was much lower than that measured at 50 °C. Compared with wax, large molecules, such as asphaltenes and resins, are the primary reason for the high viscosity of the heavy oil (Ghanavati et al. 2013). In order to elucidate the

Table 2 Viscosity and SARA compositions of heavy oil samples

Sample	Heavy oils		Saturate		Aromatic		Resin		Asphaltene	
	Viscosity, mPa s	MW, g/mol	Content, wt%[a]	MW, g/mol[b]	Content, wt%	MW, g/mol	Content, wt%	MW, g/mol	Content, wt%	MW, g/mol
Heavy crude oil	8960[c] 167,372[d]	710	21.6	561.0	26.3	708.0	46.5	1454.0	5.5	10,031
Oil sample after aquathermolysis	5590[c] 71,133[d]	697	28.3	502.0	27.7	681.0	38.9	1503.0	5.1	9078

[a] Stands for the normalized weight percent of SARA

[b] Stands for the unit of average molecular weight of SARA

[c] Determined at 80 °C

[d] Determined at 50 °C

mechanism of viscosity reduction via catalytic aquathermolysis, the oil viscosity was measured at 80 °C unless otherwise specified. The oil viscosity decreased by 24.4 % at 80 °C in the control experiment without catalysts. The results proved that the synthesized catalyst can effectively reduce the viscosity of Shengli heavy oil.

The SARA analysis shows that asphaltene and resin contents decreased by 0.44 % and 7.65 % after aquathermolysis, respectively. In contrast, the contents of saturate and aromatic fractions of the oil sample increased by 6.69 % and 1.39 % after aquathermolysis. This indicates that the resin and asphaltene fractions were partly cracked into saturate and aromatic fractions during aquathermolysis. According to Table 2, the average molecular weights of heavy oil, saturates, aromatics, and asphaltenes decreased significantly after aquathermolysis, but the molecular weight of resins increased slightly in comparison with the original oil samples. It is known that a part of the resin cracks into saturates or aromatics, and meanwhile, the asphaltenes might crack into resins during thermolysis (Wang et al. 2009). The molecular weight of new resin transferred from asphaltene might be larger than that of the original resin, while the molecular weight of the later became smaller after cracking. Thus, the overall average molecular weight only changed slightly although the contents of resin decreased significantly. Moreover, the elemental compositions of various separated asphaltene and resin fractions were characterized and listed in Table 3. It is found that the S and N contents in the asphaltene fraction decreased accompanied by an increase in the two element contents in the resin fraction after aquathermolysis. The results indicate that part of high heteroatomic compounds in asphaltene fraction was transferred into resin.

High resolution MS was performed to study the changes of heteroatomic compounds in Shengli heavy oil before and after aquathermolysis. The results are shown in Figs. 4, 5, 6, 7, and 8 and Table 4. Because of the different ionizations of different compounds, the contents in Fig. 4 only

Table 3 Elemental composition of the asphaltene and resin fractions

Elemental content, wt%	C	H	S	N
Asphaltenes	81.1	8.5	2.5	2.0
Asphaltenes after aquathermolysis	80.2	8.4	2.5	1.9
Resins	85.2	10.0	1.5	1.5
Resins after aquathermolysis	85.1	10.1	1.7	1.5

show the relative contents. For example, the sulfur compounds have a low ionization ratio and they also show low contents in Fig. 4. In the negative-ion mode of ESI FT-ICR MS, heteroatomic compounds including O_2, N_1, N_1O_2, O_2S_1, O_3, O_1, N_1S_1, N_1O_1, N_2, $N_2O_2S_2$, S_1, and $N_1O_2S_2$ were detected in the heavy oil sample. After aquathermolysis, the contents of $N_2O_2S_2$, S_1, and $N_1O_2S_2$ decreased sharply, and were hardly detected. This indicates that the sulfur compounds are apt to crack during aquathermolysis because of the relatively small bond energy of C–S bonds, which also corresponds with the element analysis results. N_1-, O_1-, and O_3-type contents increased after aquathermolysis while O_2 type decreased.

The type and content of N_1 compounds are shown in Fig. 5. The N_1 compounds in the negative-ion mode are nonbasic nitrogen compounds including pyrrole, benzopyrrole, carbazole, benzocarbazole, and dibenzocarbazole. Their DBE ranges from 9 to 16, and the carbon number is from 20 to 40 (C_{20} to C_{40}). After aquathermolysis, the DBE of the compounds ranged from 12 to 15, and the carbon number was from 20 to 40. Benzocarbazole-type compounds (DBE 12) and dibenzocarbazole (DBE 15) had the highest content. The N_1 compounds with DBE of 9–11 decreased. Almost no DBE 6 (benzopyrrole type) and DBE 3 (pyrrole type) compounds were detected.

According to the results of O_1 and O_2 compounds, the oxygen compounds in heavy crude oil are mainly phenolics and carboxylic acids. Because the carboxylic acid compounds are thermally instable, they tend to crack into

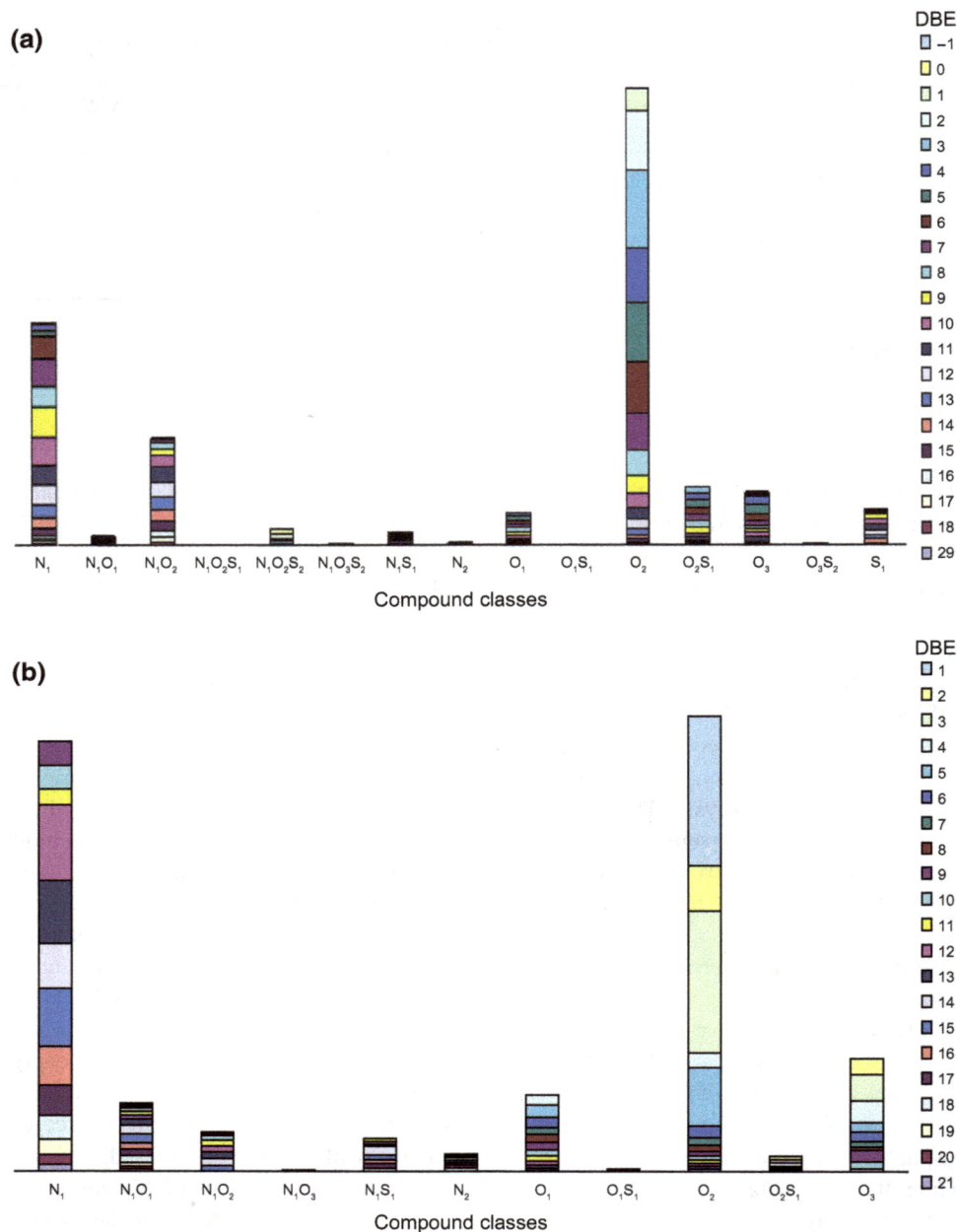

Fig. 4 Relative contents of heteroatomic compounds in Shengli heavy oil by ESI FT-ICR MS (negative-ion mode). **a** Heavy crude oil. **b** Heavy oil after aquathermolysis

hydrocarbons and carbon dioxide during aquathermolysis. The phenolic compounds mainly include alkyl phenolic and alkyl benzophenolic (from benzophenolics to quadribenzophenolics) compounds. The O_1 class of the negative-ion mode of ESI FT-ICR MS results is shown in Fig. 6. For heavy oil samples before and after aquathermolysis, the distributions of DBE and carbon number were continuous. DBE was from 4 to 9, while the carbon number was from 18 to 35 (C_{18} to C_{35}). Since only the hydroxyl compounds can be ionized under the negative ESI mode,

and the minimal DBE was 4, it would be deduced that the O_1 class compounds were mainly alkyl phenolics and their benzo-derivatives. The O_1 class compounds are relatively stable, and their content did not change obviously after aquathermolysis.

As shown in Fig. 7, the DBE of O_2 class compounds in the heavy oil sample was from 1 to 7, and their carbon numbers were from 16 to 35. After aquathermolysis, the DBE was from 1 to 3, and the carbon numbers were from 16 to 20. Furthermore, the contents also decreased

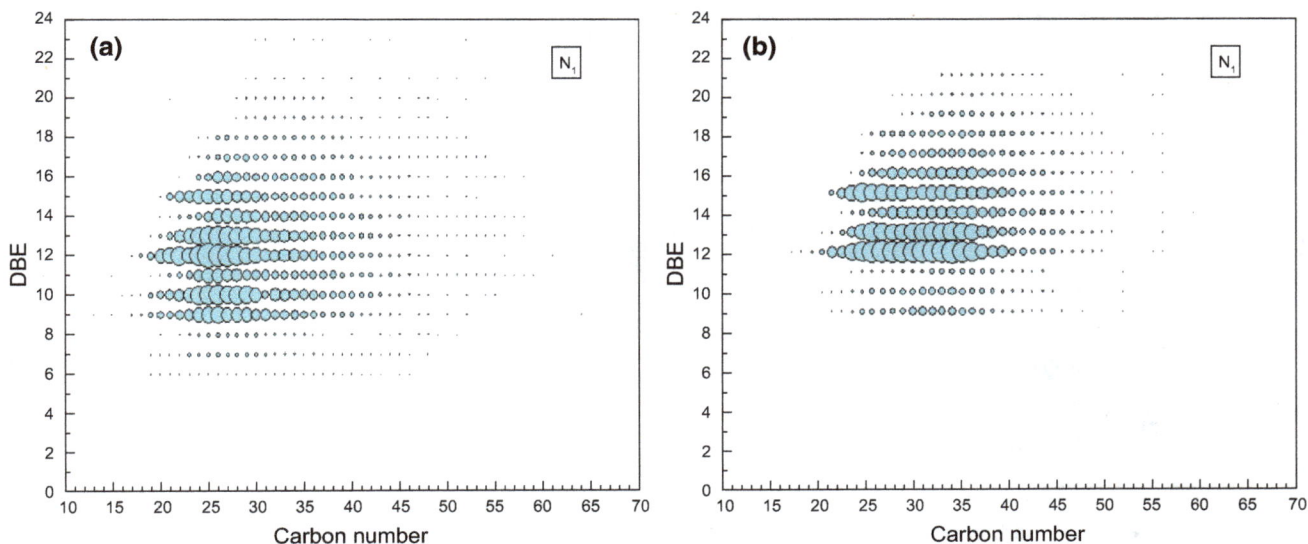

Fig. 5 Types and distributions of N_1 class compounds in Shengli heavy oil by ESI FT-ICR MS (negative-ion mode). **a** Heavy crude oil. **b** Heavy oil after aquathermolysis

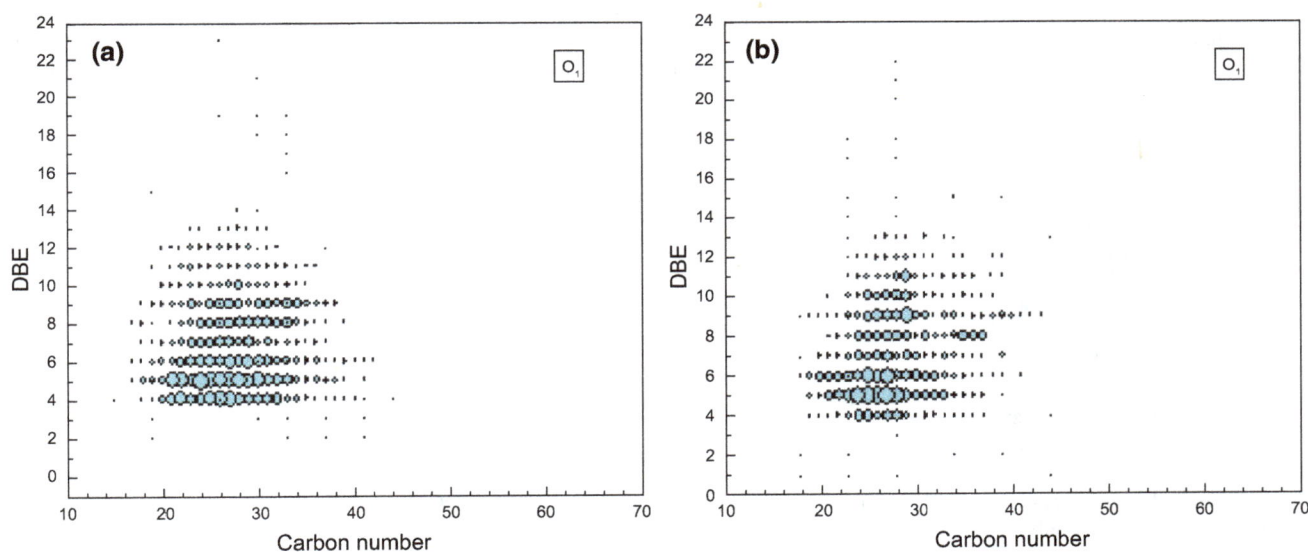

Fig. 6 Types and distributions of O_1 class compounds in Shengli heavy oil by ESI FT-ICR MS (negative-ion mode). **a** Heavy crude oil. **b** Heavy oil after aquathermolysis

obviously. The O_2 class compounds of DBE 1 should be carboxylic acid compounds, and the DBE 2 should be naphthenic acids. The DBE 3-7 should be polyring naphthenic acids. Both carboxylic acids and naphthenic acids are active and easily cracked. As a result, the contents of O_2 class compounds decreased significantly after catalytic reaction. The results of O_2S_1 class compounds are shown in Fig. 8. Like O_2 class compounds, their contents decreased noticeably. Their DBE was from 4 to 8, and the molecular structures should be thiophene rings fused with naphthenic

acids up to benzothiophene fused with dual ring naphthenic acids.

Other types of heteroatomic compounds are shown in Table 4, where N_1, O_1, O_2, and O_2S_1 classes are identified in the negative-ion ESI FT-ICR MS spectrum. The sulfur compounds were easily cracked besides O_2 class compounds for the low energy of C–S bond. Most of them were hardly detected in the heavy oil sample after aquathermolysis. For other heteroatomic compounds, their DBE and carbon number also decreased, which means that the large

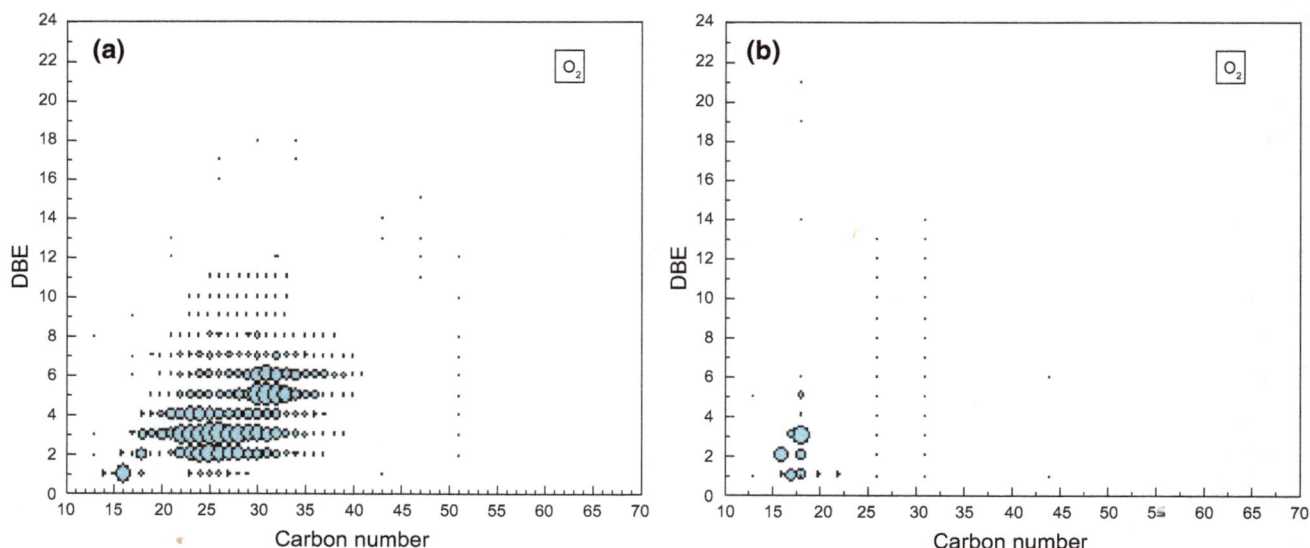

Fig. 7 Types and distributions of O_2 class compounds in Shengli heavy oil by ESI FT-ICR MS (negative-ion mode). **a** Heavy crude oil. **b** Heavy oil after aquathermolysis

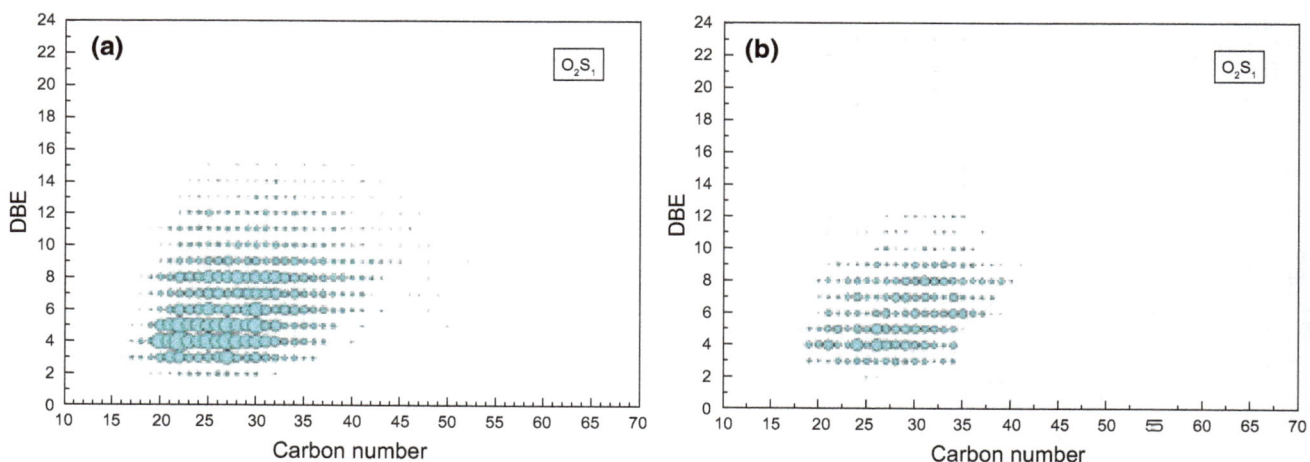

Fig. 8 Types and distributions of O_2S_1 class compounds in Shengli heavy oil by ESI FT-ICR MS (negative-ion mode). **a** Heavy crude oil. **b** Heavy oil after aquathermolysis

molecules cracked into relatively small molecules after aquathermolysis. The analytical results of gas products in Table 5 also show that H_2S, CO_2, H_2, CO, and small hydrocarbons (C_1–C_5) were also detected, which also proved that the sulfur and oxygen heteroatomic compounds had been cracked.

3.3 Changes of the molecular structures of asphaltenes and resins

The above results show that the synthesized catalyst effectively reduced the viscosity of heavy oils during catalytic aquathermolysis, and the SARA analysis also shows

decreases in the asphaltene and resin contents. Since the asphaltene and resin are the high-molecular weight components, they might be the direct reason for the high viscosity of the heavy oil. It is necessary to study the change of the molecular structures of asphaltene and resin during the aquathermolysis process. FT-IR was used to characterize the ratio of groups to that of methyl groups, n_{CH_2}/n_{CH_3} of resin and asphaltene, which means the length of the aliphatic side chains, through Eq. (1) (Wang et al. 2009). According to the FT-IR spectra in Fig. 9, the calculated results are listed in Table 6. The n_{CH_2}/n_{CH_3} of asphaltene and resin fractions after aquathermolysis were smaller than those of the original ones. This indicates that

Table 4 Various fractions of the heteroatomic compound groups in heavy oil samples before and after aquathermolysis (negative ions)

Heavy crude oil		Heavy oil after aquathermolysis		Summary of changes
Group	DBE and carbon number distribution	Group	DBE and carbon number distribution	
N_1	9–15; C_{20}–C_{35}	N_1	12–15; C_{20}–C_{35}	Some compounds decreased
N_1O_2	10–16; C_{20}–C_{40}	N_1O_2	10–15; C_{20}–C_{40}	Some compounds decreased
N_1O_1	13–18; C_{20}–C_{35}	N_1O_1	10–18; C_{20}–C_{35}	Parts of compounds decreased obviously
N_1S_1	11–17; C_{20}–C_{35}	N_1S_1	14–18; C_{20}–C_{35}	Some compounds decreased
N_2	12–18; C_{20}–C_{35}	N_2	12–18; C_{20}–C_{35}	Some compounds decreased
O_2S_1	3–9; C_{20}–C_{35}	O_2S_1	3–9; C_{20}–C_{35}	Parts of compounds decreased obviously
O_1S_1	6, 10; C_{24}–C_{26}	O_1S_1	6–8; C_{23}–C_{27}	Parts of compounds decreased obviously
O_1	4–12; C_{18}–C_{35}	O_1	4–11; C_{18}–C_{35}	Some compounds decreased
O_2	1–8; C_{15}–C_{35}	O_2	1–3; C_{15}–C_{20}	Parts of compounds decreased obviously
O_3	4–11; C_{20}–C_{35}	O_3	2–11; C_{18}–C_{35}	Parts of compounds decreased obviously
$N_1O_3S_2$	2–4; C_{15}–C_{30}			Cracked totally
O_3S_2	2–6; C_{15}–C_{30}			Cracked totally
S_1	6–15; C_{22}–C_{50}			Cracked totally
$N_1O_2S_1$	11–13; C_{25}–C_{30}			Cracked totally
$N_1O_2S_2$	4–8; C_{20}–C_{35}			Cracked totally

Table 5 Gas product compositions of heavy oil after catalytic reaction

Gas products	H_2S	H_2	CO	CO_2	CH_4	C_2	C_3	C_4	C_5
Content, vol%	0.4	2.9	1.2	41.7	29.1	10.8	9.4	7.5	3.2

parts of the side chains were cracked during aquathermolysis. The side chains of resin were more apt to be cracked than those of asphaltene.

$$n_{CH_2}/n_{CH_3} = 2.93 A_{1460}/A_{1380} - 3.70 \qquad (1)$$

where n_{CH_2} and n_{CH_3} are the number of methylene and methyl groups; A_{1460} and A_{1380} are the IR absorption intensity at 1460 and 1380 cm^{-1}, respectively.

An improved Brown–Ladner method was adopted to calculate the molecular structures of resins and asphaltenes with the data of elemental analysis, average molecular weight, and ^{1}H-NMR (Suzuki et al. 1982; Ali et al. 1990; Strausz et al. 1992). The calculated results are listed in Table 7. The ^{1}H-NMR spectra of resins and asphaltenes are shown in Figs. 10 and 11. The aromaticity (f_A) and

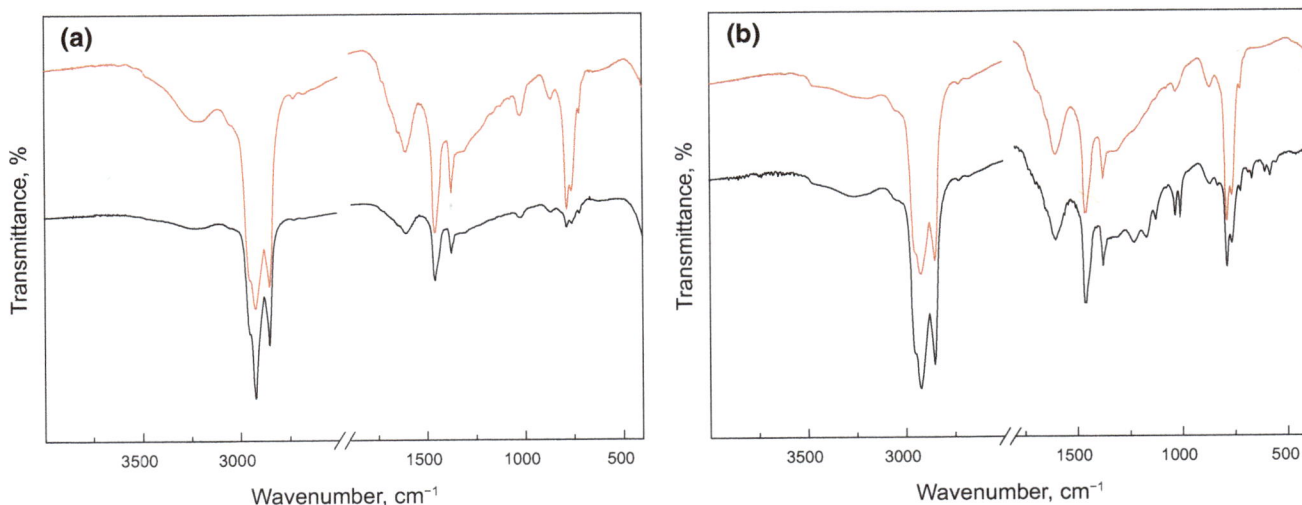

Fig. 9 IR spectra of resins (**a**) and asphaltenes (**b**) extracted from heavy crude oil (*red line*) and heavy oil after aquathermolysis (*black line*)

Table 6 n_{CH_2}/n_{CH_3} of asphaltenes and resins

Sample	n_{CH_2}/n_{CH_3}
Heavy crude oil	
Asphaltene	0.212
Resin	0.644
Heavy oil after aquathermolysis	
Asphaltene	0.177
Resin	0.171

aromatic condensation index (H_{AU}/C_A) of asphaltene and resin were calculated from Eqs. (2) and (3) (Liang 2000). The parameter H_{AU}/C_A establishes the relation between the average number of aromatic carbons and the number of carbon atoms probably incorporated into polycondensed structures of the asphaltenes and resins.

$$f_A = \frac{C/H - \left(H_\alpha^* + H_\beta^* + H_\gamma^*\right)/2H_T^*}{C/H} \tag{2}$$

$$\frac{H_{AU}}{C_{AU}} = \frac{H_A^*/H_T^* + H_\alpha^*/2H_T^*}{C/H - \left(H_\alpha^* + H_\beta^* + H_\gamma^*\right)/2H_T^*} \tag{3}$$

In Eqs. (2), (3), and Table 7, C/H is the atomic ratio of carbon to hydrogen, which is calculated from the elemental

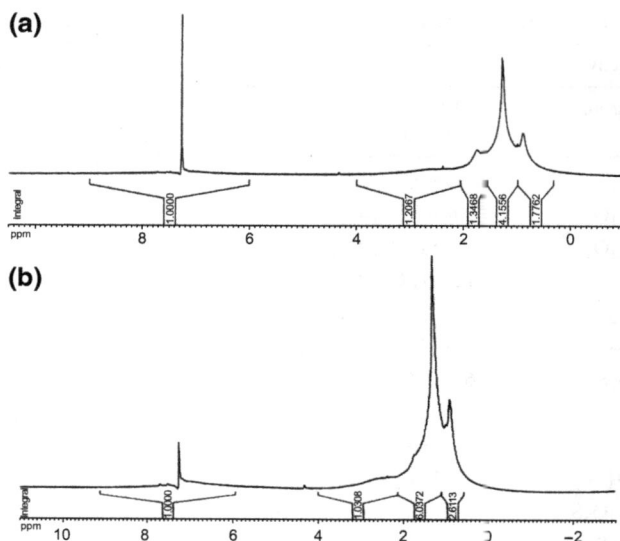

Fig. 10 ^{1}H NMR spectra of asphaltene fraction extracted from heavy crude oil (**a**) and heavy oil after aquathermolysis (**b**)

analysis data. H_A^* refers to the relative content of aromatic hydrogens; H_α^*, H_β^*, and H_γ^* refer to relative content of hydrogens which are in α, β, and γ positions relative to aromatic rings: $H_T^* = H_A^* + H_\alpha^* + H_\beta^* + H_\gamma^*$. C, C_A, C_N, and

Table 7 Average structure parameters of resin and asphaltene fractions extracted from heavy oil before and after aquathermolysis

Sample	Asphaltene		Resin	
	Heavy crude oil	Heavy oil after aquathermolysis	Heavy crude oil	Heavy oil after aquathermolysis
MW, g/mol	10,031	9078	1454	403
C content, wt%	81.1	80.2	85.2	85.1
H content, wt%	8.5	8.4	10.0	10.1
C/H	0.796	0.791	0.707	0.699
Relative content, %				
H_A^*	10.54	6.49	7.25	6.88
H_α^*	12.72	11.43	15.96	15.48
H_β^*	58.01	58.24	60.40	60.70
H_γ^*	18.73	23.84	16.39	16.95
H_{AU}/C_A	0.48	0.37	0.61	0.61
R	106.6	101.8	13.8	13.9
R_A	74.6	62.3	8.5	8.5
R_N	32.0	39.6	5.3	5.3
C	678.2	606.4	103.2	106.6
C_A	300.2	251.0	36.1	36.2
C_N	128.1	158.3	21.1	21.4
C_P	249.9	197.1	46.1	49.1
f_A	0.44	0.41	0.35	0.34
f_N	0.19	0.26	0.20	0.20
f_P	0.37	0.33	0.45	0.46

(a)

(b)

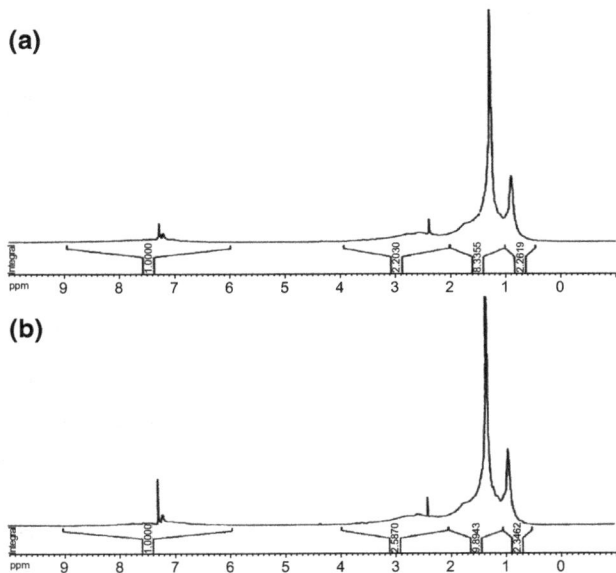

Fig. 11 ^{1}H NMR spectra of resin fraction extracted from heavy crude oil (**a**) and heavy oil after aquathermolysis (**b**)

C_P are the total carbon number, aromatic carbon number, naphthenic carbon number, and paraffinic carbon number, respectively. R is the total number of rings; R_A is the number of aromatic rings; and R_N is the number of naphthenic rings. f_A is the ratio of the number of aromatic carbons to the total number of carbons; f_N is the ratio of the number of naphthenic carbons to the total number of carbons; and f_P is the ratio of the number of paraffinic carbons to the total number of carbons. The higher the H_{AU}/C_A ratio, the lower the aromatic condensation degree.

For asphaltene, both f_A and H_{AU}/C_A decreased after catalytic aquathermolysis. The reduction in aromaticity could be attributed to the rupture of polyring structures after the removal of heteroatoms or the hydrogenation of unsaturated groups, while the increase in the aromatic condensation degree might originate from the aggregation of aromatic units after the loss of side chains. For resin, the H/C, f_A and H_{AU}/C_A changed slightly after catalytic aquathermolysis, almost the same as those of the crude resin. This is consistent with the previous results that the content of the resin changed much more but its molecular weight remained the same. It seems that during the aquathermolysis, the side chains of asphaltene might be more easily cracked than the side chains of resin, and parts of asphaltene might be transferred to resin after cracking.

The detailed structures of asphaltene and resin are also shown in Table 7. The average molecular structure of asphaltene changed significantly after aquathermolysis, while the resin remained ostensibly unchanged. Almost all the parameters of asphaltene, such as R_T, R_A, C_T, C_A, and C_P, decreased after aquathermolysis. The smaller R_T and

C_T of asphaltene proved that the cracking and ring-opening reactions occurred during aquathermolysis. The increased R_N and C_N also indicated that the aromatic rings were opened by hydrogenation. The decreased H_{AU}/C_A and increased f_N also indicated the lower aromatic condensation degree of the asphaltene after aquathermolysis. The changes of C_P and f_P further illustrated that the alkyl side chains were cracked off relatively easily over the catalyst. All these changes in the molecular structures of asphaltene benefited viscosity reduction and partially improved the quality of heavy oil. The changes of resin were more complicated than asphaltene. After aquathermolysis, part of the original resin was cracked into relatively small molecules, leading to a decrease in the average molecular weight, aromatic, and naphthenic ring numbers of the resin. Meanwhile, part of the cracked asphaltenes were of lower molecular weight and hence classified into resin fraction. After catalystic aquathermolysis, the resins might be of higher molecular weight, aromatic, and naphthenic ring numbers than the original resin. Therefore, the newly formed resin fraction had almost the same average molecular structure parameters compared with that of the original resin fraction.

In general, the catalytic aquathermolysis had more influence on the structure of the asphaltene than that of the resin. This is because the heteroatomic compounds, especially the sulfur and oxygen heteroatomic compounds, are usually found in the asphaltene fraction, where the bond energy of the C–S and C–O is lower than that of the C–C in every fractions. Therefore, the weak bonds are easier to crack by the catalyst during catalytic aquathermolysis and hence the high macromolecule content in asphaltene and resin fractions reduced. Moreover, asphaltenes are interfacially active compounds, and tend to adsorb at the water/oil interface. The addition of catalyst could reduce the interfacial tension force and improve the mixing of water and heavy oil, and meanwhile, the active metals, cobalt ions, were also concentrated at the interface. More importantly, H_2O would act as an H-donor in the catalytic reaction, where its role was same as that in heavy oil hydrotreatment. Therefore, the asphaltene could be cracked more easily.

4 Conclusion

In summary, we synthesized a complex cobalt compound with interfacial activity, cobalt dodecylbenzenesulfonate (CoDBS). It was thermally stable and suitable to be used as the catalyst in the catalytic aquathermolysis of heavy oil. The viscosity of heavy oil was significantly reduced after aquathermolysis. According to the SARA analysis and ESI FT-ICR MS results, the content of asphaltene and resin reduced as a result of cracking of sulfur- and O_2-type heteroatomic compounds. The catalyst would cause more changes of the aromatic structures in asphaltene according to the NMR results. The interfacially active complex cobalt compound catalyst is suitable for heavy oil viscosity reduction, especially for heavy oils of high-aromaticity and high oxygen content.

Acknowledgments The authors acknowledge the financial support from the Key Programs of Science and Technology of SINPOEC (Grant No. P11093).

References

Ali LH, Al-Ghannam KA, Al-Rawi JM, et al. Chemical structure of asphaltenes in heavy crude oils investigated by NMR. Fuel. 1990;69(4):519–21. doi:10.1016/0016-2361(90)90326-L.

Chao K, Chen YL, Li J, et al. Upgrading and visbreaking of super-heavy oil by catalytic aquathermolysis with aromatic sulfonic copper. Fuel Process Technol. 2012;104:174–80. doi:10.1016/j.fuproc.2012.05.010.

Chen YL, He J, Wang YQ, et al. GC-MS used in study on the mechanism of the viscosity reduction of heavy oil through aquathermolysis catalyzed by aromatic sulfonic $H_3PMo_{12}O_{40}$. Energy. 2010;35(8):3454–60. doi:10.1016/j.energy.2010.04.041.

Chen YL, Wang YQ, Lu JY, et al. The viscosity reduction of nano-keggin-$K_3PMo_{12}O_{40}$ in catalytic aquathermolysis of heavy oil. Fuel. 2009;88(8):1426–34. doi:10.1016/j.fuel.2009.03.011.

Chen YL, Wang YQ, Wu C, et al. Laboratory experiments and field tests of an amphiphilic metallic chelate for catalytic aquathermolysis of heavy oil. Energy Fuels. 2008;22(3):1502–8. doi:10.1021/ef800136.

Clark PD, Dowling NI, Hyne JB, et al. The chemistry of organosulphur compound types occurring in heavy oils: 4. the high-temperature reaction of thiophene and tetrahydrothiophene with aqueous solutions of aluminium and first-row transition-metal cations. Fuel. 1987;66(10):1353–7. doi:10.1016/0016-2361(87)9018-5.

Clark PD, Hyne JB. Chemistry of organosulphur compound types occurring in heavy oil sands: 3. reaction of thiophene and tetrahydrothiophene with vanadyl and nickel salts. Fuel. 1984;63(12):1649–54. doi:10.1016/0016-2361(84)90094-2.

Clark PD, Hyne JB, Tyrer JD. Chemistry of organosulphur compound types occurring in heavy oil sands: 1. high temperature hydrolysis and thermolysis of tetrahydrothiophene in relation to steam stimulation processes. Fuel. 1983;62(8):959–62. doi:10.1016/0016-2361(83)90170-9.

Clark PD, Hyne JB, Tyer JD. Some chemistry of organosulphur compound types occurring in heavy oil sands: 2. influence of pH on the high temperature hydrolysis of tetrahydrothiophene and thiophene. Fuel. 1984;63(1):125–8. doi:10.1016/0016-2361(84)90266-7.

Clark PD, Lesage KL, Tsang TG, et al. Reactions of benzo[b]thiophene with aqueous metal species: their influence on the production and processing of heavy oils. Energy Fuels. 1988;2(4):578–81. doi:10.1021/ef00010a027.

Colati KA, Dalmaschio GP, Castro EVR, et al. Monitoring the liquid/liquid extraction of naphthenic acids in Brazilian crude oil using electrospray ionization FT-ICR mass spectrometry (ESI FT-ICR MS). Fuel. 2013;108:647–55. doi:10.1016/j.fuel.2013.02.007.

Ghanavati M, Shojaei MJ, Ahamd RSR, et al. Effects of asphaltene content and temperature on viscosity of Iranian heavy crude oil: experimental and modeling study. Energy Fuels. 2013;27(12):7217–32. doi:10.1021/ef400776h.

Hart A. A review of technologies for transporting heavy crude oil and bitumen via pipelines. J Pet Explor Prod Technol. 2014;4(3):327–36. doi:10.1007/s13202-013-0086-6.

Hughey CA, Hendrickson CL, Rodgers RP, et al. Kendrick mass defect spectrum: a compact visual analysis for ultrahigh-resolution broadband mass spectra. Anal Chem. 2001;73(19):4676–81. doi:10.102/ac010560w.

Hughey CA, Rodgers RP, Marshall AG, et al. Identification of acidic NSO compounds in crude oils of different geochemical origins by negative ion electrospray Fourier transform ion cyclotron resonance mass spectrometry. Org Geochem. 2002;33(7):743–59. doi:10.1016/S0146-6380(02)00038-4.

Hyne JB, Clark PD, Clarke RA, et al. Aquathermolysis of heavy oils. Rev Tec Intevep. 1982;2(2):87–94.

Jeon SG, Na JG, Ko CH, et al. Preparation and application of an oil-soluble CoMo bimetallic catalyst for the hydrocracking of oil sands bitumen. Energy Fuels. 2011;25(10):4256–60. doi:10.1021/ef200703t.

Kök M. Influence of reservoir rock composition on the combustion kinetics of crude oil. J Therm Anal Calorim. 2009;97(2):397–401. doi:10.1007/s10973-008-9636-4.

Liang W. Heavy oil chemistry. Beijing: China University of Petroleum Publishing House; 2000 **(in Chinese)**.

Luo H, Deng WN, Gao JJ, et al. Dispersion of water-soluble catalyst and its influence on the slurry-phase hydrocracking of residue. Energy Fuels. 2011;25(3):1161–7. doi:10.1021/ef014378.

Qian KN, Robbins WK. Resolution and identification of elemental compositions for more than 3000 crude acids in heavy petroleum by negative-ion microelectrospray high-field Fourier transform ion cyclotron resonance mass spectrometry. Energy Fuels. 2001;15(6):1505–11. doi:10.1021/ef010111z.

Santos RG, Loh W, Bannwart AC, et al. An overview of heavy oil properties and its recovery and transportation methods. Braz J Chem Eng. 2014;31(3):571–90. doi:10.1059/0104-6632.

Shah AR, Fishwick R, Wood J, et al. A review of novel techniques for heavy oil and bitumen extraction and upgrading. Energy Environ Sci. 2010;3(6):700–14. doi:10.1039/B918960B.

Shi Q, Xu CM, Zhao SQ, et al. Characterization of basic nitrogen species in coker gas oils by positive-ion electrospray ionization Fourier transform ion cyclotron resonance mass spectrometry. Energy Fuels. 2009;24(1):563–9. doi:10.1021/ef9008983.

Smith DF, Schaub TM, Kim S, et al. Characterization of acidic species in Athabasca bitumen and bitumen heavy vacuum gas oil by negative-ion ESI FT-ICR MS with and without acid–ion exchange resin prefractionation. Energy Fuels. 2008;22(4):2372–8. doi:10.1021/ef8000345.

Strausz OP, Mojelsky TW, Lown EM, et al. The molecular structure of asphaltene: an unfolding story. Fuel. 1992;71(12):1355–63. doi:10.1016/0016-2361(92)90206-4.

Suzuki T, Itoh M, Takegami Y, et al. Chemical structure of tar-sand bitumens by ^{13}C and ^1H n.m.r. spectroscopic methods. Fuel. 1982;61(5):402–10. doi:10.1016/0016-2361(82)90062-x.

Wang JQ, Li C, Zhang LL, et al. Phase separation and colloidal stability change of Karamay residue oil during thermal reaction. Energy Fuels. 2009;23(6):3002–7. doi:10.1021/ef801149q.

Wang LT, He C, Zhang YH, et al. Characterization of acidic compounds in heavy petroleum resid by fractionation and negative-ion electrospray ionization Fourier transform ion cyclotron resonance mass spectrometry analysis. Energy Fuels. 2013;27(8):4555–63. doi:10.1021/ef400459m.

Wang YQ, Chen YL, He J, et al. Mechanism of catalytic aquathermolysis: influences on heavy oil by two types of efficient catalytic ions: Fe^{3+} and Mo^{6+}. Energy Fuels. 2010;24(3):1502–10. doi:10.1021/ef901139k.

Wu W, Song FR, Yan CY, et al. Structural analyses of protoberberine alkaloids in medicine herbs by using ESI-FT-ICR-MS and HPLC-ESI-MS[n]. J Pharm Biomed Anal. 2005;37(3):437–46. doi:10.1016/j.pba.2004.11.026.

Yi YF, Li SY, Ding FC, et al. Change of asphaltene and resin properties after catalytic aquathermolysis. Pet Sci. 2009;6(2):194–200. doi:10.1007/s12182-009-0031-y.

Zhao DW, Wang J, Gates ID, et al. Thermal recovery strategies for thin heavy oil reservoirs. Fuel. 2014;117:431–41. doi:10.1016/j.fuel.2013.09.023.

Zhao XF, Tan XH, Liu YJ, et al. Behavior of an oil-soluble catalyst for aquathermolysis of heavy oil. Ind Catal. 2008;11:31–4. doi:10.3969/j.ssn.1008-1143.

7

Local buckling failure analysis of high-strength pipelines

Yan Li[1] · Jian Shuai[1] · Zhong-Li Jin[2] · Ya-Tong Zhao[1] · Kui Xu[1]

Handling editor: Jin-Jiang Wang

Abstract Pipelines in geological disaster regions typically suffer the risk of local buckling failure because of slender structure and complex load. This paper is meant to reveal the local buckling behavior of buried pipelines with a large diameter and high strength, which are under different conditions, including pure bending and bending combined with internal pressure. Finite element analysis was built according to previous data to study local buckling behavior of pressurized and unpressurized pipes under bending conditions and their differences in local buckling failure modes. In parametric analysis, a series of parameters, including pipe geometrical dimension, pipe material properties and internal pressure, were selected to study their influences on the critical bending moment, critical compressive stress and critical compressive strain of pipes. Especially the hardening exponent of pipe material was introduced to the parameter analysis by using the Ramberg–Osgood constitutive model. Results showed that geometrical dimensions, material and internal pressure can exert similar effects on the critical bending moment and critical compressive stress, which have different, even reverse effects on the critical compressive strain. Based on these analyses, more accurate design models of critical bending moment and critical compressive stress have been proposed for high-strength pipelines under bending conditions, which provide theoretical methods for high-strength pipeline engineering.

Keywords Local buckling · High-strength pipeline · Finite element analysis · Critical bending moment · Critical compressive stress · Critical compressive strain

1 Introduction

Local buckling is an ultimate state of pipelines under complex loading conditions caused by subsidence, earthquake and landslides, etc., in geological hazard zones (Han et al. 2012; Shantanu et al. 2011). Local gross deformation in wrinkled sections can do harm to loading-carrying capacity and even damage structure integrity of pipelines (Dama et al. 2007). Technological advances have resulted in high-strength pipe steel being widely used for long-distance transmission of natural gas. However, an increase in the diameter/thickness ratio of high-strength pipe steel makes pipelines more and more susceptible to local buckling failure. Several researchers have begun to study the difference of material properties between high-strength pipe steel and traditional pipe steel (Chen et al. 2008; Igi et al. 2008; Timms et al. 2009; Suzuki et al. 2010). Traditional high-strength pipelines typically have higher yield ratio and lower strain capacity compared with medium-strength pipelines (Fathi et al. 2010). However, the high strain pipelines exhibit excellent strain capacity, which is equal or better than traditional strength pipelines, and favorable to structural behavior (Nobuhisa et al. 2008). Based on the study of high-strength pipelines crossing strike-slip faults, Liu et al. (2016) suggested that the high-strength steel has a significant influence on the critical

✉ Jian Shuai
  sjclass@126.com

1 Faculty of Mechanical and Transportation Engineering, China University of Petroleum, Beijing 102249, China

2 China ENFI Engineering Corporation, Beijing 100038, China

Edited by Yan-Hua Sun

stress along the axial direction. Therefore, more comprehensive research on local buckling mechanisms in high-strength pipelines is of great importance to prevent local buckling failure from occurring.

Critical buckling theory was introduced in the 1960s, and experiments and numerical simulations have later been conducted to investigate the local buckling of pressurized pipes (Dorey and Cheng 2001; Dorey et al. 2002, 2005). Also, several prediction methods were established to evaluate the ultimate state (Paquette and Kyriakides 2006; Limam et al. 2010) and are mostly used to calculate the critical bending moment and the critical compressive strain. Mohareb (1995) investigated the local buckling behavior through examining several plain pipes and proposed the 'Mohareb–Murray interaction equation' to determine the critical bending moment based on an ideal elastoplastic material model. ABS standard (ABS 2006) put forward a model to determine the maximum allowable moment. However, both methods have some weaknesses. Specifically, the model in the ABS standard is only applicable for pipes with a diameter/thickness ratio of 10–60, which cannot be used in pipes with larger diameter/thickness ratio. Also, it is limited by practicable materials and ignoring strain hardening effect; the 'Mohareb–Murray interaction equation' and model in ABS may overestimate or underestimate the critical bending moment (Nazemi 2009). Thus, accurately determining the critical bending moment is a crucial issue for predicting local buckling behavior. Currently, the critical compressive strain has been studied for a long time and some results have been used in practical standards, e.g., the formulation from Murphey and Langner (Gresnigt and Foeken 2001) has been used in American Bureau of Shipping (ABS 2006) and the research conducted by Gresnigt (1986) has been adopted by Canadian Standards Association (CSA 2007). Also, other standards provide similar design formulae of critical compressive strain (ASCE 1984; ALA 2001), which have been widely used in engineering design. However, critical compressive stress is currently poorly understood. Plantenma studied the buckling stress of circular cylinders and round tubes under compression conditions (Ahn et al. 2016), which are not applicable under bending loading conditions. These models mentioned above have not comprehensively considered the critical properties of strain hardening of high-strength pipelines with bending load. Thus, in this paper, the primary concern is about the local buckling of high-strength pipelines.

In this paper, a practical three-dimensional model was established using the nonlinear finite element method (FEM) and calibrated with previous experimental results (Zimmerman et al. 2004). After that, deformation behavior and different failure modes of unpressurized and pressurized pipes were studied under bending conditions. The influences of pipe diameter, pipe wall thickness, yield strength, hardening exponent and internal pressure on deformation behavior of local buckling of high-strength pipelines were analyzed in detail. Finally, methods were proposed to predict the critical bending moment and critical compressive stress of local buckling of high-strength pipelines under bending load and demonstrated with an application example.

2 Numerical simulation

Laboratory testing is an effective way to obtain first-hand data, but it is costly and time-consuming. Alternatively, numerical simulations can provide a fast, convenient and cost-efficient way to study local buckling. The bearing capacity of submarine and long-distance transmission pipelines are mainly depending on pipe dimensions, material properties and loading conditions (Dang et al. 2010; Sun 2013), which can be determined by detailed parametric analysis. So the first thing is to establish a proper finite element model to do the parametric analysis.

2.1 Finite element model

Commercial nonlinear finite element software, ABAQUS, was used to develop the three-dimensional nonlinear finite element model. Although the wrinkled section of local buckling distributes within a limited area, the pipe length is important for investigating deformation behavior of local buckling. To avoid end effects, the pipe length is supposed to be 3.5 times larger than the pipeline diameter (Ozkan and Mohareb 2009b). Therefore, a pipe with a length of $L = 6D$ was used in this study. Models were constructed using solid elements, which have been proved a reliable tool for simulating local buckling of pipes (Ozkan and Mohareb 2009a). Finally, a modified RIKS algorithm in ABAQUS has been used to perform finite element simulations.

2.2 Loading and boundary conditions

A fixed constraint is applied to one end of the pipe, and a bending moment is applied to the other end by a coupling constraint. Pressure is applied on the internal surface. Under pure bending conditions, the bending moment is the only load, which increases monotonically until local buckling occurs. Under mixed loading conditions, the finite element analysis is conducted with two steps: pressure loading step and bending step. Pressure is set as a constant internal pressure, while bending is the same as that under pure bending conditions. The finite element model is shown in Fig. 1, with element type of C3D20R and the element size of 4.7 mm × 35 mm × 35 mm.

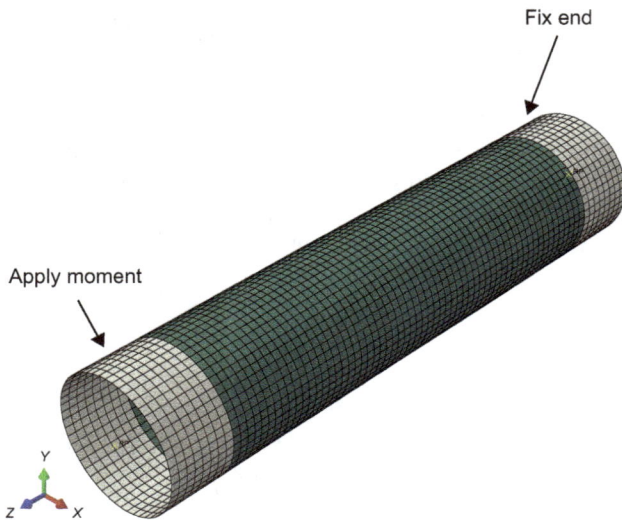

Fig. 1 Finite element model

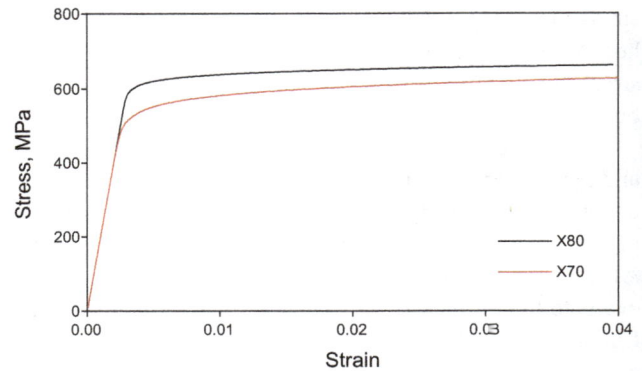

Fig. 2 Stress–strain curves of X70 and X80

2.3 Material parameters

With increasing demand for natural gas, high-strength steel pipelines have been widely used for long-distance transmission of natural gas in China. Two typical high-strength pipeline steels X70 and X80 are established in the FEM simulations. The nonlinearity of materials can be described by the Ramberg–Osgood constitutive model (Ramberg and Osgood 1943; Kamaya 2016). The constitutive relationship can be expressed as follows:

$$\varepsilon = \frac{\sigma}{E} + \alpha \left(\frac{\sigma}{\sigma_s} \right)^n \tag{1}$$

where ε is the true strain; σ is the true stress; E is the elastic modulus; σ_s is the yield strength that is defined at 0.5% strain; n is the hardening exponent; α is the "yield" offset.

2.4 Model calibration

In order to evaluate the validity of the numerical simulation, the results were calibrated with previous experimental results (Zimmerman et al. 2004). Four specimens were used to study the local buckling behavior in experiments, including two material grades X70 and X80. Figure 2 and Table 1 show the corresponding stress–strain curves and material parameters, respectively. One specimen of each material was tested under pure bending conditions, and the other was tested under bending conditions with a maximum operation pressure of 80% specified minimum yield strength (SMYS). Detailed test data are listed in Table 2.

Selecting a reasonable length is crucial for numerical simulation, because the value of the critical compressive strain varies with the effective length of the pipe. Previous studies demonstrate that buckles typically occur within the

length of $1D$ in the axial direction (Mohareb 1995; Ozkan and Mohareb 2009b). With this assumption, the relationship between moment and compressive strain was plotted in Fig. 3, which can be used to calculate the critical bending moments. The results of three methods are listed in Table 3. The critical bending moments obtained from FEM are generally correctly predicted with an error within 4.23%, which is better than the Mohareb and the ABS method.

Figure 3 suggests that the buckling trend obtained by FEM agrees well with experimental results. At first, the compressive strain increases with bending moment. When the bending moment reaches the maximum value, local buckling occurs, and then the bending moment begins to decrease, the strain still increases until the pipe loses its loading capacity completely. The maximum value of bending moment is the critical bending moment, and corresponding compressive strain is the critical compressive strain. The buckling mode exhibits two main failure shapes, which can be observed in both experimental results and FEM results (Fig. 4a–d). The unpressurized pipe buckled in a "diamond shape," while the pressurized pipe buckled in a pipe wall outward bulging shape.

3 Parametric analysis

An excessive load induced by pressure and bending leads to local buckling of a pipeline, while the failure behavior of local buckling can be described by the critical bending moment, critical compressive stress and the critical compressive strain. The critical bending moment is defined as the maximum moment at the critical point of bending vs. compressive strain curve, while the corresponding stress and strain are critical compressive stress and critical compressive strain. These three parameters primarily vary as a function of several variables, which can be divided into three types: geometrical parameters, material

Table 1 Material parameters of X70 and X80

Material grade	Elastic modulus E, MPa	Poisson's ratio v	Yield strength σ_s, MPa	Yield offset α	Hardening exponent n
X70	200,000	0.3	552	0.81	22
X80	200,000	0.3	620	0.61	45

Table 2 Test data and loading conditions

Specimen No.	Material grade	Pipe outer diameter D, mm	Pipe wall thickness t, mm	D/t	Pressure
U70	X70	762	9.4	82	0
P70	X70	762	9.4	82	80% SMYS
U80	X80	762	15.7	48	0
P80	X80	762	15.7	48	80% SMYS

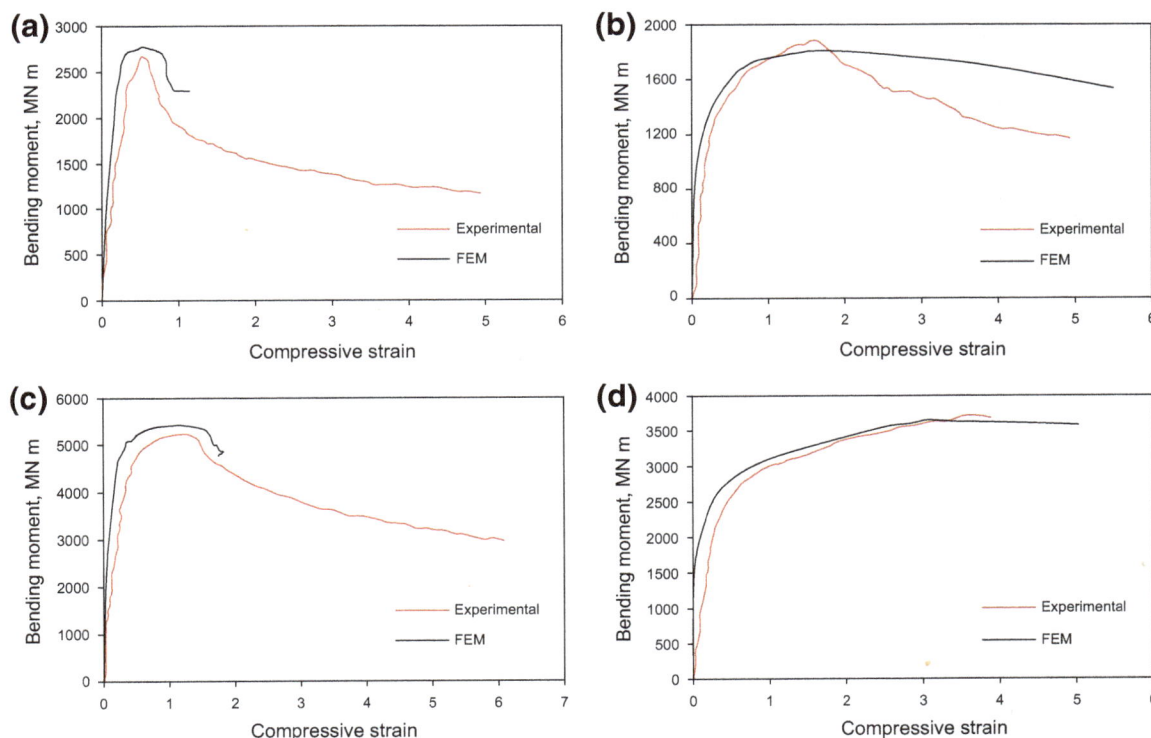

Fig. 3 Bending moment versus compressive strain curves of four specimens. **a** U70: X70 unpressurized pipe. **b** P70: X70 pressurized pipe. **c** U80: X80 unpressurized pipe. **d** P80: X80 pressurized pipe

parameters and load parameter. It is essential to analyze the influence of these parameters on local buckling behavior under bending when designing long-distance pipelines. In the analysis, only one parameter is changed at a time and others are taken as the basic values. Specifically, the basic values of geometrical parameters include the diameter and wall thickness of pipes, which were selected based on real pipeline dimensions and design code for gas pipeline engineering. The geometrical dimensions (diameter and wall thickness) of pipes used for X701 (X801), X702 (X802) and X703 (X803) are 813 mm × 11.7 mm, 1016 mm × 14.6 mm, 1219 mm × 15.3 mm, respectively. The basic values of material parameters include yield strength and hardening exponent. Specifically, the yield strengths of basic pipe materials (pipe grade X70 and X80) are 580 and 641 MPa, while the hardening exponents

Table 3 Comparison of critical bending moment obtained from test and other different methods

Specimen No.	Experimental value, MN m	FEM		Mohareb		ABS	
		Value, MN m	Error, %	Value, MN m	Error, %	Value, MN m	Error, %
U70	2667	2776	4.09	2939	10.20	2042	−23.43
P70	1845	1808	−2.01	2412	30.73	1176	−36.26
U80	5198	5417	4.22	5421	4.29	4032	−22.43
P80	3718	3646	−4.23	4408	18.56	2092	−43.73

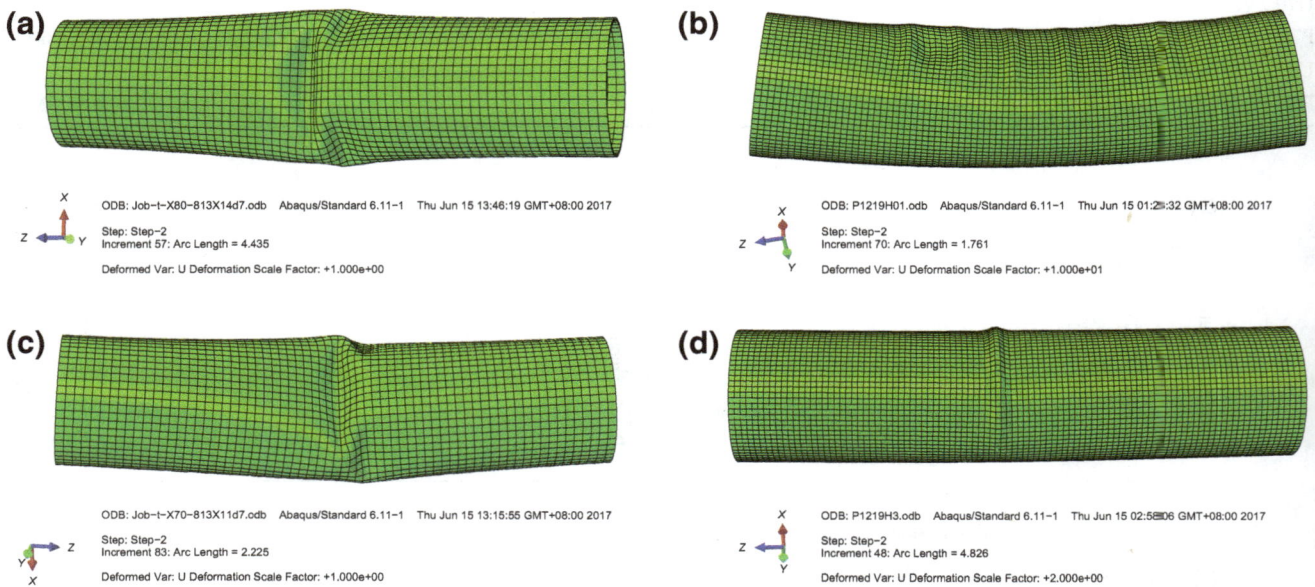

Fig. 4 Buckling shape obtained from FEM. **a** U70: X70 unpressurized pipe. **b** P70: X70 pressurized pipe. **c** U80: X80 unpressurized pipe. **d** P80: X80 pressurized pipe

are 10.0 and 17.15, respectively. The load parameter is the internal pressure that varies with the maximum operating pressure (MAOP). Detailed parameter descriptions are included in Table 4.

3.1 Critical bending moment

As a key parameter of local buckling, the critical bending moment is correlated with the pipe dimensions, pipe material properties and the internal pressure. In terms of pipe dimensions, both the pipe diameter and thickness can enhance the pipe stiffness. Figure 5 shows that the critical bending moment increases exponentially with the pipe diameter. And for a certain diameter, the higher the pipe grade, the higher the critical bending moment. Figure 6 shows that the critical bending moment increases linearly with thickness.

To analyze the effect of pipe material properties on the critical bending moment, the initial value was taken as the SMYS specified in the standard. The critical bending moment vs. yield strength (Fig. 7) exhibits that the critical bending moment increases linearly with the yield strength. With the same yield strength, the critical bending moment of X802 is lower than that of X702. It is because the hardening exponent of X80 pipes is larger than that of X70 pipes. Figure 8 shows that the critical bending moment decreases nonlinearly with hardening exponent. With the same hardening exponent, the critical bending moment of higher grade pipes is larger.

For pressurized pipes, as Fig. 9 shows, the critical bending moment decreases exponentially with the internal pressure. Compared with X70 grade pipes, the critical bending moment of X80 grade pipes decreases more rapidly.

Table 4 Ranges of parameters used in FEM

Parameter type	Basic values	Range of univariate analysis
Diameter D, mm	X701, X801: 813	813–1219
	X702, X802: 1016	
	X703, X803: 1219	
Thickness t, mm	X701, X801: 11.7	X701: 11.7–21
		X801: 10.2–21
	X702, X802: 14.6	X702: 14.6–26.2
		X802: 12.8.2–24
	X703, X803: 15.3	X703: 17.5–31.5
		X803: 15.3–31.5
Yield strength σ_s, MPa	X70: 580	X70: 490–580
	X80: 641	X80: 560–641
Hardening exponent n	X70: 10.0	5–25
	X80: 17.5	
Internal pressure P, MPa	0	0-MAOP

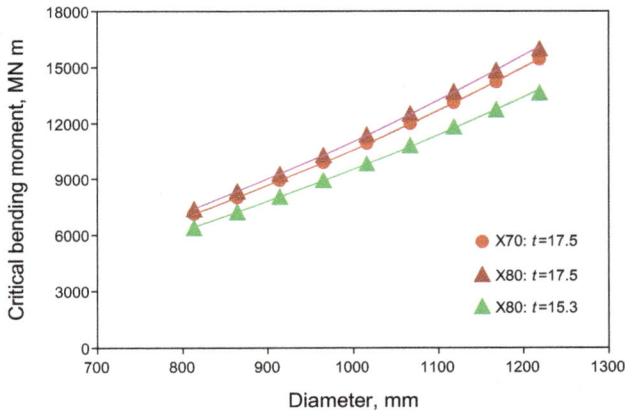

Fig. 5 Relationship between the critical bending moment and the pipe diameter

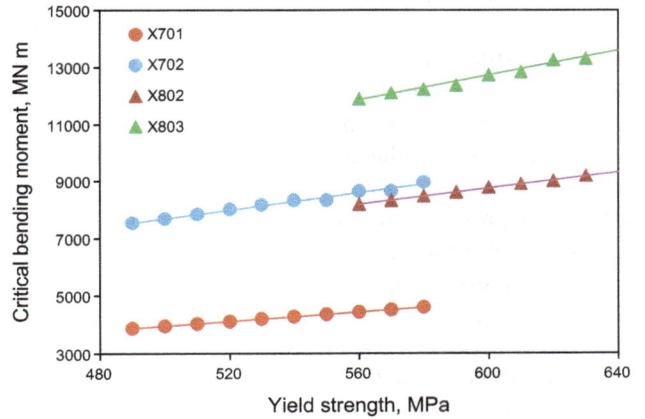

Fig. 7 Relationship between the critical bending moment and the yield strength

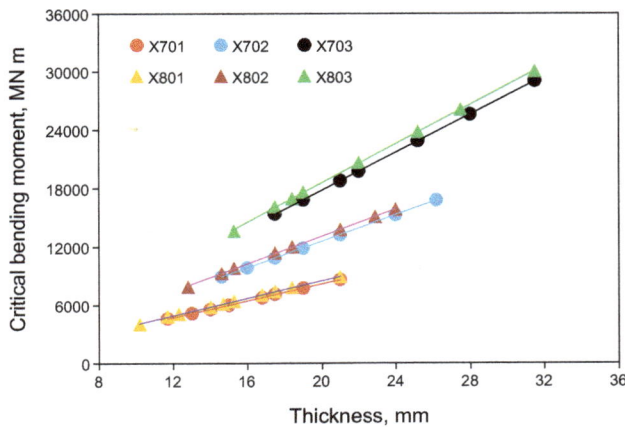

Fig. 6 Relationship between the critical bending moment and the wall thickness

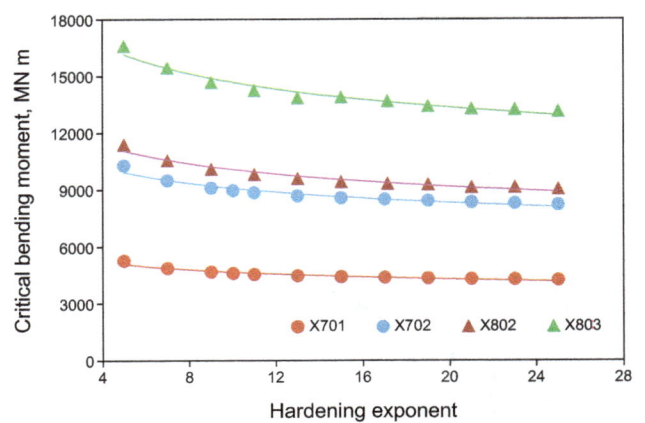

Fig. 8 Relationship between the critical bending moment and the hardening exponent

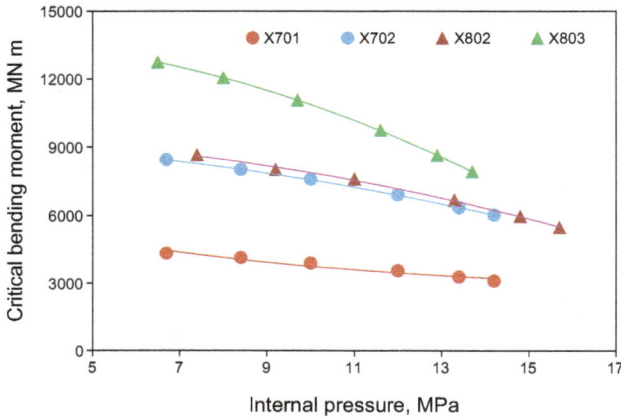

Fig. 9 Relationship between the critical bending moment and the internal pressure

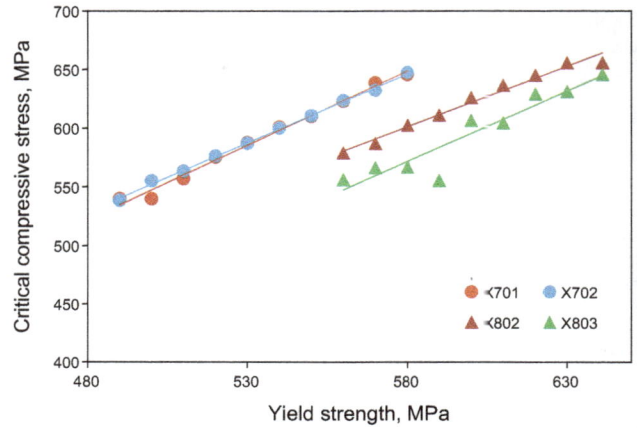

Fig. 11 Relationship between the critical compressive stress and the yield strength

3.2 Critical compressive stress

Stress and strain can be used to determine the structural behavior under loading conditions. In terms of the geometrical dimensions of pipes, the critical compressive stress increases with thickness and decreases with diameter. Thus, the effect of geometrical dimensions on the critical compressive stress can be illustrated by the thickness/diameter ratio. As Fig. 10 shows, the critical compressive stress increases near-linearly with the thickness/diameter ratio.

In terms of pipe material properties, two pipe grades, X70 and X80, were used in this study. Figure 11 shows that the critical compressive stress increases linearly with the yield strength. The hardening exponent of X80 grade pipe is higher than that of the X70 grade pipe; whereas the critical compressive stress of the former is lower than that of the latter. It should be noted that X701 and X702 almost

coincide, but there is some difference between X801 and X802. This can be explained by the fact that X701 and X702 have the same thickness/diameter ratio, which differs from that of X801 and X802. Figure 12 shows that the critical compressive stress decreases nonlinearly with the hardening exponent. With the same hardening exponent, the critical compressive stress of X80 grade pipe is larger than that of X70 grade pipes. This is because the yield strength of higher grade pipes is larger, so the stress state level of higher grade pipes is higher.

As Fig. 13 suggests, the internal pressure has a negative influence on the critical compressive stress, and at the same internal pressure, the critical compressive stress of the X80 grade pipe is lower than that of the X70 grade pipe. A comparison between X80 grade pipe and X70 grade pipe indicates that the critical compressive stress decreases rapidly as the pipe grade becomes higher.

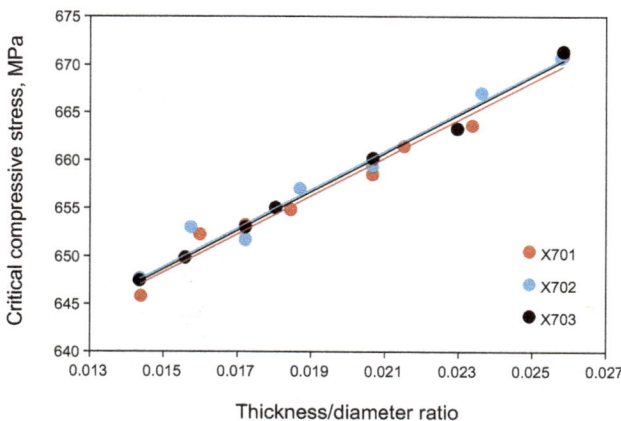

Fig. 10 Relationship between the critical compressive stress and the thickness/diameter ratio

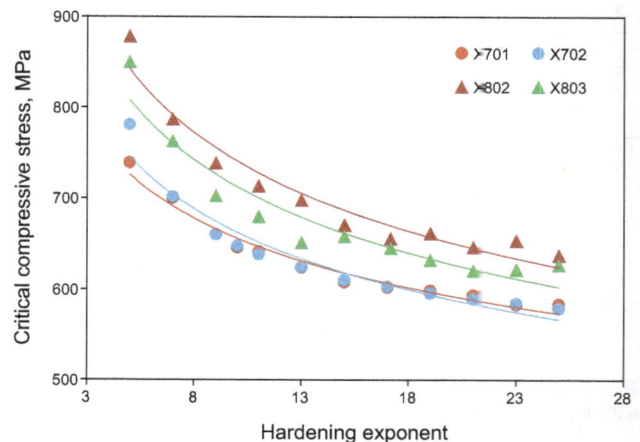

Fig. 12 Relationship between the critical compressive stress and the hardening exponent

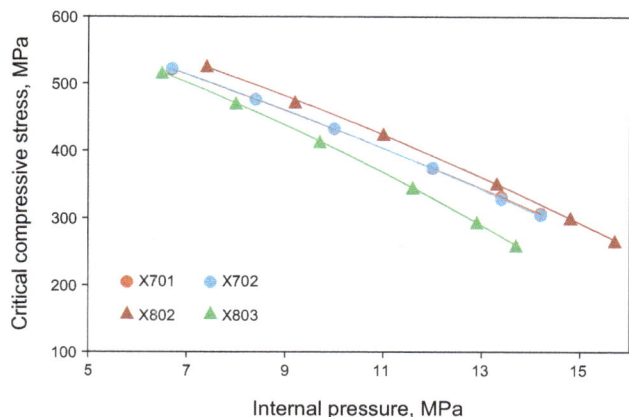

Fig. 13 Relationship between the critical compressive stress and the internal pressure

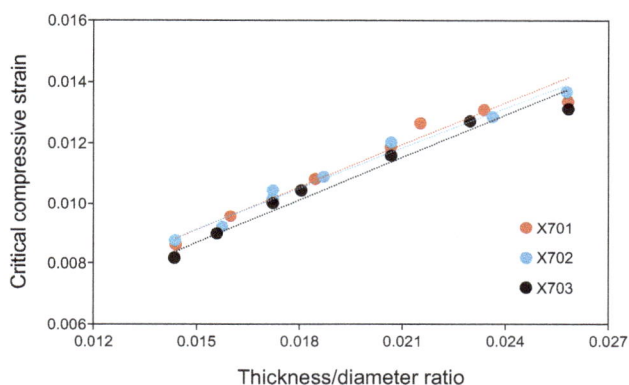

Fig. 14 Relationship between the critical compressive strain and the thickness/diameter ratio

3.3 Critical compressive strain

The critical compressive strain refers to the critical point where the structure loses its stability. It should be noted that the critical compressive strain is a strain design for local buckling, not for other failure modes. The critical compressive strain increases near-linearly with the thickness/diameter ratio, which is similar to the relationship between the critical compressive stress and the thickness/diameter ratio (Fig. 14). Figure 15 shows no significant variation of critical compressive strain with yield strength, indicating that the critical compressive strain is insensitive to the yield strength; also, the relationship between the critical compressive strain and the yield strength is significantly different from that between the critical compressive stress and the yield strength, which can be explained by the stress–strain state during the occurrence of local buckling. The local buckling typically occurs after yielding as Fig. 16 shows. The ratio of Mises

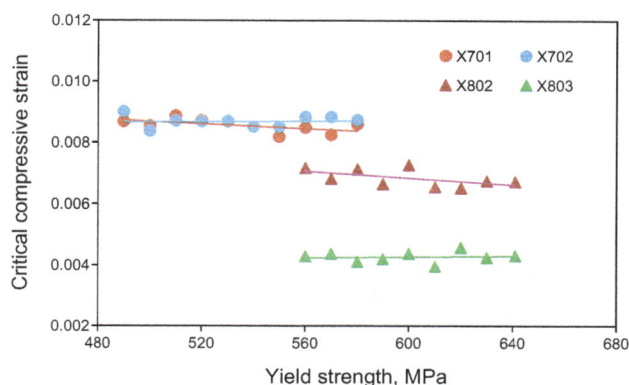

Fig. 15 Relationship between the critical compressive strain and the yield strength

stress to yield strength does not vary significantly among different materials.

The critical compressive strain decreases with an increase in the hardening exponent, which is similar to the critical compressive stress, while the critical compressive strains vary in a narrow range as Fig. 17 shows. The critical compressive strain increases nonlinearly with the internal pressure, and its value of the higher grade pipe (X802) is lower than that of the lower grade pipe (X702) (Fig. 18), which is different from the relationship between critical compressive stress and internal pressure.

4 Recommended design methods

4.1 Design methods

Over 300 finite element models with different parameters have been simulated by ABAQUS software, and the results are summarized in Sect. 3. It can be concluded that the critical bending moment of the locally buckled pipe has a highly positive correlation with the pipe diameter, pipe wall thickness, yield strength and the hardening exponent, and a negative correlation with the internal pressure. Importantly, the hardening exponent is considered as an important controlling factor of the critical bending moment. For the unpressurized pipe, the critical bending moment of local buckling can be calculated with Eq. (2), in which the hardening exponent is taken into consideration.

$$M = a_1 \cdot D^2 \cdot t \cdot \sigma_s \cdot n^{a_2} \qquad (2)$$

where M is the critical bending moment of the unpressurized pipe, MN, m; D is the outer diameter of the pipe, mm; t is the wall thickness of the pipe, mm; σ_s is the yield strength of pipe material, MPa; n is the hardening exponent of the pipe material; a_1 and a_2 are the correlation coefficients.

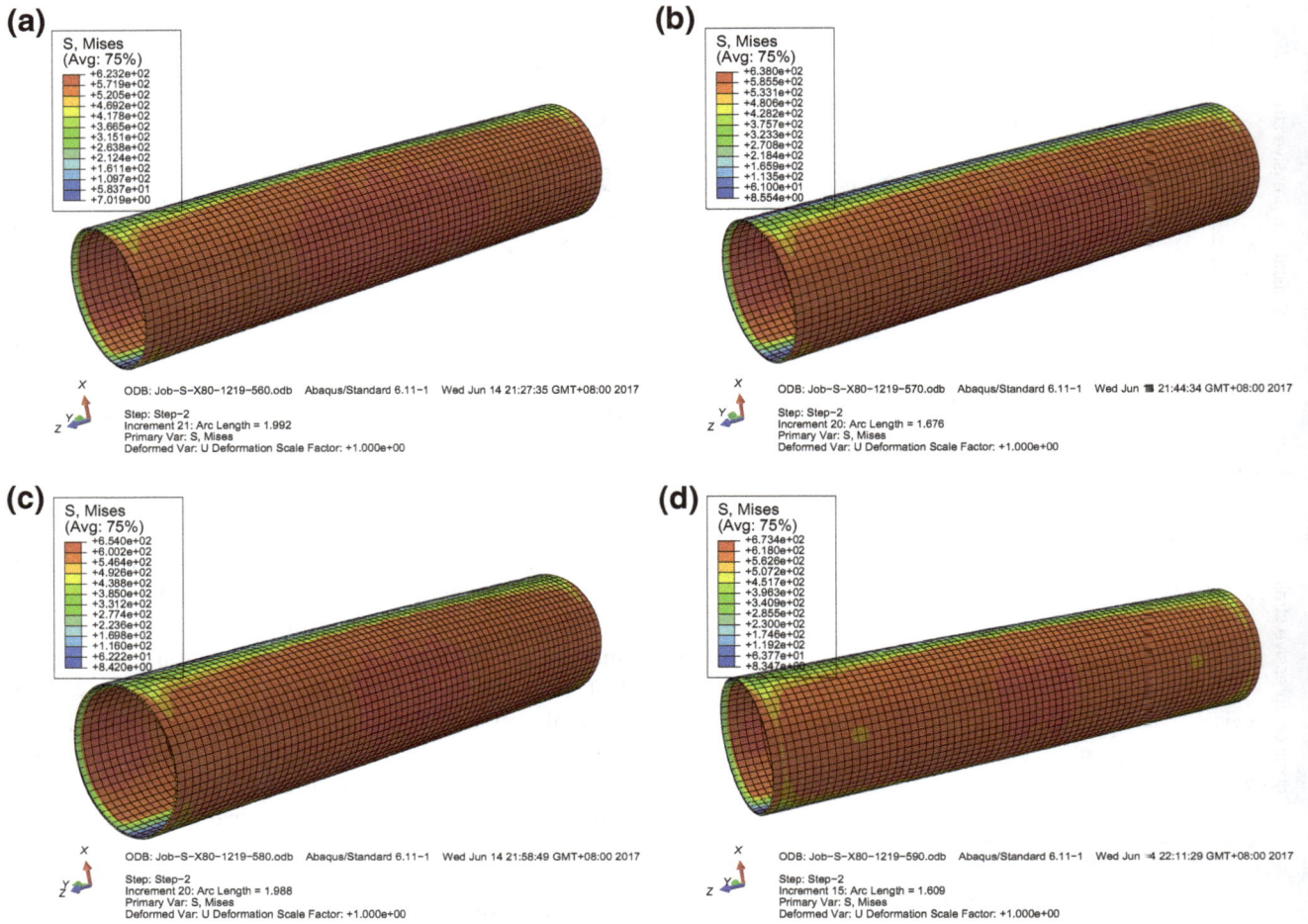

Fig. 16 Mises stress distribution at the local buckling state. **a** Yield strength $\sigma_s = 560$ MPa. **b** Yield strength $\sigma_s = 570$ MPa **c** Yield strength $\sigma_s = 580$ MPa. **d** Yield strength $\sigma_s = 590$ MPa

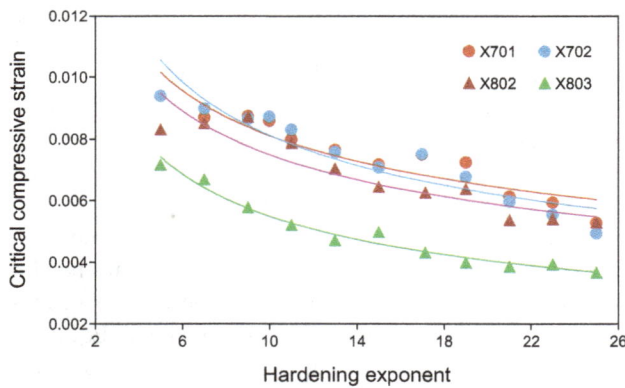

Fig. 17 Relationship between the critical compressive strain and the hardening exponent

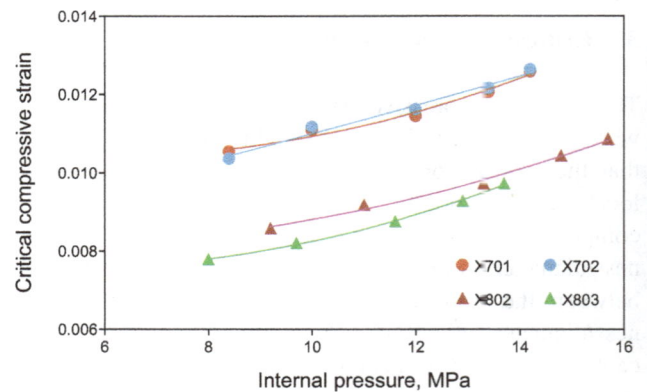

Fig. 18 Relationship between the critical compressive strain and the internal pressure

Based on Eq. (2), the critical bending moment of the pressurized pipe can be obtained by introducing the internal pressure as a quadratic polynomial, and described as follows:

$$M_p = M \cdot \left[b_1 \cdot (\sigma_\theta / \sigma_s)^2 + b_2 \right] \tag{3}$$

where M_p is the critical bending moment of the pressurized pipe, MN m; σ_θ is the hoop stress of the pipe at the internal pressure, MPa; b_1 and b_2 are the correlation coefficients. The related parameters can be obtained by fitting analysis, and the final calculation formulas are as follows:

$$M = 1.41 \cdot D^2 \cdot t \cdot \sigma_s \cdot n^{-0.13} \tag{4}$$

$$M_p = M \cdot \left[-0.576 \cdot (\sigma_\theta / \sigma_s)^2 + 0.99 \right] \tag{5}$$

Similarly, the critical compressive stress of local buckling positively correlates with the thickness/diameter ratio, yield strength and the hardening exponent, and negatively correlates with the internal pressure. Thus, the prediction model of unpressurized and pressurized pipes can be expressed as Eqs. (6) and (7). According to the fitting results, the final prediction model can be expressed as Eqs. (8) and (9).

$$\sigma_c = \sigma_s \cdot (c_1 \cdot t/D + c_2) \cdot (c_3^n + c_4) \tag{6}$$

$$\sigma_{cp} = \sigma_c \cdot [d_1 \cdot (\sigma_\theta / \sigma_s) + d_2] \tag{7}$$

$$\sigma_c = \sigma_s \cdot (4.66 \cdot t/D + 0.94) \cdot (0.8^n + 1) \tag{8}$$

$$\sigma_{cp} = \sigma_c \cdot [-0.77 \cdot (\sigma_\theta / \sigma_s) + 1.06] \tag{9}$$

where σ_c is the critical compressive stress of the unpressurized pipe, MPa; c_1, c_2 and c_3 are the correlation coefficients; σ_{cp} is the critical compressive stress of the pressurized pipe, MPa; d_1, d_2 and d_3 are the correlation coefficients.

4.2 Application of the proposed methods

X70 1016-mm and X80 1219-mm gas pipelines were used as application examples to demonstrate the proposed methods. Figures 19 and 20 show comparisons of the critical bending moment and the critical compressive stress obtained from different methods. The comparisons indicate

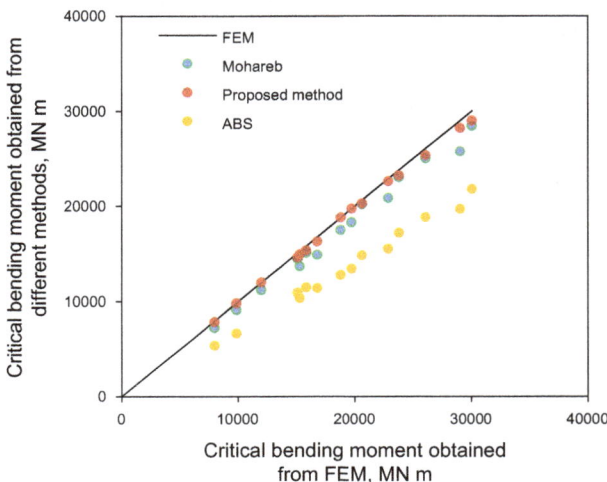

Fig. 19 Comparison of the critical bending moment obtained from different methods

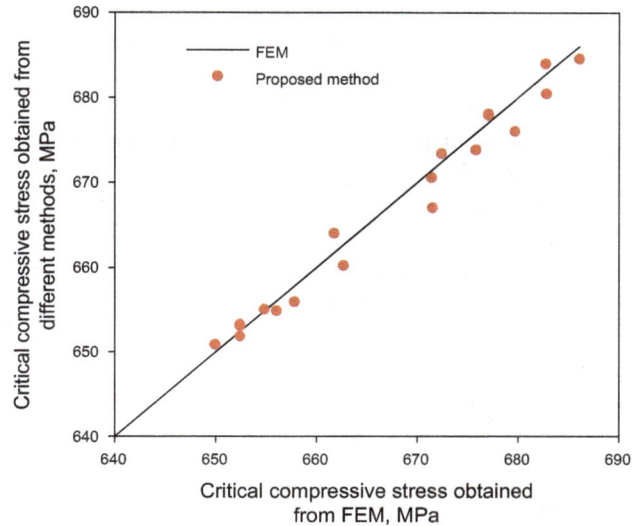

Fig. 20 Accuracy of the proposed method of critical compressive stress compared with FEM

that the proposed method can provide better estimates of both the critical bending moment and the critical compressive stress.

5 Conclusions

A finite element analysis model for local buckling was established based on the present work. The effects of key parameters including pipe diameter, pipe thickness, yield strength, hardening exponent and internal pressure, on the critical bending moment, critical compressive stress and the critical compressive strain have been investigated comprehensively through a series of models. Based on these parametric analysis results, the prediction models for critical bending moment and critical compressive stress were proposed and verified. The following conclusions can be drawn.

1. The critical bending moment of local buckling increases exponentially with the pipe diameter and increases linearly with the pipe thickness and the yield strength, while it decreases nonlinearly with an increase in the hardening exponent.
2. Local buckling occurs after the stress of the pipe reaches yield strength. Similar to the critical bending moment, the critical compressive stress increases linearly with the thickness/diameter ratio and the yield strength, and decreases exponentially with the hardening exponent and internal pressure.
3. The thickness/diameter ratio and hardening exponent have a similar influence on the critical compressive strain. However, the yield strength has no influence on

the critical compressive strain and has a linear relationship with the critical compressive stress.

4. The parameter analysis demonstrates that local buckling behavior varies with hardening exponent. The calculation methods for the critical bending moment and the critical compressive stress were proposed based on the effect of the hardening exponent on local buckling behavior. These two methods were validated by comparing the results with FEM, and the comparison results showed high accuracy.

Acknowledgements The work is supported by the National Science-Technology Support Plan Projects of China, under Award No. 2015BAK16B02.

References

ABS. Guide for building and classing subsea pipeline systems. Houston: ABS; 2006.

Ahn K, Lim IG, Yoon J, Huh H. A simplified prediction method for the local buckling load of cylindrical tubes. Int J Precis Eng Manuf. 2016;17(9):1149–56.

ALA. Guidelines for the design of buried steel pipe. Chicago: ALA; 2001.

ASCE. Guidelines for the seismic design of oil and gas pipeline systems. Reston: ASCE; 1984.

Chen H, Ji L, Gong S, Gao H. Deformation behavior prediction of X80 steel line pipe and implication on high strain pipe specification. In: International pipeline conference. Calgary, Alberta, Canada. 2008.

CSA. Oil and gas pipeline systems. CSA Z662-07, Ontario, Canada, 2007.

Dama E, Karamanos SA, Gresnigt AM. Failure of locally buckled pipelines. J Press Vessel Technol. 2007;129(2):272–9. doi:10.1115/1.2716431.

Dang XB, Gong SF, Jin WL, Li ZG, Zhao DY, Ning HE. Analysis on ultimate bearing capacity of deep-water submarine pipe. J Zhejiang Univ (Eng Sci). 2010;44(4):778–82. doi:10.3785/j.issn.1008-973X.2010.04.027 **(in Chinese)**.

Dorey AB, Cheng JJR. Critical buckling strains in energy pipelines. Structural engineering report SER 237. Edmonton, 2001.

Dorey AB, Murray DW, Cheng JJR. Material property effects on critical buckling strains in energy pipelines. In: International pipeline conference. 2002. pp. 475–84. doi:10.1115/IPC2002-27225.

Dorey AB, Murray DW, Cheng JJR. Critical buckling strain equations for energy pipelines—a parametric study. J Offshore Mech Arct Eng. 2005;128(3):248–55. doi:10.1115/1.2199561.

Fathi A, Cheng JJR, Adeeb S, Zhou J. Critical buckling strain in high strength steel pipes using isotropic-kinematic hardening. In: International pipeline conference. 2010. pp. 39–47.

Gresnigt AM. Plastic design of buried steel pipelines in settlement areas. Heron. 1986;4(31).

Gresnigt AM, Foeken RJV. Local buckling of UOE and seamless steel pipes. In: International offshore and polar engineering conference. Norway: The International Society of Offshore and Polar Engineers, 2001.

Han B, Wang ZY, Zhao HL, Jing HY, Wu ZZ. Strain-based design for buried pipelines subjected to landslides. Pet Sci. 2012;9(2):236–41. doi:10.1007/s12182-012-0204-y.

Igi S, Kondo J, Suzuki N, Zhou J, Duan DM. Strain capacity of X100 high-strain linepipe for strain-based design application. In: International pipeline conference. 2008. pp. 729–36.

Kamaya M. Ramberg-Osgood type stress–strain curve estimation using yield and ultimate strengths for failure assessments. Int J Press Vessels Pip. 2016;137:1–12. doi:10.1016/j.ijpvp.2015.04.001.

Limam A, Lee LH, Corona E, Kyriakides S. Inelastic wrinkling and collapse of tubes under combined bending and internal pressure. Int J Mech Sci. 2010;52(5):637–47. doi:10.1016/j.ijmecsci.2009.06.008.

Liu X, Zhang H, Han Y, Xia M, Zheng W. A semi-empirical model for peak strain prediction of buried X80 steel pipelines under compression and bending at strike-slip fault crossings. J Nat Gas Sci Eng. 2016;32:465–75. doi:10.1016/j.jngse.2016.04.054.

Mohareb ME. Deformational behaviour of line pipe. Doctoral dissertation. Edmonton: University of Alberta. 1995.

Nazemi N. Behavior of X60 line pipe under combined axial and transverse loads with internal pressure. Master thesis. University of Windsor, Windsor, 2009.

Nobuhisa S, Satoshi I, Katsumi M. Seismic integrity of high-strength pipelines. JFE Tech Rep. 2008;17:14–9.

Ozkan IF, Mohareb M. Moment resistance of steel pipes subjected to combined loads. Int J Press Vessel Pip. 2009a;86(4):252–64. doi:10.1016/j.ijpvp.2008.11.013.

Ozkan IF, Mohareb M. Testing and analysis of steel pipes under bending, tension, and internal pressure. J Struct Eng. 2009b;135(2):187–97. doi:10.1061/(ASCE)0733-9445(2009)135:2(187)

Paquette JA, Kyriakides S. Plastic buckling of tubes under axial compression and internal pressure. Int J Mech Sci. 2006;48(8):855–67. doi:10.1016/j.ijmecsci.2006.03.003.

Ramberg W, Osgood WR. Description of stress–strain curves by three parameters. Technical Report Archive & Image Library. 1943. NACA-TN-902.

Shantanu J, Prashant A, Arghya D, Sudhir KJ. Analysis of buried pipelines subjected to reverse fault motion. Soil Dyn Earthq Eng. 2011;31(7):930–40. doi:10.1016/j.soildyn.2011.02.003.

Sun YY. Study of perfect pipes on ultimate capacity under complex loads. Master thesis. Dalian: Dalian University of Technology, 2013. **(in Chinese)**.

Suzuki N, Tajika H, Igi S, Okatsu M, Kondo J, Arakawa T. Local buckling behavior of 48″, X80 high-strain line pipes. In: International pipeline conference. 2010.

Timms CMJ, Degeer DD, Chebaro MR, Tsuru E. Compressive strain limits of large diameter X80 UOE linepipe. In: The 19th international offshore and polar engineering conference. The International Society of Offshore and Polar Engineers. 2009; pp. 181–9.

Zimmerman T, Timms C, Xie J, Asante J. Buckling resistance of large diameter spiral welded linepipe. In: 2004 international pipeline conference. American Society of Mechanical Engineers, 2004. pp. 365–73. doi:10.1115/IPC2004-0364.

8

Drag reduction behavior of hydrolyzed polyacrylamide/xanthan gum mixed polymer solutions

Mehdi Habibpour[1] · Peter E. Clark[1]

Abstract Partially hydrolyzed polyacrylamide (HPAM) as the main component of slickwater fracturing fluid is a shear-sensitive polymer, which suffers from mechanical degradation at turbulent flow rates. Five different concentrations of HPAM as well as mixtures of polyacrylamide/xanthan gum were prepared to investigate the possibility of improving shear stability of HPAM. Drag reduction (DR) measurements were performed in a closed flow loop. For HPAM solutions, the extent of DR increased from 30% to 67% with increasing HPAM concentration from 100 to 1000 wppm. All the HPAM solutions suffered from mechanical degradation and loss of DR efficiency over the shearing period. Results indicated that the resistance to shear degradation increased with increasing polymer concentration. DR efficiency of 600 wppm xanthan gum (XG) was 38%, indicating that XG was not as good a drag reducer as HPAM. But with only 6% DR decline, XG solution exhibited a better shear stability compared to HPAM solutions. Mixed HPAM/XG solutions initially exhibited greater DR (40% and 55%) compared to XG, but due to shear degradation, DR% dropped for HPAM/XG solutions. Compared to 200 wppm HPAM solution, addition of XG did not improve the drag reduction efficiency of HPAM/XG mixed solutions though XG slightly improved the resistance against mechanical degradation in HPAM/XG mixed polymer solutions.

Keywords Slickwater · Polyacrylamide · Xanthan gum · Drag reduction · Shear stability

1 Introduction

Due to the development of horizontal drilling and hydraulic fracturing technology in recent years, oil and gas production from unconventional resources such as shale formations has played an important role in supplying energy in the USA (Larch et al. 2012; Palisch et al. 2010). Reaching economic rates of hydrocarbon production from shale formations is only possible when micro-fractures are created and connected through effective stimulation treatments such as horizontal fracturing in multiple stages (Loveless et al. 2014; Barati and Liang 2014; Wu et al. 2013).

Creation of openings in the reservoir rock involves pumping fracturing fluids into the wellbore at elevated flow rates and pressures. The viscosity of a fracturing fluid has a remarkable effect on the fracture initiation and the final size of the fracture (Kalgaonkar and Patil 2012). The main component of low-viscosity fracturing fluids is water and low concentrations of polymer ranging from 0.25 to 10 lb per thousand gallons (Bunger et al. 2013; Palisch et al. 2010). In the last decade, an amazing shift from traditional gel-based fracturing fluids toward using slickwater has occurred. Because of the relatively low viscosity of slickwater, cleanup problems and damage associated with using viscous fluids are minimized, which makes slickwater suitable for fracturing low-permeability reservoirs (Wu et al. 2013; Palisch et al. 2010). As a result of lower viscosity, slickwater cannot suspend and transport proppants as effectively as gelled fluids. To overcome poor proppant transport, higher pumping rates are applied, which in turn

Edited by Xiu-Qin Zhu.

✉ Peter E. Clark
  peter.clark@okstate.edu

[1] School of Chemical Engineering, Oklahoma State University, Stillwater, OK 74078, USA

leads to significant energy loss due to friction and turbulence in the tubular pipeline (Palisch et al. 2010; Kaufman et al. 2008). In order to lower the surface pumping pressures and compensate for the energy losses during pumping, a small amount of high molecular weight polymer is dissolved in the fluid, which acts as a drag reducer. In slickwater treatments, friction reducer is a significant component of the fluid. Long, flexible chain polyacrylamide-based polymers are known to be the most effective drag reduction polymers (Kim et al. 1998).

When a polymer solution is subjected to turbulent flow, the large molecules present in the solution undergo different conformational changes such as flow orientation, stretching, and relaxation. These conformational changes alter the flow field and dynamics of turbulent structures in the vicinity of pipe wall. In the macroscopic scale, those variations result in large drag reduction. However, mechanical degradation of polymer solutions complicates their use (Campolo et al. 2015; Moussa and Tiu 1994). Several theories have been proposed to explain mechanical degradation of polymer solutions, among which extreme stretching of polymer chains and scission of the polymer chain have been most accepted (Rodriguez and Winding 1959). Another study reports that chain scission occurs near the midpoint of the polymer which is caused by the extensional features of the turbulent flow (Lim et al. 2003; Merrill and Horn 1984). Due to the susceptibility of PAM molecules to shear degradation under turbulent flow conditions, researchers have considered other candidates such as rigid polysaccharide molecules to improve the shear stability of drag reducers. Kim et al. used a rotating disk apparatus to verify the DR behavior of 100 wppm XG solutions with four different molecular weights. They observed that XG is more resistant to shear forces than most flexible polymers (Kim et al. 1998). Other researchers compared solutions of polyethylene oxide (PEO), PAM, XG, and guar gum (GG). They observed that synthetic polymers initially induced greater DR efficiency, but as result of mechanical degradation, DR efficiency of synthetic polymers declined more quickly than the polysaccharides. The available data in the literature suggest that XG is shear resistant at turbulent flow conditions, but DR induced by XG is small compared to synthetic polymers such as PAM (Soares et al. 2015; Kulmatova 2013; Wyatt et al. 2011; Kenis 1971). According to the literature, binary solutions of polymers can give rise to DR efficiency and shear stability. Dingilian and Ruckenstein (1974) investigated the DR behavior of PAM, PEO, and carboxymethylcellulose (CMC) single and binary solutions. They observed negative deviation from additivity for binary solutions of two flexible polymer molecules (PEO/PAM) and positive deviation for PAM/CMC and PEO/CMC binary solutions, where flexible and rigid molecules coexist in the solution. Dschagarowa and

Bochossian (1978) also observed synergism in the DR behavior of poly-isobutylene and 1,4-cis-isoprene rubber binary solutions. They also concluded that flow conditions can affect the degree of deviation from additivity. Malhotra et al. (1988) verified DR of PAM/GG, GG/XG, and PAM/XG binary solutions and concluded that the degree of DR and synergism depend on polymer concentration and flow rate. They also concluded that positive deviation in additivity occurs when both polymers are rigid. But, they mentioned that their conclusions are limited to their experimental conditions and synergism might be observed in many other binary mixtures at higher concentrations. Reddy and Singh (1985) verified the same binary solutions at higher concentrations ($C > 200$ wppm) and in a larger experimental setup. Contrary to previous authors, they observed negative deviation for PAM/polysaccharide binary solutions. They also verified the shear stability of the polymer solutions and concluded that the solutions showing synergism in DR produce better shear stability (Reddy and Singh 1985). Recently Sandoval and Soares (2016) verified the DR behavior of PAM/XG and PEO/XG binary solutions in a pressure-driven flow loop. The authors reported clear synergism between polymers. They concluded that the improvement observed in the mixed solutions was related to the change in the polymeric conformation from coiled to elongated (Sandoval and Soares 2016). Regarding the DR behavior of binary polymer solutions, very limited work is reported in the literature. On the other hand, the existing published research is limited to certain conditions such as small-scale experimental setups (capillary tube), low flow rates and Reynolds numbers, and certain concentrations or molecular weights. Also, some of the available data contradict each other. As mentioned above, Reddy and Singh (1985) had a different observation from Sandoval and Soares (2016) regarding the DR behavior of PAM/XG binary solution, which we believe arises from the differences in the experimental setup and the molecular weights of the polymers.

In this work, we focus on studying the DR behavior and shear stability of anionic HPAM/XG binary polymer solution. Anionic PAM is the partially hydrolyzed form of polyacrylamide with more flexibility and relatively large molar mass. Both polymers are widely used in industrial applications such as slickwater treatments. There has been little or no work regarding the DR behavior of anionic HPAM and XG mixtures. We believe that it would be interesting to verify the DR behavior of two negatively charged molecules coexisting in a solution. Since various HPAM and XG molecules with different molecular weights are available in the market, we also study the behavior of single polymer solutions, so that we can compare them with binary solutions and assess the degree of improvement in shear stability.

2 Materials and methods

2.1 Materials and preparation

Anionic polyacrylamide as a linear flexible polymer and XG as a semirigid polymer were chosen for this investigation. The molecular weights provided for PAM (Kemira Co.) and XG (PFP Technology) by the vendors are $(6–8) \times 10^6$ kg mol^{-1} and $(4–5) \times 10^6$ kg mol^{-1}, respectively. The accurate values for molar mass were measured using light scattering (for HPAM) and viscometry (for XG) methods. In the light scattering experiments, all solvents were filtered through 0.02 μm cellulose acetate Millipore filters. The HPAM solutions were prepared by dissolution of a known amount of polymer in the 0.5 mol L^{-1} NaCl solvent. The solutions were filtered through a 0.45 μm cellulose acetate Millipore filters. A concentration range of the polymer solutions (0.2–2.0 g L^{-1}) were analyzed using the batch mode (without size separation) of a multi-angle light scattering (MALS) detector (DAWN-HELEOS II, Wyatt Technology) with a laser wavelength of 658 nm. The specific refractive index increments (dn/dc) value of the HPAM in 0.5 mol L^{-1} NaCl solution, which was determined by an OPTILAB T-reX differential refractometer (Wyatt Technology) at 633 nm and 25 °C, was 0.162 mL/g. Viscosity averaged molecular weight of XG was measured using the method described in the literature (Sohn et al. 2001). In this method, the intrinsic viscosity $[\eta]$ was measured with an Ubbelohde capillary viscometer (diameter capillary 0.46 mm, Schott-Gerate), immersed in a water bath maintained at 25.0 ± 0.1 °C. First, the specific viscosity η_{sp} was calculated from the relative viscosity, which is the ratio of the viscosity of xanthan gum solution at a certain concentration to that of the solvent. Each concentration was measured five times. The plot of η_{sp}/C versus C (C is the concentration of solutions) gave a straight line, the intercept of which was $[\eta]$. Then, using the Mark–Houwink equation the molar mass was calculated:

$$[\eta] = \lim_{c \to 0} \frac{\eta_{sp}}{c} = 6.6 \times 10^{-6} \left(\overline{M}_v\right)^{1.35} \quad [\text{mL/g}] \quad (1)$$

The measured values for the molecular weights of the polymers, HPAM and XG, were 7.2×10^6 kg mol^{-1} and 3.9×10^6 kg mol^{-1}, respectively.

To establish a baseline for further studies, deionized water was used throughout. Using an analytical balance, polymer powders were weighed with an accuracy of ±1 mg (Mettler Toledo XS603s). The preparation of the polymer solutions took place in a separate tank. First, the polymer powders were gradually sprinkled into the solvent and slowly agitated at 30 rpm, in order to prevent particles from clumping on the surface. Then the prepared solutions were stored overnight for complete hydration. The studied

concentrations of HPAM solutions were 100, 150, 200, 500, and 1000 wppm. In order to investigate the effect of XG, mixed polymer solutions of 100 wppm HPAM + 100 wppm XG (total $C = 200$ wppm), 150 wppm HPAM + 100 wppm XG (total $C = 250$ wppm), and for comparison, 600 wppm XG solutions were prepared.

2.2 Drag reduction and viscosity measurements

A closed-loop flow system was used for drag reduction and turbulent flow measurements (Fig. 1). The system is comprised of a 60-L supply tank connected to a progressive cavity pump (SEEPEX BN 10-12) with a pumping capacity of 30 GPM (113.56 L/min), and a seamless stainless steel horizontal pipe test section of $L = 8$ ft. (2.44 m) and inner diameter of 1 in. (2.54 cm). The flow rate of the system was measured using a mass flow meter (OPTIMASS 1000, KROHNE) with an accuracy of ±0.15% and a repeatability of ±0.05% as stated by the manufacturer. The pressure drop data along the measuring length of the pipe were gathered using a membrane differential pressure transducer (PX409, OMEGA). At each flow rate, the Fanning friction factor is calculated by:

$$f = \frac{D}{2\rho U^2} \left(\frac{\Delta P}{L}\right) \quad (2)$$

Here, f is the Fanning friction factor, ρ is the fluid density, U is the average velocity, ΔP is pressure drop in the measurement length of L, and D is the internal diameter of the pipe.

In the flow experiments, the drag reduction efficiency of the polymer solutions was defined as:

$$\text{DR} = \left(\frac{f_w - f_s}{f_w}\right)_{Re=\text{const}} \quad (3)$$

where subscripts "w" and "s" stand for water and polymer solution, respectively, and "$Re =$ const." signifies the fact that the comparison between the flows is made at the same Reynolds numbers. Here $Re = \rho UD/\eta_a$, and η_a is fluid viscosity. In drag reduction experiments, the flow rate was increased stepwise (and kept constant for 1 min at each flow rate), and then, the polymer solutions were sheared in the flow loop at maximum flow of 30 GPM for 2 h and sampling was performed eight times at shearing periods of $t = 0, 15, 30, 45, 60, 80, 100,$ and 120 min. After the shearing process, the flow rate was decreased stepwise in the same manner as it was increased.

Viscosity measurements were carried out in a DHR controlled stress rheometer. The instrument is equipped with a Peltier system to control sample temperature. In our experiments, sample temperature was maintained at 25 ± 0.01 °C. The geometry used in the apparent viscosity

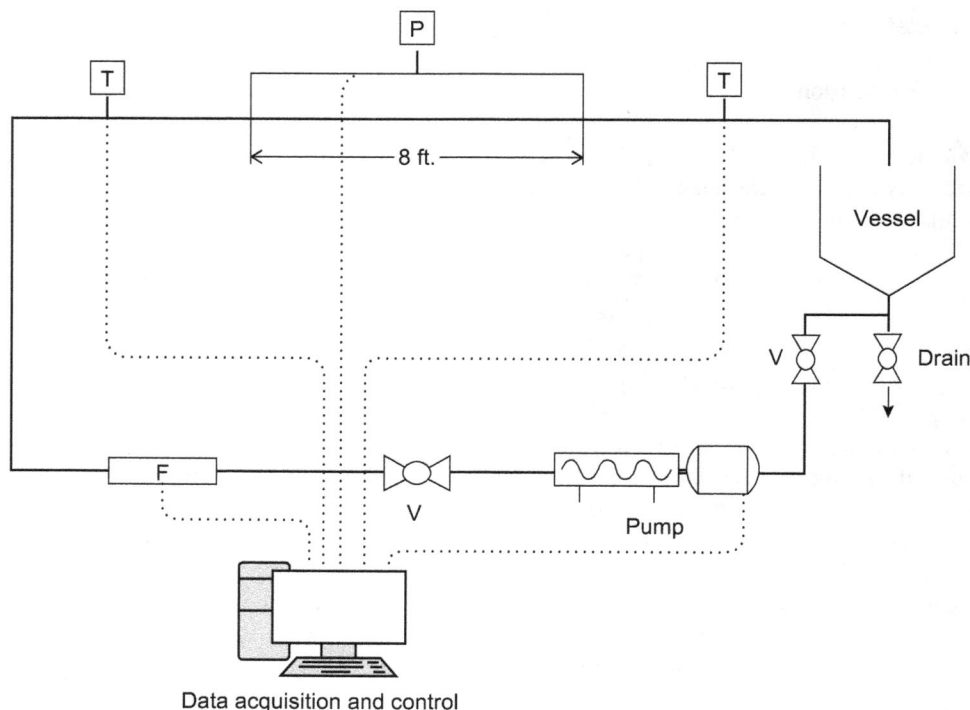

Fig. 1 Schematic illustration of the flow loop; instrumentation consists of a digital flow meter (F), differential pressure transducer (P), valves (V), and temperature sensors (T)

measurements was cone and plate. The cone diameter and angle were 60 mm and 2°, respectively. Approximately, 2 mL of the solution was placed between the cone (rotating plate) and the fixed plate and the instrument was set to the strain control mode. In this mode, shear rate ($\dot{\gamma}$) was logarithmically increased from 0.01 to 1000 s^{-1} and shear stress was measured simultaneously, and then, apparent viscosity was calculated using $\eta_a = \sigma/\dot{\gamma}$ correlation, where σ and η_a are shear stress and apparent viscosity, respectively. The power-law and Carreau–Yasuda viscosity models were used for fitting the viscosity data.

In the power-law model, n (behavior index) is a measure of deviation from Newtonian behavior and K (consistency index) is a measure of average viscosity. K and n known as power-law parameters were used in friction factor calculations.

$$\eta_a = K(\dot{\gamma})^{n-1} \tag{4}$$

$$\frac{\eta - \eta_\infty}{\eta_0 - \eta_\infty} = [1 + (\lambda\dot{\gamma})^a]^{\frac{n-1}{a}} \tag{5}$$

In the Carreau–Yasuda equation, η is the shear viscosity, and η_0 and η_∞ are viscosities at zero-shear and infinite-shear plateaus, while λ, n, and a represent the inverse shear rate at the onset of shear thinning, power-law index, and transition index, respectively.

3 Results and discussion

3.1 Polyacrylamide solutions

In order to assess the degree of drag reduction and polymer degradation, several samples were taken at specified intervals over the shearing process. Shear stress and apparent viscosity of the samples were measured in the rheometer, and the power-law and Carreau–Yasuda parameters were calculated. For Newtonian fluids, the value of n is 1 and for shear-thinning fluids, such as polymer solutions, n is <1. As the value of n deviates from 1, the degree of non-Newtonian behavior increased. The Carreau–Yasuda and power-law parameters calculated for different concentrations of the anionic polyacrylamide are shown in Table 1 and Fig. 2a, b, respectively. Comparison of power-law parameters of fresh (zero shearing time) HPAM solutions showed that by increasing concentration, values of K increased and n values decreased, indicating increased non-Newtonian behavior. It can also be found that, as the shearing time increased, the values of power-law parameters changed for all the polymer solutions. Figure 2a shows that, over the shearing time, K values of HPAM solutions decreased and n values increased and get close to one (Newtonian behavior). The difference in the n values of the solutions before and after the shearing was

Table 1 Parameters of Carreau–Yasuda model for HPAM, XG, and HPAM/XG solutions

Concentration of solution	η_0, Pa s	η_∞, Pa s	λ, s	a	n
100 wppm HPAM	0.0835	0.0035	4.96	1.37	0.37
150 wppm HPAM	0.2668	0.0041	5.46	3.15	0.26
200 wppm HPAM	0.3717	0.0042	5.80	3.14	0.27
500 wppm HPAM	1.2105	0.0052	8.70	3.74	0.26
1000 wppm HPAM	3.5071	0.0067	10.1	3.52	0.21
600 wppm XG	0.2689	0.0040	5.34	1.92	0.43
100 wppm HPAM + 100 wppm XG	0.1331	0.0039	4.18	2.51	0.36
150 wppm HPAM + 100 wppm XG	0.2873	0.0038	7.2	2.28	0.33

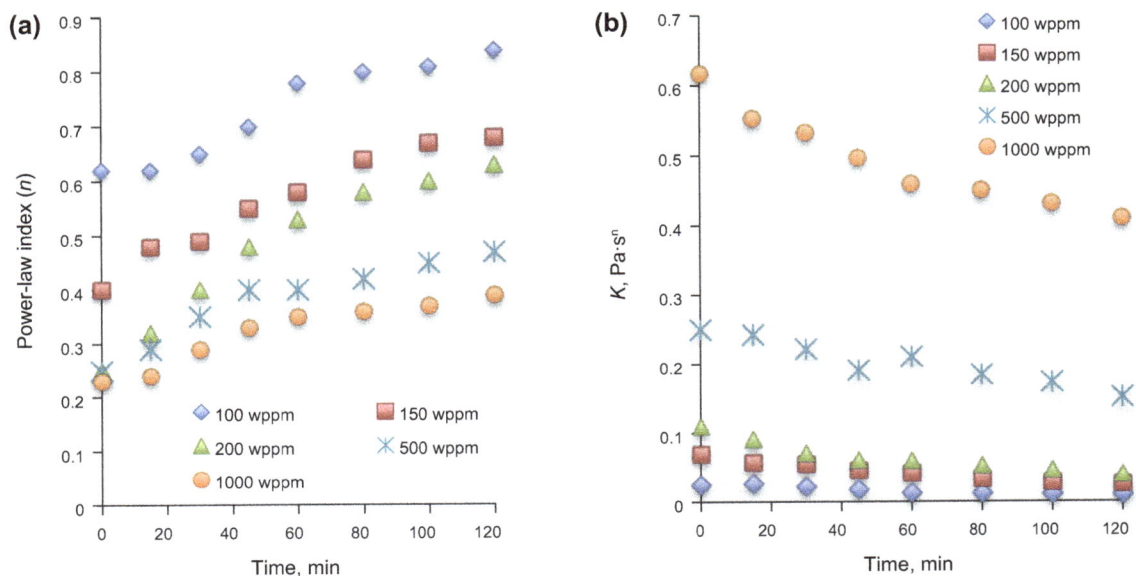

Fig. 2 Power-law model parameters variation over shearing time for HPAM solutions

large (35%–152% increase for various concentrations). The overall change in n values (after the shearing period) was 35% for a 100 wppm solution, and by increasing concentration, the maximum change in n values (152%) appeared at 200 wppm. As the concentration increased above 200 wppm, n dropped to 69% for the 1000 wppm solution. As a result of shear degradation and consequently viscosity reduction, K values declined. By increasing concentration from 100 wppm to 1000 wppm, K values reached a maximum decline of 64% at 200 wppm, but with further concentration increase a 33% decline in K value was found for 1000 wppm HPAM solution.

Another indication of shear degradation is shown in Fig. 3, where the apparent viscosity of the 200 wppm HPAM solution, at shear rates under 10 s^{-1}, undergoes drastic reduction as the circulation time increases. The reason that we only saw changes at lower shear rates might be related to the scission of the molecules that belong to the high tale of the molecular weight distribution of the

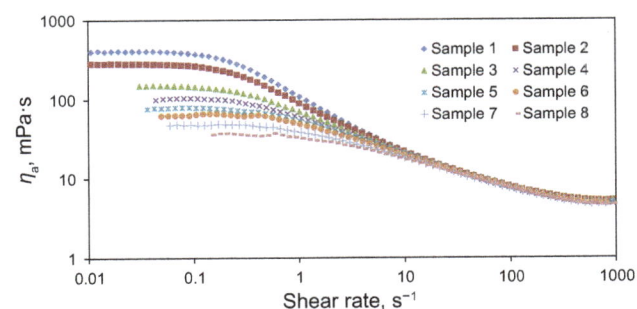

Fig. 3 Apparent viscosity of 200 wppm sheared HPAM samples (samples 1–8: fresh, 15, 30, 45, 60, 80, 100, and 120 min shearing time)

polymer (Liberatore et al. 2004). Those molecules contributed to the formation of the zero-shear viscosity plateau.

Prandtl–Karman coordinates are a semilog graph of $f^{-1/2}$ versus $Re.f^{1/2}$, where f and Re are Fanning friction factor and Reynolds number, respectively. Using Prandtl–Karman coordinates, the degree of DR of polymer solutions can be compared with respect to the boundaries of drag reduction; the onset of drag reduction as the point of departure from Prandtl–Karman law and maximum drag reduction (MDR) or Virk's asymptote (White and Mungal 2008; Virk 1975). Prandtl–Karman plot for different concentrations of HPAM is shown in Fig. 4. Three flow regimes can be detected:

(a) $Re\sqrt{f} < 200$; laminar flow. All polymer solutions obey Poiseuille's law.
(b) $200 < Re\sqrt{f} < 350$; laminar to turbulent transition.
(c) $Re\sqrt{f} > 350$; turbulent flow.

It is observed that except for the laminar to turbulent transition region, the turbulent friction factor data for the solutions are bounded in the area between the two universal asymptotes and the data points are linear for all concentrations.

Results of linear fitting of the data points in the turbulent region (Table 2) revealed that by increasing the concentration of polymer, the slope of the lines tended to increase. According to the literature, the slope increment with respect to the Prandtl–Karman law (δ) is proportional to the square root of polymer concentration ($\delta \propto \sqrt{C}$) with a proportionality constant that is the characteristic of the polymer (Virk and Baher 1970). The slope increment values are reported in Table 2. The proportionality constant for the HPAM used in this work is calculated to be

0.27 ± 0.05. Figure 4 shows that by increasing HPAM concentration, the extent of DR increases and the data points approach Virk's asymptote. Among different concentrations, the extent of DR and slope of data points for 100 wppm HPAM solution were closest to the Prandtl–Karman line (the onset of DR). On the other hand, the greatest DR belongs to the 1000 wppm solution. For solutions with polymer content greater than 200 wppm, in $200 < Re\sqrt{f} < 500$ range, the data points were nearly tangential to the MDR line, but by increasing the flow rates (Reynolds number), the data points tended to deviate from MDR. Results indicated that further increase in concentration did not have a large effect on the degree of DR in the experimental conditions of this work.

The effect of polymer concentration and mechanical degradation on the Fanning friction factor of HPAM solutions is shown in Fig. 5a, b, respectively. It can be seen that even at low concentrations of polymer (100 wppm), Fanning friction factor values were much lower than those of water. Increasing polymer concentration resulted in further reduction in friction factors (data points get close to the maximum drag reduction asymptote), and consequently, drag reduction efficiency increased The smallest values of friction factors at different Reynolds numbers belong to 1000 wppm solution. It was also observed that the difference between turbulent friction factors of 200, 500, and 1000 wppm solutions was very small, which is in agreement with the results shown in Prandtl–Karman coordinates, indicating that at the studied range of Reynolds numbers, DR efficiency was close to its maximum value (MDR); further increase in concentration did not change friction factors. We can also observe that as the Reynolds numbers increased, the data points tended to deviate from MDR.

Shearing had a large impact on reducing the DR ability of polyacrylamide solutions. Comparing the friction factors of 200 wppm HPAM solution at different Reynolds numbers (Fig. 5b) showed a 30%–50% decline after 2 h of shearing. It can be observed that as a result of shearing, friction factor values moved away from MDR line and shifted toward water friction factors at different Reynolds numbers. Although flow rate was constant for all the experiments, a shift in the data points of the sheared samples toward lower Reynolds numbers was observed.

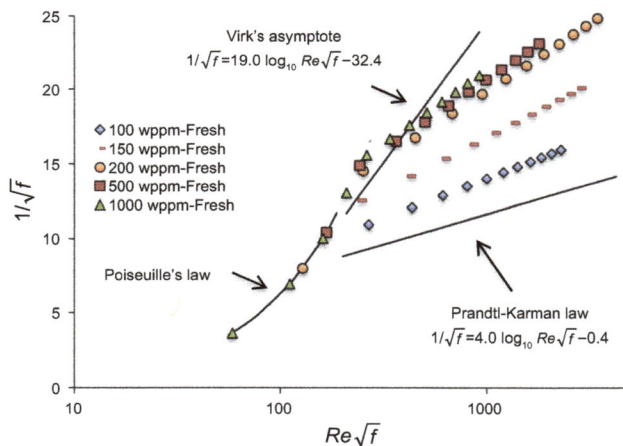

Fig. 4 Prandtl–Karman coordinates for HPAM solutions

Table 2 Linear fit results of turbulent regime

Concentration, wppm	Linear fit equation	R^2	Slope increment (δ)
100	$1/\sqrt{f} = 6.0 \log Re\sqrt{f} - 4.1$	0.99	2.05
150	$1/\sqrt{f} = 7.1 \log Re\sqrt{f} - 4.4$	0.99	3.09
200	$1/\sqrt{f} = 9.3 \log Re\sqrt{f} - 7.3$	0.99	5.23
500	$1/\sqrt{f} = 9.6 \log Re\sqrt{f} - 8.2$	0.99	5.60
1000	$1/\sqrt{f} = 12.8 \log Re\sqrt{f} - 15.9$	0.99	8.81

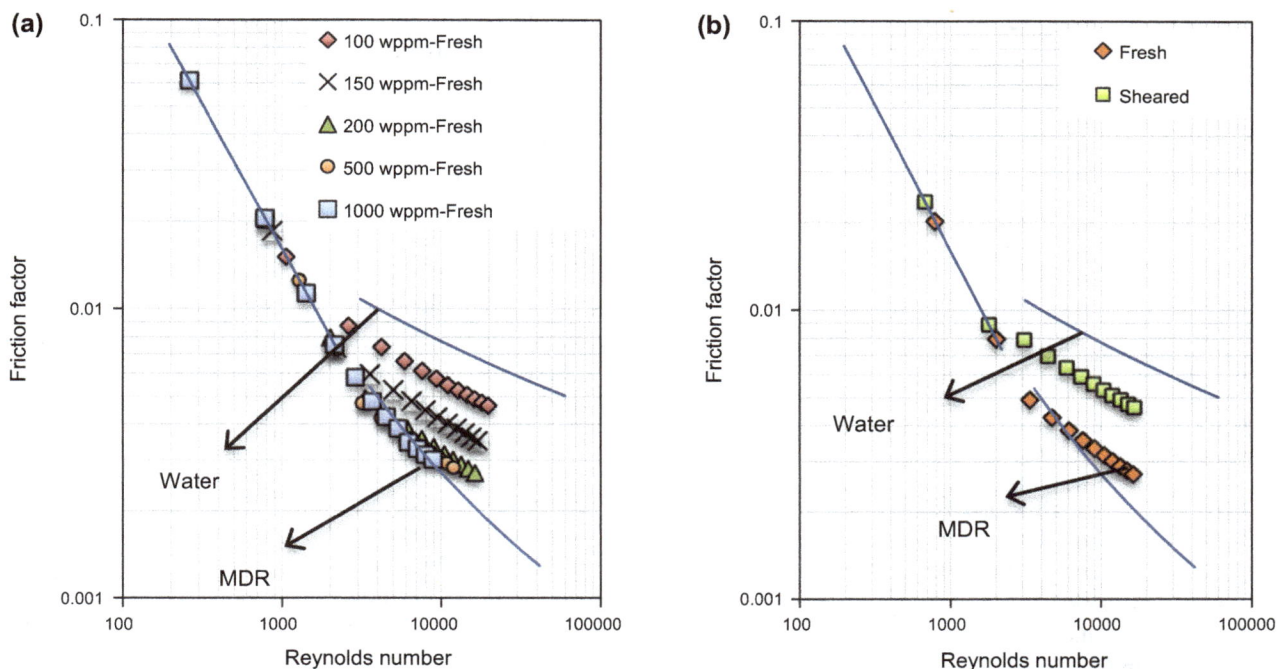

Fig. 5 Friction factor versus *Re*: **a** effect of HPAM concentration and **b** effect of shearing on 200 wppm HPAM

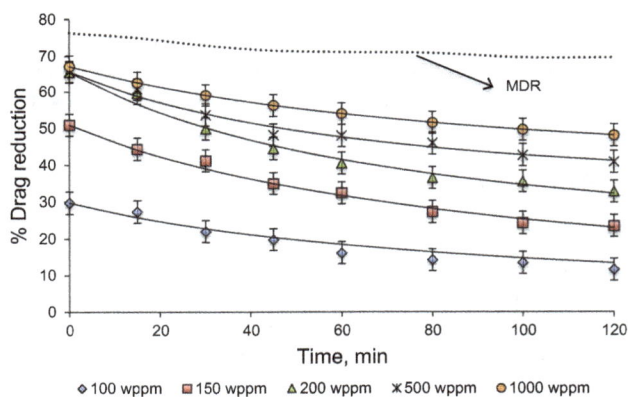

Fig. 6 %Drag reduction for different concentrations of HPAM at 30 GPM (113.56 L/min)

The shift was an indication of alteration in the flow regime, which occurred due to the change in the rheological properties of the solutions as a result of shear degradation. Figure 6 shows the changes in the DR% over shearing time for different concentrations of HPAM at constant flow rate. It was observed that, for fresh samples (*t* = 0), the extent of DR increased from 30% to 67% by increasing polymer concentration from 100 to 1000 wppm. Results also indicated that solutions above 200 wppm produced nearly identical DR at the early stages of shearing (*t* < 20 min), which corresponds to the results shown in Prandtl–Karman coordinates (Fig. 4). But beyond 20 min, the DR curves

tended to diverge, indicating that the degree of shear degradation was different for each polymer concentration. Ptasinski et al. (2003) reported that in flexible polymers such as HPAM, as shear force reached a certain level, the molecules stretch and consequently effective viscosity increased, the turbulent buffer layer thickened, and due to dissipation of the energy from turbulent fluctuations drag was reduced. It is also known that drag is reduced when turbulent flow interacts with polymer networks. When polymer concentration is high enough (above critical overlap concentration, *C**), polymer chains are packed closer and begin to interact with each other and form entangled networks (Skelland and Meng 1996).

The decline in the degree of DR was observed at all the concentrations of HPAM solution. Resistance to mechanical degradation over the shearing period is shown in Fig. 7.

Results showed that, as the concentration increased, the resistance to mechanical degradation also increased. This made the 100 wppm solution the least resistant and 1000 wppm solution the most resistant to mechanical degradation. These results agreed with those reported by as Soares et al. (2015), indicating that increasing polymer concentration (PAM or PEO) in the solution could increase the resistance to mechanical degradation. It was also observed that the shear stabilities of 100, 150, 200 wppm solutions were very close and followed the same trends. In other words, 200 ppm solution was only slightly more resistant to degradation ($DR_{t=120}/DR_{max} = 0.5$) than

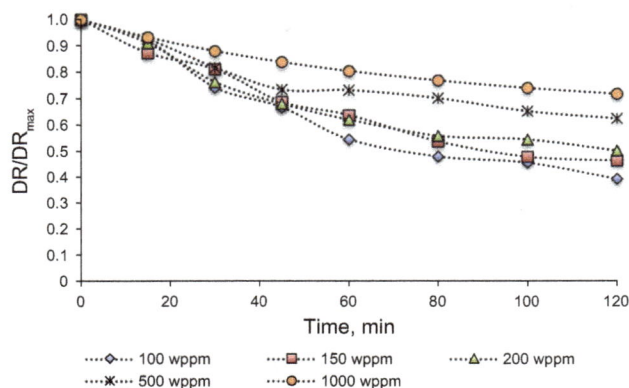

Fig. 7 Resistance to mechanical degradation (DR/DR$_{max}$) for HPAM solutions at 30 GPM (113.56 L/min)

100 wppm solution (DR$_{t=120}$/DR$_{max}$ = 0.39). But, above 200 wppm a sudden rise in the shear resistance of the solutions was observed. Both 500 and 1000 wppm solutions showed superior shear stability compared to other solutions. We found that although 200 wppm HPAM solution had similar initial DR% to 500 and 1000 wppm solutions (Fig. 6), it had a larger decline in DR (Fig. 7). The values of DR$_{t=120}$/DR$_{max}$ for 500 and 1000 wppm solutions were 0.62 and 0.72 at the end of 2-h shearing period, respectively. It is also interesting to note that the DR/DR$_{max}$ declined at a slower rate after $t = 60$ min. For example, the reduction in DR/DR$_{max}$ value for 1000 wppm solution from $t = 60$ min to $t = 120$ min was only 0.06.

From viscosity measurements, the critical overlap concentration (the concentration that marks the beginning of polymer–polymer interaction) of the HPAM solution was determined to be 200 wppm. At $C < 200$ wppm, there was no polymer network, so there was no decline due to polymer network breakup, but at $C > 200$ wppm, the networks were more entangled and upon deformation can recover quickly. Also, it seems that at higher concentrations ($C > 200$ wppm), the extensional force from the turbulent flow was distributed among a larger number of molecules, which resulted in a lower number of chain scissions and higher DR stability. But at $C < 200$ wppm polymer networks just began to form and were weak.

Therefore, at the beginning of the shearing process DR efficiency increased, but gradually as shear forces acted, unstable polymer networks, as well as individual polymer molecules broke up, which resulted in a large decline in the DR efficiency of $C < 200$ wppm HPAM solutions compared to higher concentrations.

Another result was that the rate of DR decline was fast in the first 60–80 min of the shearing process. This is probably due to the presence of longer chains of polymer in the fresh solutions, which are more susceptible to chain scissions (require less energy to break). After 60 min, the gap between DR lines (Fig. 6) remained nearly constant, which indicated that polymer chains were broken up at a similar rate at all the concentrations.

Several models for correlating the DR behavior of polymer solutions can be found in the literature. Two of the most common models are the exponential decay model (Bello et al. 1996) and Brostow's model (Brostow 1983). Based on the exponential decay model proposed by Bello et al. (1996), DR follows an exponential decline with time for polyacrylamide and polysaccharide solutions:

$$DR\%(t)/DR\%(0) = e^{(-t/\lambda)} \tag{6}$$

where DR% (t) and DR% (0) (or DR$_{max}$) are percent drag reduction at times t and $t = 0$, respectively and λ is an adjustable parameter. Brostow's equation is a more general model, which has been applied to various shear degradation applications (Lim et al. 2003) and has been developed to quantitatively describe DR and its changes with time:

$$DR\%(t)/DR\%(0) = [1 + W(1 - e^{(-ht)})]^{-1} \tag{7}$$

Here W is the average number of vulnerability points per chain and h is called rate constant.

The two models were used in this work to correlate the DR data of HPAM and HPAM/XG solutions. Calculated model parameters and goodness of fit (R^2) for both models are reported in Table 3. The solid lines in Fig. 6 represent the fitting results. It can be found that for the studied concentrations of HPAM solution, both models describe the degradation behavior well, though Brostow's model gives better fits and its predicted values are more accurate.

Table 3 Parameters of Brostow and exponential decay models for HPAM solutions

Concentration, wppm	Exponential decay model		Brostow's model		
	λ, min	R^2	W	h, min^{-1}	R^2
100	87.32	0.96	374.2	32.7e−5	0.94
150	146.7	0.97	43.43	2.3e−4	0.98
200	162.3	0.95	2.723	3.9e−3	0.99
500	245.0	0.90	0.8628	9.6e−3	0.98
1000	312.1	0.88	0.6351	8.1e−3	0.96

3.2 HPAM/XG mixed solutions

In order to verify the possibility of stabilizing HPAM with XG, mixed solutions were prepared and sheared in the same manner as the HPAM solutions. Based on the results obtained from the previous section, the DR behavior of binary polymer solutions was compared to that of 200 wppm HPAM solution. Several authors have mentioned that the degree of DR in XG solutions depends on the concentration and molar mass of XG (Soares et al. 2015; Pereira et al. 2013). So higher concentrations of XG are required to reach the same level of DR as HPAM. Our preliminary experiments showed that the DR efficiency of 600 wppm XG was close to the DR values of 200 wppm HPAM.

The Carreau–Yasuda model fitting results of the mixed solutions and a solution containing XG are included in Table 1. Also, friction factors of the mixed polymer solutions and solutions of 600 wppm XG and 200 wppm HPAM are shown in Fig. 8a. It was found that, although polymer concentration in the 600 wppm XG solution was greater than that of 200 wppm HPAM solution, the 600 wppm XG solution possessed greater friction factors. Higher flexibility of HPAM polymer chains with respect to semirigid XG chains resulted in higher energy adsorption (stretching) due to interaction with dynamic turbulent flow and subsequently, superior drag reduction capability of flexible polymers. Also, it was found that in the fresh binary solutions, there is no outstanding synergetic effect between HPAM and XG molecules. Even the 150 wppm HPAM + 100 wppm XG solution, which has a higher total polymer concentration, showed larger friction factors than 200 wppm HPAM.

Figure 8b compares the friction factors of the solutions after 120 min of shearing. Because of high shear stability, XG solution maintained its initial friction reduction

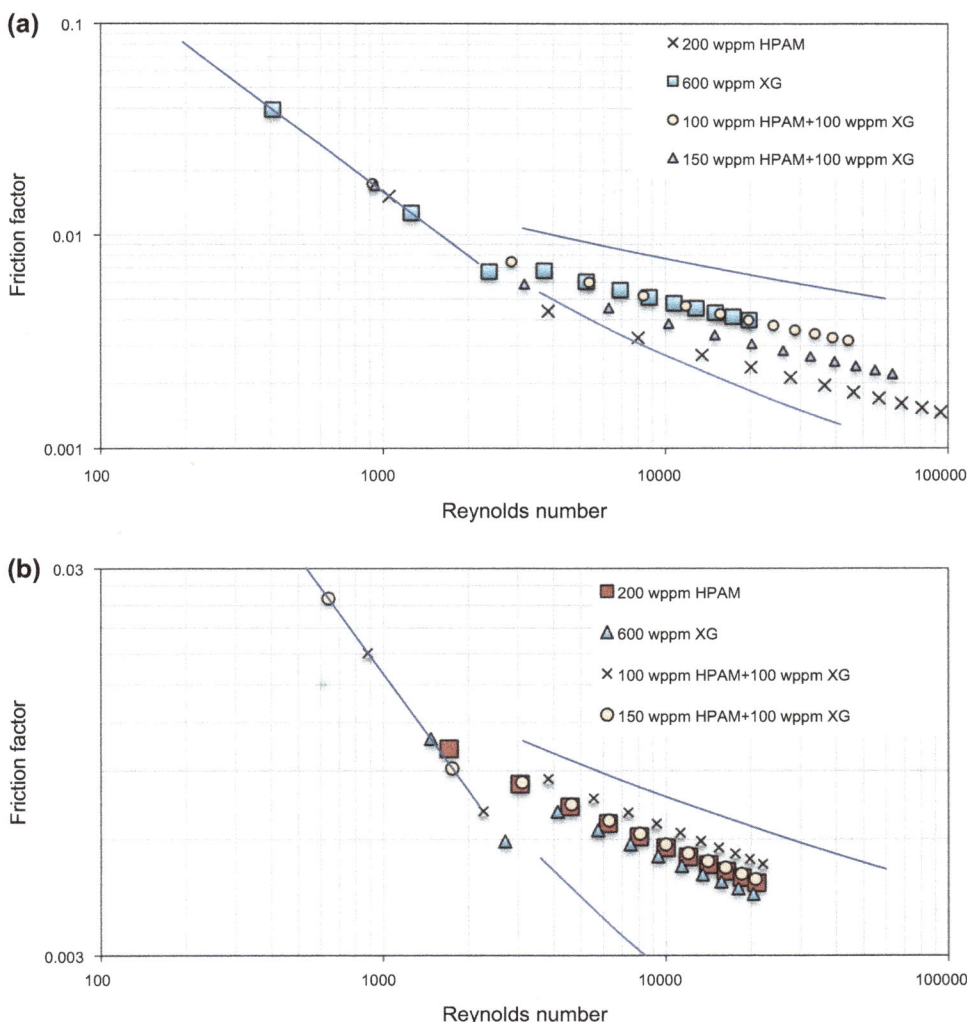

Fig. 8 Fanning friction factor of mixed solutions in **a** fresh form and **b** after shearing for 120 min

efficiency, with a relatively small change in friction factor values. But for the mixed HPAM/XG solutions, the friction factor values largely increased as a result of shearing. It is interesting to note that 200 wppm HPAM solution had a larger increase in its friction factors than mixed solutions (Fig. 8b). This suggested that addition of XG was beneficial in controlling the shear degradation of HPAM. The decline in the DR efficiency and shear resistance of the samples over shearing time is shown in Figs. 9 and 10. Also, as mentioned above, Brostow's and exponential

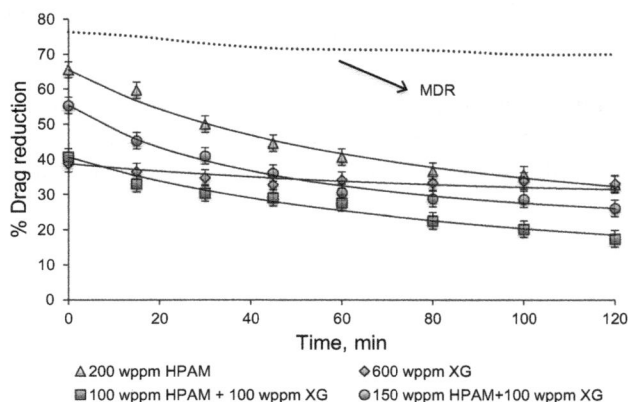

Fig. 9 %Drag reduction for mixed HPAM/XG solutions at 30 GPM (113.56 L/min)

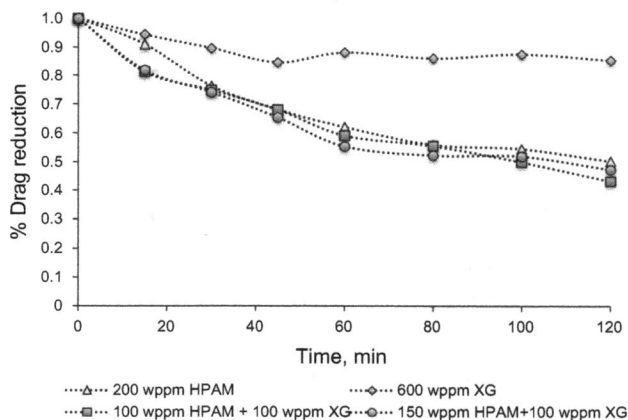

Fig. 10 Effect of shearing on DR decline for mixed HPAM/XG solutions

decay models were used to correlate DR with time (solid lines in Fig. 9; Table 4).

The results indicated that the DR efficiencies of both HPAM and XG binary solutions were smaller than that of 200 wppm HPAM solution, indicating that partially replacing HPAM with XG in the binary solutions had reduced the DR efficiency of the single HPAM solution. Regarding the DR behavior of HPAM/XG binary mixtures, different views can be found in the literature. Sandoval and Soares (2016) recently reported that the DR efficiency of HPAM/XG mixture is close to the DR values of the single HPAM solution at the same total concentration. However, they confirmed that the synergetic effect in PAM/XG mixture is not as notable as PEO/XG. Reddy and Singh (1985) also studied PAM/XG mixtures and reported that the single PAM solution has a higher DR efficiency than a binary mixture. Their data were limited to fresh solutions. We believe that the differences in the reported data might be attributed to the differences in the experimental conditions. In our work, since we were dealing with two negatively charged polymers in the solution, we would expect to obtain different results. We also believed that due to the higher flexibility (higher molecular weight) of the anionic HPAM used in our experiments, HPAM molecules were more susceptible to scission and the interaction between HPAM and XG was reduced. Compared to 600 wppm XG solution, the DR% of 100 wppm HPAM/150 wppm XG solution was high only in the first 60 min of the shearing process and after that, as a result of mechanical degradation the DR values declined. Also, it was found that the 100 wppm HPAM/100 wppm XG solution was less efficient in DR than the 600 wppm XG solution from the beginning till end of the shearing period. Comparison of the shear resistance (DR/DR$_{max}$) of the solutions revealed that, as expected, 600 wppm XG solution had the best shear stability among all the solutions and the least decline in DR over the shearing period (0.15 decline in DR/DR$_{max}$ value). Contrary to XG, 200 wppm HPAM and the binary HPAM/XG solutions suffered from poor shear resistance and a large decline in their DR efficiencies. These results are in agreement with the results reported by Sandoval and Soares (2016). It is interesting to note that the shear resistance of both binary solutions and 200 wppm HPAM solution was nearly identical.

Table 4 Parameters of Brostow and exponential decay models for mixed HPAM/XG solutions

Concentration, wppm	Exponential decay model		Brostow's model		
	λ, min	R^2	W	h, min^{-1}	R^2
200 (HPAM)	145	0.92	1.792	6.8e−3	0.99
600 (XG)	571	0.94	0.3853	8.8e−3	0.96
100 (HPAM) + 100 (XG)	137	0.92	44.49	2.2e−4	0.96
150 (HPAM) + 100 (XG)	127	0.89	1.768	8.4e−3	0.99

From the obtained friction factor results, it can be inferred that addition of XG reduced the DR capability of mixed solutions (DR% = 40% and 55%) with respect to 200 wppm HPAM solution. Also results suggested that increasing HPAM concentration in the mixed solutions does not change the DR efficiency decline significantly. Since there was only a small amount of chain degradation in semirigid polymers, the extent of DR depended only on the interaction between polymer chains and entanglements, which is directly related to polymer concentration. It should be considered that the concentration of XG was 100 wppm for mixed solutions, which is below the critical concentration of XG ($C^* \approx 300$ ppm). Hence, as suggested by other authors (Wyatt et al. 2011; Berman 1978), mixing HPAM with higher concentrations of XG would be more beneficial in maintaining high DR efficiency and even better control of shear degradation of polyacrylamide solutions, which is the subject of our future work. Same as for the HPAM solutions, Brostow and exponential decay models were used to correlate the drag reduction behavior over time for the HPAM/XG mixed solutions. The result is shown in Fig. 9 and Table 4. It can be found that both models give good fitting, but similar to HPAM solutions, the Brostow's model correlated the DR reduction data better.

4 Conclusions

The results from this work show that among different concentrations of HPAM, 1000 wppm solution has the highest drag reduction efficiency and the lowest decline in DR%. Increasing the concentration above 200 wppm does not change the DR% of the fresh samples significantly, though increasing concentration increases resistance to mechanical degradation. Results also indicate that mixing flexible HPAM with XG did not improve the degree of DR significantly, but slightly improved the DR stability of HPAM solutions. Since the drag reduction of rigid polymers depends only on concentration, mixing HPAM with higher concentrations of XG might be beneficial in maintaining the DR efficiency of HPAM as well as improving its shear stability. Brostow and exponential decay models were used to correlate the drag reduction decline data. Results indicate that both models can correlate the DR behavior of HPAM, as well as mixed HPAM/XG solutions. Brostow's model gave slightly better fits.

Acknowledgements The authors would like to thank the Research Partnership to Secure Energy for America (RPSEA) and Oklahoma State University Chemical Engineering Department for partial support of this project.

References

Barati R, Liang J-T. A review of fracturing fluid systems used for hydraulic fracturing of oil and gas wells. J Appl Polym Sci. 2014;131(16). doi:10.1002/app.40735.

Bello JB, Müller AJ, Sáez AE. Effect of intermolecular cross links on drag reduction by polymer solutions. Polym Bull. 1996;36(1):111–8. doi:10.1007/BF00296015.

Berman NS. Drag reduction by polymers. Annu Rev Fluid Mech. 1978;10(1):47–64.

Brostow W. Drag reduction and mechanical degradation in polymer solutions in flow. Polymer. 1983;24(5):631–8.

Bunger AP, McLennan J, Jeffrey R. Effective and sustainable hydraulic fracturing. InTech, Janeza Trdine. 2013;9:51000.

Campolo M, Simeoni M, Lapasin R, Soldati A. Turbulent drag reduction by biopolymers in large scale pipes. J Fluids Eng. 2015;137(4):041102.

Dingilian G, Ruckenstein E. Positive and negative deviations from additivity in drag reduction of binary dilute polymer solutions. AIChE J. 1974;20(6):1222–4.

Dschagarowa E, Bochossian T. Drag reduction in polymer mixtures. Rheol Acta. 1978;17(4):426–32.

Kalgaonkar RA, Patil PR. Performance enhancements in metal-crosslinked fracturing fluids. 1 January 2012. SPE: Society of Petroleum Engineers; 2012.

Kaufman PB, Penny GS, Paktinat J. Critical evaluation of additives used in shale slickwater fracs. 1 January 2008. SPE: Society of Petroleum Engineers; 2008.

Kenis PR. Turbulent flow friction reduction effectiveness and hydrodynamic degradation of polysaccharides and synthetic polymers. J Appl Polym Sci. 1971;15(3):607–18.

Kim CA, Choi HJ, Kim CB, Jhon MS. Drag reduction characteristics of polysaccharide xanthan gum. Macromol Rapid Commun. 1998;19(8):419–22. doi:10.1002/(SICI)1521-3927(19980801)19:8<419:AID-MARC419>3.0.CO;2-0.

Kulmatova D. Turbulent drag reduction by additives. Phys Rev Lett. 2013;84:4765–8.

Larch KL, Aminian K, Ameri S. The impact of multistage fracturing on the production performance of the horizontal wells in shale formations. 1 January 2012. SPE: Society of Petroleum Engineers; 2012.

Liberatore MW, Baik S, McHugh AJ, Hanratty TJ. Turbulent drag reduction of polyacrylamide solutions: effect of degradation on molecular weight distribution. J Nonnewton Fluid Mech. 2004;123(2):175–83.

Lim S, Choi H, Lee S, So J, Chan C. λ-DNA induced turbulent drag reduction and its characteristics. Macromolecules. 2003;36(14):5348–54.

Loveless D, Holtsclaw J, Weaver JD, Ogle JW, Saini RK. Multifunctional boronic acid crosslinker for fracturing fluids. 19 January 2014. IPTC: International Petroleum Technology Conference; 2014.

Malhotra J, Chaturvedi P, Singh R. Drag reduction by polymer–polymer mixtures. J Appl Polym Sci. 1988;36(4):837–58.

Merrill E, Horn A. Scission of macromolecules in dilute solution: extensional and turbulent flows. Polym Commun. 1984;25(5):144–6.

Moussa T, Tiu C. Factors affecting polymer degradation in turbulent pipe flow. Chem Eng Sci. 1994;49(10):1681–92.

Palisch TT, Vincent M, Handren PJ. Slickwater fracturing: food for thought. SPE Prod Oper. 2010;. doi:10.2118/115766-PA.

Pereira AS, Andrade RM, Soares EJ. Drag reduction induced by flexible and rigid molecules in a turbulent flow into a rotating cylindrical double gap device: comparison between Poly (ethylene oxide), Polyacrylamide, and Xanthan Gum. J Nonnewton Fluid Mech. 2013;202:72–87. doi:10.1016/j.jnnfm.2013.09.008.

Ptasinski P, Boersma B, Nieuwstadt F, Hulsen M, Van den Brule B, Hunt J. Turbulent channel flow near maximum drag reduction: simulations, experiments and mechanisms. J Fluid Mech. 2003;490:251–91.

Reddy G, Singh R. Drag reduction effectiveness and shear stability of polymer–polymer and polymer–fibre mixtures in recirculatory turbulent flow of water. Rheol Acta. 1985;24(3):296–311.

Rodriguez F, Winding C. Mechanical degradation of dilute poly-isobutylene solutions. Ind Eng Chem. 1959;51(10):1281–4. doi:10.1021/ie50598a034.

Sandoval GA, Soares EJ. Effect of combined polymers on the loss of efficiency caused by mechanical degradation in drag reducing flows through straight tubes. Rheol Acta. 2016;55(7):559–569.

Skelland A, Meng X. The critical concentration at which interaction between polymer molecules begins in dilute solutions. Polym Plast Technol Eng. 1996;35(6):935–45.

Soares EJ, Sandoval GA, Silveira L, Pereira AS, Trevelin R, Thomaz F. Loss of efficiency of polymeric drag reducers induced by high Reynolds number flows in tubes with imposed pressure. Phys Fluids (1994–present). 2015;27(12):125105.

Sohn J-I, Kim C, Choi H, Jhon M. Drag-reduction effectiveness of xanthan gum in a rotating disk apparatus. Carbohydr Polym. 2001;45(1):61–8.

Virk P, Baher H. The effect of polymer concentration on drag reduction. Chem Eng Sci. 1970;25(7):1183–9.

Virk PS. Drag reduction fundamentals. AIChE J. 1975;21(4):625–56. doi:10.1002/aic.690210402.

White CM, Mungal MG. Mechanics and prediction of turbulent drag reduction with polymer additives. Annu Rev Fluid Mech. 2008;40:235–56.

Wu Q, Sun Y, Zhang H, Ma Y, Bai B, Wei M. Experimental study of friction reducer flows in microfracture during slickwater fracturing. 8 April 2013. SPE: Society of Petroleum Engineers; 2013.

Wyatt NB, Gunther CM, Liberatore MW. Drag reduction effectiveness of dilute and entangled xanthan in turbulent pipe flow. J Nonnewton Fluid Mech. 2011;166(1–2):25–31. doi:10.1016/j.jnnfm.2010.10.002.

9

Carbon isotopic fractionation during vaporization of low molecular weight hydrocarbons (C6–C12)

Qian-Yong Liang[1,2] · Yong-Qiang Xiong[2] · Jing Zhao[1] · Chen-Chen Fang[3] ·
Yun Li[2]

Abstract Three series of laboratory vaporization experiments were conducted to investigate the carbon isotope fractionation of low molecular weight hydrocarbons (LMWHs) during their progressive vaporization. In addition to the analysis of a synthetic oil mixture, individual compounds were also studied either as pure single phases or mixed with soil. This allowed influences of mixing effects and diffusion though soil on the fractionation to be elucidated. The LMWHs volatilized in two broad behavior patterns that depended on their molecular weight and boiling point. Vaporization significantly enriched the ^{13}C present in the remaining components of the C_6–C_9 fraction, indicating that the vaporization is mainly kinetically controlled; the observed variations could be described with a Rayleigh fractionation model. In contrast, the heavier compounds (n-C_{10}–n-C_{12}) showed less mass loss and almost no significant isotopic fractionation during vaporization, indicating that the isotope characteristics remained sufficiently constant for these hydrocarbons to be used to identify the source of an oil sample, e.g., the specific oil field or the origin of a spill. Furthermore, comparative studies suggested that matrix effects should be considered when the carbon isotope ratios of hydrocarbons are applied in the field.

Keywords Low molecular weight hydrocarbons · Gas chromatography–isotope ratio mass spectrometry · Isotope fractionation · Vaporization

1 Introduction

The C_6–C_{12} low molecular weight hydrocarbons (LMWHs) are an important part of petroleum, consisting of different compound classes (n-, iso-, $cyclo$-alkanes, and aromatics). Various parameters based on the chemical and isotopic compositions of these LMWHs have been widely utilized to make oil/source correlations (Bjorøy et al. 1994; Ten Haven 1996; Odden et al. 1998; Whiticar and Snowdon 1999; Obermajer et al. 2000; Wever 2000), assess the thermal maturity of oils and condensates (Thompson 1983; Mango 2000), determine the source allocation of mixed oils (Chung et al. 1998; Rooney et al. 1998), and identify various secondary alterations of crude oils (George et al. 2002; Pasadakis et al. 2004; Zhang et al. 2005). The characterization of these light compounds is also a powerful tool for tracing the source of petroleum-related contaminants and understanding the environmental processes that control the transport and fate of these contaminants (Dempster et al. 1997; Kelley et al. 1997; Gray et al. 2002; Kolhatkar et al. 2002; Mancini et al. 2002, 2008; Smallwood et al. 2002; Zwank et al. 2003).

Liquid petroleum hydrocarbons, particularly LMWHs, lose mass through vaporization, which can occur in a wide variety of settings, including during the weathering of oil spills (i.e., before sampling), and during sampling, transportation, and storage. The different vaporization behavior

✉ Yong-Qiang Xiong
  xiongyq@gig.ac.cn

[1] Key Laboratory of Marine Mineral Resources, Guangzhou Marine Geological Survey, Ministry of Land and Resources, Guangzhou 510760, Guangdong, China

[2] State Key Laboratory of Organic Geochemistry, Guangzhou Institute of Geochemistry, Chinese Academy of Sciences, Guangzhou 510640, Guangdong, China

[3] PetroChina Research Institute of Petroleum Exploration and Development, Beijing 100083, China

Edited by Jie Hao

of each component of an oil sample will alter the sample's chemical and isotopic composition, thus likely influencing the application of some identification methods based on these properties (Cañipa-Morales et al. 2003). It is therefore essential to clarify the possible effects of vaporization on the composition of oil samples prior to the interpretation of the data.

Numerous studies have examined the effects of evaporation on the composition of LMWHs (Thompson 1987, 1988; Cañipa-Morales et al. 2003). Strict collection and preservation procedures are required to avoid the evaporation of a crude oil sample to facilitate the accurate determination of its LMWHs distribution, because even minor evaporation would affect the quality of the data (Cañipa-Morales et al. 2003). Although compound-specific isotope analysis (CSIA) has become a powerful tool for oil characterization and correlation (Chung et al. 1998; Odden et al. 1998; Harris et al. 1999; Whiticar and Snowdon 1999), few studies have considered the effect of evaporation on the $\delta^{13}C$ values of LMWHs (Harrington et al. 1999; Shin and Lee 2010).

In addition, different volatile organic compounds have shown different carbon isotope fractionation trends during evaporation. For example, the enrichment of ^{13}C in the vapor fraction was reported for the evaporation of benzene, toluene, ethylbenzene, and xylene (collectively known as BTEX) ($\Delta^{13}C_{vapor-liquid} \approx +0.2‰$) (Harrington et al. 1999), trichloroethylene ($\Delta^{13}C_{vapor-liquid} = +0.1‰$ to $+0.7‰$) (Poulson and Drever 1999), chlorinated aliphatic hydrocarbons ($\Delta^{13}C_{vapor-liquid} = +0.31‰$ for trichloroethene and $\Delta^{13}C_{vapor-liquid} = +0.65‰$ for dichloromethane) (Huang et al. 1999), and MTBE (tert-butyl methyl ether, $\Delta^{13}C_{vapor-liquid} = +0.2-0.5‰$ in different physical contexts) (Kuder et al. 2009). The enrichment of ^{13}C in the vapor phase could be explained by higher vapor pressure of ^{13}C-substituted organic compounds relative to ^{12}C-substituted organic compounds (Baertschi et al. 1953; Narten and Kuhn 1961; Jancso and Van Hook 1974). In contrast, some evaporation experiments have shown that progressive evaporation considerably enriches the remaining liquid fraction in ^{13}C, with $\Delta^{13}C_{vapor-liquid} = -0.58‰$ and $-0.41‰$ for benzene and toluene, respectively (Shin and Lee 2010). Kinetic fractionation was evidently dominant in controlling the carbon isotopic fractionation during these evaporation experiments.

Previous experimental studies have investigated the evaporation of LMWHs mainly by simulating the evaporation of one or two pure components in each experiment and determining their composition and isotope fractionation at different stages during the evaporation (Huang et al. 1999; Poulson and Drever 1999; Shin and Lee 2010). The effect of the matrix in which the vaporization of a hydrocarbon is studied has seldom been discussed in these experimental studies. However, crude oils and oil products generally comprise a complicated mixture of hydrocarbons, and the matrix effect of the other components may to some extent influence the evaporation behavior of each individual compound.

Additionally, evaporation in natural environments commonly involves other media such as water and soils. A few studies have explored the effects of mixture with water and adsorption to soil on carbon isotope fractionation (Harrington et al. 1999; Slater et al. 1999; Höhener et al. 2003; Schüth et al. 2003; Bouchard et al. 2008a, b). No significant carbon isotope fractionation was observed during the equilibrium vaporization of aqueous solution of toluene and trichloroethylene (Slater et al. 1999), the soil adsorption of BTEX (Harrington et al. 1999) and the sorption of halogenated hydrocarbon compounds (trichloroethene, cis-dichloroethene, vinylchloride) and BTEX compounds onto activated carbon, lignite coke, and lignite (Schüth et al. 2003). However, significant fractionation has been observed after passing some volatile organic compounds across alluvial sand (e.g., $\Delta^{13}C_{vapor-liquid}$ is $-2.14 \pm 0.22‰$, $-1.73 \pm 0.52‰$, and $-1.55 \pm 0.45‰$ for n-pentane, n-hexane, and benzene, respectively) (Bouchard et al. 2008b), and an unsaturated soil zone (Bouchard et al. 2008a). Therefore, the matrix effects (of both mixing and soil diffusion) on the evaporation fractionation should be better understood prior to utilizing the $\delta^{13}C$ value of volatile organic compounds.

The main purpose of this research is to gain better insights into carbon isotope fractionation during the vaporization of C_6-C_{12} LMWHs and to determine the influences of multi-component mixtures and soil diffusion on the vaporization of these hydrocarbons. Three series of laboratory vaporization experiments were conducted at room temperature. The first investigated the vaporization of a mixture of C_6-C_{12} LMWHs by assessing their vaporization characteristics in a mixture compounds. The second series compared the vaporization of individual pure single-phase compounds. The third series examined vaporization in soil.

2 Experimental

2.1 Reagents and chemicals

n-Hexane (n-C_6, 99%), n-heptane (n-C_7, HPLC grade, 99+%), n-octane (n-C_8, 98+%), n-nonane (n-C_9, 99%), n-decane (n-C_{10}, 99%), n-undecane (n-C_{11}, 99%), n-dodecane (n-C_{12}, 99+%), benzene (99%), ethylbenzene (99%), o-xylene (99%), methylcyclohexane (MCH, 99%), and deuterated n-octane (n-C_8D_{18}, 99%) were purchased from Alfar Aesar China (Tianjin) Co., Ltd. n-Pentane (n-

C_5, 99%) and toluene (99%) were purchased from Qianhui Chemicals and Glassware Co. Ltd. (Guangzhou, China).

The soil used here was first freeze-dried, pulverized with an agate mortar, and then sieved. The fraction with particle sizes less than 100 mesh was heated at 250 °C in an oven for 4 h to eliminate any naturally present LMWHs and microbes. The main minerals in the soil were quartz, chlorite, feldspar, illite, calcite, and dolomite. The total organic carbon of the soil was 0.12%.

2.2 Vaporization experiments

Three series of vaporization experiments were designed in which three pure compounds (n-hexane, n-nonane, and n-dodecane) and a mixture of compounds were progressively volatilized in the laboratory. The mixture was of twelve LMWHs, including C_6–C_{12} n-alkanes, MCH, and BTEX. The mixture was prepared by adding the 12 pure compounds (n-C_6, 700 μL; benzene, 700 μL; n-C_7, 400 μL; MCH, 400 μL; toluene, 400 μL; n-C_8, 300 μL; ethylbenzene, 300 μL; o-xylene, 300 μL; n-C_9, 200 μL; n-C_{10}, 200 μL; n-C_{11}, 200 μL; and n-C_{12}, 200 μL) to a 4-mL glass vial capped with an aluminum–rubber seal.

In the first vaporization experiment, aliquots (approximately 200 μL) of the mixture of compounds were delivered into a series of 4-mL glass vials and weighed. Each vial was then placed in a fume cupboard to allow open vaporization without any agitation, and an air conditioner was used to control the room temperature at 24 ± 1 °C. At intervals up to 72 h, vaporization was measured. For the GC measurement, the vials were then weighed and filled with n-pentane. They were tightly capped with aluminum–rubber seals, shaken in an ultrasonicator for 10 min to increase the dissolution of the residue hydrocarbons into the n-pentane solvent, and then kept in a freezer prior to analysis.

The second series of vaporization experiments was conducted similar to the first, although instead of the artificial oil mixture, pure compounds were individually studied (n-hexane, n-nonane, or n-dodecane). Aliquots (about 100 μL) of each individual pure compound were added to a series of 4-mL glass vials, and the following procedures for the vaporization experiment were identical to those of the first series.

The third series of vaporization experiments was designed to reveal the effect of diffusion through soil on the vaporization of LMWHs. A certain amount of soil (1 or 2 g) was added to 4-mL glass vials containing a pure hydrocarbon (n-hexane or n-nonane). Subsequent procedures were as in the preceding series. Finally, the residual compounds were ultrasonically extracted into pentane solvent from the soils.

After vaporization, the concentrations and $\delta^{13}C$ values of the target compounds in the unaltered original sample and the evaporated residual aliquots were measured by directly injecting n-pentane solutions containing the target compounds into gas chromatography (GC) and gas chromatography–isotope ratio mass spectrometry (GC–IRMS) apparatus. The remaining fraction (F) of each component was calculated by measuring the weight (pure compound) or concentration (artificial oil) of the corresponding compound before and after vaporization.

2.3 Gas chromatography (GC)

GC analyses were performed on an Agilent 7890 GC instrument equipped with a split/splitless injector, an HP-PONA fused silica capillary column (50 m × 0.20 mm i.d. × 0.50 μm), and a flame ionization detector. The temperatures for both injection and detection were set at 300 °C. Nitrogen (99.999%) was used as the carrier gas at a maintained constant flow rate of 1.0 mL/min. The injection was operated in split mode (10:1). The GC oven temperature was programmed to rise for 5 min from 35 to 50 °C at a rate of 4 °C/min, and then to 180 °C at 8 °C/min. C_6–C_{12} LMWHs were quantified by integration of the peak areas. The response factors of these hydrocarbons relative to the internal standard (n-C_8D_{18}) were calculated based on the peak area ratios of each C_6–C_{12} hydrocarbon compared with the internal standard.

2.4 Gas chromatography–isotope ratio mass spectrometry (GC–IRMS)

Carbon isotopic compositions of LMWHs were measured with a gas chromatograph (Agilent 6890) equipped with a DB-5MS column (50 m × 0.25 mm i.d. × 0.25 μm) coupled to an isotope ratio monitoring mass spectrometer (GV IsoPrime). Helium was used as the carrier gas with a maintained constant head pressure of 8.5 psi. The GC oven temperature was programmed to be initially held at 35 °C for 5 min, then raised to 50 °C at 1.5 °C/min, held for 3 min, increased to 53 °C at 0.5 °C/min, and finally increased to 200 °C at 25 °C/min and held for 2 min. The combustion by-product (H_2O) was removed by passing the analyte stream through a selectively permeable membrane (Nafion^{TM}) with a dry He counter flow. Carbon isotope ratios were computed by five pulses of CO_2 reference gas with known $\delta^{13}C$ values (−22.5‰, VPDB), which were injected via the interface to the IRMS instrument at the beginning and end of each analysis. A standard mixture of n-alkanes (C_{12}–C_{32}) from Indiana University with known isotopic composition was used daily to monitor the performance of the instrument. The reported isotopic data represent the arithmetic means of at least two replicate analyses, and the repeatability is better than ±0.3‰ (VPDB).

2.5 Quantification of isotope fractionation during vaporization and diffusion across soil

The experimental isotope factors can be determined using the following Rayleigh equations:

$$R = R_o \cdot F^{\alpha-1} \qquad (1)$$

$$\ln A = (\alpha-1) \cdot \ln F \qquad (2)$$

$$A = (\delta^{13}C_R + 1000)/(\delta^{13}C_I + 1000) \qquad (3)$$

where F is the mass fraction of the original compound remaining, R and R_o are the $^{13}C/^{12}C$ value of the individual compound at a specific F ($F < 1$) and at $F = 1$, respectively, and α (equal to R_{vapor}/R_{liquid}) is the vapor–liquid fractionation factor. $\delta^{13}C_R$ and $\delta^{13}C_I$ are the $\delta^{13}C$ values of the residual and initial compound, respectively. For each compound of the artificial oil, the α values were calculated by linear regression of $\ln F$ versus $\ln A$.

The corresponding isotope enrichment factors can be calculated according to:

$$\varepsilon(‰) = (\alpha-1) \cdot 1000 \qquad (4)$$

The uncertainty was characterized using the standard uncertainty of the slope obtained by linear regression using the least-squares method, which was performed by the data analysis tool installed in Microsoft Excel.

The theoretical fractionation factor between the diffusion coefficients is given by the following equation (Craig 1953; Cerling et al. 1991; Bouchard et al. 2008c; Jeannottat and Hunkeler 2012):

$$\alpha_t = \frac{D_h}{D_l} = \sqrt{\frac{(MW_h + MW_a) \cdot MW_l}{(MW_l + MW_a) \cdot MW_h}} \qquad (5)$$

where D represents the diffusion coefficient, MW the molecular weight, and the subscripts l, h, and a represent light isotopes only, molecules with one heavy isotope, and air, respectively ($MW_a = 28.8$ g/mol in this case).

3 Results and discussion

An evaporation system with constant boundary conditions will usually have a constant evaporation rate for a single liquid (one component) with respect to time (Stiver and Mackay 1984). In contrast to the linear evaporation of a pure compound, the evaporative loss of a mixture by total weight or volume is either logarithmic (approximately seven or more components) or a square root function (between about five and seven components) with time (Fingas

1997). This implies that the evaporation behavior of a given component is probably different between its pure state (single-component liquid) and when it is in a mixture due to the occurrence of intermolecular interactions. Therefore, the carbon isotope fractionations of LMWHs during evaporation from a mixture and from soil were investigated to explore their evaporation behavior under conditions resembling practical situations.

3.1 Carbon isotope fractionations of LMWHs during the progressive vaporization of artificial oil

To eliminate the possible influence of co-elution and other factors on the measurement of the LMWHs, a mixture consisting of twelve pure standards (n-C_6, benzene, n-C_7, MCH, toluene, n-C_8, ethylbenzene, o-xylene, n-C_9, n-C_{10}, n-C_{11}, and n-C_{12}) was selected here to replace a real oil.

The progression of the remaining mass fractions (F) and carbon isotope compositions of the individual C_6–C_{12} LMWHs in the volatilized residues of the mixture are summarized in Table 1. The mass losses are being observed against time (h). n-C_6, benzene, n-C_7, MCH, toluene, n-C_8, ethylbenzene, and o-xylene in the mixture showed expected mass losses, with the lighter compounds being the most volatile, and these compounds in the residual liquid were enriched in ^{13}C. However, the heaviest compounds (n-decane, n-undecane, and n-dodecane) after 72 h showed mass losses of 66%, 16%, and 1% and $\delta^{13}C_{R-I}$ values of 0.6, 0.4, and 0.2‰ (within the CSIA error), respectively.

Based on their vaporization rates and variations in $\delta^{13}C$ values, the considered LMWHs fall into two categories: a lighter C_6–C_9 fraction and a heavier C_{10}–C_{12} fraction. The lighter fraction volatilized more quickly and showed considerable carbon isotope fractionation. Plots of F versus vaporization time (Fig. 1a) were used to evaluate the vaporization rates of individual LMWHs: The steeper the curve, the faster the vaporization rate. The plots show that the vaporization rates of the LMWHs were inversely related to their boiling points or carbon number and that the individual components of the lighter fraction rapidly evaporated. The residues of individual LMWHs in the lighter fraction became gradually enriched in ^{13}C during vaporization (Table 1; Fig. 2). Their $\delta^{13}C_{R-I}$ values reached up to 0.5‰ (beyond the analytical error of CSIA) once 20%–60% of the compounds were removed, and over 3‰ after the evaporation of about 90% of the component.

The heavier fraction evaporated more slowly (Fig. 1a). After 72-h vaporization, the amounts of n-C_{10}, n-C_{11}, and n-C_{12} remaining in the mixture compounds sample were 34%, 84%, and 99% of the original, respectively (Table 1; Fig. 2). These compounds showed relatively less isotope

Table 1 Remaining fraction and the corresponding carbon isotopic composition of individual LMWHs in the residual mixture compounds during progressive vaporization (24 ± 1 °C)

T^a	F^b	$\delta^{13}C \pm SD$ (‰)^c	$\delta^{13}C_{R-I}$ (‰)^d	T^a	F^b	$\delta^{13}C \pm SD$ (‰)^c	$\delta^{13}C_{R-I}$ (‰)^d	T^a	F^b	$\delta^{13}C \pm SD$ (‰)^c	$\delta^{13}C_{R-I}$ (‰)^d
$n\text{-}C_6$				12	0.19	−26.9 ± 0.16	2.4	12	0.87	−49.4 ± 0.07	0.0
0	1	−46.1 ± 0.02	0.0	15	0.06	−24.4 ± 0.05	4.8	15	0.75	−49.0 ± 0.12	0.4
1	0.81	−45.6 ± 0.03	0.5	$n\text{-}C_8$				20	0.66	−49.4 ± 0.00	0.0
2	0.45	−44.9 ± 0.21	1.2	0	1	−45.9 ± 0.15	0.0	24	0.56	−49.3 ± 0.05	0.2
4	0.17	−44.0 ± 0.08	2.1	2	0.96	−45.9 ± 0.10	0.1	33	0.33	−48.4 ± 0.24	1.0
6	0.03	−42.1 ± 0.00	4.0	4	0.92	−45.8 ± 0.13	0.1	48	0.13	−47.3 ± 0.28	2.2
Ben				6	0.85	−45.7 ± 0.11	0.3	64	0.04	−44.5 ± 0.06	4.9
0	1	−24.6 ± 0.17	0.0	12	0.54	−45.6 ± 0.01	0.3	$n\text{-}C_{10}$			
1	0.89	−24.1 ± 0.22	0.5	15	0.34	−44.5 ± 0.25	1.4	0	1	−35.6 ± 0.10	0.0
2	0.55	−23.6 ± 0.03	0.9	20	0.18	−44.2 ± 0.00	1.7	2	0.96	−35.8 ± 0.11	−0.2
4	0.29	−22.3 ± 0.03	2.3	24	0.12	−43.6 ± 0.06	2.3	6	0.97	−35.4 ± 0.10	0.1
6	0.10	−20.6 ± 0.17	4.0	33	0.02	−39.4 ± 0.00	6.5	12	0.94	−35.7 ± 0.00	−0.1
$n\text{-}C_7$				EB				24	0.83	−35.5 ± 0.10	0.0
0	1	−39.4 ± 0.12	0.0	0	1	−28.3 ± 0.18	0.0	33	0.76	−35.3 ± 0.05	0.3
1	0.97	−39.3 ± 0.10	0.1	2	0.98	−28.3 ± 0.09	0.0	48	0.57	−35.0 ± 0.00	0.6
2	0.82	−39.3 ± 0.07	0.2	4	0.95	−28.3 ± 0.05	0.0	64	0.40	−35.0 ± 0.03	0.6
4	0.66	−39.0 ± 0.04	0.5	6	0.90	−28.1 ± 0.03	0.2	72	0.34	−34.9 ± 0.05	0.6
6	0.46	−38.2 ± 0.07	1.2	12	0.61	−28.1 ± 0.06	0.2	$n\text{-}C_{11}$			
9	0.21	−37.4 ± 0.24	2.0	15	0.41	−27.0 ± 0.25	1.3	0	1	−27.8 ± 0.28	0.0
12	0.10	−36.6 ± 0.04	2.8	20	0.22	−27.2 ± 0.05	1.1	2	0.97	−27.6 ± 0.07	0.2
15	0.02	−33.5 ± 0.13	5.9	24	0.16	−26.0 ± 0.05	2.3	6	0.96	−27.6 ± 0.04	0.2
MCH				33	0.03	−25.1 ± 0.15	3.2	12	0.97	−27.7 ± 0.03	0.2
0	1	−29.3 ± 0.07	0.0	o-Xy				24	0.97	−27.5 ± 0.03	0.3
1	0.97	−29.2 ± 0.07	0.0	0	1	−28.0 ± 0.13	0.0	33	0.94	−27.7 ± 0.00	0.2
2	0.81	−29.2 ± 0.03	0.0	2	0.99	−28.1 ± 0.08	0.0	48	0.89	−27.6 ± 0.00	0.3
4	0.66	−29.0 ± 0.09	0.3	4	0.97	−28.0 ± 0.16	0.0	64	0.86	−27.5 ± 0.01	0.3
6	0.45	−28.6 ± 0.07	0.7	6	0.94	−28.1 ± 0.00	−0.1	72	0.84	−27.5 ± 0.03	0.4
9	0.19	−27.9 ± 0.01	1.4	12	0.72	−28.0 ± 0.08	0.0	$n\text{-}C_{12}$			
12	0.10	−27.3 ± 0.07	2.0	15	0.55	−27.5 ± 0.21	0.6	0	1	−31.9 ± 0.07	0.0
15	0.02	−25.6 ± 0.05	3.7	20	0.37	−27.4 ± 0.03	0.6	2	0.98	−31.9 ± 0.05	0.0
Tol				24	0.29	−27.0 ± 0.03	1.1	6	0.96	−31.7 ± 0.01	0.1
0	1	−29.3 ± 0.25	0.0	33	0.08	−24.9 ± 0.00	3.1	12	0.97	−31.8 ± 0.03	0.0
1	0.99	−29.1 ± 0.23	0.1	$n\text{-}C_9$				24	0.99	−31.7 ± 0.03	0.1
2	0.88	−28.9 ± 0.09	0.3	0	1	−49.4 ± 0.04	0	33	1.00	−31.8 ± 0.00	0.0
4	0.77	−28.8 ± 0.12	0.5	2	1.00	−49.5 ± 0.19	−0.1	48	0.97	−31.9 ± 0.00	0.0
6	0.59	−28.2 ± 0.14	1.1	4	0.98	−49.3 ± 0.21	0.1	64	0.99	−31.7 ± 0.03	0.2
9	0.36	−27.4 ± 0.18	1.8	6	0.94	−49.3 ± 0.06	0.2	72	0.99	−31.7 ± 0.03	0.2

^{a} t, time of vaporization (h)

^{b} F, fraction of original compound remaining

^{c} $\delta^{13}C \pm SD$, average ± standard deviation of carbon isotopic composition obtained by three parallel measurements

^{d} $\delta^{13}C_{R-I}$, carbon isotope difference between the residual and initial composition

fractionations over the entire process of vaporization, particularly $n\text{-}C_{11}$ and $n\text{-}C_{12}$, which showed variations of $\delta^{13}C$ of less than 0.6‰ (Table 1). A similar result was reported in a previous study (Wang and Huang 2003): the $\delta^{13}C$ values of residual C_{10}, C_{11}, C_{12}, C_{13}, and C_{14} n-alkanes changed by less than ±0.3‰ when 45%, 29%,

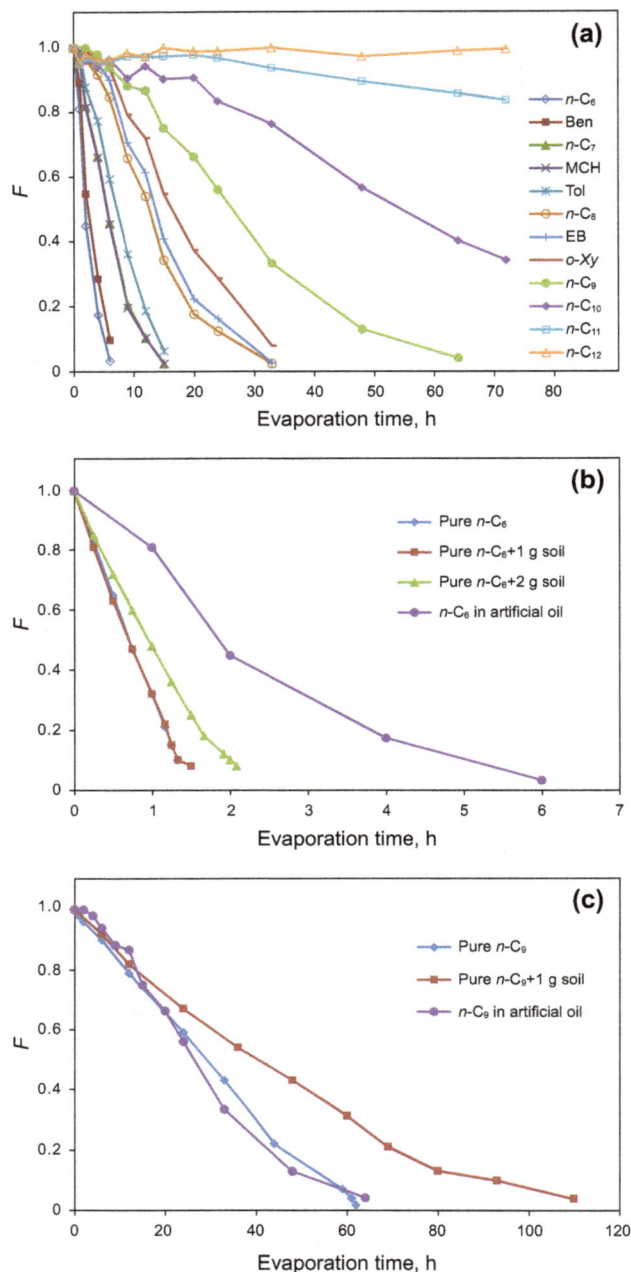

Fig. 1 Variation of the remaining mass fraction (F) as a function of vaporization time. **a** The artificial oil; **b** *n*-hexane; **c** *n*-nonane

30%, 37%, and 51% of the starting compounds remained in the vial, respectively. Therefore, the $\delta^{13}C$ values of the heavier *n*-alkanes varied little enough to make them useful identifiers of oil that has been evaporated to some extent.

Figure 3 shows the good linear correlation between $\ln[(\delta^{13}C_R + 1000)/(\delta^{13}C_I + 1000)]$ and $\ln F$ for the compounds of the lighter fraction, whose regression coefficients (R^2) were 0.98 (*n*-C_6), 0.99 (*n*-C_7), 0.97 (*n*-C_8), 0.91 (*n*-C_9), 0.996 (MCH), 0.97 (benzene), 0.97 (toluene), 0.91

Fig. 2 Carbon isotope fractionation of individual LMWHs in the ▶ residual artificial oil ($\delta^{13}C_{R-I}$) versus remaining fraction (F) of the corresponding components during progressive vaporization

(ethylbenzene), and 0.93 (*o*-xylene). These results suggest that the carbon isotope fractionations of these compounds during vaporization from a multi-component system (the artificial oil) followed the Rayleigh fractionation model.

The vapor–liquid carbon isotope enrichment factor (ε), also sometimes noted by $\Delta^{13}C_{vapor-liquid}$, is considered the best way to express the isotope fractionation effect (Hayes, 1993). All the ε values observed here were negative, ranging from -0.87 to $-1.74‰$ ($\varepsilon = $ slope $\times$ 1000, Fig. 3), indicating that the progressive vaporization of these compounds was dominated by kinetic fractionation, i.e., the preferential removal of molecules containing the lighter isotope. The same trend was observed by Shin and Lee (2010), who reported enrichment factors for benzene and toluene of -0.58 and $-0.41‰$, respectively. The magnitude of carbon isotope fractionation during the vaporization of a pure liquid phase appears to be considerably less than that from a multi-component system.

3.2 Carbon isotope fractionations of LMWHs during the progressive vaporization of single pure liquid and diffusion through soil

To understand better matrix effects on the carbon isotope fractionation of individual LMWHs during vaporization from a mixture, vaporization experiments of three pure compounds (*n*-hexane, *n*-nonane, and *n*-docecane) were conducted under the same conditions as the assessment of the artificial oil. Only 7.4% mass loss and no obvious $\delta^{13}C$ variation (<0.5‰, VPDB) were observed for *n*-dodecane after 72 h of vaporization. Consequently, only *n*-hexane and *n*-nonane are discussed.

Table 2 lists the progressive vaporization results for *n*-hexane, both its pure single phase and when in 1 g soil and 2 g soil. The pure single phase lost mass approximately as quickly as when it was mixed with 1 g soil: Both showed mass losses of about 90% after 80 min, with the $\delta^{13}C$ of the residue shifted more than 2.0‰ in both cases. *n*-Hexane in 2 g soil lost about 90% of its mass after 120 min of vaporization, and the $\delta^{13}C$ of the residue shifted 3.3‰. Its maximal $\delta^{13}C$ shift was 3.5‰ after 125 min with a corresponding mass loss of 92%, which represents a much slower mass loss than observed for the pure liquid and the *n*-hexane in 1 g soil. Increasing the mass of soil slowed the vaporization rate and lessened the change of $\delta^{13}C$ of the residue.

Figure 1b shows that *n*-C_6 added to 1 g soil volatilized at a similar rate to the pure liquid phase, indicating that

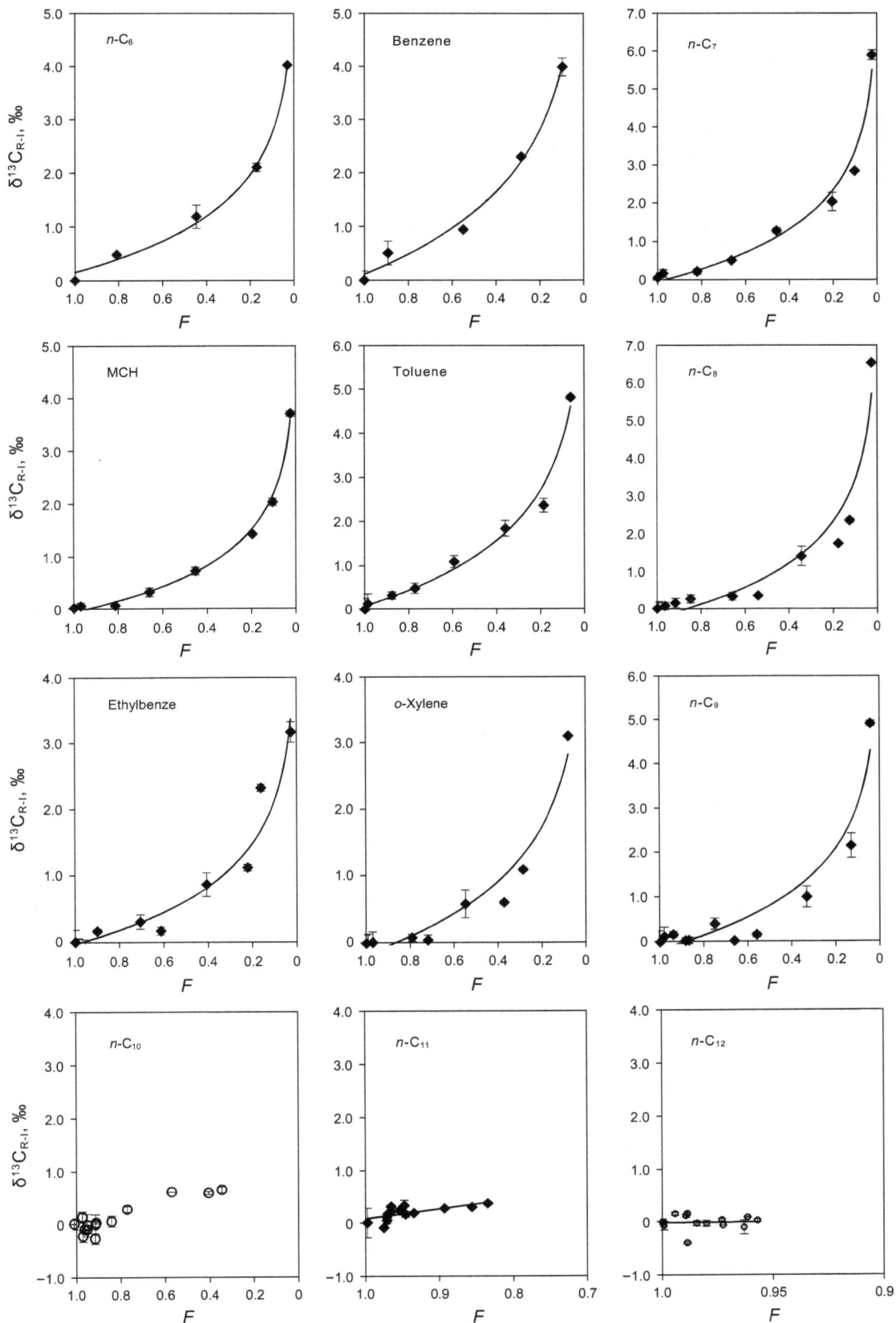

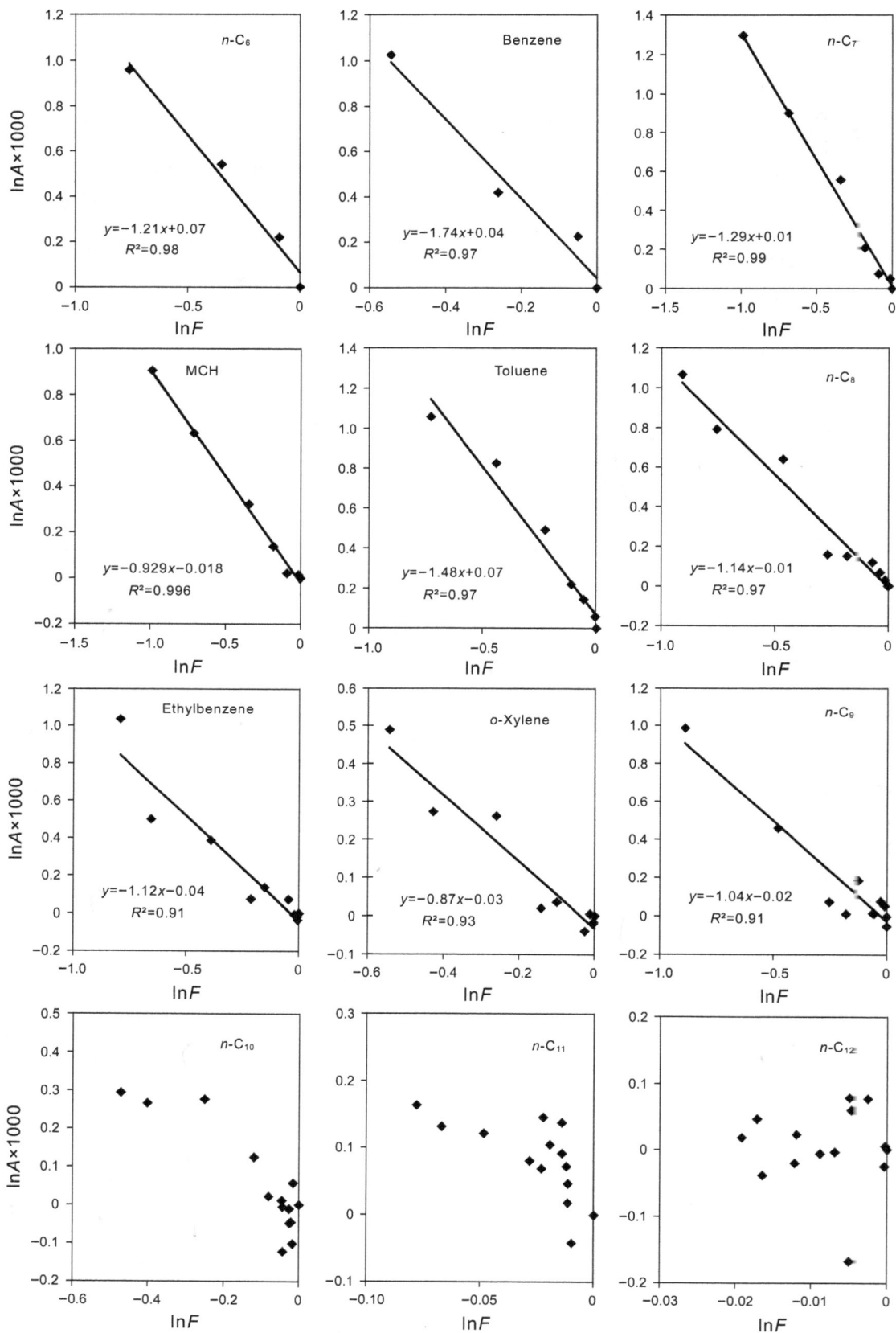

n-C$_6$

$y = -1.21x + 0.07$
$R^2 = 0.98$

Benzene

$y = -1.74x + 0.04$
$R^2 = 0.97$

n-C$_7$

$y = -1.29x + 0.01$
$R^2 = 0.99$

MCH

$y = -0.929x - 0.018$
$R^2 = 0.996$

Toluene

$y = -1.48x + 0.07$
$R^2 = 0.97$

n-C$_8$

$y = -1.14x - 0.01$
$R^2 = 0.97$

Ethylbenzene

$y = -1.12x - 0.04$
$R^2 = 0.91$

o-Xylene

$y = -0.87x - 0.03$
$R^2 = 0.93$

n-C$_9$

$y = -1.04x - 0.02$
$R^2 = 0.91$

n-C$_{10}$

n-C$_{11}$

n-C$_{12}$

◀**Fig. 3** Carbon isotope fractionation of individual compounds, as ln$A$, versus fraction of residual liquid, presented as ln$F$, during the progressive vaporization of artificial oil

diffusion through the 1 g soil did not have a remarkable effect on the vaporization of the n-C_6 owing to its relatively high volatility. Both n-C_9 in 1 g soil and n-C_6 in 2 g soil evaporated less quickly than their respective pure compounds. The results show that a LMWH's diffusion through

soil can slow its rate of vaporization, with the effect depending on the volatility of the compound.

Table 3 lists the progressive vaporization results of n-nonane, both its pure single phase and in 1 g soil. The compound evaporated more slowly in the soil than alone, but it showed a greater $\delta^{13}C$ shift in the soil (Fig. 1c).

The strong correlations between ln$A$ and ln$F$ for n-C_6 ($R^2 = 0.998$, 0.99, and 0.98 for the pure single phase, in 1 g soil, and in 2 g soil, respectively) and n-C_9 ($R^2 = 0.99$

Table 2 Carbon isotopic composition of residual n-hexane evaporated as a free phase and from soil (24 ± 1 °C)

Pure n-C_6				Pure n-C_6 + 1 g soil				Pure n-C_6 + 2 g soil			
t^a	F^b	$\delta^{13}C$ ± SD (VPDB, ‰)^c	$\delta^{13}C_{R-I}$ (‰)^d	t^a	F^b	$\delta^{13}C$ ± SD (VPDB, ‰)^c	$\delta^{13}C_{R-I}$ (‰)^d	t^a	F^b	$\delta^{13}C$ ± SD (VPDB, ‰)^c	$\delta^{13}C_{R-I}$ (‰)^d
0	1	−46.2 ± 0.08	0	0	1	−46.2 ± 0.05	0	0	1	−46.1 ± 0.10	0
15	0.83	−46.0 ± 0.04	0.2	15	0.81	−45.9 ± 0.12	0.2	15	0.85	−45.9 ± 0.03	0.2
30	0.65	−45.8 ± 0.08	0.5	30	0.63	−45.8 ± 0.07	0.3	30	0.72	−43.9 ± 0.01	0.3
45	0.47	−45.6 ± 0.05	0.7	45	0.47	−45.6 ± 0.04	0.6	45	0.60	−45.6 ± 0.04	0.5
60	0.32	−45.1 ± 0.05	1.1	60	0.32	−45.3 ± 0.11	0.9	60	0.48	−45.5 ± 0.06	0.6
70	0.21	−44.8 ± 0.09	1.4	70	0.22	−45.0 ± 0.02	1.2	75	0.36	−45.1 ± 0.05	1.1
75	0.15	−44.5 ± 0.12	1.8	75	0.15	−44.5 ± 0.06	1.7	90	0.25	−44.6 ± 0.05	1.6
80	0.10	−44.1 ± 0.08	2.1	80	0.10	−44.2 ± 0.11	2.0	100	0.18	−44.1 ± 0.01	2.1
				85	0.08	−44.0 ± 0.05	2.2	115	0.12	−43.2 ± 0.08	2.9
								120	0.10	−42.8 ± 0.09	3.3
								125	0.08	−42.7 ± 0.07	3.5

^{a} t, time of vaporization (min)

^{b} F, fraction of original compound remaining

^{c} $\delta^{13}C$ ± SD, average ± standard deviation of carbon isotopic composition obtained by three parallel measurements

^{d} $\delta^{13}C_{R-I}$, carbon isotope difference between the residual and initial composition

Table 3 Carbon isotopic composition of residual n-nonane evaporated as a free phase and from soil

Pure n-C_9				Pure n-C_9 + 1 g soil			
t^a	F^b	$\delta^{13}C$ ± SD (VPDB, ‰)^c	$\delta^{13}C_{R-I}$ (‰)^d	t^a	F^b	$\delta^{13}C$ ± SD (VPDB, ‰)^c	$\delta^{13}C_{R-I}$ (‰)^d
0	1	−49.4 ± 0.08	0	0	1	−49.6 ± 0.06	0
1	0.98	−49.5 ± 0.04	−0.1	6	0.92	−49.5 ± 0.04	0.1
2	0.96	−49.4 ± 0.05	0.0	12	0.82	−49.4 ± 0.09	0.2
6	0.9	−49.5 ± 0.11	0.0	24	0.67	−49.3 ± 0.05	0.2
12	0.79	−49.4 ± 0.06	0.1	36	0.54	−49.2 ± 0.06	0.4
24	0.59	−49.1 ± 0.06	0.3	48	0.43	−49.0 ± 0.09	0.5
33	0.43	−48.8 ± 0.04	0.6	60	0.31	−48.7 ± 0.05	0.9
44	0.22	−48.6 ± 0.04	0.9	69	0.21	−48.5 ± 0.02	1.0
59	0.07	−48.0 ± 0.08	1.4	80	0.13	−48.1 ± 0.09	1.5
61	0.039	−47.6 ± 0.03	1.8	93	0.098	−49.6 ± 0.15	1.8
62	0.017	−47.1 ± 0.03	2.3	110	0.038	−47.2 ± 0.19	2.4

^{a} t, time of vaporization (h)

^{b} F, fraction of original compound remaining

^{c} $\delta^{13}C$ ± SD, average ± standard deviation of carbon isotopic composition obtained by three parallel measurements

^{d} $\delta^{13}C_{R-I}$, carbon isotope difference between the residual and initial composition

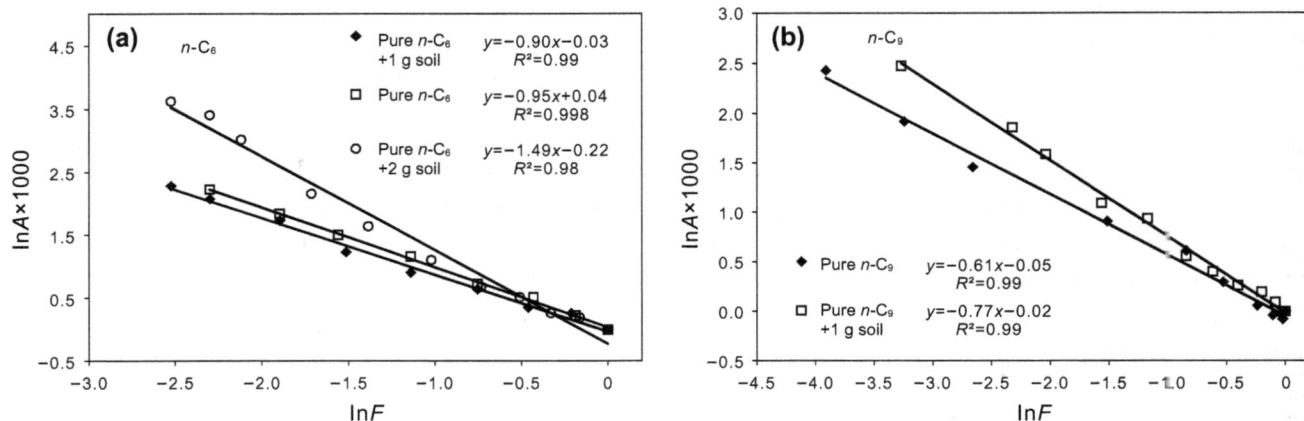

Fig. 4 Carbon isotope fractionation of vaporization for individual n-C_6 (**a**) and n-C_9 (**b**), as ln$A$, versus fraction of residual liquid, presented as ln$F$, during progressive vaporization

and 0.99 for the pure single and in 1 g soil, respectively) (Fig. 4) indicate that the vaporization in each case followed the Rayleigh trend.

3.3 Possible mechanism of carbon isotope fractionation of LMWHs during vaporization

Theoretically, the effects of equilibrium vapor pressure (evaporation-controlled) and kinetics (diffusion-controlled) are the two main factors that influence the fractionation of stable isotopes of organic compounds during vaporization, and the competition between them directly determine the direction of the fractionation. The evaporation-controlled process usually results in "inverse isotope fractionation," characterizing of enriching ^{13}C in the vapor phase (Baertschi et al. 1953; Balabane and Letolle 1985; Huang et al. 1999; Poulson and Drever 1999; Wang and Huang 2001; Jeannottat and Hunkeler 2012; Xiao et al. 2012), whereas diffusion-controlled vaporization, which depends on the system itself and intermolecular free energy due to the van der Waals attractive forces among molecules, results in the "normal isotope fractionation," characterizing of enriching ^{13}C in the residual liquids (Shin and Lee 2010; Xiao et al. 2012; Kuder et al. 2009; Bouchard et al. 2008a, b, c; Jeannottat and Hunkeler 2012; Hayes 1993; Wang and Huang 2001).

Table 4 lists the carbon isotope enrichment factors of the LMWHs considered here and in previous studies, along with values calculated using Eq. (5) (Craig 1953; Cerling et al. 1991; Bouchard et al. 2008c; Jeannottat and Hunkeler 2012). All the experimental values of ε (ε_e) for the C_6–C_9 LMWHs are negative, indicating the enrichment of ^{13}C in the residual liquids and "normal isotope fractionation" during vaporization. The vaporization of these compounds is thus diffusion-controlled, and the equilibrium vapor pressure has little effect on their natural vaporization.

The n-alkanes in the mixture compounds showed decreasing experimental values of ε with their increasing carbon number (Table 4): Values of -1.21 ± 0.06, -1.29 ± 0.14, -1.14 ± 0.23, and -1.04 ± 0.17‰ were observed with 95% confidence limits for n-C_5, n-C_7, n-C_8, and n-C_9, respectively. Benzene, toluene, ethylbenzene, and o-xylene, respectively, showed values of -1.74 ± 0.08, -1.48 ± 0.10, -1.12 ± 0.12, and -0.87 ± 0.09‰ with 95% confidence limits. These results confirm that the isotope enrichment factor of a compound is controlled by its molecular weight and boiling point. Therefore, the intermolecular binding energies (van der Waals attraction forces) are the main factor controlling the isotope fractionation during the vaporization of the mixture compounds and the single compounds (n-C_6 and n-C_9) both as a pure single phase and when in soil.

In a diffusion-limited vaporization system, it is well known that the higher the vapor saturation is above that of volatilizing water, the lower the isotope effects (Craig and Gordon 1965). The results of this study combined with previous results indicate that this rule holds for the vaporization of LMWHs. The carbon isotope enrichment factor of volatilizing pure single-phase n-hexane was -0.95 ± 0.04‰ (95% confidence limit, Table 4). While that for pure n-hexane volatilizing across 1 g soil, 2 g soil, and a soil column became gradually higher with the progression along the series of matrices in which the vapor space become increasingly unsaturated and the vapor pressures gradually decreased. A similar trend was observed during the vaporization of n-C_9. This demonstrates the effects of the matrix: diffusion materials like soil can decrease the vapor saturation and make the remaining liquid enriched in ^{13}C, further increasing the isotope enrichment factor.

n-C_{10}, n-C_{11}, and n-C_{12}, on the other hand, showed little change in isotopic composition with mass loss (or lack

Table 4 Comparison of experimental carbon isotope enrichment factors of this work, previous studies, and the theoretical enrichment factors of LMWHs during vaporization

Compound	ε_e^a (‰)	95% CI[b]	Experimental material	α_t^c	ε_t^d (‰)	References
n-C_6	−1.21	±0.06	Artificial oil	0.99856	−1.44	This study
	−0.95	±0.04	Pure liquid			
	−0.90	±0.07	In 1 g soil			
	−1.49	±0.18	In 2 g soil			
	−1.73	±0.52	Soil column			Bouchard et al. (2008b)
	−1.02	/	Natural oil			Xiao et al. (2012)
n-C_7	−1.29	±0.14	Artificial oil	0.99889	−1.11	This study
	−1.01	/	Natural oil			Xiao et al. (2012)
n-C_8	−1.14	±0.23	Artificial oil	0.99912	−0.88	This study
	−0.57	/	Natural oil			Xiao et al. (2012)
n-C_9	−1.04	±0.17	Artificial oil	0.99929	−0.71	This study
	−0.61	±0.08	Pure liquid			
	−0.77	± 0.06	In 1 g soil			
Benzene	−1.74	±0.08	Artificial oil	0.99829	−1.71	This study
	−1.55	±0.45	Soil column			Bouchard et al. (2008b)
	−0.58	±0.04	Pure liquid			Shin and Lee (2010)
MCH	−0.93	±0.04	Artificial oil	0.99885	−1.15	This study
	−0.60	/	Natural oil			Xiao et al. (2012)
Toluene	−1.48	±0.10	Artificial oil	0.99872	−1.28	This study
	−0.41	±0.04	pure liquid			Shin and Lee (2010)
	−1.48	/	Natural oil			Xiao et al. (2012)
Ethylbenzene	−1.12	±0.12	Artificial oil	0.99900	−1.00	This study
o-Xylene	−0.87	±0.09	Artificial oil	0.99900	−1.00	This study

[a] ε_e, experimental enrichment factors calculated in this work

[b] CI, confidence interval used to elucidate enrichment factors; this was performed by the data analysis tool installed in Microsoft Excel

[c] α_t, theoretical fractionation factor calculated by Eq. (5)

[d] ε_t, theoretical enrichment factor

/, not mentioned

thereof) during vaporization. The isotopic composition of the residual liquids varied almost within the CSIA error for these compounds (Table 1; Fig. 1a). This may be because (1) these hydrocarbons are heavier than those in the lighter fraction, and their strong intermolecular binding energies reduced their evaporation rates and (2) the vapors of these hydrocarbons approached close to saturation, thus greatly impeding their vaporization, which resulted in them showing greatly lower mass loss than the lighter fraction.

4 Conclusions

The effect of vaporization on the carbon isotopic compositions of LMWHs was investigated through three series of experiments examining a mixture of compounds and pure compounds both alone in a single state and when diffusing across soil. Most of the mixture compounds showed obvious mass loss during vaporization, with the rate of vaporization decreasing with the increasing carbon number of the compounds, indicating that molecular weight and boiling point were the main regulator of that vaporization.

Isotope analysis showed that the vaporization patterns of the C_6–C_{12} LMWHs could be classified into two types: one for the lighter C_6–C_9 fraction and another for the heavier C_{10}–C_{12} fraction. The remaining portion of the lighter fraction was significantly enriched in ^{13}C by vaporization, with the vaporization fractionation of each hydrocarbon following the Rayleigh model, indicating that kinetic isotope effects controlled the natural vaporization of the molecules and their diffusion through soils. Additionally, significant isotope enrichments (more than 3‰) were apparent in the $\delta^{13}C_{R-I}$ values of the corresponding compounds when more than 90% of each components of the lighter fraction had evaporated. In contrast, the heavier fraction remained isotopically consistent due to its lower mass loss during the vaporization, indicating that the isotopic characteristics of these heavier hydrocarbons could be extremely useful for identifying the source of a given oil sample, even one slightly evaporated.

Comparison of all the series of studies conducted here suggests that both mixing a given hydrocarbon and its diffusion through soil could slow its vaporization and increase its carbon isotope enrichment factors, because both the mixture and the soil decreased the vapor pressure in the vapor–liquid system. The values of carbon isotope enrichment factors for the LMWHs are quite close to those calculated from theory and reflect a diffusion-controlled vaporization process during the natural vaporization of the LMWHs.

The C_6–C_{12} LMWHs are widely used to identify the source of oil samples, to assess the thermal maturity of oils and condensates, to determine the source allocation of mixed oils, to identify various secondary alterations of crude oils, and to trace the source of petroleum-related contaminants. This study shows that there is significant isotope fractionation during the natural vaporization of the lighter fraction of these hydrocarbons, which means that isotope monitoring using the C_6–C_9 LMWHs should be used carefully. However, as natural vaporization has little influence on the isotopic compositions of the heavier hydrocarbons in a short time (i.e. within 72 h), these molecules can provide reliable carbon isotope data better than C_6–C_9 LMWHs for use in petroleum and environmental science.

Acknowledgements We are grateful to Chen H. S. for the technical assistance. This work is financially supported by the National "863" Project (Grant No. 2012AA0611401) and the program of the Chinese Academy of Sciences (Grant No. KZCX2-YW-JC103). This is contribution number IS-2343 from Guangzhou Institute of Geochemistry, Chinese Academy of Sciences (GIGCAS). We also acknowledge three anonymous reviewers for their helpful comments and suggestions.

References

Baertschi P, Kuhn W, Kuhn H. Fractionation of isotopes by distillation of some organic substances. Nature. 1953;171:1018–20.

Balabane M, Letolle R. Inverse overall isotope fractionation effect through distillation of some aromatic molecules (anethole, benzene and toluene). Chem Geol. 1985;52:391–6.

Bjorøy M, Hall PB, Moe RP. Stable carbon isotope variation of n-alkanes in Central Graben oils. Org Geochem. 1994;22(3–5): 355–81.

Bouchard D, Höhener P, Hunkeler D. Carbon isotope fractionation during volatilization of petroleum hydrocarbons and diffusion across a porous medium: a column experiment. Environ Sci Technol. 2008a;42(21):7801–6.

Bouchard D, Hunkeler D, Gaganis P, et al. Carbon isotope fractionation during diffusion and biodegradation of petroleum hydrocarbons in the unsaturated zone: field experiment at Værløse airbase, Denmark, and modeling. Environ Sci Technol. 2008b;42(2):596–601.

Bouchard D, Hunkeler D, Höhener P. Carbon isotope fractionation during aerobic biodegradation of n-alkanes and aromatic com-

pounds in unsaturated sand. Org Geochem. 2008c;39:23–33.

Cañipa-Morales NK, Garlán-Vidal CA, Guzmán-Vega MA, et al. Effect of evaporation on C_7 light hydrocarbon parameters. Org Geochem. 2003;34(6):813–26.

Cerling TE, Solomon DK, Quade J, et al. On the isotopic composition of carbon in soil carbon dioxide. Geochim Cosmochim Acta. 1991;55(11):3403–5.

Chung HM, Walters CC, Buck S, et al. Mixed signals of the source and thermal maturity for petroleum, accumulations from light hydrocarbons: an example of the Beryl field. Org Geochem. 1998;29(1–3):381–96.

Craig H. The geochemistry of the stable carbon isotopes. Geochim Cosmochim Acta. 1953;3(2):53–92.

Craig H, Gordon LI. Stable isotopes in oceanographic studies and paleotemperatures. In: Tongiorgi, E, editor. Conference on stable isotopes in oceanographic studies and paleotemperatures. Spoleto, Italy; 1965. p. 9–130.

Dempster HS, Lollar BS, Feenstra S. Tracing organic contaminants in groundwater: a new methodology using compound–specific isotope analysis. Environ Sci Technol. 1997;31 3193–7.

Fingas MF. Studies on the evaporation of crude oil and petroleum products: I. The relationship between evaporation rate and time. J Hazard Mater. 1997;56(3):227–36.

George SC, Boreham CJ, Minifie SA, et al. The effect of minor to moderate biodegradation on C_5–C_9 hydrocarbons in crude oils. Org Geochem. 2002;33(12):1293–317.

Gray JR, Lacrampe-Couloume G, Gandhi D, et al. Carbon and hydrogen isotopic fractionation during biodegradation of methyl tert–butyl ether. Environ Sci Technol. 2002;36(9):1931–8.

Harris SA, Whiticar MJ, Eek MK. Molecular and isotopic analysis of oils by solid phase microextraction of gasoline range hydrocarbons. Org Geochem. 1999;30:721–37.

Harrington RR, Poulson SR, Drever JI, et al. Carbon isotope systematics of monoaromatic hydrocarbons: vaporization and adsorption experiments. Org Geochem. 1999;30(8A):765–75.

Huang L, Sturchio NC, Abrajano T, et al. Carbon and chlorine isotope fractionation of chlorinated aliphatic hydrocarbons by evaporation. Org Geochem. 1999;30(8A):777–85.

Höhener P, Duwig C, Pasteris G, et al. Biodegradation of petroleum hydrocarbon vapors: laboratory studies on rates and kinetics in unsaturated alluvial sand. J Contam Hydrol. 2003;56(1–2):93–115.

Hayes JM. Factors controlling ^{13}C contents of sedimentary organic-compounds: principles and evidence. Mar Geol. 1993;113(1–2): 111–25.

Jancso G, Van Hook WA. Condensed phase isotope effects (especially vapor pressure isotope effects). Chem Rev. 1974; 74(6):689–750.

Jeannottat S, Hunkeler D. Chlorine and carbon isotopes fractionation during volatilization and diffusive transport of trichloroethene in the unsaturated zone. Environ Sci Technol. 2012;46(6):3169–76.

Kelley CA, Hammer BT, Coffin RB. Concentrations and stable isotope values of BTEX in gasoline-contaminated groundwater. Environ Sci Technol. 1997;31(9):2469–72.

Kolhatkar R, Kuder T, Philp P, et al. Use of compound-specific stable carbon isotope analyses to demonstrate anaerobic biodegradation of MTBE in groundwater at a gasoline release site. Environ Sci Technol. 2002;36(23):5139–46.

Kuder T, Philp P, Allen J. Effects of volatilization on carbon and hydrogen isotope ratios of MTBE. Environ Sci Technol. 2009;43(6):1763–8.

Mancini SA, Lacrampe-Couloume G, Jonker H, et al. Hydrogen isotopic enrichment: an indicator of biodegradation at a petroleum hydrocarbon contaminated field site. Environ Sci Technol. 2002;36(11):2464–70.

Mancini SA, Devine CE, Elsner M, et al. Isotopic evidence suggests different initial reaction mechanisms for anaerobic benzene biodegradation. Environ Sci Technol. 2008;42(22):8290–6.

Mango FD. Light hydrocarbons as oil maturity indicators. Abstr Pap Am Chem Soc. 2000;219:U687–8.

Narten A, Kuhn W. Genaue bestimmung kleiner dampfdruckunterschiede isotoper verbindungen II. Der ^{13}C/^{12}C–Isotopieeffekt in tetrachlorkohlenstoff und in Benzol. Helv Chim Acta. 1961;44(6):1474–9.

Obermajer M, Osadetz KG, Fowler MG, et al. Light hydrocarbon (gasoline range) parameter refinement of biomarker-based oil–oil correlation studies: an example from Williston Basin. Org Geochem. 2000;31(10):959–76.

Odden W, Patience RL, van Graas GW. Light hydrocarbon (gasoline range) parameter refinement of biomarker-based oil–oil correlation studies: an example from Williston Basin. Org Geochem. 1998;28(12):823–47.

Pasadakis N, Obermajer M, Osadetz KG. Definition and characterization of petroleum compositional families in Williston Basin, North America using principal component analysis. Org Geochem. 2004;35(4):453–68.

Poulson SR, Drever JI. Stable isotope (C, Cl, and H) fractionation during vaporization of trichloroethylene. Environ Sci Technol. 1999;33(20):3689–94.

Rooney MA, Vuletich AK, Griffith CE. Compound-specific isotope analysis as a tool for characterizing mixed oils: an example from the west of Shetlands area. Org Geochem. 1998;29(1–3):241–54.

Schüth C, Taubald H, Bolaño N, et al. Carbon and hydrogen isotope effects during sorption of organic contaminants on carbonaceous materials. J Contam Hydrol. 2003;64:269–81.

Shin WJ, Lee KS. Carbon isotope fractionation of benzene and toluene by progressive evaporation. Rapid Commun Mass Spectrom. 2010;24(11):1636–40.

Slater GF, Dempster HS, Lollar BS, et al. Headspace analysis: a new application for isotopic characterization of dissolved organic contaminants. Environ Sci Technol. 1999;33(1):190–4.

Smallwood BJ, Philp RP, Allen JD. Stable carbon isotopic composition of gasolines determined by isotope ratio monitoring gas chromatography mass spectrometry. Org Geochem. 2002;33(2):149–59.

Stiver W, Mackay D. Evaporation rate of spills of hydrocarbons and petroleum mixtures. Environ Sci Technol. 1984;18(11):834–40.

Ten Haven HL. Applications and limitations of Mango's light hydrocarbon parameters in petroleum correlation studies. Org Geochem. 1996;24(10–11):957–76.

Thompson KFM. Classification and thermal history of petroleum based on light hydrocarbons. Geochim Cosmochim Acta. 1983;47(2):303–16.

Thompson KFM. Fractionated aromatic petroleums and the generation of gas-condensates. Org Geochem. 1987;11(6):573–90.

Thompson KFM. Gas-condensate migration and oil fractionation in deltaic systems. Mar Pet Geol. 1988;5(3):237–46.

Wang Y, Huang YS. Hydrogen isotopic fractionation of low molecular weight n-alkanes during progressive vaporization. Org Geochem. 2001;32(8):991–8.

Wang Y, Huang YS. Hydrogen isotopic fractionation of petroleum hydrocarbons during vaporization: implications for assessing artificial and natural remediation of petroleum contamination. Appl Geochem. 2003;18(10):1641–51.

Wever HE. Petroleum and source rock characterization based on C$_7$ star plot results: examples from Egypt. Bull Am Assoc Pet Geol. 2000;84(7):1041–54.

Whiticar MJ, Snowdon LR. Geochemical characterization of selected Western Canada oils by C$_5$–C$_8$ compound specific isotope correlation (CSIC). Org Geochem. 1999;30(9):1127–61.

Xiao QL, Sun YG, Zhang YD, et al. Stable carbon isotope fractionation of individual light hydrocarbons in the C$_6$–C$_8$ range in crude oil as induced by natural evaporation: experimental results and geological implications. Org Geochem. 2012;50:44–56.

Zhang CM, Li ST, Zhao HJ, et al. Applications of Mango's light hydrocarbon parameters to petroleum from Tarim basin, NW China. Appl Geochem. 2005;20(3):545–51.

Zwank L, Berg M, Schmidt TC, et al. Compound-specific carbon isotope analysis of volatile organic compounds in the low-microgram per liter range. Anal Chem. 2003;75(20):5575–83.

10

In situ catalytic upgrading of heavy crude oil through low-temperature oxidation

Hu Jia[1] · Peng-Gang Liu[1] · Wan-Fen Pu[1] · Xian-Ping Ma[2] ·
Jie Zhang[2] · Lu Gan[1]

Abstract The low-temperature catalytic oxidation of heavy crude oil (Xinjiang Oilfield, China) was studied using three types of catalysts including oil-soluble, water-soluble, and dispersed catalysts. According to primary screening, oil-soluble catalysts, copper naphthenate and manganese naphthenate, are more attractive, and were selected to further investigate their catalytic performance in in situ upgrading of heavy oil. The heavy oil compositions and molecular structures were characterized by column chromatography, elemental analysis, and Fourier transform infrared spectrometry before and after reaction. An Arrhenius kinetics model was introduced to calculate the rheological activation energy of heavy oil from the viscosity–temperature characteristics. Results show that the two oil-soluble catalysts can crack part of heavy components into light components, decrease the heteroatom content, and achieve the transition of reaction mode from oxygen addition to bond scission. The calculated rheological activation energy of heavy oil from the fitted Arrhenius model is consistent with physical properties of heavy oil (oil viscosity and contents of heavy fractions). It is found that the temperature, oil composition, and internal molecular structures are the main factors affecting its flow

ability. Oil-soluble catalyst-assisted air injection or air huff-n-puff injection is a promising in situ catalytic upgrading method for improving heavy oil recovery.

Keywords In situ catalytic oxidation · Heavy oil · Upgrading · Low-temperature oxidation · Mechanism

1 Introduction

The resources of heavy oil in the world are more than twice as great as those of conventional light crude oil (Shah et al. 2010). Heavy crude oils, with an API gravity less than 20 (Hascakir et al. 2009), have been found in different regions in the world, such as the Orinoco Belt in Venezuela, Alberta in Canada, some regions of the Gulf of Mexico, and northeastern China (Galarraga et al. 2007; Rana et al. 2007; Hinkle et al. 2008; Chen et al. 2009; Martínez-Palou 2011). However, the compositional complexity and poor flow properties of heavy crude oils bring out many problems and challenges in oil production, transportation, and refining (Saniere et al. 2004). High contents of unsaturated fractions, mainly resins and asphaltenes in heavy oil are the main factors affecting the viscosity. The recovery of heavy oil faces a challenge because of its high viscosity. Among a variety of enhanced oil-recovery methods, the application of in situ combustion (ISC) process has been regarded as one of the most promising strategies for heavy oil reservoirs. Different from recovery mechanisms of the traditional thermal recovery of hot fluid injection and steam stimulation, the injection of high-pressure air for ISC is the process of injecting air into a reservoir to oxidize a small portion of hydrocarbons in situ for upgrading. The oil viscosity is significantly reduced to initiate heavy oil to flow through porous media. The oil is moved toward the

✉ Hu Jia
   jiahuswpu@swpu.edu.cn; tiger-jia@163.com

✉ Wan-Fen Pu
   tiger-jia@163.com

[1] State Key Laboratory of Oil and Gas Reservoir Geology and Exploitation, Southwest Petroleum University, Chengdu 610500, Sichuan, China

[2] Oil Production Technology Institute of Dagang Oilfield, CNPC, Tianjin 300280, China

Edited by Yan-Hua Sun

production wells by a vigorous drive of flue gases, steam, and hot water. Since it is difficult to reestablish the front, the success of an ISC process mainly depends on the combustion front stability and its rapid propagation. Also oil from the combustion zone is upgraded in situ as the heaviest components burn and the lighter crude oil components evaporate ahead of the combustion front (Zhang et al. 2013). Hence, the quality of the produced oil is improved. Similarly, the aggregation–dissociation phenomenon was detected by Ekulu et al. (2005), showing that high temperatures can degrade the molecular structure of heavy components.

This has stimulated great interest in the application of catalysts to the ISC process in order to accelerate reaction and promote upgrading of reservoir crude oils. Catalysts have the ability to restructure the strongly associating structures as well as rearrange oxygen-containing groups that will effectively reduce the oil viscosity. It is also accepted that the addition of catalysts may promote the breaking of C–S, C–O, and C–N bonds, and hence the chemical properties of heavy oil are changed in favor of liquid flow. Rezaei et al. (2013) and Murillo-Hernández et al. (2009) studied the thermocatalytic influence of nanoparticles and ionic liquids on the oxidation reaction. Lower activation energies were observed for various reactions involved in the combustion process. Kök and Bagci (2004) investigated catalytic effects of metallic salts on the combustion properties of crude oil by reaction kinetic cell experiments. The reaction had been positively catalyzed and the molar ratio of CO_2/CO of effluent gases increased by additives. Wang et al. (2012) conducted a comparative study of the catalytic effect of tungsten oxide supported on zirconia via different methods to reduce the viscosity of heavy crude oil. With the heavy components converted into light ones and the removal of heteroatoms, the powdered catalyst prepared by the hydrothermal method reduced the viscosity of oil from 5740 to 1020 mPa s, a reduction of 82 %. Upgrading and viscosity reduction of heavy oil by a catalytic ionic liquid and its modified version were investigated by Shaban et al. (2014). With the asphaltenes dissolved by ionic liquids, the oil viscosity reduction increased from 26.8 % to 78.6 % and the sulfur reduction increased to 20 % when the reaction was performed with the modified catalyst at 90 °C. Moore et al. (1999) conducted a combustion tube experiment in the presence of a nickel/alumina catalyst, an 8-point gravity increase was detected for the test Middle East crude of 18° API gravity in addition to a 50 % sulfur removal in the produced oil. Pu et al. (2015) used oxidation tube experiments to study the low-temperature oxidation and characteristics of heavy oil under reaction conditions of 120 °C,

72 h, and 30 MPa. With the addition of the oil-soluble catalyst, cobalt naphthenate to promote the crude oil oxidation, the viscosity of the oxidized oil decreased dramatically. Abuhesa and Hughes (2009) compared the conventional and catalytic ISC process for oil recovery using a low-pressure combustion cell. They found that the presence of a catalyst advanced the combustion reactions and the resultant oil was upgraded by up to 11 points due to several processes such as pyrolysis, distillation, and cracking, which were governed by temperature. Weissman et al. (1996) applied a combustion tube apparatus to down-hole catalytic upgrading of heavy crude oil in the presence and absence of alumina-supported nickel molybdenum, catalytic upgrading was due to hydrogenation and the API of oxidized oil increased from 15.3 to 23.2 with about a 50 % sulfur removal.

Many available published papers state that the presence of catalyst could significantly accelerate the aquathermolysis reaction to further upgrade the quality of heavy oil (Maity et al. 2010; Qin and Xiao 2013; Wu et al. 2013; Shokrlu and Babadagli 2014; Wang et al. 2014). In aquathermolysis, water at high temperatures or in the form of steam plays a crucial role, and the oil viscosity reduces only slightly if water is absent (Wang et al. 2010; Maity et al. 2010). However, the reservoir temperature after steam injection will gradually reduce with an increase in the distance to the wellbore and it is difficult to maintain a relative high temperature, leading to an insufficient energy supply for the catalysis (Liu et al. 2005; Xu et al. 2012b). The production of superheated water or steam is very energy-intensive with a large environmental burden and the necessary process equipment is expensive, resulting in poor economic performance. More to the point, some heavy oils are waterless or oil production areas have extreme water shortages, making it difficult to implement aquathermolysis. Whereas, it is hoped that the oil viscosity can be reduced without adding water and an idea is to apply air injection into heavy oil reservoirs in the presence of catalyst for ISC. Besides, the catalytic oxidation is superior to a single thermal oxidation since it combines the functions of heteroatom reduction, upgrading, splitting of asphaltene and resin, high yields of light components, and accelerated reaction rate.

In this paper, the Xinjiang heavy crude oil was used for the research. We primarily screened different types of catalysts including dispersed, water-soluble, and oil-soluble catalysts by the use of static oxidation tube experiments from the observation of upgrading and viscosity reduction of the target oil. More in-depth analysis was conducted to investigate catalytic oxidation mechanisms at low temperatures.

2 Experimental

2.1 Experimental materials

Heavy crude oil obtained from a reservoir block of the Xinjiang Oilfield (Junggar Basin, China) was used in this study. The reservoir temperature is about 90 °C, the oil viscosity is 3523 mPa s at atmospheric pressure and at 50 °C, and the oil density is 0.938 g/cm^3 at atmospheric pressure and at 25 °C.

Three types of catalysts were purchased commercially from the Chengdu Kelong Chemical Reagent Co., Ltd. (Sichuan province, China) and were used as received with the American Chemical Society (ACS) grade. The oil-soluble catalysts are copper naphthenate ($Cu(C_{11}H_7O_2)_2$) and manganese naphthenate ($Mn(C_{11}H_7O_2)_2$). The organic ligand is naphthenic acid ($C_{22}H_{14}O_4$) which serves as the carrier of metal ions to reduce the viscosity of heavy oil and is also crucial to the catalytic performance, and the transition metal ions that serve as catalytic centers are, respectively, Cu^{2+} and Mn^{2+}. The dispersed catalyst is copper oxide (CuO), and the water-soluble catalyst is copper dichloride ($CuCl_2$). The fine quartz used in the experiment was obtained from the Pixian Zhitai Chemical (Sichuan province, China) with 60–80 mesh size. Prior to experiments, the heavy crude oil was separated from the sand to eliminate any catalytic effects arising from other metal salts that might exist in the crude oil. All oil samples were dehydrated to eliminate the effect of dilution and emulsification on viscosity measurement.

2.2 Isothermal oxidation

2.2.1 Experimental setup

Upgrading and viscosity reduction of heavy oil was conducted in a stainless steel reactor. The main body of the reactor was made of a stainless steel tube of 8 mm inside diameter (I.D.), 14 mm outside diameter (O.D.), and 100 cm in length and with an effective volume of 50.3 mL for loading samples. The other equipment used was described in detail in a previous study (Zhao et al. 2012). Figure 1 shows a schematic diagram of the experimental setup for isothermal oxidation of heavy oil.

2.2.2 Experimental procedures

All runs were conducted at 90 °C and at 20 MPa (reservoir conditions) to investigate the catalytic oxidation of heavy oil in porous media. The experimental procedures are as follows: (1) A fixed amount of fine quartz was loaded into the oxidation tube to simulate the porous medium with the approximate porosity of 38 % and permeability of 2100 mD. (2) The cleaned oil (around 12 mL) mixed with and without different catalysts (catalyst content was 0.5 wt% for each case) was pumped into the oxidation tube at a constant flow rate of 0.5 mL/min to saturate the porous medium. The remaining pore volume was used for compressed air injection, and the outlet of the oxidation tube was shut off. (3) High pressure air was injected into the oxidation tube at a constant rate of 1.0 mL/min until the inlet pressure achieved the required test pressure. Then, the inlet was shut off for performing experiments. At the end of experiments, the oxidation tube was first cooled to room temperature and the oil plus sand mixtures were put into a filter centrifuge for further characterization. All the runs were performed twice to ensure the repeatability and the accuracy of the experimental data.

2.3 Viscosity measurement

The viscosity measurement of the oil sample before and after reaction was conducted with a programmable viscometer (Brookfield DV-III, USA) at 50 °C for screening the most efficient catalysts.

The accuracy of the measurement of oil viscosity is ±1.0 % for a specific spindle running at a specific speed. The viscosity reduction ($\Delta\eta$) of the heavy oil is defined as

$$\Delta\eta = \frac{\eta_0 - \eta}{\eta_0} \times 100\%,$$

where $\Delta\eta$ is the percentage of reduction in the oil viscosity; η_0 and η are the oil viscosity before and after reaction, respectively.

The target oil samples were heated to 100 °C and held constant for a few minutes. Then these oils were cooled down from 100 to 40 °C at a rate of 0.5 °C/min, and the viscosity at different temperatures was measured during the cooling process.

2.4 Oil sample separation and characterization

To obtain enough material for further analysis, four group compositions including saturates, aromatics, resins, and asphaltenes were separated from oil samples by a chromatography column according to the industrial standard of China Petroleum SY/T 5119. The elemental composition of oils (C, H, O, N, and S) was measured by an elemental analyzer (Elementar Vario EL-III, Germany). The structural changes of heavy oil before and after reaction were characterized by Fourier transform infrared spectroscopy (FT-IR WQF-520, China).

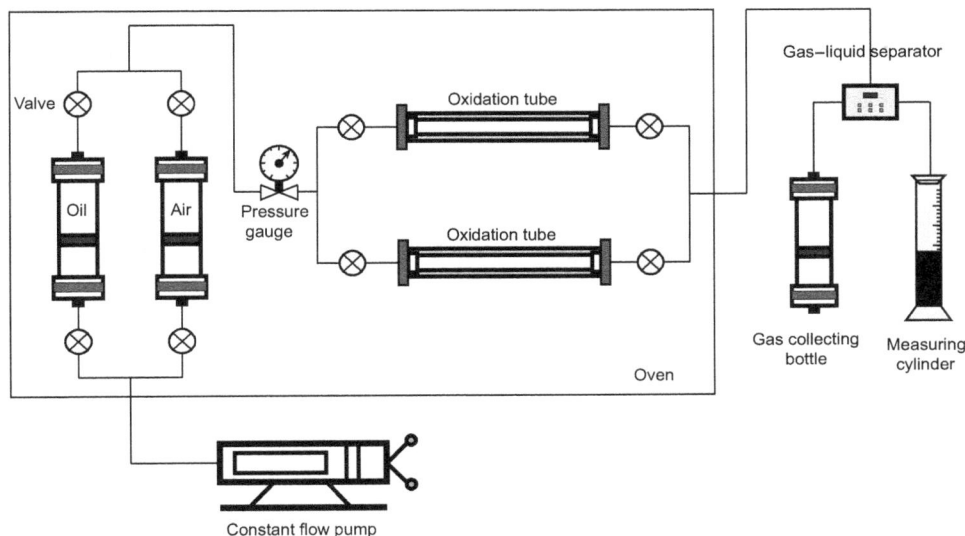

Fig. 1 Schematic diagram of the setup for isothermal oxidation of heavy oil

3 Results and discussion

3.1 Analyses of viscosity reduction mechanisms and oxidation models

It is widely accepted that the high content of resins and asphaltenes in heavy oil is responsible for high viscosity (Chao et al. 2012). The viscosity enhancing effect of the resins on heavy oil is due to a large number of heteroatoms contained. This promotes the formation of complex structures between family molecules, with little relationship with the size of the resin itself. In the asphaltene case, in addition to its own high viscosity, its viscosity enhancing effect is more reflected in the impact of micelles formed by interaction on the viscosity (Li et al. 2010). When the resins and aromatics in heavy oil are not enough to surround the asphaltenes to form micelles, the asphaltene molecules are likely to further associate with each other to form large clumps and coagulations, which greatly increase the viscosity of heavy oil. The approaches of viscosity reduction upon treatment of heavy crude oil with catalysts are mainly from two aspects. Intermolecular forces with weak bond energies expressed such as hydrogen bond, coordination bond, and van der Waals force are easy fractured, resulting in that the tight association structure in heavy oil becomes loose. The small molecules wrapped inside the heavy oil molecular aggregates are released and the viscosity undergoes reversible reduction. On the other hand, part of the chemical bonds with weak bond energies (C–C, C–O, and C–S bonds) are ruptured, leading to scission reactions for alkyl side chains and generating chain bridges, heterocyclic rings, etc. With heavy substances converting to light components, the molecular aggregation is weakened along with a decrease in the average relative molecular weight, and the viscosity reduces irreversibly. It is acknowledged that condensed aromatic rings form the main structure of resin and asphaltene. Rich fatty structural units connected to the aromatic rings, complex branched structures, and a high degree of aromatization make the molecular structure exhibit strong polarity and make it easy to be attacked by oxygen atoms, which contribute to a fast oxidation reaction rate and large reactivity for resin and asphaltene at low temperatures (Castro and Vazquez 2009; Chen et al. 2013b).

In general, there will be a series of reactions between crude oil and oxygen once they are in contact. Oxygen atoms are first bound into the carbon chains and the hydrocarbon compounds are converted into ketones, aldehydes, alcohols, and other partially oxidized materials (Kapadia et al. 2013). After that, the chain scission reaction occurs. In this process, hydrocarbon products keep reacting with oxygen, whilst chains with functional groups are split into carbon dioxide, carbon monoxide, water, and shorter chain hydrocarbon compounds. A two-step reaction mode can be used to describe the oxygen addition reaction and decarboxylation process described elsewhere (Zhao et al. 2012).

Step 1 (oxygen addition reaction):

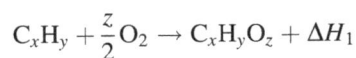

$$C_xH_y + \frac{z}{2}O_2 \rightarrow C_xH_yO_z + \Delta H_1$$

Step 2 (decarboxylation reaction):

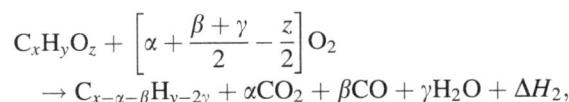

$$C_xH_yO_z + \left[\alpha + \frac{\beta + \gamma}{2} - \frac{z}{2}\right]O_2$$
$$\rightarrow C_{x-\alpha-\beta}H_{y-2\gamma} + \alpha CO_2 + \beta CO + \gamma H_2O + \Delta H_2,$$

where ΔH is the reaction enthalpy and x, y, z, α, β, and γ are the stoichiometric coefficients of the chemical reaction. The oxygen addition reactions can cause a dramatic increase in the oil viscosity with the formation of heavier, less desirable oil fractions. On the contrary, the decarboxylation reactions can reduce the viscosity of crude oil with the generation of shorter chain hydrocarbon compounds. It seems that there is a competition for the predominant role between the two types of reactions at low temperatures (Barzin et al. 2010).

3.2 Viscosity reduction with different types of catalysts

The viscosity of oil samples before and after oxidation reaction with different types of catalysts are shown in Fig. 2. This shows that the viscosities of oil samples after reaction increased from the original value 3523 mPa s, respectively, to 4731 and 4196 mPa s for the dispersed and water-soluble catalysts, while to 1238 and 1705 mPa s for the oil-soluble catalysts ($Cu(C_{11}H_7O_2)_2$) and ($Mn(C_{11}H_7O_2)_2$). The dispersed and water-soluble catalysts failed to reduce the viscosity of heavy oil, with thickening ratios of 34.3 % and 19.1 %, respectively, while the oil-soluble catalysts exhibited excellent capacity to reduce the oil viscosity. The compared results indicate that the oil-soluble catalysts are the most active catalysts for catalytic oxidation of heavy crude oils, aiming to reduce the oil viscosity.

The viscosity variation of heavy oil catalyzed by different types of catalysts can be explained as follows. According to low-temperature oxidation models, it is believed that the dispersed and water-soluble catalysts, CuO or $CuCl_2$ can accelerate the chemical combination of oxygen with liquid hydrocarbons to form hydroperoxides,

which tend to further react and polymerize with each other (Niu et al. 2011), and the oxygen addition reaction plays a predominant role. As well, the molecular size of hydrocarbons increases and the structure becomes more complicated. Under the action of molecular forces, the macromolecules with oxygen-containing functional groups aggregate and associate mutually to form supermolecular structures with various levels, leading to an increase in viscosity. As to the oil samples catalyzed by oil-soluble catalysts, experimental results indicate that the reaction mode may be changed easily from oxygen addition switching to bond scission. Homogeneous catalysis may be another reason for the superiority of the oil-soluble catalysts $Cu(C_{11}H_7O_2)_2$ and $Mn(C_{11}H_7O_2)_2$. To better understand the role of oil-soluble catalysts in reducing the viscosity of heavy oil, a blank experiment was carried out under the same conditions without a catalyst. Figure 2 shows that the oil viscosity increased from 3523 mPa s (original value) to 4047 mPa s (after oxidation), which further proves that the introduction of oil-soluble catalysts $Cu(C_{11}H_7O_2)_2$ and $Mn(C_{11}H_7O_2)_2$ can significantly promote the viscosity reduction of heavy oil.

The viscosity of these oil samples was measured at 50 °C. Since the wax contents in crude oil, and oil catalyzed by oil-soluble catalysts $Cu(C_{11}H_7O_2)_2$ and $Mn(C_{11}H_7O_2)_2$ were 2.11 %, and 1.46 % and 1.42 %, respectively, and the melting point of microcrystalline wax is usually around 70 °C, viscosity measurement at 80 °C is meaningful for heavy oil. It will help us to focus on studying the effects of asphaltene and resin, which are usually regarded as the primary cause of high viscosity for heavy oil (Ghanavati et al. 2013; Wang et al. 2014). In order to avoid the influence of wax during viscosity measurement, the viscosity of oil samples catalyzed by oil-soluble catalysts was subsequently measured at 80 °C. The viscosity reduction of oil samples is summarized in Fig. 3. After catalytic oxidation at low temperatures, the viscosity of the oil sample, measured at 50 °C, reduced by 64.9 % and 48.2 %, respectively, for the oil-soluble catalysts $Cu(C_{11}H_7O_2)_2$ and $Mn(C_{11}H_7O_2)_2$; while the oil viscosity reduced by 51.6 % and 40.4 % for the two catalysts when the viscosity was measured at 80 °C. The viscosity of the original crude oil was 3523 and 441.36 mPa s at 50 and 80 °C.

However, the viscosity reduction of oil samples measured at 80 °C was lower than that tested at 50 °C, which also confirms the effect of microcrystalline wax on the heavy oil viscosity. The thickening ratio at 80 °C was 4.9 % in the controlled experiment. The oil-soluble catalysts could drastically reduce the viscosity of heavy oil, and the oil-soluble catalyst $Cu(C_{11}H_7O_2)_2$ was more effective than the oil-soluble catalyst $Mn(C_{11}H_7O_2)_2$. In the subsequent study, we selected the two oil-soluble catalysts to

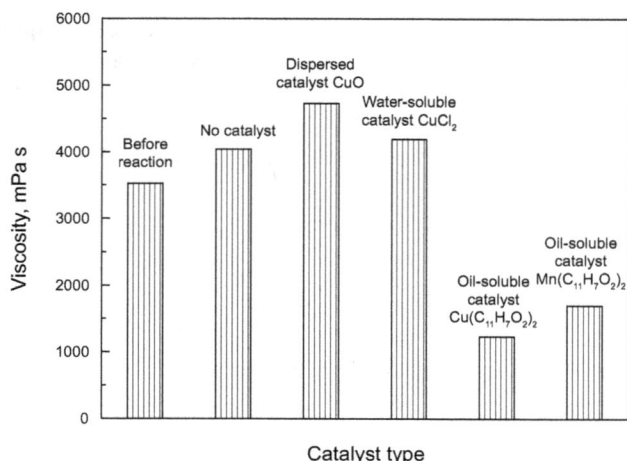

Fig. 2 Viscosity variations of oil samples before and after catalytic oxidation over different catalysts

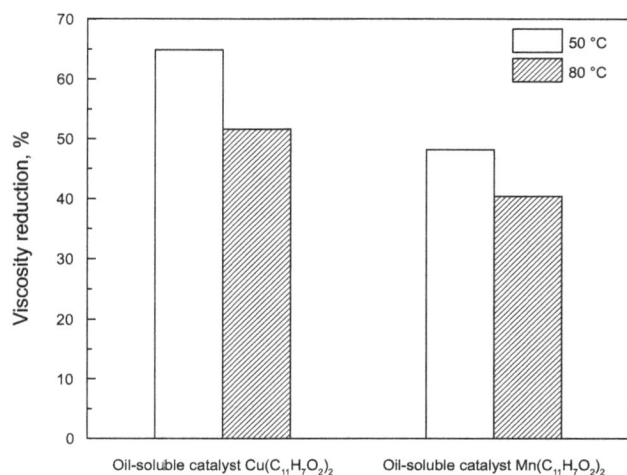

Fig. 3 Viscosity reduction of heavy oil after catalytic oxidation

further investigate the oxidation performance of heavy oil over these oil-soluble catalysts.

3.3 Composition variation of heavy oil before and after oxidation

Table 1 summarizes the group compositions of heavy oil before and after catalytic oxidation with oil-soluble catalysts $Cu(C_{11}H_7O_2)_2$ and $Mn(C_{11}H_7O_2)_2$, which were determined based on the method for determining the contents of wax, resins, and asphaltenes in crude oil (SY/T 7550-2004). It should be noted that the content of each unsaturate is a relative weight content without saturates. The composition changes of oil samples show an increase in the contents of saturates and aromatics and a loss in resins and asphaltenes after reaction, verifying that the oil-soluble catalysts have the ability to crack part of heavy components (asphaltenes and resins) into light components (saturates and aromatics) (Qin and Xiao 2013). As we know, the resins and asphaltenes consist primarily of condensed aromatic rings, naphthenic rings, alkyl side chains, and heteroatoms. The macroscopic decrease in the content of heavy components implies the probable pyrolysis of C–R (R=S, N, and O) bonds in terms of microscopic molecular structure. The C–S bond is believed to be one of the weakest bonds in heavy oil molecules. It is also found that catalysts can lead to the splitting of a few C–O and C–

N bonds owing to the emergence of a variety of oxygen-containing and nitrogen-containing compounds in the reaction products (Chen et al. 2010, 2011b). The participation of transition metal ions can attack heteroatoms of macromolecules of heavy components to perform the catalytic process, such as pyrolysis, decarboxylation, hydrodesulfurization, depolymerization, isomerization, and ring opening. The C–R (R=S, N, and O) bonds will be weakened and break easily at high temperatures, resulting the heavy components to be depolymerized to form fragments with different sizes. Another reason for composition changes in this study is the generation and accumulation of a large number of free radicals with high reactivity during oxidation according to Sarma et al. (2002). The exothermic reaction pathways are opened, wherein highly reactive free radicals are generated. The catalysis is believed to be related to these highly reactive free radicals to improve the quality of heavy crude oil.

In terms of low-temperature oxidation models, oxygen is firstly in combination with liquid hydrocarbons to form oxygen-containing functional groups, mainly ketones, aldehydes, and alcohols (Chen et al. 2013a). Generally, these medium products are unstable and are vulnerable to bond scission reactions. Hydrocarbon products keep reacting with oxygen, and thus the chains with oxygen-containing functional groups are further broken down to carbon monoxide, carbon dioxide, water, and shorter chain hydrocarbon compounds. Zhang et al. (2015) investigated the characteristics of low-temperature oxidation (LTO) of heavy oils through laboratory experiments and found that the heavy components served as the main reactant in the LTO process. During the oxidation, the carbon chains of heavy components were broken, producing some light components and leading to a relative decrease in the total content of heavy components. Similar phenomena have also been found in other published studies (Yu et al. 2008; Tang et al. 2011; Li et al. 2013b). Because the branched chains are more easily broken than linear chains, and chain scission happens more easily than dehydrogenation and oxygenation in the same carbon number alkane, oxygen addition and chain scission will play a predominant role in this stage despite the carbon chains undergoing a polycondensation process at the same time. Meanwhile, with the effect of oil-soluble catalysts, the transition of the predominant reaction mode from oxygen addition to bond

Oil sample	Catalyst	Content, %		
		Saturates + aromatics	Resins	Asphaltenes
Before reaction	–	45.07	46.62	8.31
After reaction	Oil-soluble catalyst $Mn(C_{11}H_7O_2)_2$	51.49	44.77	3.74
After reaction	Oil-soluble catalyst $Cu(C_{11}H_7O_2)_2$	54.12	43.73	2.15

Table 1 Variation of group compositions of heavy oil

scission can be easily achieved, which is verified by the viscosity reduction of 65 % and 52 % for oil-soluble catalysts $Cu(C_{11}H_7O_2)_2$ and $MnC_{11}H_7O_2)_2$, respectively, at 50 °C. The catalytic oxidation of crude oil is from the attack of oxygen atoms. The long alkyl side chains and naphthene groups located on the polycyclic aromatics in resin and asphaltene are likely to cleave and open the rings, leading to a decrease in resins and asphaltenes as well as an increase in saturates and aromatics (Li et al. 2013a).

Table 1 shows that the content of heavy components (resins and asphaltenes) decreased by 6.4 % and 9.0 %, respectively, after oxidation of heavy oil with catalysts $Mn(C_{11}H_7O_2)_2$ and $Cu(C_{11}H_7O_2)_2$. Similar to viscosity reduction, the experimental results demonstrate that the catalyst $Cu(C_{11}H_7O_2)_2$ shows high activity in the aspect of upgrading heavy components. Table 1 also indicates that the resin content, respectively, decreased by 2.89 % and 1.85 % after oxidation over catalysts $Cu(C_{11}H_7O_2)_2$, and $Mn(C_{11}H_7O_2)_2$, while asphaltene content decreased by 6.16 % and 4.57 %, implying that the two catalytic ions mainly react with asphaltene. The different catalytic performances of additives are related to the molecular structure differences between resin and asphaltene. This is due to the fact that asphaltenes have longer alkyl branch sides, more layer structures, and higher average molecular weight than the resins. From another perspective, it is known that some of the resins can crack into saturate or aromatic fractions, and asphaltenes can split into resin during thermolysis (Wang et al. 2009). Intuitively, the cracking of original resins can decrease their content only to a limited degree, but the new resins generated by asphaltene splitting can compensate for the cracking loss of original resins. Hence, the overall content of resins only changes slightly.

3.4 Elemental analysis of heavy oil before and after reaction

Elemental analysis results of heavy oil before and after oxidation are summarized in Table 2. It can be seen that the contents of carbon and heteroatoms (S, O, and N) decreased after low-temperature catalytic oxidation, while for the hydrogen content and the atomic ratio of H to C, N_H/N_C, of oil samples a reverse tendency was observed. Among these changes, the removal of heteroatoms is related to the cleavage of C–R (R=S, O, N) bonds. It is not

surprising that the change of content of S is more evident than those of O and N after oxidation. It can be explained as the bond energy of C–S is relatively low (Hyne and Greidanus 1982). Meanwhile, during the upgrading process, the combination of a metallic ion in the catalyst with a S atom can weaken C–S bonds and cause the breakage of heavy oil molecules. By contrast, since the C–N bond is the most stable bond among other bonds of molecules in the oil, most of nitrogen remains in the heavy component molecules after reaction, resulting in a lesser decrease in its content compared with other heteroatoms. Also, the atomic ratio of H to C of reacted oil over the oil-soluble catalysts increased, which is due to the cracking of weak C–C bonds in heavy hydrocarbons.

API gravities of original heavy crude oil and oil samples catalyzed by oil-soluble catalysts $Mn(C_{11}H_7O_2)_2$ and $Cu(C_{11}H_7O_2)_2$, were 19.2, 23.3, and 24.7, respectively. Because even a small fraction of bond splitting can lead to a significant improvement of the flow properties of heavy oil, the cleavage of some C–R (R=S, N, and O) bonds favors reducing crude oil viscosity eventually with the quality slightly upgraded after oxidation. In addition, when many heavy oil molecules split into fragments, the total number of rings is reduced and some of saturated fragments will release light components, leading to an increase in the content of saturates and aromatics. In general, the changes of heavy oil compositions and molecular structures are responsible for viscosity reduction and quality improvement. The changes of elemental contents after catalytic oxidation over the catalyst $Cu(C_{11}H_7O_2)_2$ are more obvious than that over the catalyst $Mn(C_{11}H_7O_2)_2$, which means that the catalyst $Cu(C_{11}H_7O_2)_2$ is superior to the catalyst $Mn(C_{11}H_7O_2)_2$ for splitting heavy oil molecules.

A comparison of elemental composition between the Xinjiang heavy oil and other heavy oils in previous studies has also been made in terms of catalytic reaction (Wang et al. 2012; Xu et al. 2012a; Qin and Xiao 2013). It is easy to find from Table 3 that the changing trends of elements are consistent with the Xinjiang heavy oil catalyzed by oil-soluble catalysts. In these experiments, the heteroatoms were removed and the values of N_H/N_C increased after catalytic reaction, which is a positive sign for heavy oil upgrading. For example, Qin and Xiao (2013) investigated the effects of the oil-soluble catalyst (Fe^{2+}) on elemental

Table 2 Elemental analysis of heavy crude oil	Oil sample	Catalyst	Content, %					N_H/N_C
			C	H	O	N	S	
	Before reaction	–	81.69	12.95	1.76	0.93	2.64	1.90
	After reaction	Oil-soluble catalysts $Mn(C_{11}H_7O_2)_2$	80.21	15.23	1.31	0.73	1.59	2.28
	After reaction	Oil-soluble catalysts $Cu(C_{11}H_7O_2)_2$	79.94	15.81	1.19	0.72	1.36	2.37

Table 3 Comparison of elemental composition of heavy oil from the literature

	Catalyst	Before reaction						After catalytic reaction					
		C, %	H, %	O, %	N, %	S, %	N_H/N_C	C, %	H, %	O, %	N, %	S, %	N_H/N_C
Wang et al. (2012)	20-W/Zr-HD-650	86.63	11.47	–	1.00	0.90	1.59	86.06	12.25	–	0.99	0.70	1.71
Qin and Xiao (2013)	Fe^{2+}-catalyst	81.20	13.70	1.26	0.95	0.28	2.02	79.50	16.70	0.42	0.91	0.05	2.52
Xu et al. (2012a)	Cu^{2+}-catalyst	80.23	11.77	2.76	1.96	3.28	1.76	81.24	12.64	2.16	1.75	2.21	1.87

compositions of heavy oil during steam huff and puff. The results showed that the Fe^{2+}-catalyst could effectively decrease the amounts of heteroatoms (S, O, and N) after the upgrading process. Meanwhile, the atomic ratio of H to C, N_H/N_C increased and the quality of the heavy oil was also improved after treatment. The comparison indicates that the oil-soluble catalyst should be a favorable catalyst for viscosity reduction and upgrading of heavy oil.

3.5 FT-IR spectra of heavy oil before and after reaction

Figure 4 shows the FT-IR spectra of heavy oil before and after reaction. An absorption peak at 3344 cm^{-1} was observed to become stronger after low-temperature oxidation without a catalyst, indicating the occurrence of alcohol addition reactions. As well, the bands between 1705 and

Fig. 4 FT-IR spectra of heavy oil. **a** Before reaction. **b** After reaction without a catalyst. **c** After reaction with the oil-soluble catalyst $Cu(C_{11}H_7O_2)_2$

1597 cm^{-1} were enhanced, which could be attributed to the formation of new carbonyl groups (C=O), such as carboxylic acids, ketones, aldehydes, and others favored by the degradation of hydroperoxides (Chao et al. 2012). It is believed that there could be numerous reactions between crude oil and oxygen once they are in contact. Under the attack of oxygen atoms, carbon–hydrogen bonds in hydrocarbon components split to generate oxygen-containing groups. Accordingly, comparative results of Fig. 4a, b indicate that the oxygen addition reaction plays a predominant role during the low-temperature oxidation without a catalyst. By comparing Fig. 4b, c the absorption peak at 3344 cm^{-1} became weaker, indicating that some of alcohol functional groups were removed; the observed bands in 1705–1597 cm^{-1} were weakened, which could be attributed to breakage of C=O bonds and the occurrence of hydrogenation reaction. Due to the catalytic effect of the oil-soluble catalyst, the contents of oxygen-containing groups decreased. Bond scission reactions are responsible for the further reaction of oil oxides and dominates the reaction modes. The reaction schemes can be shown through to Chen et al. (2013a), as shown below:

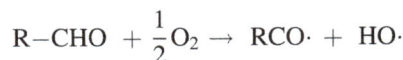

$$R-COOH \rightarrow CO_2 + RH$$

$$R-CHO + \frac{1}{2}O_2 \rightarrow RCO\cdot + HO\cdot$$

$$RCO\cdot \rightarrow CO + R\cdot$$

$$R-CHO + O_2 \rightarrow R-CO_3H$$

$$R-CO_3H \rightarrow CO_2 + ROH.$$

It is concluded that the oil-soluble catalyst has the capability to cause the transition of the reaction mode from oxygen addition to bond scission. In comparison with Fig. 4a, absorption peaks in Fig. 4c at 1023 and 1162 cm^{-1} were weakened, and the absorption peak at 1267 cm^{-1} has nearly disappeared, indicating the breakage of C–N, C–O, and C–S bonds; the absorption peak at 796 cm^{-1} was also weakened, showing the removal of the existing alkyl side chain; the absorption peak at 868 cm^{-1} became weaker, implying the occurrence of ring-opening reactions (Castro and Vazquez 2009). As well, the absorption peak at 1374 cm^{-1} was enhanced in Fig. 4b while being reduced in Fig. 4c, showing the aggregation of condensed aromatic

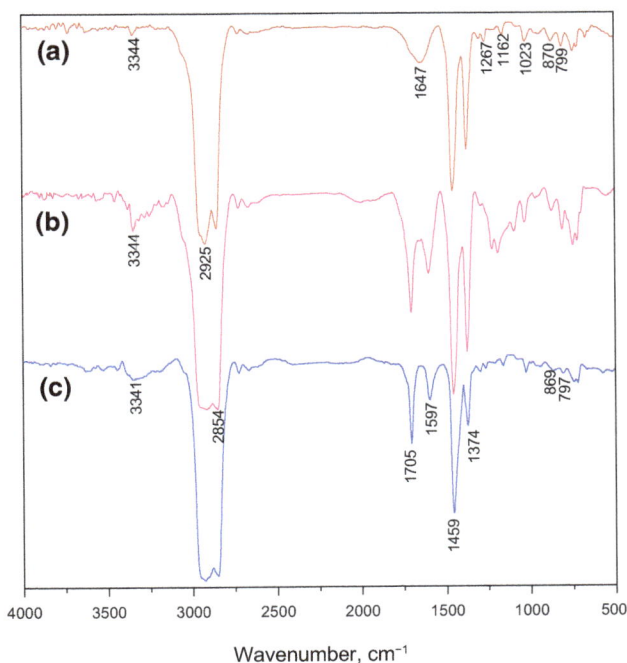

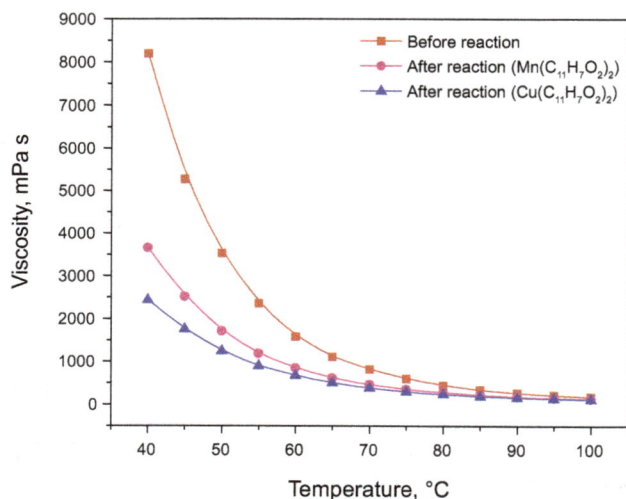

Fig. 5 Viscosity–temperature curves for oil samples

rings during the oxidation without a catalyst and the addition of the oil-soluble catalyst could inhibit the polymerization. All these changes also indicate that some of heteroatoms were removed and the tight macromolecular rings were reduced to fragments of different sizes, which is consistent with the result of the elemental analysis.

3.6 Analysis of viscosity–temperature characteristics

3.6.1 Temperature sensitivity

The complexity of the microstructure of heavy oil results in viscosity–temperature characteristics different from its macroperformance. The viscosities of heavy oil before and after catalytic reaction were measured and analyzed from the microscopic point of view. Figure 5 shows that the viscosity of heavy oil exhibited strong temperature sensitivity. The viscosity of oil samples changed smoothly when the temperature was higher than 60 °C during the cooling process, while rapidly increasing when the temperature was below 60 °C, and the lower the temperature, the more dramatic the change in viscosity. This is due to heavy oil being a multiphase liquid mixture. The relative size, distribution, and content of macromolecular solid particles (e.g., asphaltenes) in the liquid, as well as the momentum exchange between different components, agglomerating by association and arrangement mode, significantly affect the viscosity of heavy oil. Most crude oil is a relatively stable colloidal dispersed system, in which the asphaltene is the core of the dispersed phase (Ye et al. 2010). The compositions gradually change from the centers of micelles to the dispersion medium. Connection or disassembly of these micelles loosens the internal structure and lowers the

cohesive force in crude oil, which is the essence of viscosity–temperature characteristics.

Regression analysis was applied to the viscosity–temperature relationship of heavy oil before and after catalytic oxidation, and the regression equation conforms to the Arrhenius equation perfectly. The Arrhenius equation is of the form (Zhu et al. 2011, 2012):

$$\eta = 1000 A \exp(E/RT),$$

where η is the oil viscosity, mPa s; E is the activation energy, kJ mol^{-1}; R is the universal gas constant; T is the measured temperature, K; A is a constant. Taking the logarithm of both sides and the final form of the kinetic model used for analysis is as follows:

$$\log \eta = \log 1000 + \log A + E/(2.303RT).$$

The viscosity–temperature curve is converted into a $\log\eta - 1/T$ curve and the results are shown in Fig. 6. The curves are not straight lines in the strict sense. With an increase in $1/T$, $\log\eta$ gradually deviates upward. This indicates that not only the van der Waals force plays a role between micelles, but also the molecular structure has changed inside the crude oil, both of which lead to the viscosity increasing rapidly during the cooling process (Huang et al. 2014). There is no apparent inflection point in the curve and the curve is curved upward smoothly, indicating that the internal molecular structure of heavy oil gradually changes during the cooling process and no obvious phase transformation point occurs. The $1/T$ is roughly divided into three intervals according to the variation tendencies of the curve. When $\log\eta$ is plotted vs. $1/T$ for each interval, a straight line is obtained, and the values of E can be calculated from the slope of the linear

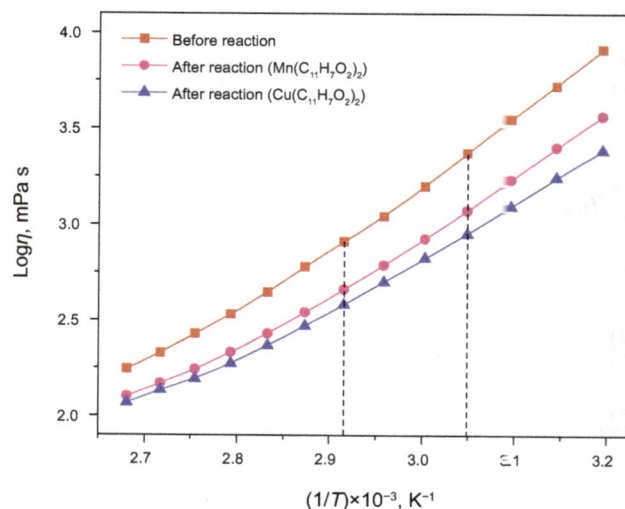

Fig. 6 Relationship curves between $\log\eta$ and $1/T$ during the cooling process

fit. Kinetic parameters of all oil samples are summarized in Table 4 in detail.

The correlation coefficient of linear fit in each temperature interval is more than 0.9889, indicating that the curves conform to the Arrhenius equation. E is the value of activation energy which refers to the energy barrier that must be overcome to form a cavity near the particle and large enough for particle to move before the fluid begins to flow. It is a measure of the friction force between particles inside the fluid, which depends on the polarity of particles, molecular mass, and molecular configurations. In the high temperature interval (from 70 to 100 °C), the activation energies of oil samples decrease with the following sequence, i.e., oil before reaction, oil after reaction with the catalyst $Mn(C_{11}H_7O_2)_2$, and oil after reaction with the catalyst $Cu(C_{11}H_7O_2)_2$. These results correspond to the variation of four group compositions of heavy oil in Table 1. In the other two temperature interval cases, this consistency is also detected. The reason may be that the unreacted crude oil contains more heavy components (resins and asphaltenes) and less light components (saturates and aromatics) than the catalytically oxidized heavy oil. Actually, a higher content of heavy components can provide larger molecules, and the interaction force is stronger between various molecules, hence, higher activation energy is needed for molecule flow (Chen et al. 2011a). It will lead to the macrophenomena of high viscosity for crude oil. Besides, intermolecular hydrogen bonds can also increase the viscosity of crude oil. The rheological activation energy of the oil sample after reaction with the catalyst $Cu(C_{11}H_7O_2)_2$ is the lowest, which further demonstrates that the catalyst $Cu(C_{11}H_7O_2)_2$ is more favorable for heavy oil viscosity reduction compared to the catalyst $Mn(C_{11}H_7O_2)_2$.

It can be seen from Table 4 that the activation energies are different in various temperature intervals for the same oil sample, and the activation energy at low temperatures is higher than that at high temperatures. There are several reasons contributing to the increase in the activation energy at low temperatures. (1) During the cooling process, the degree of the thermal motion of particles is weakened and the solubility of resin in heavy oil reduces, resulting in resin molecules continuously precipitating from crude oil. These can absorb and aggregate on the surfaces of asphaltene particles to form a solvent layer of varying thickness (Ye et al. 2010). This adsorption effect becomes significant as well as an increase in the solvent layer at lower temperatures. The aggregated asphaltene adsorbs the solvent layer during the cooling process, resulting in increases in the micelle volume and the interaction force. (2) At low temperatures, with an increase in the volume of resin–asphaltene aggregations, the distance between micelles is reduced. Adjacent micelles may be connected

Table 4 Kinetic parameters of oil samples before and after reaction

Temperature intervals, °C	Before reaction			After reaction (catalyst $Mn(C_{11}H_7O_2)_2$)			After reaction (catalyst $Cu(C_{11}H_7O_2)_2$)		
	Regression equation	R^{2a}	E, kJ/mol	Regression equation	R^2	E, kJ/mol	Regression equation	R^2	E, kJ/mol
70–100	$\log \eta = -5.4 + 2868.7/T$	0.9960	54.93	$\log \eta = -4.3 + 2393.6/T$	0.9916	45.83	$\log \eta = -3.8 + 2184.8/T$	0.9889	41.83
55–70	$\log \eta = -7.2 + 3467.2/T$	0.9966	66.38	$\log \eta = -6.3 + 3085.2/T$	0.9993	59.07	$\log \eta = -5.5 + 2772.2/T$	0.9994	53.08
40–55	$\log \eta = -7.8 + 3690.9/T$	0.9997	70.67	$\log \eta = -7.2 + 3370.9/T$	0.9996	64.54	$\log \eta = -6.1 + 2994.3/T$	0.9998	57.33

[a] R^2 stands for the correlation coefficient of linear fit line

with each other through the action of hydrogen bonds and coordinative bonds to form a spatial network structure and to wrap a lot of liquid oil in it (Zhu et al. 2011, 2012). (3) A small amount of paraffin contained in crude oil crystallizes and precipitates at low temperatures, and interacts with resin and asphaltene in the precipitation process, giving the heavy oil significant structural strength at low temperatures. All of these lead to a rapid increase in viscosity at low temperatures, and this is why the $\log\eta$ versus $1/T$ curve has greater deviation upward at low temperatures. Hence, it is crucial for production of heavy crude oil to maintain the reservoir at relatively high temperatures if possible.

On the basis of analysis of activation energy for various oil samples, it is found that the surrounding temperature, oil composition, and internal molecular structure are the main factors affecting the flow capability of heavy crude oil under reservoir conditions. As is known, in situ combustion is a thermally enhanced recovery method that promotes combustion reactions between a small portion of hydrocarbons in place and the injected air to generate heat in the reservoir. Viscosity is significantly reduced due to heat release in exothermic reactions, resulting in some asphaltene molecules splitting to small molecules. Hence, the flow properties of heavy oil are improved (Xu et al. 2012b). Moreover, the generated heat provides the driving force for viscous oil to flow easily to achieve high displacement efficiency. Besides, compared with traditional steam-based recovery processes, heat loss is avoided as energy generation is performed within the oil-bearing zone. On the other hand, the elemental content and molecular structure of oil samples have been changed after low-temperature catalytic oxidation with oil-soluble catalysts. Part of the heavy components, mainly resin and asphaltene fractions are converted into light components, such as saturated and aromatic fractions, leading to in situ upgrading and viscosity reduction of heavy crude oil. We believe that catalytic assisted air injection (or huff and puff) should be a favorable thermally enhanced oil recovery method for heavy crude oil.

4 Conclusions

(1) Oil-soluble catalysts are found to be the most effective catalysts for reducing the viscosity of heavy crude oil and upgrading the heavy crude oil when implementing low-temperature catalytic oxidation. Copper naphthenate $Cu(C_{11}H_7O_2)_2$ shows more favorable catalytic effects than manganese naphthenate $Mn(C_{11}H_7O_2)_2$.

(2) The combined analyses of group composition, elemental content, and FT-IR spectra indicate that part of heavy components (asphaltenes and resins) may

split into light components (saturates and aromatics) during low-temperature oxidation with oil-soluble catalysts and then heavy crude oil is upgraded. The catalyst plays an important role in promoting the oxidation-cracking reaction and has the ability to switch the reaction mode from oxygen addition to bond scission.

(3) The viscosity of heavy crude oil is intensively sensitive to temperature, and the oil viscosity increases rapidly with a decrease in temperature when the temperature is lower than 60 °C. A regression analysis is applied to study the viscosity–temperature relationship of oil samples, exhibiting the goodness fit of an Arrhenius model. The calculated activation energies correspond to the physical characteristics of viscosity and heavy component content.

(4) The surrounding temperature, oil composition, and internal molecular structure are the main factors affecting heavy crude oil flow ability. Implementing oil-soluble catalyst assisted air injection or air huff-n-puff injection should be a favorable strategy to improve heavy crude oil recovery.

Acknowledgments This work is supported by the National Natural Science Foundation of China (No. 51404202), Sichuan Youth Science and Technology Fund (No. 2015JQ0038), and the Scientific Research Starting Project of Southwest Petroleum University (No. 2014QHZ001).

References

Abuhesa MB, Hughes R. Comparison of conventional and catalytic in situ combustion processes for oil recovery. Energy Fuels. 2009;23(1):186–92. doi:10.1021/ef800804a.

Barzin Y, Moore RG, Mehta SA, et al. Role of vapor phase in oxidation/combustion kinetics of high-pressure air injection (HPAI). In: SPE annual technical conference and exhibition, Florence, Italy, September 19–22, 2010. doi:10.2118/135641-MS.

Castro LV, Vazquez F. Fractionation and characterization of Mexican crude oils. Energy Fuels. 2009;23(3):1603–9. doi:10.1021/ef8008508.

Chao K, Chen YL, Liu HC, et al. Laboratory experiments and field test of a difunctional catalyst for catalytic aquathermolysis of heavy oil. Energy Fuels. 2012;26(2):1152–9. doi:10.1021/ef2018385.

Chen L, Li JH, Sun C. Dispersing effect of resins on asphaltenes in crude oil. J Xi'an Shiyou Univ. 2011a;26(1):64–70 (in Chinese).

Chen Y, He J, Wang Y, et al. GC–MS used in study on the mechanism of the viscosity reduction of heavy oil through aquathermolysis catalyzed by aromatic sulfonic $H_3PMo_{12}O_{40}$. Energy. 2010;35(8):3454–60. doi:10.1016/j.energy.2010.04.041.

Chen Y, Li P, Wang Q, et al. Change of carbazole compounds in heavy oil by catalytic aquathermolysis and the catalytic mechanism of viscosity reduction. J Fuel Chem Technol. 2011b; 39(4):271–7.

Chen Y, Wang Y, Lu J, et al. The viscosity reduction of nano-keggin-$K_3PMo_{12}O_{40}$ in catalytic aquathermolysis of heavy oil. Fuel. 2009;88(8):1426–34. doi:10.1016/j.fuel.2009.03.011.

Chen Z, Wang L, Duan Q, et al. High-pressure air injection for improved oil recovery: low-temperature oxidation models and thermal effect. Energy Fuels. 2013a;27(2):780–6. doi:10.1021/ef301877a.

Chen ZY, Niu BL, Tang LZ, et al. Experimental study of low temperature oxidation mechanisms and activity of oil components. J Fuel Chem Technol. 2013b;41(11):1336–42 (in Chinese).

Ekulu G, Nicolas C, Achard C, et al. Characterization of aggregation processes in crude oils using differential scanning calorimetry. Energy Fuels. 2005;19(4):1297–302. doi:10.1021/ef049813p.

Galarraga F, Márquez G, Reategui K, et al. Comparative study of crude oils from the Machete area in the Eastern Venezuelan Basin by pyrolysis of asphaltenes. J Anal Appl Pyrolysis. 2007;80(2):289–96. doi:10.1016/j.jaap.2007.03.006.

Ghanavati M, Shojaei MJ, Ramazani SAA. Effects of asphaltene content and temperature on viscosity of Iranian heavy crude oil: experimental and modeling study. Energy Fuels. 2013;27(12): 7217–32. doi:10.1021/ef400776h.

Hascakir B, Acar C, Akin S. Microwave-assisted heavy oil production: an experimental approach. Energy Fuels. 2009;23(12): 6033–9. doi:10.1021/ef9007517.

Hinkle A, Shin EJ, Liberatore MW, et al. Correlating the chemical and physical properties of a set of heavy oils from around the world. Fuel. 2008;87(13–14):3065–70. doi:10.1016/j.fuel.2008. 04.018.

Huang J, Li BG, Qin B, et al. Study of viscosity reduction mechanisms of Tahe heavy oil by catalytic aquathermolysis. Pet Process Petrochem. 2014;45(5):16–20 (in Chinese).

Hyne JB, Greidanus JW. Aquathermolysis of heavy oil. In: The second international conference on heavy crude and tar sands, Caracas Venezuela, 1982; p. 25–30.

Kapadia P, Gates ID, Mahinpey N, et al. Kinetic models for low temperature oxidation subranges based on reaction products. In: SPE heavy oil conference, June 11–13, Calgary, Alberta, 2013. doi:10.2118/165527-MS.

Kök MV, Bagci S. Characterization and kinetics of light crude oil combustion in the presence of metallic salts. Energy Fuels. 2004;18(3):858–65. doi:10.1021/ef0301755.

Li J, Yuan Y, Chen D. Morphology analysis of Tahe super heavy resin and asphaltene. Chem Eng Oil Gas. 2010;39(5):454–6 (in Chinese).

Li J, Chen YL, Liu HC, et al. Influences on the aquathermolysis of heavy oil catalyzed by two different catalytic ions: Cu^{2+} and Fe^{3+}. Energy Fuels. 2013a;27(5):2555–62. doi:10.1021/ef400328s.

Li Y, Pu W, Wang A, et al. Experimental study on low temperature high pressure oxidation in air injection in light oil reservoir. Oil Drill. Prod. Technol. 2013b;34(6):90–2 (in Chinese).

Liu Y, Chen E, Wen S, et al. The preparation and evaluation of oil-soluble catalyst for aquathermolysis of heavy oil. Chem Eng Oil Gas. 2005;34(6):511–2 (in Chinese).

Maity SK, Ancheyta J, Marroquín G. Catalytic aquathermolysis used for viscosity reduction of heavy crude oils: a review. Energy Fuels. 2010;24(5):2809–16. doi:10.1021/ef100230k.

Martínez-Palou R. Transportation of heavy and extra-heavy crude oil by pipeline: a review. J Pet Sci Eng. 2011;75(3–4):274–82. doi:10.1016/j.petrol.2010.11.020.

Moore RG, Laureshen CJ, Mehta SA, et al. A downhole catalytic upgrading process for heavy oil using in situ combustion. J Can Pet Technol. 1999;38(13):72–96. doi:10.2118/99-13-44.

Murillo-Hernández JA, López-Ramírez S, Domínguez JM, et al. Survey on ionic liquids effect based on metal anions over the thermal stability of heavy oil. J Therm Anal Calorim. 2009;95(1):173–9. doi:10.1007/s10973-007-8919-5.

Niu BL, Ren SR, Liu YH, et al. Low-temperature oxidation of oil components in an air injection process for improved oil recovery. Energy Fuels. 2011;25(10):4299–304. doi:10.1021/ef200891u.

Pu W, Yuan C, Jin F, et al. Low-temperature oxidation and characterization of heavy oil via thermal analysis. Energy Fuels. 2015;29(2):1151–9. doi:10.1021/ef502135e.

Qin W, Xiao Z. The researches on upgrading of heavy crude oil by catalytic aquathermolysis treatment using a new oil-soluble catalyst. Adv Mater Res. 2013;608–609:1428–32. doi:10.4028/www.scientific.net/AMR.608-609.1428.

Rana MS, Sámano V, Ancheyta J, et al. A review of recent advances on process technologies for upgrading of heavy oils and residua. Fuel. 2007;86(9):1216–31. doi:10.1016/j.fuel.2006.08.004.

Rezaei M, Schaffie M, Ranjbar M. Thermocatalytic in situ combustion: influence of nanoparticles on crude oil pyrolysis and oxidation. Fuel. 2013;113(5):516–21. doi:10.1016/j.fuel.2013.05.062.

Saniere A, Henaut I, Argillier J. Pipeline transportation of heavy oils, a strategic, economic and technological challenge. Oil Gas Sci Technol. 2004;59(5):455–66.

Sarma HK, Yazawa N, Moore RG, et al. Screening of three light-oil reservoirs for application of air injection process by accelerating rate calorimetric and TG/PDSC tests. J Can Pet Technol. 2002;41(3):50–61. doi:10.2118/02-03-04.

Shaban S, Dessouky S, El Fatah Badawi A, et al. Upgrading and viscosity reduction of heavy oil by catalytic ionic liquid. Energy Fuels. 2014;28(10):6545–53. doi:10.1021/ef500993d.

Shah A, Fishwick R, Wood J, Leeke G, Rigby S, Greaves M. A review of novel techniques for heavy oil and bitumen extraction and upgrading. Energy Environ Sci. 2010;3(6):700–14. doi:10.1039/B918960B.

Shokrlu YH, Babadagli T. Viscosity reduction of heavy oil/bitumen using micro- and nano-metal particles during aqueous and non-aqueous thermal applications. J Pet Sci Eng. 2014;119(7):210–20. doi:10.1016/j.petrol.2014.05.012.

Tang X, Su X, Cui Y, et al. Effect of low temperature air oxidation on the composition of heavy oil. J Southwest Pet Univ (Sci Technol Ed). 2011;33(4):149–52 (in Chinese).

Wang H, Wu Y, He L, et al. Supporting tungsten oxide on zirconia by hydrothermal and impregnation methods and its use as a catalyst to reduce the viscosity of heavy crude oil. Energy Fuels. 2012;26(11):6518–27. doi:10.1021/ef301064b.

Wang J, Li C, Zhang L, et al. Phase separation and colloidal stability change of Karamay residue oil during thermal reaction. Energy Fuels. 2009;23(6):3002–7. doi:10.1021/ef801149q.

Wang J, Liu L, Zhang L, et al. Aquathermolysis of heavy crude oil with amphiphilic nickel and iron catalysts. Energy Fuels. 2014;28(12):7440–7. doi:10.1021/ef502134p.

Wang Y, Chen Y, He J, et al. Mechanism of catalytic aquathermolysis: influences on heavy oil by two types of efficient catalytic ions: Fe^{3+} and Mo^6. Energy Fuels. 2010;24(3):1502–10. doi:10.1021/ef901339k.

Weissman JG, Kessler RV, Sawicki RA, et al. Downhole catalytic upgrading of heavy crude oil. Energy Fuels. 1996;10(4):883–9. doi:10.1021/ef9501814.

Wu C, Su J, Zhang R, et al. The use of a nano-nickel catalyst for upgrading extra-heavy oil by an aquathermolysis treatment under steam injection conditions. Pet Sci Technol. 2013;31(21): 2211–8. doi:10.1080/10916466.2011.644016.

Xu H, Pu C, Wu F. Low frequency vibration assisted catalytic aquathermolysis of heavy crude oil. Energy Fuels. 2012a;26(6): 5655–62. doi:10.1021/ef301057t.

Xu H, Pu C, Wu F. Mechanism of underground heavy oil catalytic aquathermolysis. J Fuel Chem Technol. 2012b;40(10):1206–11 (in Chinese).

Ye ZB, Cheng L, Xiang WT, et al. Characterization of resins and the adsorption of resins on asphaltene particles. J Southwest Pet Univ. 2010;32(6):147–54 (in Chinese).

Yu H, Yang B, Xu G, et al. Air foam injection for IOR: from laboratory to field implementation in Zhongyuan oilfield China. In: SPE/DOE symposium on improved oil recovery, April 20–3, Tulsa, OK, 2008. doi:10.2118/113913-MS.

Zhang L, Deng J, Wang L, et al. Low-temperature oxidation characteristics and its effect on the critical coking temperature of heavy oils. Energy Fuels. 2015;29(2):538–45. doi:10.1021/ ef502070k.

Zhang X, Liu Q, Che H. Parameters determination during in situ combustion of Liaohe heavy oil. Energy Fuels. 2013;27(6): 3416–26. doi:10.1021/ef400095b.

Zhao J, Jia H, Pu W, et al. Sensitivity studies on the oxidation behavior of crude oil in porous media. Energy Fuels. 2012;26(11): 6815–23. doi:10.1021/ef301044c.

Zhu J, Li CX, Xin PG. Analysis of viscosity-temperature characteristics and rheology behavior for heavy oil. J Petrochem Univ. 2011;24(2):66–8 (in Chinese).

Zhu J, Li CX, Xin PG, et al. Study on microstructure and rheology of heavy oil. J Xi'an Shiyou Univ. 2012;27(2):54–7 (in Chinese).

11

Estimation of the water–oil–gas relative permeability curve from immiscible WAG coreflood experiments using the cubic B-spline model

Dai-Gang Wang[1,2] · Yong-Le Hu[1] · Jing-Jing Sun[3] · Yong Li[1]

Abstract Immiscible water-alternating-gas (WAG) flooding is an EOR technique that has proven successful for water drive reservoirs due to its ability to improve displacement and sweep efficiency. Nevertheless, considering the complicated phase behavior and various multiphase flow characteristics, gas tends to break through early in production wells in heterogeneous formations because of overriding, fingering, and channeling, which may result in unfavorable recovery performance. On the basis of phase behavior studies, minimum miscibility pressure measurements, and immiscible WAG coreflood experiments, the cubic B-spline model (CBM) was employed to describe the three-phase relative permeability curve. Using the Levenberg–Marquardt algorithm to adjust the vector of unknown model parameters of the CBM sequentially, optimization of production performance including pressure drop, water cut, and the cumulative gas–oil ratio was performed. A novel numerical inversion method was established for estimation of the water–oil–gas relative permeability curve during the immiscible WAG process. Based on the quantitative characterization of major recovery mechanisms, the proposed method was validated by interpreting coreflood data of the immiscible WAG experiment. The proposed method is reliable and can meet engineering requirements. It provides a basic calculation theory for implicit estimation of oil–water–gas relative permeability curve.

Keywords Cubic B-spline model · Immiscible · WAG flooding · Relative permeability · Numerical inversion

1 Introduction

The relative permeability curve is essential to describe the complicated multiphase flow characteristics in porous media (Masihi et al. 2011; Chen et al. 2013). In general, water–oil or oil–gas relative permeability data can be obtained from steady- or unsteady-state displacement experiments with core samples. Such experimental data can be interpreted using analytical methods, e.g., the Johnson–Bossler–Naumann (JBN) method. However, due to the idealized hypothesis, the precision usually cannot be guaranteed when using analytical methods to calculate the water–oil or oil–gas relative permeability curve. To improve the precision of the estimated result, Sigmund and McCaffery applied a nonlinear regression to the problem of history matching laboratory coreflood data for the first time and proposed a numerical inversion method for the water–oil relative permeability curve (Sigmund and McCaffery 1979). In contrast to the existing analytical methods, when the numerical inversion methods are adopted to interpret laboratory coreflood data, production performance prior to and after water breakthrough can be utilized comprehensively. The estimated result is not only complete but also highly precise (Daoud and Velasquez 2006; Barroeta and Thompson 2010). In recent decades, a variety of numerical inversion methods have been developed to implicitly estimate the relative permeability curve for water–oil or oil–gas systems (Chen et al. 2008; Li et al. 2009; Eydinov et al.

✉ Dai-Gang Wang
wdg@petrochina.com.cn; dgwang@pku.edu.cn

[1] Research Institute of Petroleum Exploration & Development, PetroChina, Beijing 100083, China

[2] School of Earth and Space Sciences, Peking University, Beijing 100871, China

[3] College of Petroleum Engineering, China University of Petroleum, Qingdao, Shandong 266580, China

Edited by Yan-Hua Sun

2009; Wang et al. 2010; Wang and Li 2011; Li and Yang 2011; Abdollahzadeh et al. 2011; Zhang and Yang 2013; Xu et al. 2013; Miao et al. 2014).

As planar and vertical reservoir heterogeneity escalates, it is a great challenge to recover the remaining oil from mature waterflooded oilfields which suffer from extremely high water cut and unfavorable recovery performance of original oil in place (Li 2009; Han 2010). Due to the highly scattered and relatively enriched distribution of remaining oil, efficient enhanced oil recovery (EOR) techniques have become imperative. Water-alternating-gas (WAG) injection has been identified as a cost-effective EOR process yielding high recovery in some oilfields (Luo et al. 2013; Salehi et al. 2014; Laochamroonvorapongse et al. 2014; Sheng 2015). Nevertheless, considering the complicated phase behavior and various flow characteristics in heterogeneous formations, gas tends to break through early in production wells due to overriding, fingering, and channeling, which may result in unfavorable recovery performance. So far, few attempts have been made to implicitly estimate water–oil–gas relative permeability curves during immiscible WAG injection. Taking a synthetic reservoir as an example, Li et al. proposed a numerical inversion method for estimation of the three-phase relative permeability curve using the ensemble Kalman filter algorithm for assisted history matching. However, due to the inherent limitations confronted by the relative permeability representation model, no significant recognitions were achieved (Li et al. 2012; Chen and Reynolds 2015). Using the Levenberg–Marquardt (LM) algorithm for automatic history matching, Hou et al. addressed the optimization of production performance and relative permeability representation models and finally proposed a numerical inversion method for estimation of the radial water–oil relative permeability curve, which attracts great interest by researchers in petroleum engineering (Hou et al. 2012a, b, 2015).

As a result of the above-mentioned problems, this paper presents a novel numerical inversion method for estimation of the water–oil–gas relative permeability curve during immiscible WAG processes. The structure of this paper is organized as follows. In Sect. 2, the formulation and architecture of the relative permeability representation model are presented. Section 3 provides a brief description of the proposed numerical inversion method using the LM based history matching techniques. In Sect. 4, laboratory tests including phase behavior studies, minimum miscibility pressure (MMP) measurement, and immiscible WAG coreflood experiments are conducted to understand the major recovery mechanisms and thus generate accurate fluid properties under reservoir conditions. Finally, the reliability and robustness of the proposed method are validated by interpreting coreflood data of the immiscible

WAG experiment to implicitly estimate the water–oil–gas relative permeability curve.

2 Relative permeability representation model

According to whether it is required to assume the shape of the relative permeability curve, there are two main categories in the representation model: the parametric model and the nonparametric model (Kulkarni and Datta-Gupta 2000). The parametric model uses explicit equations to generate the two- or three-phase relative permeability curve, assuming that the relative permeability curve fits into the shape of a certain type of the functional model (e.g., the power law model). Nevertheless, the number of degrees of freedom of the parametric model is not enough for all types of relative permeability curves for actual reservoirs. Due to its simplicity, the power law model is widely used (Lee and Seinfeld 1987; Reynolds et al. 2004). The nonparametric model is far more general and flexible because there is no assumption regarding the shape of the relative permeability curves [e.g., the cubic B-spline model (CBM)], which has the advantage of being able to accurately represent any set of relative permeability curves. Thus far, the nonparametric model most widely used is the CBM (Chen et al. 2008; Eydinov et al. 2009). For this study, the CBM is adopted to describe the water–oil–gas relative permeability curve during the WAG recovery process.

First, the dimensionless fluid saturation is defined as

$$S_{wD} = \frac{S_w - S_{iw}}{1 - S_{iw} - S_{orw}} \tag{1}$$

$$S_{owD} = 1 - S_{wD} \tag{2}$$

$$S_{gD} = \frac{S_g - S_{gc}}{1 - S_{iw} - S_{gc} - S_{org}} \tag{3}$$

$$S_{ogD} = 1 - S_{gD}, \tag{4}$$

where S_{wD} and S_{owD} are the dimensionless water and oil saturation for the water–oil system, respectively; S_{gD} and S_{ogD} are the dimensionless gas and oil saturation for the oil–gas system, respectively; S_w is the water saturation; S_{iw} is the initial water saturation; S_{orw} is the residual oil saturation for the water–oil system; S_g is the gas saturation; S_{gc} is the critical gas saturation; and S_{org} is the residual oil saturation for the oil–gas system.

The CBM model for the water–oil–gas relative permeability curve is given by

$$k_{rp}(S_{pD}) = \sum_{j=-3}^{n-1} C_{j+2}^p B_{j+3}(S_{pD}), \quad p = w, ow, g, og, \tag{5}$$

where k_{rp} is the p-phase relative permeability at the dimensionless fluid saturation S_{pD}; n is the number of the controlling knots; C_{j+2}^p is the controlling points for the B-spline approximations of the p-phase relative permeability curve; and $B_{j,3}(S_{pD})$ is the basis function for the CBM model. For detail, please refer to de Boor (de Boor 1978).

It should be noted that the traditional B-spline curve is "attached" to the controlling knots but does not normally pass through them. Therefore, to ensure two of its controlling knots C_0^p and C_n^p for the p-phase relative permeability curve being traversed, two phantom knots C_{-1}^p and C_{n+1}^p are introduced for each curve to clamp the endpoints of three-phase relative permeability curves. Equation (6) is proposed to describe the correlation:

$$C_{-1}^p = 2C_0^p - C_1^p \quad \text{and} \quad C_{n+1}^p = 2C_n^p - C_{n-1}^p. \tag{6}$$

Since the water–oil relative permeability and oil–gas relative permeability are, respectively, normalized by $K_o(S_{iw})$ and $K_o(S_{gc})$, it leads to $C_0^{ow} = C_0^{og} = 1$. Moreover, the endpoints of the fluid saturation S_{iw}, S_{orw}, S_{gc}, and S_{org} are regarded as known values, i.e., $C_n^{ow} = C_n^{og} = C_0^w = C_0^g = 0$. There are $n - 1$ parameters to be estimated for the oil relative permeability of the water–oil or oil–gas system, and there are n parameters to be estimated for the water or gas relative permeability. Finally, there are total $4n - 2$ parameters to be estimated for the water–oil–gas system.

To enforce the monotonicity and convexity of the three-phase relative permeability curve, a log transformation from the controlling knots to pseudo-controlling knots is carried out. For the water-phase or gas-phase relative permeability curve,

$$\begin{cases} x_1^u = \ln\left(\dfrac{C_1^u}{\frac{1}{2}(C_2^u + 0) - C_1^u}\right); \\[2ex] x_i^u = \ln\left(\dfrac{C_i^u - (2C_{i-1}^u - C_{i-2}^u)}{\frac{1}{2}(C_{i+1}^u + C_{i-1}^u) - C_i^u}\right), \ 2 \le i \le n-1; \quad u = w, g. \\[2ex] x_n^u = \ln\left(\dfrac{C_n^u - (2C_{n-1}^u - C_{n-2}^u)}{1 - C_n^u}\right). \end{cases} \tag{7}$$

For the oil-phase relative permeability curve of the water–oil or oil–gas system,

$$\begin{cases} y_1^v = \ln\left(\dfrac{C_1^v - (2C_2^v - C_3^v)}{\frac{1}{2}(C_2^v + 1) - C_1^v}\right); \\[2ex] y_i^v = \ln\left(\dfrac{C_i^v - (2C_{i+1}^v - C_{i+2}^v)}{\frac{1}{2}(C_{i+1}^v + C_{i-1}^v) - C_i^v}\right), \ 2 \le i \le n-2; \quad v = ow, og. \\[2ex] y_{n-1}^v = \ln\left(\dfrac{C_{n-1}^v - 0}{\frac{1}{2}C_{n-2}^v - C_{n-1}^v}\right). \end{cases} \tag{8}$$

To sum up, the vector $\mathbf{m}$ of unknown model parameters of the CBM can be expressed as

$$\mathbf{m} = \left[x_1^w, x_2^w, \ldots, x_n^w, y_1^{ow}, y_2^{ow}, \ldots, y_{n-1}^{ow}, x_1^g, x_2^g, \ldots, x_n^g, y_1^{og}, y_2^{og}, \ldots, y_{n-1}^{og} \right]. \tag{9}$$

With regard to the actual estimation of the three-phase relative permeability curve, the vector $\mathbf{m}$ of unknown model parameters, which is composed of the pseudo-controlling knot vectors ($\mathbf{x}$ and $\mathbf{y}$) mentioned above, is adjusted subsequently using the optimization algorithm. After each iteration, the controlling knot vectors ($\mathbf{C}^u$ and $\mathbf{C}^v$) are calculated by inverse transforming the pseudo-controlling parameters. Then, the water–oil and oil–gas relative permeabilities satisfying the monotonicity and convexity rule are obtained from the cubic B-spline model. For this study, the number of controlling knots n is equal to 7.

Once the water–oil and oil–gas relative permeabilities are obtained after each iteration, the modified Stone's Model II (Aziz and Settari 1979) is further adopted to calculate the oil relative permeability curve when all three phases are mobile, which takes the form of

$$K_{ro}(S_w, S_g) = K_{row}^0 \left[\left(\frac{K_{row}(S_w)}{K_{row}^0} + K_{rw}(S_w) \right) \times \left(\frac{K_{rog}(S_g)}{K_{row}^0} + K_{rg}(S_g) \right) - \left(K_{rw}(S_w) + K_{rg}(S_g) \right) \right], \tag{10}$$

where K_{rw} and K_{row} are the water and oil relative permeability for the water–oil system, respectively; K_{rg} and K_{rog} are the gas and oil relative permeability for the oil–gas system, respectively; and K_{row}^0 is the oil relative permeability at the connate water saturation (in oil–water flow), and at the critical saturation (in oil–gas flow).

3 Methodology

3.1 Least-squares objective function

Provided that the predicted production performance is in accordance with the observed values, a least-squares objective function needs to be established for estimation of the water–oil–gas relative permeability curve and can be described as follows:

$$O(\mathbf{m}) = \frac{1}{2}(\mathbf{g}(\mathbf{m}) - \mathbf{d}_{obs})^T \mathbf{C}_D^{-1}(\mathbf{g}(\mathbf{m}) - \mathbf{d}_{obs}), \tag{11}$$

where $O(\mathbf{m})$ is the least-squares objective function; $\mathbf{m}$ is a ($m \times 1$) vector of the unknown model parameters; T is a symbol denoting the transpose of a vector or matrix; $\mathbf{d}_{obs}$ is a ($n \times 1$) vector of the observed (or measured) data; $\mathbf{g}(\mathbf{m})$ is a ($n \times 1$) vector of the predicted data; and $\mathbf{C}_D$ is the ($n \times n$) covariance matrix. With regard to the actual history matching problems, the objective function $O(\mathbf{m})$ is usually nonlinear and the vector of model parameters $\mathbf{m}$ should be confined to a reasonable range according to reservoir conditions. In this study, the pressure drop, water cut, and the cumulative gas–oil ratio are considered as the observed production performance to establish the least-squares objective function.

3.2 LM algorithm

The LM algorithm (Oliver and Chen 2011), one of the gradient-based algorithms most widely used, has high computational efficiency and a quick convergence speed. When using the LM algorithm to solve the inverse history matching problems, the smooth transition can be addressed successively between the steepest descent algorithm and the Newton algorithm (Barua et al. 1988). The optimization procedure should satisfy the following principle: if the least-squares objective function is far from the minimum point, the convergence direction should be identical to that of the steepest descent algorithm; if the objective function is close to the minimum point, the convergence direction is the same as that of the Newton algorithm. Optimization of production performance in this paper is performed using the LM algorithm. In addition, a finite difference method is adopted to compute the sensitivity matrix of the least-squares objective function at the unknown model parameters. The generalized form of the LM algorithm is depicted as Eq. (12):

$$(\lambda \mathbf{I} + \mathbf{H}(\mathbf{m}^k))\delta \mathbf{m}^{k+1} = -\nabla O(\mathbf{m}^k), \qquad (12)$$

where $\mathbf{H}(\mathbf{m}^k)$ is the Hessian matrix for the kth iteration; $\mathbf{I}$ is an $(n \times n)$ identity matrix; λ is the damping factor to guarantee the half-positive definitiveness of the Hessian matrix; $O(\mathbf{m}^k)$ is the least-squares objective function for the kth iteration; $\mathbf{m}^k$ and $\mathbf{m}^{k+1}$ are, respectively, the vector of unknown model parameters obtained by the kth and $(k + 1)$th iterations, and $\delta \mathbf{m}^{k+1} = \mathbf{m}^{k+1} - \mathbf{m}^k$; and ∇ is the Hamiltonian operator.

The following is the specific calculation procedure of using the LM algorithm for automatic history matching. First, input the initial damping factor λ_0. After each iteration, it is necessary to adjust the values of the damping factor. The principle for adjustment is summarized as follows: (1) Calculate the vector of unknown model parameters $\mathbf{m}^{k+1}$. If $O(\mathbf{m}^{k+1}) \geq O(\mathbf{m}^k)$, the iteration is regarded as a failure, and then $\lambda = \lambda \times 10$. If $O(\mathbf{m}^{k+1}) < O(\mathbf{m}^k)$, the iteration is regarded as a success, and then $\lambda = \lambda \times 10$. (2) Submit the damping factor λ adjusted to Eq. (12) and carry out the next iteration. The iteration described previously is repeated until the termination condition is satisfied.

The termination condition of iteration utilized takes the form of Eq. (13):

$$|O(\mathbf{m}^{k+1}) - O(\mathbf{m}^k)| < \varepsilon_1 \quad \text{or} \quad count > count_{max}, \qquad (13)$$

where ε_1 is the convergence precision (fraction); $count$ is the iteration times (integer); and $count_{max}$ is the maximum iteration times (integer). For this study, $\varepsilon_1 = 10^{-6}$ and $count_{max} = 100$.

3.3 Procedure of parameter estimation

The procedure for implicitly estimating the three-phase relative permeability curve using LM based history matching techniques is briefly described as follows: (1) initialize the unknown controlling parameters of the CBM to generate prior knowledge of the three-phase relative permeability curve; (2) implement reservoir simulation using the prior curves to generate the predicted production data; (3) establish a least-squares objective function to reflect the discrepancy between the predicted and observed values of production performance; and (4) advance the minimization of the least-squares objective function using the LM algorithm to subsequently adjust the unknown controlling parameter vector of the CBM until all the observed production data are assimilated. Finally, the water–oil–gas three-phase relative permeability curve is achieved and evaluated.

4 Laboratory tests

Using synthetic core samples, laboratory tests including phase behavior studies, MMP measurements, and immiscible WAG coreflood experiments were conducted to understand the major recovery mechanisms of immiscible WAG injection and thus generate accurate fluid properties under reservoir conditions.

4.1 Phase behavior studies

Well product and dissolved gas were collected from a low-permeability classic sandstone reservoir in the Jidong Oil-field, China. The density and viscosity of the well product were measured to be 0.871 g/cm^3 and 8.477 mPa s at 20 °C and at atmospheric pressure, respectively. The injection gas with a molar mass of 20.6 g/mol and a relative density of 0.71 came from a neighboring reservoir. The compositions of the well product, dissolved gas, and the injection gas are listed in Table 1. For this study, a synthetic oil sample was prepared from the well product and the dissolved gas according to the initial dissolved gas–oil ratio of 70.58 m^3/sm^3.

Single-phase flash and saturated pressure measurements were performed on the synthetic oil sample at the reservoir temperature of 120.8 °C. The dissolved gas–oil ratio and the saturated pressure were measured to be 69.0 m^3/sm^3 and 14.1 MPa, respectively, which are so close to those determined under the initial reservoir conditions that the synthetic oil sample satisfies the requirement of phase behavior studies.

4.1.1 Influence of hydrocarbon gas injected on saturated pressure

The integrated experimental apparatus (Guo et al. 2000) provided by the Ruska Instrument Corporation, United States, was used to understand the influence of hydrocarbon gas injected on saturated pressure at a reservoir temperature of 120.8 °C. This apparatus consisted of a fluid property measurement device and a high-pressure falling-ball viscometer. Figure 1 shows the correlation of relative volume with pressure under various mole fractions of hydrocarbon gas injected.

Figure 1 demonstrates that there is a distinct breaking point indicating the saturated pressure, especially when the mole fraction of hydrocarbon gas injected is relatively low. Meanwhile, the fluid phase behavior changed significantly as the experimental pressure dropped. With an increase in the mole fraction of hydrocarbon gas injected, the correlation of relative volume with pressure gradually shifted toward the right accompanied by the disappearance of the breaking point, which indicates that the continuous solution of injected hydrocarbon gas results in an increase in the saturated pressure and a reduction of discrimination between gas–liquid phases to a large extent.

4.1.2 Influence of hydrocarbon gas injected on fluid properties

Using the constant composition expansion test, differential liberation test, and the swelling test under the reservoir

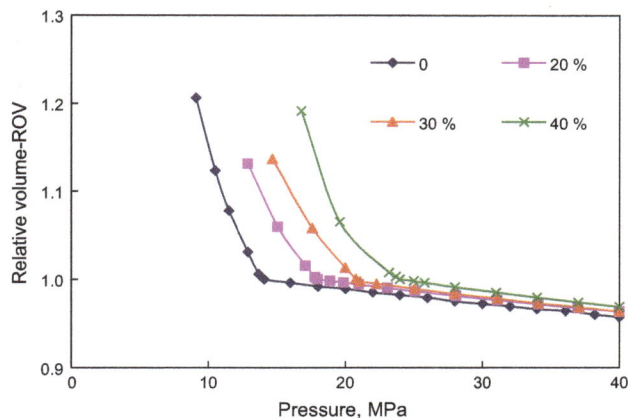

Fig. 1 Correlation of relative volume with pressure under various mole fractions of hydrocarbon gas injected

temperature of 120.8 °C, the influence of various mole fractions of hydrocarbon gas injected on variation of fluid properties was investigated, as presented in Figs. 2 and 3. The results show that, with the escalation of hydrocarbon gas injected, the saturated pressure increased gradually along with a decrease in oil density and viscosity, which shows a closer similarity between gas and liquid phases, and as such better fluid properties will be achieved.

4.2 MMP measurement

A long slim-tube displacement experiment was conducted to determine the minimum miscibility pressure (MMP) between the synthetic oil sample and hydrocarbon gas. Generally, if the recovery factor is greater than 80 % when hydrocarbon gas breakthrough occurs or the ultimate oil recovery reaches 90 %–95 % after 1.2 pore volume (PV) hydrocarbon gas is injected, it is treated as a state of miscibility. The experimental setup was provided by the Ruska Company, United States, and consisted of an

Table 1 Compositions of the well product, dissolved gas, and the injection gas

Component	Composition, wt%		
	Well product	Dissolved gas	Injection gas
CO_2	0.06	0.12	0.32
N_2	0.07	0.14	0.19
CH_4	30.67	61.79	83.62
C_2H_6	8.76	17.64	7.64
C_3H_8	5.96	11.98	3.78
i-C_4H_{10}	0.72	1.46	0.97
n-C_4H_{10}	1.55	3.13	1.52
i-C_5H_{12}	0.27	0.54	0.66
n-C_5H_{12}	0.26	0.53	0.56
C_{6+}	51.69	2.67	0.74

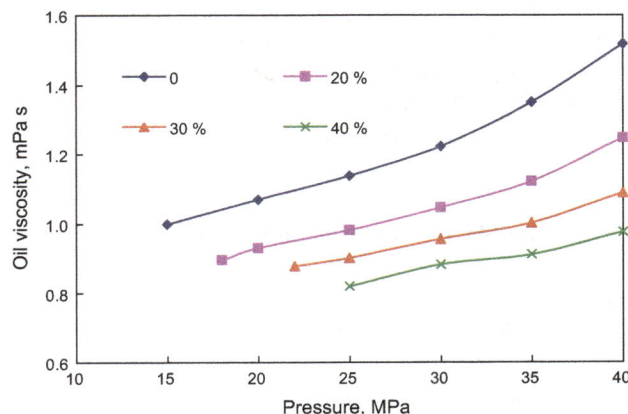

Fig. 2 Variation of oil viscosity with pressure with various mole fractions of hydrocarbon gas injected

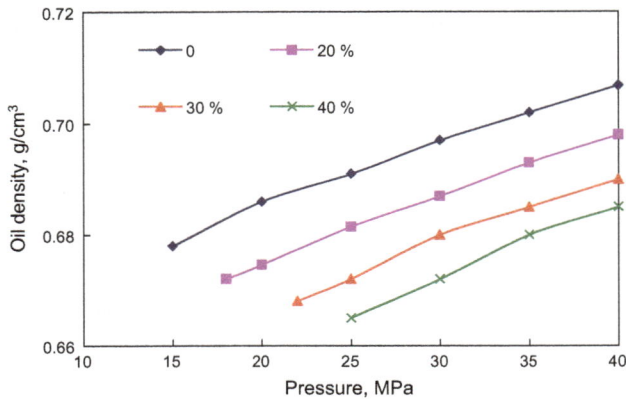

Fig. 3 Variation of oil density with pressure with various mole fractions of hydrocarbon gas injected

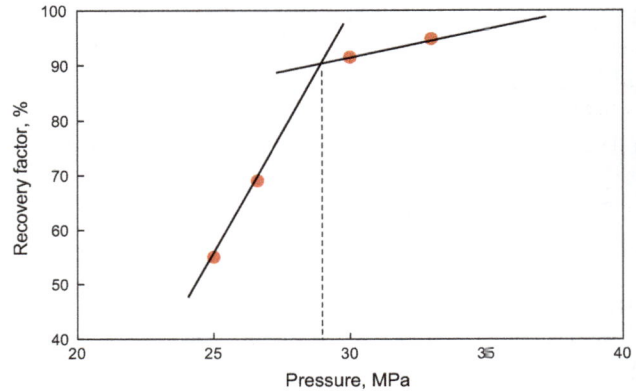

Fig. 4 Observed recovery factor versus different experimental pressures

injection system with a positive displacement pump, a slim tube, a backpressure regulator, a differential pressure transducer, a temperature-controlling system, a sample collection system, and a gas chromatograph. The long slim-tube model was approximately 18 m in length and 4 mm in diameter, with a pore volume of 125 cm³ at 20 °C and at atmospheric pressure.

Prior to displacement, the long slim-tube model was fully saturated with the synthetic oil sample at the reservoir temperature of 120.8 °C and under the ambient pressure above the bubble point. The experimental pressures were 25.0, 26.6, 30.0, and 33.0 MPa, respectively. The long slim-tube displacement experiments were performed at a constant gas injection rate of 0.167 cm³/min until 1.2 PV of hydrocarbon gas was injected. It should be noted that the pressure regulator must be adjusted sequentially during the displacement process in order to retain the ambient pressure close enough to the experimental pressure with its fluctuation range less than 0.05 MPa. As shown in Fig. 4, miscibility between the synthetic crude oil and hydrocarbon gas would be achieved as the experimental pressure reached 30.0 MPa. The MMP was further obtained using the interpolation method and its value was 29.0 MPa, which is significantly higher than the current reservoir pressure 27.0 MPa. That is to say, miscibility cannot be achieved under the current reservoir conditions.

4.3 Immiscible WAG coreflood experiment

Core samples were prepared to conduct the immiscible WAG coreflood experiments using a total of 12 representative samples taken from the same low-permeability sandstone reservoir. The physical properties of the actual reservoir core samples are listed in Table 2. The synthetic core sample was approximately 66.8 cm in length and

2.5 cm in diameter with a pore volume of 71.6 cm³, while the corresponding average porosity, average permeability, and rock compressibility were determined to be 21.95 %, 39.35×10^{-3} μm^2, and 5.2×10^{-6} MPa^{-1} respectively. The brine was composed of $NaHCO_3$ and distilled water with a salinity of 4664 mg/L. Oil samples and hydrocarbon gas injected were the same as those used in the phase behavior studies. Figure 5 presents the schematic diagram of the WAG coreflood experiment. The core sample holder had one inlet located in the left side and one outlet in the right side. The production was performed at a constant outlet pressure, and the injection was performed at a constant surface injection rate. In this displacement experiment, the initial oil saturation, residual oil saturation of the water–oil system, critical gas saturation, and the residual oil saturation of the oil–gas system were accurately measured to be 0.60, 0.44, 0, and 0.40, respectively.

Based on the measured MMP, an immiscible WAG coreflood experiment was carried out under the reservoir temperature of 120.8 °C and a backpressure of 27 MPa. The production performance, such as pressure drop, displacement efficiency, water cut, and cumulative gas–oil ratio, was simultaneously recorded with the advancing of the displacement. For the water flooding stage, the gas was injected at a constant rate of 0.3 cm³/min. As the water cut reached 85.7 % at the water flooding stage, WAG injection was initiated with a WAG ratio of 3:1 (0.24 PV water vs. 0.08 PV gas) and with four WAG cycles. Soon afterward, the subsequent water flooding was further carried out. Figures 6 and 7 show the production performance of the immiscible WAG displacement experiment. The recovery efficiency was significantly improved during the immiscible WAG processes. With regard to the four slugs of immiscible WAG injection, it was mainly the second WAG cycle that resulted in the greatest improvement of production performance.

Table 2 Properties of the actual reservoir core samples

Core number	Length, cm	Diameter, cm	Porosity, %	Absolute permeability, 10^{-3} μm^2
1	6.90	2.50	17.2	70.0
2	7.09	2.52	20.4	75.3
3	5.65	2.51	19.7	94.7
4	5.10	2.51	21.2	49.7
5	5.71	2.50	23.7	86.9
6	4.05	2.52	20.0	49.5
7	5.62	2.52	22.9	62.4
8	5.75	2.51	17.8	33.2
9	4.65	2.52	18.3	38.6
10	5.90	2.51	21.8	98.0
11	5.65	2.50	22.2	107.1
12	4.76	2.50	23.4	127.0

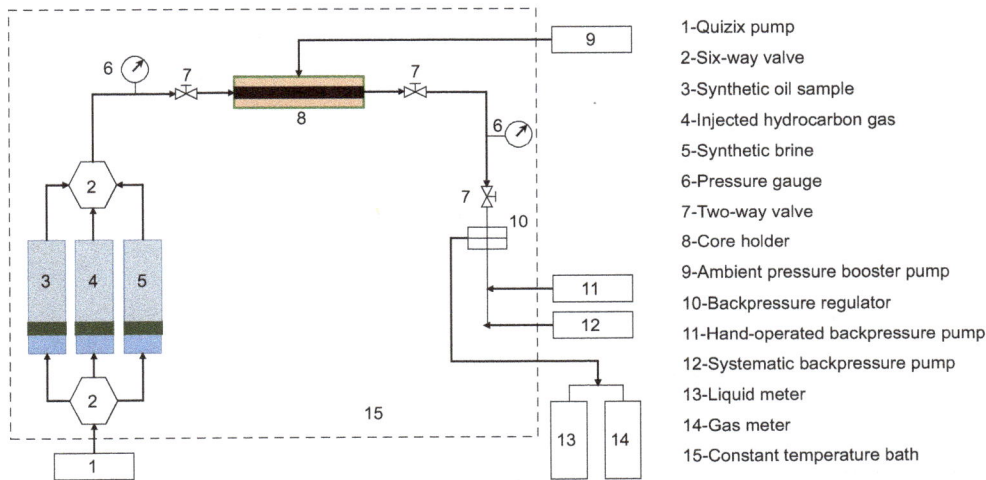

1-Quizix pump
2-Six-way valve
3-Synthetic oil sample
4-Injected hydrocarbon gas
5-Synthetic brine
6-Pressure gauge
7-Two-way valve
8-Core holder
9-Ambient pressure booster pump
10-Backpressure regulator
11-Hand-operated backpressure pump
12-Systematic backpressure pump
13-Liquid meter
14-Gas meter
15-Constant temperature bath

Fig. 5 Schematic of the immiscible WAG coreflood experiment system

5 Estimation of water–oil–gas relative permeability curve

Considering the above-mentioned immiscible WAG coreflood experiment and history matching results of fluid properties as well as MMP, a one-dimensional compositional model was established using the reservoir simulator CMG. A grid system of $50 \times 1 \times 1$ was selected to represent the physical model in the experiment, resulting in a grid block size of 1.34 cm $\times$ 2.23 cm $\times$ 2.23 cm. The controlling conditions of compositional simulation were the same as those of laboratory WAG coreflood experiments most widely used. Moreover, the influence of capillary pressure was neglected.

On this basis, the proposed numerical inversion method was employed to interpret coreflood data of the immiscible WAG experiment and to implicitly estimate the water–oil–gas relative permeability curve. Figures 8 and 9 display the estimated water–oil and oil–gas relative permeability

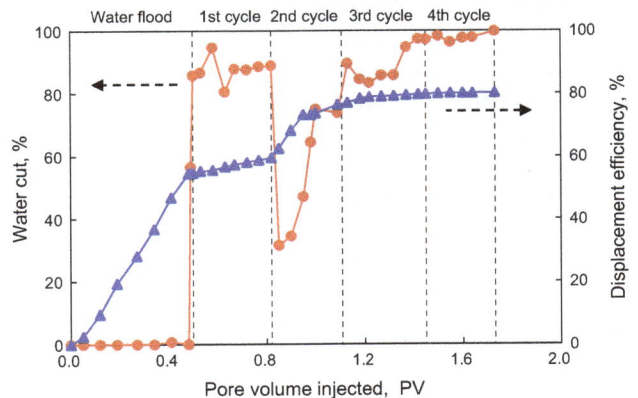

Fig. 6 Variation of the water cut and displacement efficiency during the immiscible WAG processes

curves, respectively, and the dashed lines denote the presumed water–oil and oil–gas relative permeability curves when the iteration was initialized using the LM algorithm

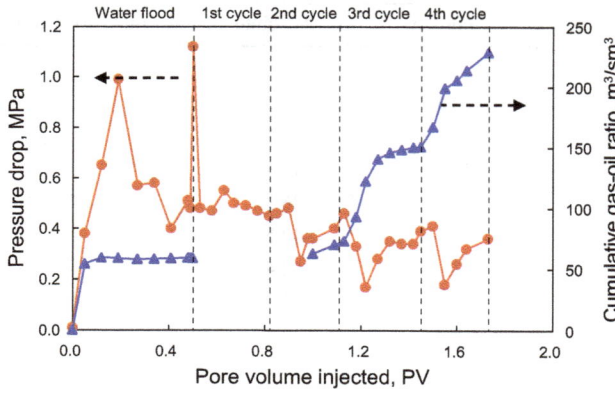

Fig. 7 Variation of the displacement pressure drop and the cumulative gas–oil ratio during the immiscible WAG process

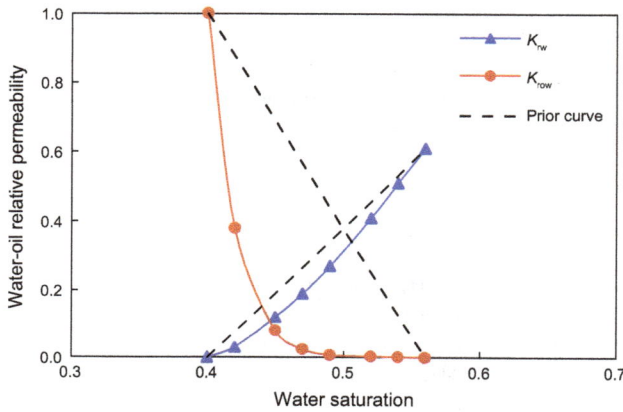

Fig. 9 Estimated oil–gas relative permeability curve for the oil–gas two-phase system

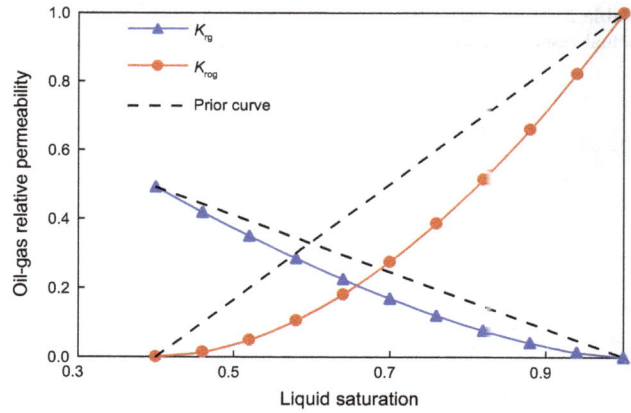

Fig. 8 Estimated water–oil relative permeability curve for the water–oil two-phase system

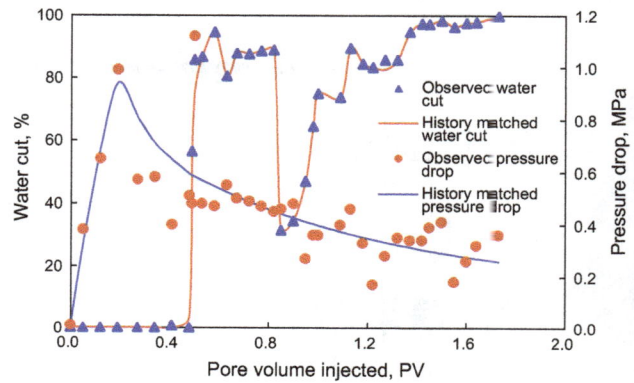

Fig. 10 Fitting results of the observed displacement pressure drop and water cut

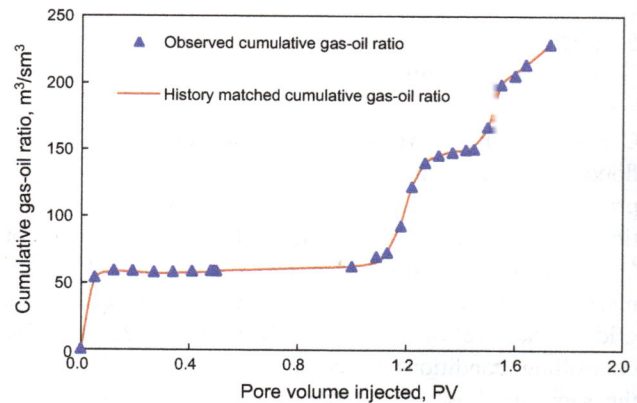

Fig. 11 Fitting results of the observed cumulative gas–oil ratio

for automatic history matching. Fitting results of production performance including pressure drop, water cut, and cumulative gas–oil ratio are plotted in Figs. 10 and 11, respectively. To compare the estimation accuracy, Eq. (14) was used to calculate the average absolute error between the predicted result and the observed production data including water cut, pressure drop, and cumulative gas–oil ratio:

$$R = \frac{1}{n_p} \sum_{i=1}^{n_p} |\zeta_i| \quad (\zeta_i = s_i - s_i'), \tag{14}$$

where R is the average absolute error; s_i is the real value of the ith data point (fraction); s_i' denotes the estimated value of the ith data point (fraction); n_p is the total number of data points; and ζ_i is the absolute error of the ith data point (fraction).

From Figs. 8, 9, 10, and 11, it can be found that the predicted production data including pressure drop, water cut, and cumulative gas–oil ratio are in good agreement with the observed values, with an average absolute error of

13.6 %, 2.3 %, and 1.7 %, respectively, which indicates that the estimated three-phase relative permeability curve is reliable. It also demonstrates that the proposed numerical inversion method is reliable and can meet engineering requirements, which provides a basic calculation theory for implicit estimation of water–oil–gas relative permeability curve during immiscible WAG processes.

6 Conclusions

(1) The cubic B-spline model (CBM) was used to describe the three-phase permeability curve. The optimization of production performance including pressure drop, water cut, and cumulative gas–oil ratio was performed by adopting the LM algorithm to subsequently adjust the vector of unknown model parameters of the CBM. Finally, a novel numerical inversion method was proposed to implicitly estimate the water–oil–gas relative permeability curves during immiscible WAG flooding processes.

(2) Actual core samples were used for phase behavior studies, MMP measurements, and immiscible WAG coreflood experiments to understand the major recovery mechanisms and thus to generate the fluid properties under reservoir conditions. Based on history matching results of fluid phase behavior and MMP, the proposed method was used to interpret coreflood data from the immiscible WAG experiment in order to implicitly estimate the water–oil–gas relative permeability curve. Results indicate that the proposed method is reliable and can meet engineering requirements. It provides a basic calculation theory for implicit estimation of three-phase relative permeability curve during immiscible WAG processes.

Acknowledgments The authors greatly appreciate the financial support of the Important National Science and Technology Specific Projects of China (Grant No. 2011ZX05010-002), and the Important Science and Technology Specific Projects of PetroChina (Grant No. 2014E-3203).

References

Abdollahzadeh A, Reynolds A, Christie MA, et al. Estimation of distribution algorithms applied to history matching. In: SPE reservoir simulation symposium, 21–3 Feb, The Woodlands; 2011. doi:10.2118/141161-MS.

Aziz K, Settari A. Petroleum reservoir simulation. Essex: Elsevier Applied Science Publishers; 1979.

Barroeta RG, Thompson LG. Importance of using pressure data while history matching a waterflooding process. In: Trinidad and Tobago energy resources conference, 27–30 June, Port of Spain, Trinidad; 2010. doi:10.2118/132347-MS.

Barua J, Horne RN, Greenstadt JL, et al. Improved estimation algorithms for automated type-curve analysis of well test. SPE Form Eval. 1988;3(1):186–96. doi:10.2118/14255-PA.

Chen BL, Reynolds AC. Ensemble-based optimization of the water-alternating-gas-injection process. In: SPE reservoir simulation symposium, 23–25 Feb, Houston; 2015. doi:10.2118/173217-MS.

Chen L, Zhang G, Ge J, et al. Research of the heavy oil displacement mechanism by using alkaline/surfactant flooding system. Colloids Surf A. 2013;434(19):63–71. doi:10.1016/j.colsurfa.2013.05.035.

Chen S, Li G, Peres A, et al. A well test for in situ determination of relative permeability curves. SPE Reserv Eval Eng. 2008;11(1):95–107. doi:10.2118/96414-PA.

Daoud AM, Velasquez LV. 3D field-scale automatic history matching using adjoint sensitivities and generalized travel time inversion. In: SPE annual technical conference and exhibition, 24–27 Sept, San Antonio; 2006. doi:10.2118/101779-MS.

de Boor C. A practical guide to splines. New York: Springer-Verlag; 1978.

Eydinov D, Gao G, Li G, et al. Simultaneous estimation of relative permeability and porosity/permeability fields by history matching production data. J Can Pet Technol. 2009;48(12):13–25. doi:10.2118/132159-PA.

Guo XQ, Yan W, Ma QL. Experimental study of the phase behavior of the reservoir fluid-CO_2 system. J Univ Pet China (Ed Nat Sci). 2000;24(3):12–5 (**in Chinese**).

Han DK. Discussion of concepts, countermeasures, and technical routines for the redevelopment of high water-cut oilfields. Pet Explor Dev. 2010;37(5):583–91 (**in Chinese**).

Hou J, Wang DG, Luo FQ, et al. Estimation of the water-oil relative permeability curve from radial displacement experiments. Part 1: numerical inversion method. Energy Fuels. 2012a;26(7):4291–9. doi:10.1021/ef300018w.

Hou J, Luo FQ, Wang DG, et al. Estimation of the water-oil relative permeability curve from radial displacement experiments. Part 2: reasonable experimental parameters. Energy Fuels. 2012b;26(7):4300–9. doi:10.1021/ef3005866.

Hou J, Zhou K, Zhang XS, et al. A review of closed-loop reservoir management. Pet Sci. 2015;12(1):114–28. doi:10.1007/s12182-014-0005-6.

Kulkarni KN, Datta-Gupta A. Estimating relative permeability from production data: a streamline approach. SPE J. 2000;5(4):402–11. doi:10.2118/66907-PA.

Laochamroonvorapongse R, Kabir CS, Lake LW. Performance assessment of miscible and immiscible water-alternating gas floods with simple tools. J Pet Sci Eng. 2014;122:18–30. doi:10.1016/j.petrol.2014.08.012.

Lee TY, Seinfeld JH. Estimation of absolute and relative permeabilities in petroleum reservoirs. Inverse Problem. 1987;3(4):711–28. doi:10.1088/0266-5611/3/4/015.

Li H, Chen S, Yang D, et al. Estimation of relative permeability by assisted history matching using the ensemble Kalman filter method. In: Canadian international petroleum conference, 16–8 Jun, Calgary; 2009. doi:10.2118/2009-052.

Li H, Chen SN, Yang DY, et al. Estimation of relative permeability by assisted history matching using the ensemble Kalman filter method. J Can Pet Technol. 2012;51(3):1–10. doi:10.2118/156027-PA.

Li H, Yang DYT. Estimation of multiple petrophysical parameters for the PUNQ-S3 model using ensemble-based history matching. In: SPE EUROPEC/EAGE annual conference and exhibition, 23–26 May, Vienna; 2011. doi:10.2118/143583-MS.

Li Y. Study of enhancing oil recovery of continental reservoirs by water drive technology. Acta Pet Sin. 2009;30(3):396–9 (**in Chinese**).

Luo P, Zhang YP, Huang S. A promising chemical-augmented WAG process for enhanced heavy oil recovery. Fuel. 2013;104:333–41. doi:10.1016/j.fuel.2012.09.070.

Masihi M, Javanbakht L, Bahaloo HF, et al. Experimental investigation and evaluation of three phase relative permeability models. J Pet Sci Eng. 2011;79(1–2):45–53. doi:10.1016/j.petrol.2011.08.017.

Miao TJ, Yu BM, Duan YG, et al. A fractal model for spherical seepage in porous media. Int Commun Heat Mass Transfer. 2014;58:71–8. doi:10.1016/j.icheatmasstransfer.2014.08.023.

Oliver DS, Chen Y. Recent progress on reservoir history matching: a review. Computational Geosciences. 2011;15(1):185–221. doi:10.1007/s10596-010-9194-2.

Reynolds PC, Li R, Oliver DS. Simultaneous estimation of absolute and relative permeability by automatic history matching of three-phase flow production data. J Can Pet Technol. 2004;43(3):37–46. doi:10.2118/04-03-03.

Salehi MM, Safarzadeh MA, Sahraei E, et al. Comparison of oil removal in surfactant alternating gas with water alternating gas, water flooding and gas flooding in secondary oil recovery process. J Pet Sci Eng. 2014;120:86–93. doi:10.1016/j.petrol.2014.05.017.

Sheng JJ. Enhanced oil recovery in shale reservoirs by gas injection. Journal of Natural Gas Science and Engineering. 2015;2:252–9. doi:10.1016/j.jngse.2014.12.002.

Sigmund PM, McCaffery FG. An improved unsteady-state procedure for determining the relative-permeability characteristics of heterogeneous porous media. SPE Journal. 1979;19(1):15–28. doi:10.2118/6720-PA.

Wang YD, Li GM, Reynolds AC. Estimation of depths of fluid contacts and relative permeability curves by history matching using iterative ensemble Kalman smoothers. SPE Journal. 2010;15(2):509–25. doi:10.2118/119056-PA.

Wang YD, Li M. Reservoir history matching and inversion using an iterative ensemble Kalman filter with covariance localization. Pet Sci. 2011;8(3):316–27. doi:10.1007/s12182-011-0148-7.

Xu P, Qiu SX, Yu BM, et al. Prediction of relative permeability in unsaturated porous media with a fractal approach. Int J Heat Mass Transf. 2013;64:829–37. doi:10.1016/j.ijheatmasstransfer.2013.05.003.

Zhang Y, Yang DY. Simultaneous estimation of relative permeability and capillary pressure for tight formations using ensemble-based history matching method. Comput Fluids. 2013;71:446–60. doi:10.1016/j.compfluid.2012.11.013.

12

Characteristics and accumulation model of the late Quaternary shallow biogenic gas in the modern Changjiang delta area, eastern China

Xia Zhang[1] · Chun-Ming Lin[1]

Abstract The Changjiang (Yangtze) is one of the largest rivers in the world. It formed a huge incised valley at its mouth during the Last Glacial Maximum; the incised-valley fill, approximately 80–110 m thick, supplies an important foundation for the generation of shallow biogenic-gas reservoirs. Two cores and 13 cone penetration tests were used to elaborate the characteristics, formation mechanism, and distribution of the shallow biogenic-gas reservoirs in the study area. The natural gas is mainly composed of CH_4 (generally >95%) with a $\delta^{13}C_{CH4}$ and $\delta^{13}C_{CO2}$ of −75.8 to −67.7‰ and −34.5 to −6.6‰, respectively, and a δD_{CH4} of −215 to −185‰, indicating a biogenic origin by the carbon dioxide reduction pathway. Commercial biogenic gas occurs primarily in the sand bodies of fluvial-channel, floodplain, and paleo-estuary facies with a burial depth of 50–80 m. Gas sources as well as cap beds are gray to yellowish-gray mud of floodplain, paleo-estuary, and offshore shallow marine facies. The organic matter in gas sources is dominated by immature type III kerogen (gas prone). The difference in permeability (about 4–6 orders of magnitude) between cap beds and reservoirs makes the cap beds effectively prevent the upward escape of gas in the reservoirs. This formation mechanism is consistent with that for the shallow biogenic gas in the late Quaternary Qiantang River incised valley to the south. Therefore, this study should provide further insight into understanding the formation and distribution of shallow biogenic gas in other similar postglacial incised-valley systems.

Keywords Biogenic gas · Formation mechanism · Late Quaternary · Modern Changjiang delta · Eastern China

1 Introduction

Biogenic gas is of significant importance because it is clean energy and an abundant resource, accounting for ~20% of the conventional natural gas reserves in the world (Rice and Claypool 1981). Recently, researchers have drawn considerable attention to the shallow biogenic gas in the near-surface marine and coastal sediments such as bay or estuarine deposits and late Quaternary incised-valley fills (García-García et al. 2007; Xu et al. 2009; Lin et al. 2010; Zhang et al. 2013; Jones et al. 2014; Okay and Aydemir 2016), in terms of the shallow-burial depth ranging from several tens of meters to hundreds of meters and easy exploitation with low investment and high benefit. Commercial reservoirs of shallow biogenic gas have been widely found in the world, including the North Sea (Heggland 1997; Vielstädte et al. 2015), Ria de Vigo incised valley, Spain (Garcia-Gil et al. 2002), western Gulf of Maine, USA (Rogers et al. 2006), and Lawrence Estuary, Canada (Pinet et al. 2008).

Shallow biogenic-gas reservoirs are principally distributed along the eastern and southern coasts of China, especially the Jiangsu–Zhejiang coastal plain area, where the postglacial Changjiang and Qiantang River incised valleys are located (Wang 1982; Zheng 1998; Lin et al. 2004; Liu et al. 2008; Li et al. 2010a; Zhang et al. 2013, 2014). Incised valleys generally have high preservation potential for their fill deposits (Dalrymple et al. 1994) and provide a fundamental and important background for the formation of the shallow biogenic-gas reservoirs therein, i.e., gas-source beds, cap beds, and sand

✉ Xia Zhang
zhangxiananjing@163.com

[1] State Key Laboratory for Mineral Deposits Research, School of Earth Sciences and Engineering, Nanjing University, Nanjing 210023, Jiangsu, China

Edited by Jie Hao

reservoirs can occur together in close geographic and stratigraphic proximity (Xu et al. 2009; Lin et al. 2004; Zhang et al. 2014). Considerable biogenic-gas accumulations with depth <120 m have been discovered and well documented in the late Quaternary Qiantang River incised valley with a predicted total gas amount of $244.5 \times 10^9 \text{ m}^3$ (Lin et al. 2004, 2010; Zhang et al. 2013). Also, the produced gas was used as gas supply for local villages and factories (Li and Lin 2010). Nevertheless, there are relatively few examples in the scientific literature regarding the shallow biogenic gas in the Changjiang incised valley area after >60 years' research and development (cf. Wang 1982; Zheng 1998). Previous research indicates that shallow biogenic gas is mainly distributed in the northern margin of the modern Changjiang delta (i.e., the M area in Fig. 1a, $\sim 2.5 \times 10^4 \text{ km}^2$; Wang 1982; Zheng 1998), and there are generally three sets of gas-bearing intervals with burial-depth ranges of 7–15, 25–35, and >50 m, respectively, involving prodelta–shallow marine, delta front, coastal plain, and floodplain facies (Zheng 1998).

Obviously, a poor understanding of the geological background, distribution, and formation of the shallow biogenic-gas reservoirs in the study area has seriously hampered the exploration and exploitation processes.

This paper is an extension of the previous work based primarily on the detailed observations and analyses of newly acquired cores ZK01 and ZK02, as well as 13 surrounding cone penetration tests (CPTs) in the Qidong and Haimen areas (Fig. 1), and the further correlation with more than 600 boreholes, most of which have been reported by Li et al. (2000, 2002) and Hori et al. (2002). The objective of this study is to discuss the conditions required for the formation of the shallow biogenic-gas reservoirs and to summarize the regularities of their distribution in the modern Changjiang delta area. This study will provide a useful insight into the exploration and exploitation of the shallow biogenic gas in similar incised-valley systems, more importantly for the modern Changjiang delta area, where there is dense population and tremendous economic development.

2 Geological setting

The Changjiang originating from the Qinghai–Tibet Plateau annually discharges water and sediment of 924 km³ and 4.8×10^{11} kg, respectively (Milliman and Syvitski 1992), and provides the primary sediment source for the modern Changjiang delta. The present-day Changjiang delta is situated at a coastal subsidence zone, with the altitude generally <5 m above mean sea level (Stanley and Chen 1996). It is bounded by hills in the west (Fig. 1a) and slopes gently toward the east with ~ 250 km in length from the apex around Zhenjiang–Yangzhou area to the modern river mouth. The modern Changjiang delta covers an area of $\sim 5.2 \times 10^4 \text{ km}^2$, with $2.3 \times 10^4 \text{ km}^2$ subaerial and $2.9 \times 10^4 \text{ km}^2$ subaqueous (Li et al. 2002). The former can be divided into three major units: the main delta body (M), and the southern (Sf) and northern (Nf) flanks (Li et al. 2002; Fig. 1a). The main body is characterized by the combination of distributary channels and three active river-mouth sand bars, which are elongated and extend southeastward (Li et al. 2002). The subaqueous part of the delta can be classified into subtidal flats with water depths <5–10 m, delta front ranging from 5–10 to 15–30 m, and prodelta (water depth of >30 and <50 m).

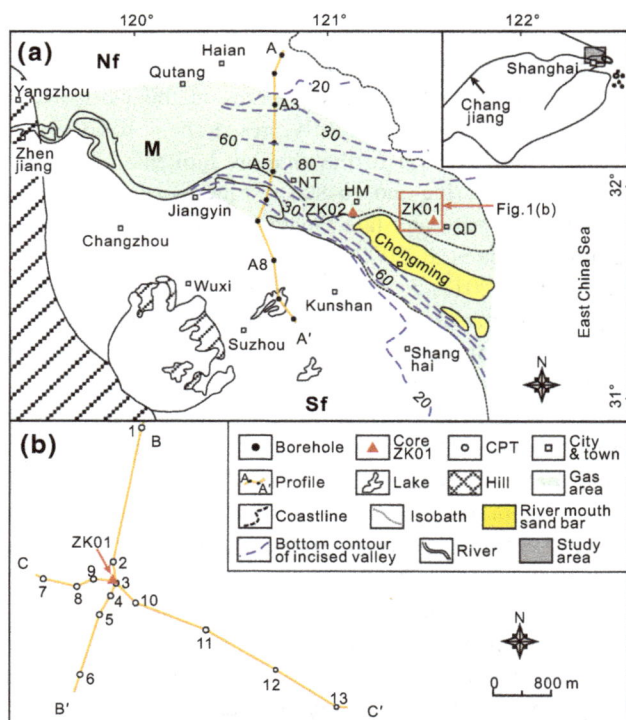

Fig. 1 a Schematic map of the modern Changjiang (Yangtze River) delta showing the locations of cores ZK01 and ZK02, boreholes, cross section, and gas-bearing area, as well as the basal topography of the Changjiang incised valley during the Last Glacial Maximum (modified from Wang 1982; Zheng 1998; Li et al. 2002; Zhang et al. 2015). b Locations of core ZK01, cone penetration tests (CPTs), and transects in the Qidong area from the northern margin of the modern Changjiang delta (see location in Fig. 1a). M Main delta area, Sf Southern flank, Nf Northern flank, QD Qidong area, HM Haimen area, NT Nantong

3 Materials and methods

Cores ZK01 (112 m penetration depth, 0.1 m diameter) and ZK02 (128 m penetration depth, 0.1 m diameter) were taken from Qidong (31°50′26.74″N, 121°33′24.08″E) and

Haimen (31°52'47.12"N, 121°09'30.69"E) areas where shallow gas is abundant, respectively, in 2014 by rotary drilling and with almost 100% recovery (see locations in Fig. 1a). In the laboratory, they were split, photographed, described, and subsampled. The grain size of 271 samples (0.5–1.0 m interval) was analyzed at 0.25 Φ spacing according to a standard method (cf. Zhou and Gao 2004), with grain-size parameters determined using the GRADI-STAT software (Blott and Pye 2001). Eighty-three samples were collected for foraminifera analysis using the method described by Zhang et al. (2014) and Wang et al. (1988). The permeabilities of 16 samples were measured by falling-head permeability test apparatus (cf. Zhang et al. 2013). The total organic carbon (TOC) and chloroform bitumen contents of 41 samples were analyzed following the method of Stax and Stein (1993). Eight mud samples were obtained for pyrolysis analysis (cf. Zou et al. 2006). Four ^{14}C ages were determined on shells or organic sediments by using accelerator-based mass spectroscopy in the Beta Analytic Radiocarbon Dating Laboratory (Lab No. Beta) in Miami, USA, and calibrated using the Calib Rev 7.1 (beta) program (Reimer et al. 2013; Table 1). Two ^{14}C ages on marine shells were calibrated utilizing the Marine13 model with the $\triangle R$ value of 135 ± 42 to deal with the marine reservoir effect (cf. Yoneda et al. 2007; Wang et al. 2012).

In addition, 13 CPTs were taken around the ZK01 core (Fig. 1(b)), with the tested parameters being cone tip resistance (q_c) and sleeve friction (f_s), in order to explore the distribution pattern of shallow gas, i.e., potential sand reservoirs (cf. Moran et al. 1989; Li and Lin 2010). During exploration, almost every CPT inevitably encountered gas with an original gas pressure of ~ 0.5 MPa and a flame height reaching up to 2 m when the gas was ignited. In this study, a total of six gas samples were collected and analyzed for chemical composition, stable carbon isotope ratios of CH_4 and CO_2, and stable hydrogen isotope ratios of CH_4 according to a standard method (cf. Ni et al. 2013). Furthermore, CPT-3 and the nearby ZK01 core were compared to calibrate the q_c and f_s curves allowing for distinguishing lithology, especially potential sandy

reservoirs (Fig. 1b). In general, q_c and f_s values increase as grain size increases (cf. Li and Lin 2010; Lin et al. 2015). Therefore, the q_c and f_s curves of sand and silt sediments show high values and large separation of curves, and the q_c curve lies generally to the left of f_s curve (cf. Li and Lin 2010; Lin et al. 2015).

4 Stratigraphic architecture

More than 600 cores were drilled in the present-day Changjiang delta area in the last five decades, laying a solid foundation for understanding the stratigraphic architecture of the coastal depositional system (Li et al. 2002; Hori et al. 2001; Wang et al. 2012). There are at least three stages for the formation of the Quaternary Changjiang incised-valley fills (Hori et al. 2002; Li and Wang 1998; Li et al. 2006; Figs. 1a, 2). Most of the incised-valley fills generated during the preceding two stages are missing resulted from the following strong down-cutting erosion and in many cases are characterized by a superposition of fluvial-channel sediments composed mainly of sandy and gravelly sediments. However, the incised-valley fill formed since the Last Glacial Maximum, the objective of this paper, is relatively complete (Fig. 2). It can be classified into five sedimentary facies that were deposited during the sea-level rise and subsequent stillstand (Figs. 2, 3). In the following text, we will use the newly drilled ZK01 core to describe each facies (Fig. 3).

Facies V (Fluvial Channel Sediments) is bounded by an erosional basal surface and shows a generally upward-fining trend (Figs. 2, 3, 4a). Sediments at the bottom consist mainly of gray or grayish-yellow gravelly sand interbedded by fine sand, silty sand, and gravels (Figs. 3, 4a). The gravels are angular to subangular, with a diameter of 2–50 mm (Fig. 4a), and the sand sediments are primarily composed of coarse (av. 43.4%) and medium (av. 28.8%) sands. The sediments at the top are characterized by an alternation of gray or grayish-yellow silty sand and silty fine sand (Fig. 3). The sediments are well-moderately sorted with a sorting coefficient of 1.14–2.52. Massive,

Table 1 AMS ^{14}C ages of organic sediments and shell samples from the newly drilled ZK01 core in the modern Changjiang delta area, China

Lab. code	Depth, m	Materials	Measured ^{14}C age, yr B.P.	Conventional ^{14}C age, yr B.P.	Calibrated ^{14}C age, cal. yr B.P.			
					Age, 1δ	Prob.	Age, 2δ	Prob.
Beta-409604	29.90	Organic sediments	5500 ± 30	5510 ± 30	6300 ± 18	0.95	6310 ± 40^{a}	0.83
Beta-409605	34.60	Shell	3800 ± 30	4200 ± 30	4090 ± 85	1	4100 ± 165	1
Beta-409606	60.00	Organic sediments	$11,080 \pm 40$	$11,110 \pm 40$	$13,000 \pm 60$	1	$12,960 \pm 120$	1
Beta-409607	75.60	Shell	$11,790 \pm 60$	$12,170 \pm 60$	$13,470 \pm 90$	1	$13,500 \pm 180$	1

In this paper, the 2δ calibrated ages are adopted and labeled in Fig. 3

[a] The age is not used because it does not follow the general trend

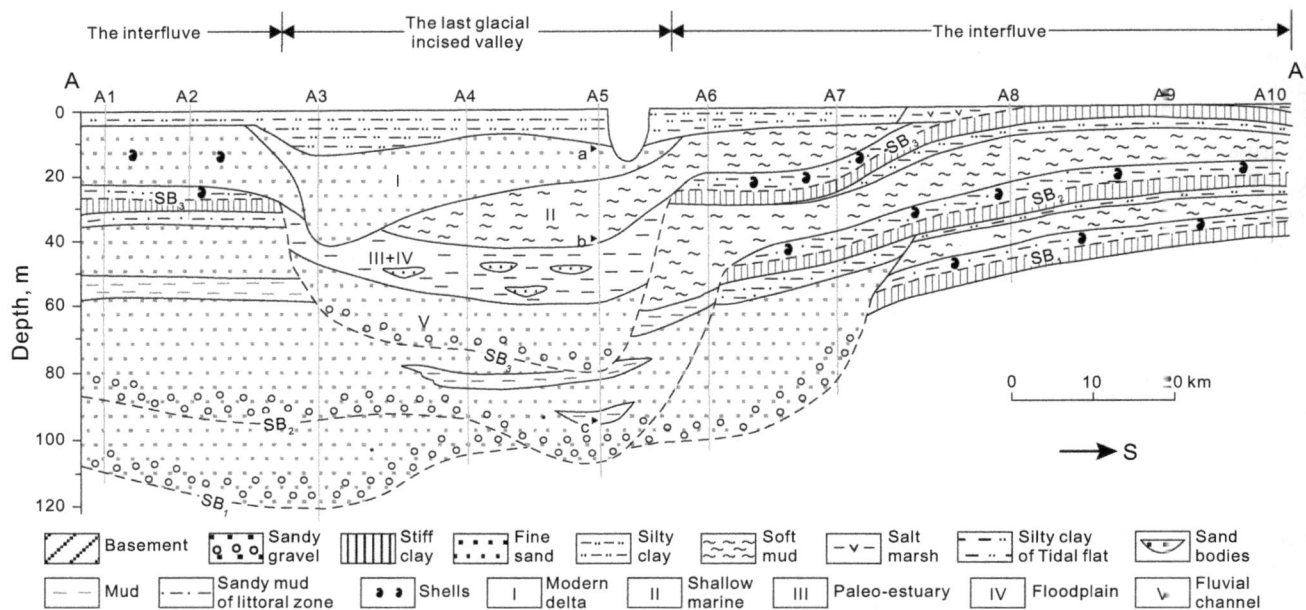

Fig. 2 Stratigraphic transect (*A–A''*) in the modern Changjiang delta region (modified from Zhang et al. 1998; see Fig. 1a for location). SB: sequence boundary with the subscript indicating the distinct stages for the formation of the Quaternary Changjiang incised valleys. [14]C data were obtained from Li et al. (2002) and calibrated by using the Calib Rev 7.1 (beta) program (Reimer et al. 2013): *a*-6595 ± 320 cal. yr B.P., 11.70 m depth; *b*- 12,900 ± 190 cal. yr B.P., 38.80 m depth; *c*- 39,200 ± 2200 cal. yr B.P., 94.50 m depth

graded, and parallel beddings, iron oxide spots, and shells are common, and there is a lack of tide-influenced sedimentary structures. Benthic foraminifera (BF) dominant by *Ammonia beccarii* are identified at the top with nine species and 15 individuals per 50 g (dry weight) sample size. A ^{14}C date at the burial depth of 75.6 m is 13,500 ± 180 cal. yr B.P. (Fig. 3; Table 1). This facies represents part of the river system and may have been deposited in a channel thalweg to bar environment (cf. Nittrouer et al. 2011; Zhang et al. 2014).

Facies IV (Floodplain Sediments) consists mainly of an alternation of gray mud and grayish-yellow sandy silt, silt, and gravelly sand (Figs. 3 and 4b). The gravels occupying 5%–10% of the coarse sediments have a diameter of 2–5 mm, up to 20 mm. Massive to graded beddings are common in coarse sediments, whereas silty blebs, and massive and lenticular beddings are abundant in mud sediments (Fig. 4b). Only one sample at 68.80 m depth contains some BF with four species and eight individuals per 50 g dry sample size.

Facies III (Paleo-estuary Sediments) is dominated by gray or yellowish-gray mud interbedded by silt and coarse sand with the lamina thickness ranging from 2 mm to 1 m (Fig. 4c). The structureless sand beds are usually typified by an erosional basal surface and present as a fining-upward succession with numerous irregular mud pebbles (Fig. 4d). They are well sorted with a sorting coefficient of 1.52–2.46 and mean grain size of 4.27–6.23 Φ. Wavy,

horizontal, and massive beddings are common. In addition, a set of tide-influenced sedimentary structures including sand-mud couplet and lenticular bedding are common. Foraminiferal fossils are abundant and mainly composed of BF which consists principally of *Ammonia beccarii* vars., *Florilus decorus*, *Elphidium magellanicum*, *Cribrononion vitreum* Wang, and *Elphidium advenum* (Cushman). The number of BF is 14–42 species and 61–10,304 individuals per 50 g dry sample size. A ^{14}C date of shells at the burial depth of 60.0 m is 12,960 ± 120 cal. yr B.P. (Fig. 3; Table 1). Facies II has been recorded as a macro-tidal system like the Qiantang River estuary with the maximum tidal range located in the Yangzhou area (Li et al. 2006), which is further testified by numerical simulations (Yang and Sun 1988; Uehara et al. 2002).

Facies II (Offshore Shallow Marine Sediments) consists mainly of gray or yellowish-gray soft mud interbedded with gray silt, fine sand, and clayey silt stripes (0.001–0.03 m thick) and blebs (Fig. 4e). Massive, horizontal and lenticular beddings, sand-filled burrows, bioturbation, and seriously broken shells are common (Fig. 4e). Foraminifera are also abundant in this facies and are mainly composed of BF. There are >60 BF species present, including *Ammonia beccarii* vars., *Elphidium magellanicum*, and *Cribrononion vitreum* Wang. The number of BF is 27–44 species and 139–5883 individuals, respectively, based on 50 g (dry weight) sample size. The foraminiferal fossils in Facies II resemble the living groups

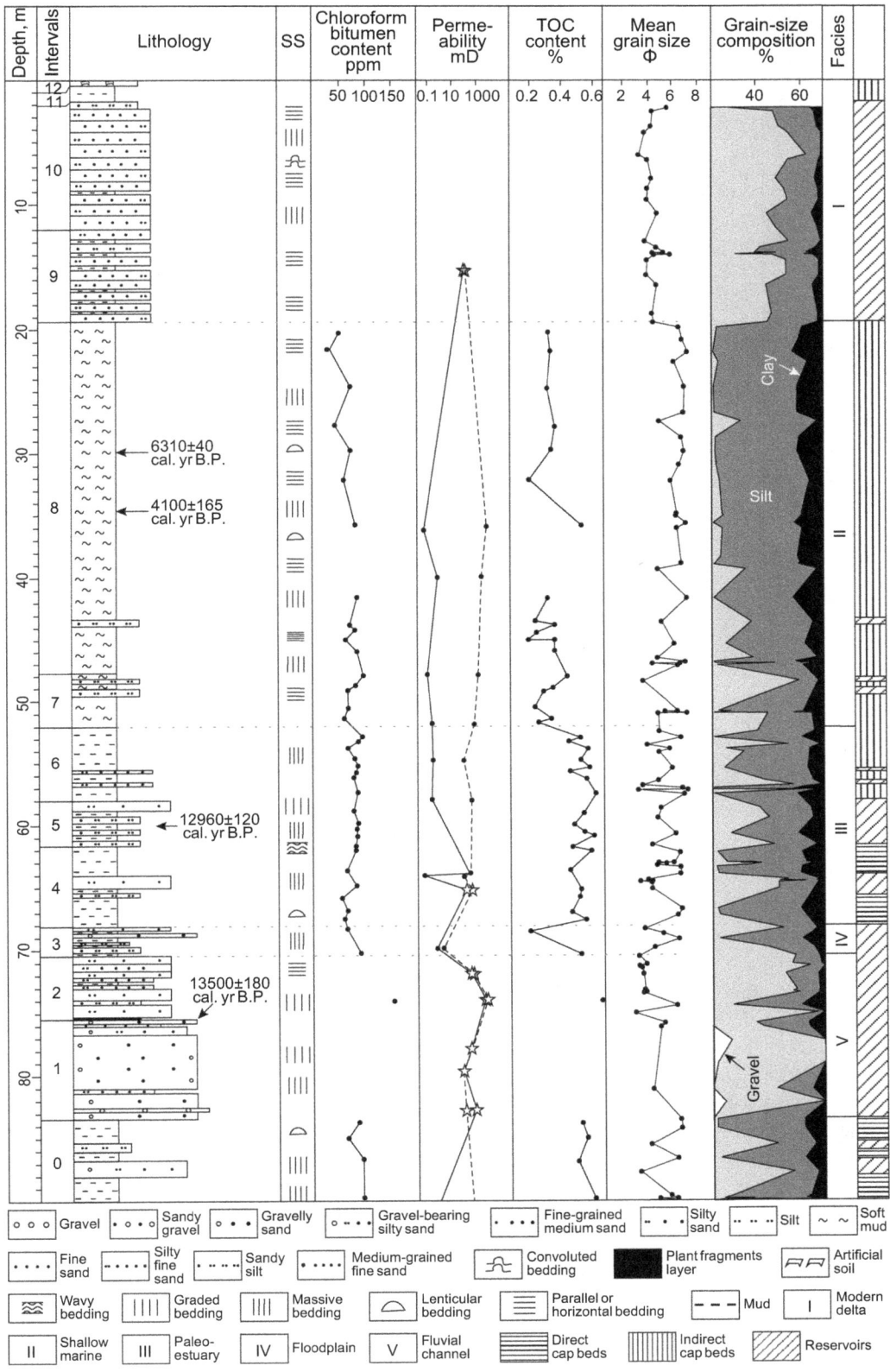

Fig. 3 Columnar section of core ZK01 in the modern Changjiang delta area (see Fig. 1 for location). *Black circles* indicate the depths at which various features were observed, and *light stars* show the sediment samples of reservoirs. In the column of permeability, the *dashed line* indicates horizontal permeability, while the *solid line* is vertical permeability. SS: sedimentary structure

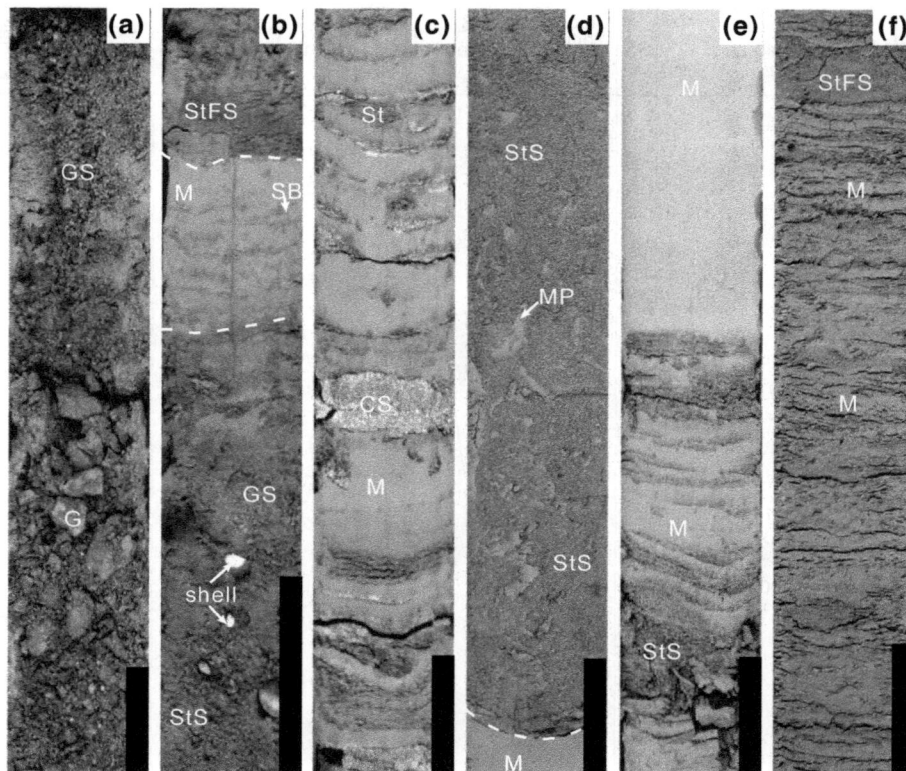

Fig. 4 Photographs of typical sedimentary characteristics in the ZK01 core. *Black scale bar* = 10 cm. **a** Facies V, 82.30–82.80 m depth: *grayish-yellow* gravelly sand (GS) with graded bedding. **b** Facies IV, 68.45–68.80 m depth: alternation of gray mud (M) and *grayish-yellow* silty fine sand (StFS), silty sand (StS), and gravelly sand with the lithological unconformity surface indicated by a *white dotted line*. Lenticular bedding and silty blebs (SB) are common in the mud sediments, while graded and massive bedding and shells are present in the sand sediments. **c** Facies III, 66.35–66.70 m depth: gray mud interbedded by thin silt (St) and coarse sand (CS) layers. **d** Facies III, 58.40–58.85 m depth: gray silty sand with an erosional basal surface (*white dashed line*) and numerous irregular mud pebbles (MP), present as a fining-upward succession. **e** Facies II, 57.30–57.80 m depth: alternation of gray mud and silty sand at the *bottom* and then massive gray soft mud at the *top*. **f** Facies I, 17.30–17.80 m depth: alternation of gray silty fine sand and mud. G: gravel

in the offshore shallow water areas (<20–55 m) of the East China Sea, South Yellow Sea, Changjiang delta, and Bohai Bay (Wang et al. 1981; Li and Wang 1998; Zhuang et al. 2002; Li et al. 2002; Li et al. 2010b).

Facies I (Modern Delta Sediments) shows an upward-coarsening sequence. The sediments at the bottom are characterized by an alternation of silty fine sand and mud (Figs. 3, 4f). The sand beds (0.3–7 cm thick) consist mainly of fine sand (52%–67%) and silt (29%–42%) with a mean grain size of 3.8–4.5 Φ and are well sorted with a sorting coefficient <2, while the mud sediments (0.1–3 cm thick) dominated by silt and clay. The sediments at the top consist predominantly of gray sand interbedded by gray or brown mud stripes (0.5–3 cm thick) or muddy gravels (Fig. 3). The sand sediments are mainly composed of fine sand (50.3%–75.8%) and silt (13.6%–37.7%), with a mean grain size of 3.3–4.4 Φ and sorting coefficient of 1.3–2.2. Parallel, massive, and convolute beddings, as well as seriously broken shells, are common (Fig. 3).

Foraminiferal fossils dominated by BF (>58 species; >42,688 in a 50-g dry sample) are also abundant in this facies. Most BF individuals are juveniles with small and/or seriously abraded shells. The BF assemblage of this facies, including *Ammonia beccarii* vars., *Elphidium naraensis*, *Florilus decorus*, *Protelphidium tuberculatum* (d'Orbigny), and *Elphidium magellanicum*, is similar to that of the modern Changjiang delta (cf. Li and Wang 1998).

5 Characteristics of shallow biogenic-gas reservoirs

5.1 Biogenic origin of shallow gas

Table 2 presents the chemical and isotopic compositions of the shallow gas in the study area. Results show that most of the gas samples are dominated by CH_4 (generally >95%), with minor N_2 (0.75%–9.07%) and CO_2 (0.98%–3.23%).

The gas sample from CPT-1 (see Fig. 1b for location) is special in consisting mainly of N_2 (64.0%) and CH_4 (23.5%). The carbon isotope values of CH_4 and CO_2 are -75.8 to $-67.7‰$ and -34.5 to $-6.6‰$, respectively, and the hydrogen isotope values of CH_4 are -215 to $-185‰$. These results indicate a biogenic origin for the shallow gas (cf. Whiticar et al. 1986; Whiticar 1999; Humez et al. 2016; Tao et al. 2016; Sun et al. 2016). The nitrogen-rich biogenic gas may be derived from the degradation of nitrogen-rich organic matter, indicating that the organic matter of the source sediments in the study area is heterogeneous (cf. Wang 1982). In addition, all the gases plot within "Bacterial Carbonate Reduction" zone in Fig. 5, indicating that the methanogenesis is predominant from carbon dioxide reduction (cf. Whiticar et al. 1986).

5.2 Gas-source sediments

There are three potential kinds of gas-source sediments in this area, including the gray and light-brown mud of Facies IV, gray and yellowish-gray mud of Facies III, gray and yellowish-gray soft mud of Facies II (Figs. 2, 3). Mud beds of Facies I can be excluded because they are thin (<0.5 m) and close to the surface (Figs. 2, 3).

Systematic analysis of core ZK01 indicates that TOC content increases with depth, and the argillic sediments of Facies III and IV have higher TOC contents than those of Facies II (Fig. 3). The TOC contents of Facies III and IV exceed the lower limits for both terrestrial (0.18%, Zhou et al. 1994) and marine gas sources (0.5%; Rice and Claypool 1981), whereas those of Facies II are below the lower limit for potential marine gas sources (Table 3). The chloroform bitumen content of source sediments shows a similar variation trend among distinct sedimentary facies (Fig. 3; Table 3).

Pyrolysis results show that the Tmax values are generally <435 °C, the gas generation potential index ranges from 0.09 to 0.19 mg/g sediment, and the hydrogen index

values vary from 15.97 to 42.48 mg/g TOC (Table 4), implying that the organic materials are substantial at the immature stage and the biogas is now still being formed at a massive generation stage (cf. Lin et al. 2004; Zhang et al. 2013). H/C and O/C ratios of kerogen vary in a range of 0.89–1.25 and 0.27–0.38, respectively, and plot in the type III kerogen area (gas prone; cf. Peters et al. 1986; Fig. 6), which are supported by the correlation of Tmax and HI indexes.

In summary, the argillic sediments of Facies III and IV are more likely to act as effective gas-source sediments (cf. Zhang and Chen 1983; Lu and Hai 1991; Wu et al. 2014). However, the TOC contents and chloroform bitumen values for the argillic sediments in the study area are significantly lower than those from the nearby Qiantang River

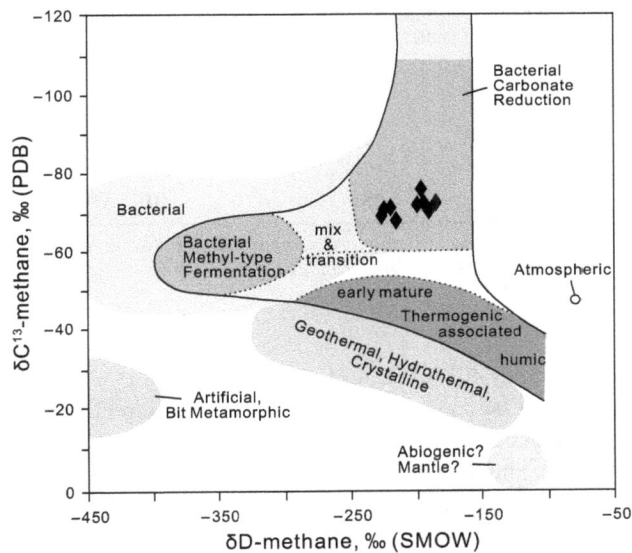

Fig. 5 Cross plot of $\delta^{13}C$ and δD of the methane for the shallow gas in the modern Changjiang delta area, eastern China, implying gases generated through the pathway of H_2 reduction of CO_2 (base diagram is from Whiticar 1999)

Table 2 Geochemical characteristics of the shallow gas in the Qidong area of the modern Changjiang delta region, China

CPT	Depth, m	Chemical composition, %					$\delta^{13}C_{CH4}$ PDB, ‰	$\delta^{13}C_{CO2}$ PDB, ‰	δD_{CH4} SMOW, ‰
		CH_4	C_2H_6	CO_2	CO	N_2			
1	70	23.6	u.d.	1.5	10.9	64.0	-67.7	-19.3	-215
3	20	87.7	u.d.	3.2	u.d.	9.1	-71.7	-34.5	-199
	70	96.2	u.d.	1.8	u.d.	2.0	-72.2	-20.4	-194
4	62	98.1	u.d.	1.2	u.d.	0.8	-70.2	-6.6	-190
13	40	97.7	u.d.	1.3	u.d.	1.0	-72.3	-21.4	-185
	70	95.1	u.d.	1.1	u.d.	3.7	-75.8	-21.4	-196

Note: *u.d.* under detection limit

Table 3 Organic matter abundance of the mud sediments from the different sedimentary facies of cores ZK01 and ZK02 in the modern Changjiang delta area, China

Facies	Depth, m	Lithology	TOC content, wt.%			Chloroform bitumen content, ppm		
			Min.	Max.	Av.	Min.	Max.	Av.
II	20.30–51.40	Gray soft mud	0.19	0.52	0.32 (17)	32	95	67 (17)
III	52.90–67.60	Gray or yellowish-gray mud	0.44	0.61	0.52 (21)	53	93	76 (21)
IV + V	68.40–74.10	Light-brown or gray mud	0.20	0.65	0.46 (3)	63	153	102 (3)

Numbers in parentheses = sample number of analyses

Note: *Min.* minimum, *Max.* maximum, *Av.* average

Table 4 Pyrolysis results of the gas-source sediments from different sedimentary facies of cores ZK01 and ZK02 in the modern Changjiang delta area, China

Facies	Lithology	Depth, m	S_1, mg/g	S_2, mg/g	Gas generation potential ($S_1 + S_2$), mg/g	Hydrogen index, mg/g	T_{max}, °C
II	Gray soft mud	24.7	0.02	0.07	0.09	22.74	463
		29.2	0.01	0.07	0.08	15.97	429
		48.0	0.03	0.13	0.16	30.15	394
III	Gray mud	63.7	0.05	0.14	0.19	31.25	424
		40.9	0.03	0.15	0.18	35.30	387
		45.1	0.03	0.15	0.18	42.48	439

Note: S_1 the amount of free hydrocarbons volatilized from the sediment sample, S_2 the hydrocarbons produced by cracking of organic matter in the sediment sample, *Hydrogen index* (S_2/TOC) × 100, T_{max} = the temperature of the maximum S_2 yield

incised-valley fill (cf. Lin et al. 2004; Zhang et al. 2013), which may be caused by the remarkably different amount of terrigenous sediment inputs. The mean annual suspended sediment load is ~4.8 × 10⁸ t/yr for the Changjiang, generally two orders of magnitude higher than that of the Qiantang River (~0.09 × 10⁸ t/yr). This huge terrigenous sediment input dilutes the organic matter abundance, as a result implying that the mud sediments in the postglacial Changjiang incised-valley fill have a relatively lower biogas generation potential than those in the Qiantang River incised-valley fill.

5.3 Reservoirs

The potential shallow biogenic-gas reservoirs in the study area can be classified into five types (Figs. 3, 7): (a) beds of sandy gravel, gravelly sand, medium-grained fine sand, and silty sand of Facies V; (b) sand bodies of Facies IV composed mainly of silty sand, sandy silt, and silt; (c) sand bodies of Facies III characterized by silty sand and sandy silt; (d) lenses of sand (silty fine sand, sandy silt, silt, and clayey silt) intercalated in the soft mud of Facies II; and (e) silty fine sand and sandy silt in Facies I. Based on the available exploration data, commercial gas was

encountered mainly in the sand bodies of Facies III and IV and secondarily in the top parts of the sand beds of Facies V (Figs. 3, 7). The sand bodies of Facies II are thin with a thickness of 0.005–0.95 m (generally <0.5 m; Figs. 3, 4e, 7). Although the sand sediments of Facies I are thick and composed mainly of coarse grains, the attributes including being close to the surface and lack of effective cap beds (0.1–3 m thick) make them unsuitable to be effective reservoirs in the study area (Fig. 3).

Therefore, determining the size, shape, and permeability of the sand bodies in Facies III, IV, and V is essential for exploration prediction and exploitation of shallow biogenic gas in the study area. The sand bodies in Facies III and IV vary significantly in thickness (0.5–4.0 m) and burial depth (50–70 m). Even in neighboring boreholes, the depth difference may be over 3–4 m (Fig. 7). In some cases, a borehole can go through up to >10 layers of sand with a total thickness of ~20 m (Fig. 7), but no sand layers are penetrated by the neighboring borehole. All the sand bodies are encased entirely by mud and vary in size (Fig. 7). Small bodies are distributed locally, but large ones can extend up to several hundreds of meters. The vertical permeabilities of the sand bodies in Facies III are of 122.9–211.5 mD, generally lower than the horizontal permeabilities

5.4 Cap beds

In the study area, commercial gas-bearing pools mainly occur as sand bodies capped directly by the mud beds of Facies III and IV, which are restricted within the incised valley (burial depth of 30–90 and thickness of 10–18 m) and are called direct or local cap beds (Figs. 2, 3, 9). By contrast, the soft mud covering the whole incised valley and deposited in an offshore shallow marine environment is called regional or indirect cap beds with a burial depth of 10–60 m and thickness of 5–20 m (Figs. 2, 3, 9). This is similar to that of the Qiantang River incised valley. Zhang et al. (2013) proposed that capillary sealing, pore-water pressure sealing, and hydrocarbon concentration sealing are considered as the main mechanisms for the conservation of the shallow biogenic gas in the Qiantang River incised-valley area. In this paper, we use vertical permeability, which has a negative relationship with the capillary sealing capacity of cap beds, to indirectly illustrate the sealing mechanism of cap beds. The vertical permeabilities of cap beds (0.05–0.52 mD) are significantly lower than those of sand reservoirs (66.99–15037 mD) with a difference of 4–6 orders of magnitude, which makes cap beds effective in preventing gas escaping from reservoirs (Table 5).

The vertical permeabilities have a similar variation trend with the contents of sand and silt, but an inverse correlation with the mud content (Fig. 3). The cap beds composed mainly of mud have the lowest permeability, and the vertical permeability is generally equal to the horizontal one, whereas those with sand bands or inclusions are marked by higher permeabilities, and the horizontal permeabilities are significantly higher than the vertical ones, by about 3–4 orders of magnitude (Figs. 3, 8c, d, e, f; Table 5). This result indicates that mud content plays a significant role in the capillary sealing ability of cap beds, namely the massive mud has much stronger sealing ability.

5.5 Gas migration and accumulation model

Methane is predominantly dissolved in the water within gas-source beds, with less being absorbed by clay minerals, and free gas is present only when saturation is attained with increasing depth of burial (cf. Lin et al. 2004; Gao et al. 2010, 2012). As a result, most gas in the study area is regarded as being transported from gas-source beds to sand beds by formation water with the differential compaction between sand and mud beds. Also, the capillary pressure difference between gas-source beds and sand reservoirs indicated by the huge difference in vertical permeability (Table 5) also drives gas migrating from mud beds to sandy reservoirs (cf. Magara 1987). After gas is released from mud sediments, it can migrate toward the overlying,

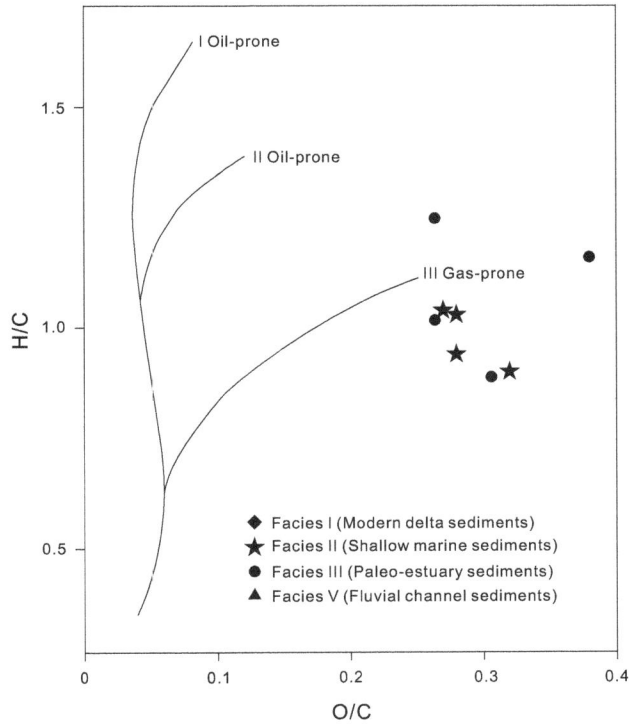

Fig. 6 Plot of H/C *vs.* O/C ratios (according to Peters et al. 1986) implying the kerogen character for the mud sediments in the late Quaternary Changjiang incised valley

(273.3–352.5 mD, Table 5), which may be influenced by the heterogeneity of sand bodies, for instance, the presence of parallel bedding and mud clasts (Figs. 4d, 8a). Generally speaking, the size, thickness, and permeability of the sand bodies in the study area are considerably lower than those of the sand bodies from the same facies in the late Quaternary Qiantang River incised-valley fill (Zhang et al. 2013, 2014). The sand bodies in Facies III of the Qiantang River incised-valley fill are characterized by a thickness of 3–20 m, burial depth of 28–78 m, width of 0.4–2 km, and permeability of 577–4,590 mD (Zhang et al. 2013, 2014).

The potential shallow biogenic-gas reservoirs in Facies V occur in the local highs, which are capped directly by the mud of Facies IV, and have a burial depth of 58–70 m (Fig. 7). The local highs, which are surrounded by fluvial-channel sand sediments formed in the previous incised-valley period however, cannot become effective gas reservoirs (Fig. 9). Permeabilities of these sand reservoirs (66.99–15,037.03 mD) are generally higher than those of sand bodies in Facies III and IV, which may be caused by the coarser grain size and lower contents of mud (Fig. 8b); nevertheless, they are characterized by a complex difference between the vertical and horizontal permeabilities, i.e., sometimes vertical permeability is higher than the horizontal, and *vice versa* (Table 5).

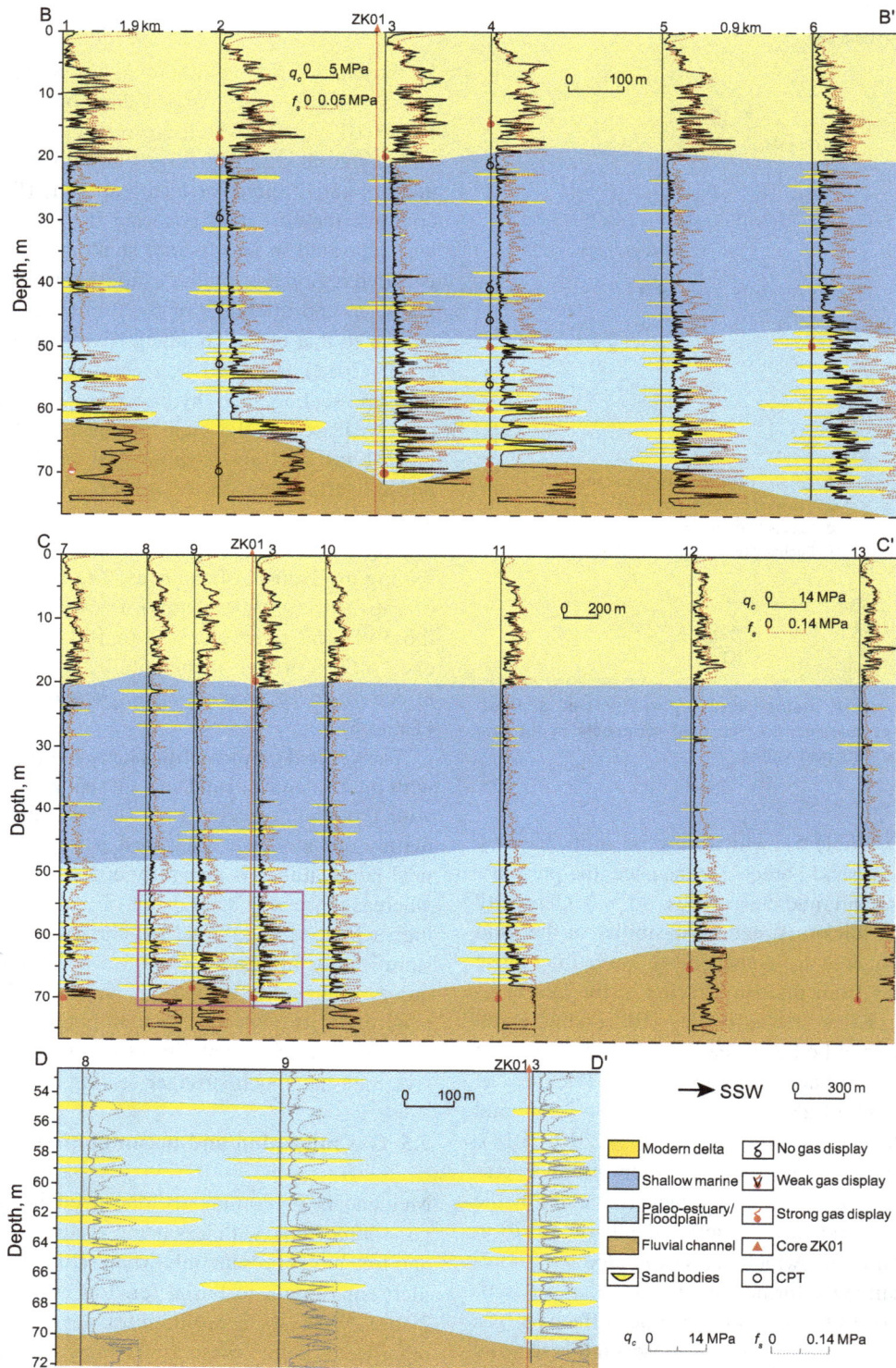

Fig. 7 Contrast of different CPTs in the Qidong area from the modern Changjiang delta region showing the distribution pattern of the sand beds in different facies. See Fig. 1 for locations. Profile **D–D′** is a close-up view of the purple box area in the profile **C–C′**

underlying, or lateral sand bodies (Fig. 9). Within sand reservoirs, the methane-filled space is initially restricted to the top, but it expands downward when methane is

abundant, and the formation water is expelled along the gas–water interface. In addition, gas migrates more frequently and easily along the bedding surface than in the

Table 5 Permeabilities of the cap beds and reservoirs of core ZK01 in the modern Changjiang delta area, China

Facies	Depth, m	Lithology	Vertical permeability, mD	Horizontal permeability, mD	
I	13.80–14.00	Alternation of silty fine sand and sandy silt	82.14–101.80/89.78(6[a])	115.37–136.25/124.46(6)	Reservoirs
II	34.35–35.00	Soft mud	0.04–0.05/0.05(3)		Indirect cap beds
	38.80–39.00	Soft mud with thin-bedded silt and silt blocks	0.43–0.80/0.59(3)	2,245.92–2,365.12/2,312.70(6)	
	46.80–47.00	Alternation of silty fine sand and soft mud	0.08–0.09/0.08(3)	1,059.74–1,294.41/1,171.16(6)	
	50.80–51.00	Soft mud with thin-bedded sandy silt	0.20–0.21/0.20(3)	456.23–640.49/552.44(6)	Direct cap beds
III	53.80–54.00	Mud with horizontal bedding	0.22–0.24/0.23(3)	58.34–113.20/75.88(6)	
	57.00–57.20	Mud with sandy and silty blocks	0.19–0.21/0.20(3)	236.73–409.30/333.07(7)	
	63.00–63.20	Sandy silt with parallel bedding	169.08–234.65/211.54(6)	256.43–302.87/273.32(6)	Reservoirs
	63.20–63.40	Mud with horizontal bedding	0.05–0.05/0.05(3)	8.82–207.28/70.17(6)	Direct cap beds
	64.35–64.55	Massive silty fine sand with mud blocks	99.87–170.37/122.88(7)	224.15–538.04/352.49(6)	Reservoirs
IV	69.05–69.25	Mud	0.45–0.53/0.48(3)	1.43–1.62/1.49(3)	Direct cap beds
V	71.10–71.30	Massive silty sand	237.19–425.86/279.66(7)	232.86–874.42/463.67(6)	
	73.15–73.35	Massive silty sand	3933.75–4235.73/4039.75(6)	5,723.47–6,287.39/6,019.80(6)	Reservoirs
	77.20–77.40	Gravelly sand	253.29–272.28/263.40(6)	272.28–316.94/304.09(6)	
	79.00–79.20	Slightly gravelly sand	62.91–75.38/66.99(6)	54.78–67.03/60.18(6)	
	82.10–82.30	Gravelly sand	548.25–798.87/652.45(6)	86.07–109.68/94.18(6)	

Note: 82.14–101.80/89.78 shows the minimum–maximum/average values

[a] Numbers in parentheses = number of analyses for one sample

(a)   **(b)**   **(c)**

(d)   **(e)**   **(f)**

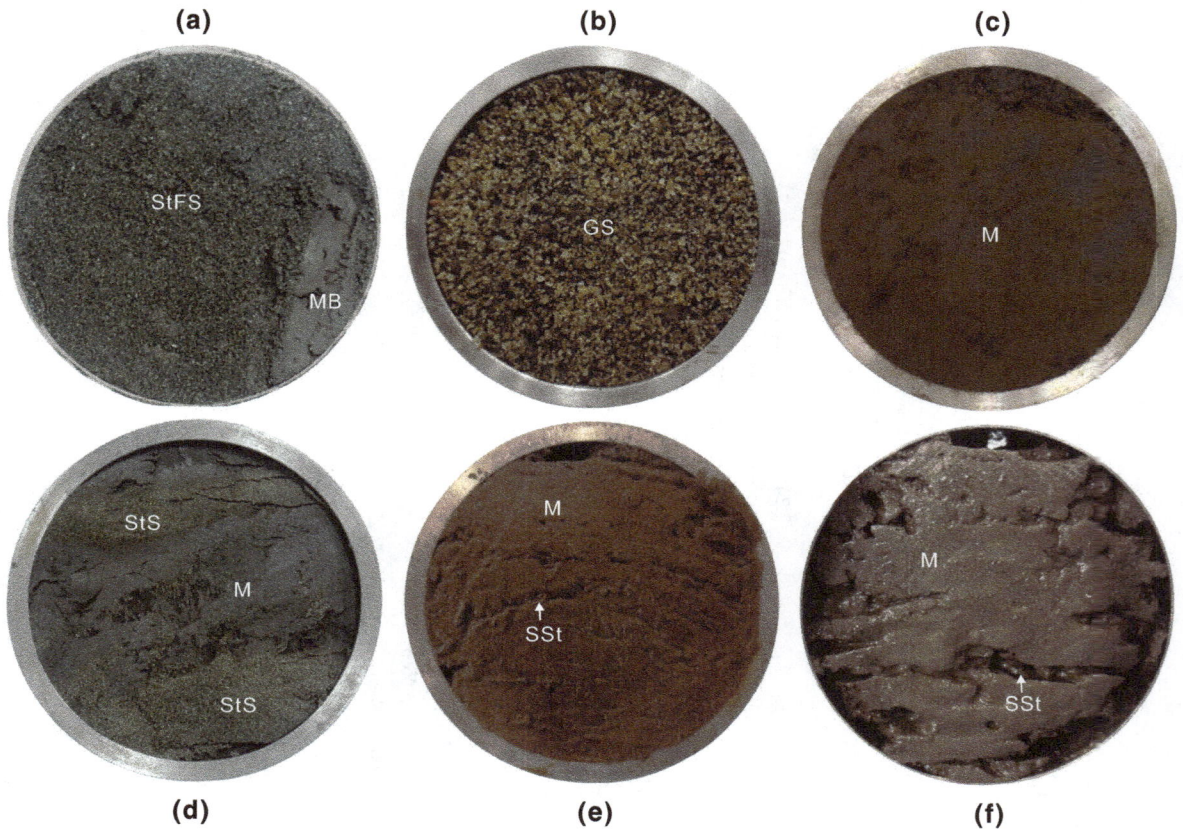

Fig. 8 Photographs showing the typical sedimentary characteristics of permeability samples. **a** 64.35–64.55 m depth: massive silty fine sand (StFS) with mud blocks (MB), *vertical* permeability of 123 mD, and *horizontal* permeability of 352 mD. **b** 77.20–77.40 m depth: structureless gravelly sand (GS) with *vertical* and *horizontal* permeabilities of 263 and 304 mD, respectively. **c** 69.05–69.25 m depth: massive mud (M) with *vertical* permeability of 0.48 mD and *horizontal* permeability of 1.49 mD. **d** 89.80–90.00 m depth: alternation of mud and silty sand (StS) with *vertical* and *horizontal* permeabilities of 0.52 and 395.6 mD, respectively. **f** and **e** 50.80–51.00 m depth: mud with thin-bedded sandy silt (SSt), *vertical* permeability of 0.20 mD, and *horizontal* permeability of 552 mD; the sand beds have been washed away after experiment

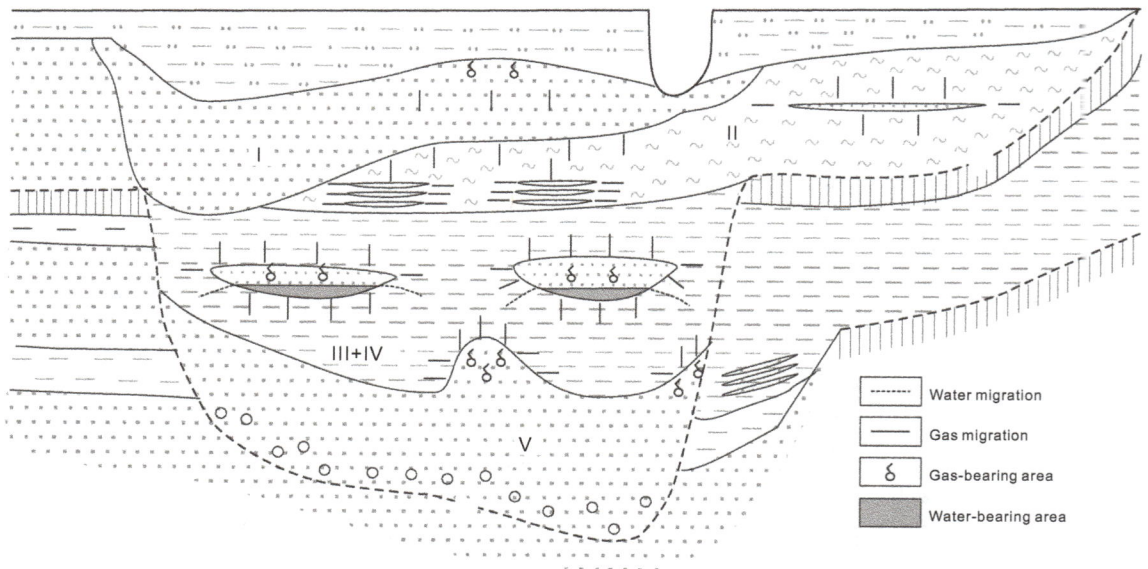

Fig. 9 Accumulation model of the shallow biogenic gas in the postglacial Changjiang incised-valley fill. See legends in Fig. 2

direction perpendicular to it within gas-source layers indicated by the difference between vertical and horizontal permeabilities (Table 5; Fig. 3).

As neotectonism in the study area is only characterized by uplift of the hilly lands and subsidence (1–3 mm/yr) in the coastal plain area (Chen and Stanley 1993), gas-bearing strata in the Changjiang delta area remain horizontal; therefore, gas migration and accumulation are mainly controlled by the lithology of gas-source layers and reservoirs, and there are no significant structural traps. As a consequence, sand bodies of Facies III and IV provide optimum conditions for in situ stratigraphic entrapment of biogenic gas and secondly in local highs of Facies V.

6 Conclusions

The natural gas in the modern Changjiang delta area consists primarily of CH_4 (generally >95%) and has a biogenic origin with carbon isotope ratios of CH_4 and CO_2 of −75.8 to −67.7‰ and −34.5 to −6.6‰, respectively, and hydrogen isotope ratios of CH_4 of −215 to −185‰. It is mainly distributed in the postglacial Changjiang incised-valley fill, which consists principally of five sedimentary facies in ascending order, i.e., fluvial channel (Facies V), floodplain (Facies IV), paleo-estuary (Facies III), offshore shallow marine (Facies II), and modern delta (Facies I). The sand bodies of Facies III and IV and local highs of Facies V are primary potential gas reservoirs. The former vary significantly in thickness (0.5–4.0 m) and burial depth (50–70 m), and the latter with a thickness larger than 10 m and burial depth of 58–70 m (Fig. 7). The main gas sources are gray or yellowish-gray mud of Facies III and IV, and the organic matter is dominated by type III kerogen (gas prone) at an immature stage. Meanwhile, the gas sources occur as cap beds, and the mud sediments of Facies III and IV that encase sand reservoirs directly are referred as direct cap beds, while the soft mud of Facies II called indirect ones. The huge difference in vertical permeability, about 4–6 orders of magnitude, between cap beds and reservoirs allows cap beds to effectively reduce the upward escape of gas in reservoirs. Therefore, it is notable that the shallow biogenic-gas reservoirs in the study area are of classic "self-generated and self-reserved" lithological entrapment type, and the sand bodies of Facies III, IV, and V should be considered as promising targets for exploration.

Acknowledgements This work was supported by the National Natural Science Foundation of China under Grant Numbers 41402092 and 41572112, the Natural Science Foundation (Youth Science Fund Project) of Jiangsu Province (BK20140604), and the Scholarship under State Scholarship Fund sponsored by the China Scholarship Council (File No. 201506195035). We thank C. Liu, Y.M., Jiang, H. Wang, J. Yu, C.W. Deng, Q.C. Yin, C. Lu, and Q. Wang for their helpful discussions, and assistance in field and core observations, and sample analyses. Special thanks should be extended to Dr. X.W. Sun of Sun Petroleum Geoservices, Australia, and Dr. J. Cao of Nanjing University for checking the English presentation. We also thank Dr. Y.H. Shuai of the PetroChina Research Institute of Petroleum Exploration and Development, China, anonymous reviewers and Petroleum Science editors for their constructive suggestions and comments.

References

Blott SJ, Pye K. Gradistat: a grain size distribution and statistics package for the analysis of unconsolidated sediments. Earth Surf Proc Land. 2001;26(11):1237–48.

Chen ZY, Stanley DJ. Yangtze delta, eastern China: 2. Late Quaternary subsidence and deformation. Mar Geol. 1993;112:13–21.

Dalrymple RW, Boyd R, Zaitlin BA. History of research, types and internal organization of incised-valley systems: introduction to the volume. In: Dalrymple RW, Boyd R, Zaitlin BA, editors. Incised-valley systems: origin and sedimentary sequences. SEPM Spec. Publ.; 1994; 51:3–10.

Gao G, Huang ZL, Huang BJ, et al. The solution and exsolution characteristics of natural gas components in water at high temperature and pressure and their geological meaning. Pet Sci. 2012;9(1):25–30.

Gao Y, Jin Q, Zhu GY. Genetic types and distribution of shallow-buried natural gases. Pet Sci. 2010;7(3):347–54.

García-García A, Orange DL, Miserocchi S, et al. What controls the distribution of shallow gas in the Western Adriatic Sea? Cont Shelf Res. 2007;27(3–4):359–74.

Garcia-Gil S, Vilas F, Garcia-Garcia A. Shallow gas features in incised-valley fills (Ria de Vigo, NW Spain): a case study. Cont Shelf Res. 2002;22(16):2303–15.

Heggland R. Detection of gas migration from a deep source by the use of exploration 3D seismic data. Mar Geol. 1997;137(1–2):41–7.

Hori K, Saito Y, Zhao QH, et al. Architecture and evolution of the tide-dominated Changjiang (Yangtze) River delta, China. Sedim Geol. 2002;146(3–4):249–64.

Hori K, Saito Y, Zhao QH, et al. Sedimentary facies and Holocene progradation rates of the Changjiang (Yangtze) delta, China. Geomorphology. 2001;41(2–3):233–48.

Humez P, Mayer B, Ing J, et al. Occurrence and origin of methane in groundwater in Alberta (Canada): gas geochemical and isotopic approaches. Sci Total Environ. 2016;541:1253–68.

Jones KL, Lindsay MBJ, Kipfer R, et al. Atmospheric noble gases as tracers of biogenic gas dynamics in a shallow unconfined aquifer. Geochim Cosmochim Acta. 2014;128:144–57.

Li CX, Chen QQ, Zhang JQ, et al. Stratigraphy and paleoenvironmental changes in the Yangtze Delta during the Late Quaternary. J Asian Earth Sci. 2000;18(4):453–69.

Li CX, Wang PX. Researches on Stratigraphy of the Late Quaternary Period in Yangtze River Mouth. Beijing: Science Press; 1998 (**in Chinese**).

Li CX, Wang P, Fan DD, et al. Characteristics and formation of late Quaternary incised-valley-fill sequences in sediment-rich deltas and estuaries: case studies from China. In: Dalrymple RW, Leckie DA, Tillman RW, editors. Incised valleys in time and space, vol. 85. SEPM Spec. Publ; 2006. p. 141–60.

Li CX, Wang P, Sun HP, et al. Late Quaternary incised-valley fill of the Yangtze delta (China): its stratigraphic framework and evolution. Sedim Geol. 2002;152(1–2):133–58.

Li P, Du J, Liu LJ, et al. Distribution characteristics of the shallow gas in Chinese offshore seabed. Chin J Geol Hazard Control. 2010a;21(1):69–74 (in Chinese).

Li XY, Shi XF, Cheng ZB, et al. Distribution of benthic foraminifera in surface sediments of the Laizhou Bay, Bohai Sea and its environmental significance. Acta Micropalaeontol Sin. 2010b;27:38–44 (in Chinese).

Li YL, Lin CM. Exploration methods for late Quaternary shallow biogenic gas reservoirs in the Hangzhou Bay area, eastern China. AAPG Bull. 2010;94:1741–59.

Lin CM, Gu LX, Li GY, et al. Geology and formation mechanism of late Quaternary shallow biogenic gas reservoirs in the Hangzhou Bay area, eastern China. AAPG Bull. 2004;88(5):613–25.

Lin CM, Li YL, Zhuo HC, et al. Features and sealing mechanism of shallow biogenic gas in incised valley fills (the Qiantang River, eastern China): a case study. Mar Pet Geol. 2010;27(4):909–22.

Lin CM, Zhang X, Xu ZY, et al. Sedimentary characteristics and accumulation conditions of shallow-biogenic gas for the Late Quaternary sediments in the Changjiang River delta area. Adv Earth Sci. 2015;30(5):589–601 (in Chinese).

Liu CL, Zhu J, Che CB, et al. Potential recoverable natural gas resources in China. Pet Sci. 2008;5(1):83–6.

Lu WW, Hai XZ. Simulation experiments on biogenic gas generation and estimation of generation amount of biogenic gas in strata. Exp Pet Geol. 1991;13:65–75 (in Chinese).

Magara K. Compaction and fluid migration. Amsterdam-Oxford, New York: Elsevier Scientific Publishing Company; 1987. p. 319.

Milliman JD, Syvitski JPM. Geomorphic/tectonic control of sediment discharge to the ocean: the importance of small mountainous rivers. J Geol. 1992;100(5):525–44.

Moran K, Hill PR. Blasco SM. Interpretation of piezocone penetrometer profiles in sediment from the Mackenzie Trough, Canadian Beaufort Sea. J Sedim Res. 1989;59:88–97.

Ni YY, Dai JX, Zou CN, et al. Geochemical characteristics of biogenic gases in China. Int J Coal Geol. 2013;113:76–87.

Nittrouer JA, Mohrig D, Allison MA, et al. The lowermost Mississippi River: a mixed bedrock-alluvial channel. Sediment. 2011;58(7):1914–34.

Okay S, Aydemir S. Control of active faults and sea level changes on the distribution of shallow gas accumulations and gas-related seismic structures along the central branch of the North Anatolian Fault, Southern Marmara shelf, Turkey. Geodinamica Acta. 2016;28:328–46.

Peters KE, Moldowan JM, Schoell M, et al. Petroleum isotopic and biomarker composition related to source rock organic matter and depositional environment. Org Geochem. 1986;10(1–3):17–27.

Pinet N, Duchesne M, Lavoie D, et al. Surface and subsurface signatures of gas seepage in the St. Lawrence Estuary (Canada): significance to hydrocarbon exploration. Mar Pet Geol. 2008;25(3):271–88.

Reimer PJ, Bard E, Bayliss A, et al. IntCal13 and Marine-13 radiocarbon age calibration curves 0–50,000 years cal BP. Radiocarbon. 2013;55(4):1869–87.

Rice DD, Claypool GE. Generation, accumulation, and resource potential of biogenic gas. AAPG Bull. 1981;65(1):5–25.

Rogers JN, Kelley JT, Belknap DF, et al. Shallow-water pockmark formation in temperate estuaries: a consideration of origins in the western gulf of Maine with special focus on Belfast Bay. Mar Geol. 2006;225(1–4):45–62.

Stanley DJ, Chen ZY. Neolithic settlement distributions as a function of sea level-controlled topography in the Yangtze delta, China. Geology. 1996;24(12):1083–6.

Stax R, Stein R. Long-term changes in accumulation of organic carbon in Neogene sediments of ODP-Leg 130 (Ontong Java Plateau). Proc ODP Sci Results. 1993;130:573–84.

Sun PA, Wang Y, Leng K, et al. Geochemistry and origin of natural gas in the eastern Junggar Basin, NW China. Mar Pet Geol. 2016;75:240–51.

Tao KY, Cao J, Wang Y, et al. Geochemistry and origin of natural gas in the petroliferous Mahu sag, northwestern Junggar Basin, NW China: carboniferous marine and Permian lacustrine gas systems. Org Geochem. 2016;100:62–79.

Uehara K, Saito Y, Hori K. Paleotidal regime in the Changjiang (Yangtze) estuary, the East China Sea, and the Yellow Sea at 6 ka and 10 ka estimated from a numerical model. Mar Geol. 2002;183(1–4):179–92.

Vielstädte L, Karstens J, Haeckel M, et al. Quantification of methane emissions at abandoned gas wells in the Central North Sea. Mar Pet Geol. 2015;68:848–60.

Wang MY. Shallow natural gas in the modern Changjiang delta area. Nat Gas Ind. 1982;3:3–9 (in Chinese).

Wang PX, Min QB, Bian YH, et al. Strata of Quaternary transgressions in east China: a preliminary study. Acta Geol Sin. 1981;55:1–13 (in Chinese).

Wang PX, Zhang JJ, Zhao QH, et al. Foraminifera and Ostracod in Surface sediments of the East China Sea. Beijing: China Ocean Press; 1988. p. 1–438 (in Chinese).

Wang ZH, Zhuang CC, Saito Y, et al. Early mid-Holocene sea-level change and coastal environmental response on the southern Yangtze delta plain, China: implications for the rise of Neolithic culture. Quat Sci Rev. 2012;35:51–62.

Whiticar MJ, Faber E, Schoell M. Biogenic methane formation in marine and freshwater environments: CO_2 reduction vs. acetate fermentation-Isotope evidence. Geochim Cosmochim Acta. 1986;50(5):693–709.

Whiticar MJ. Carbon and hydrogen isotope systematics of bacterial formation and oxidation of methane. Chem Geol. 1999;161(1–3):291–314.

Wu M, Cao J, Wang XL, et al. Hydrocarbon generation potential of Triassic mudstones in the Junggar Basin, northwest China. AAPG Bull. 2014;98(9):1885–906.

Xu ZY, Yue DL, Wu SH, et al. An analysis of the types and distribution characteristics of natural gas reservoirs in China. Pet Sci. 2009;6:38–42.

Yang CS, Sun JS. Tidal sand ridges on the East China Sea shelf. In: de Boer PL, van Gelder A, Nio SD, editors. Tide-influenced sedimentary environments and facies. Dordrecht: D. Reidel Publ; 1988. p. 23–38.

Yoneda M, Uno H, Shibata Y, et al. Radiocarbon marine reservoir ages in the western Pacific estimated by pre-bomb molluscan shells. Nucl Instrum Methods Phys Res Sect B. 2007;259(1):432–7.

Zhang JQ, Zhang GJ, Li CX. Characteristics of the late Quaternary stratigraphic sequence in the Changjiang River delta area. J Tongji Univ. 1998;26(4):438–42 (in Chinese).

Zhang X, Dalrymple RW, Yang SY, et al. Provenance of Holocene sediments in the outer part of the Paleo-Qiantang River estuary, China. Mar Geol. 2015;366(1):1–15.

Zhang X, Lin CM, Dalrymple RW, et al. Facies architecture and depositional model of a macrotidal incised-valley succession (Qiantang River estuary, eastern China), and differences from other macrotidal systems. GSA Bull. 2014;126(3–4):499–522.

Zhang X, Lin CM, Li YL, et al. Sealing mechanism for cap beds of shallow-biogenic gas reservoirs in the Qiantang River incised valley, China. Cont Shelf Res. 2013;69:155–67.

Zhang YG, Chen HJ. Concepts on the generation and accumulation of biogenic gas. Oil Gas Geol. 1983;4(2):160–70 (in Chinese).

Zheng KF. Distribution and exploration prospects of the shallow gas in the Quaternary in Jiangsu Province. Nat Gas Ind. 1998;18:20–4 (in Chinese).

Zhou XJ, Gao S. Spatial variability and representation of seabed sediment grain sizes: an example from the Zhoushan-Jinshanwei transect, Hangzhou Bay, China. Chin Sci Bull. 2004;49(23):2503–7.

Zhou ZH, Zhou RN, Guan ZQ. Geochemical properties of source materials and biogas prospects of the Quaternary gas in east of Qaidam basin. Pet Explor Dev. 1994;21:30–6 (in Chinese).

Zhuang LH, Chang FM, Li TG, et al. Foraminiferal faunas and Holocene sedimentation rates of Core EY 02-2 in the South Yellow Sea. Mar Geol Quatern Geol. 2002;22:7–14 (in Chinese).

Zou YR, Zhao CQ, Wang YP, et al. Characteristics and origin of natural gases in the Kuqa Depression of Tarim Basin, NW China. Org Geochem. 2006;37(3):280–90.

13

Characteristics of fault zones and their control on remaining oil distribution at the fault edge: a case study from the northern Xingshugang Anticline in the Daqing Oilfield, China

Xiao-Fei Fu[1,2,3,4] · **Xiao Lan**[1,4] · **Ling-Dong Meng**[1,2,3,4] · **Hai-Xue Wang**[1,2,3,4] · **Zong-Bao Liu**[1,2,3,4] · **Zhi-Qiang Guo**[5] · **Zai-He Chen**[5]

Abstract Most major oil zones in the Daqing Oilfield have reached a later, high water cut stage, but oil recovery is still only approximately 35 %, and 50 % of reserves remain to be recovered. The remaining oil is primarily distributed at the edge of faults, in poor sand bodies, and in insufficiently injected and produced areas. Therefore, the edge of faults is a major target for remaining oil enrichment and potential tapping. Based on the dynamic change of production from development wells determined by the injection–recovery relationship at the edge of faults, we analyzed the control of structural features of faults on remaining oil enrichment at the edge. Our results show that the macroscopic structural features and their geometric relationship with sand bodies controlled remaining oil enrichment zones like the edges of NNE-striking faults, the footwalls of antithetic faults, the hard linkage segments (two faults had linked together with each other to form a bigger through-going fault), the tips of faults, and the oblique anticlines of soft linkages. Fault edges formed two types of forward micro-amplitude structures: (1) the tilted uplift of footwalls controlled by inverse fault sections and (2) the hanging-wall horizontal anticlines controlled by synthetic fault points. The remaining oil distribution was controlled by micro-amplitude structures. Consequently, such zones as the tilted uplift of the footwall of the NNW-striking antithetic faults with a fault throw larger than 40 m, the hard linkage segments, the tips of faults, and the oblique anticlines of soft linkage were favorable for tapping the remaining oil potential. Multi-target directional drilling was used for remaining oil development at fault edges. Reasonable fault spacing was determined on the basis of fault combinations and width of the shattered zone. Well core and log data revealed that the width of the shattered zone on the side of the fault core was less than 15 m in general; therefore, the distance from a fault to the development target should be larger than 15 m. Vertically segmented growth faults should take the separation of the lateral overlap of faults into account. Therefore, the safe distance of remaining oil well deployment at the fault edge should be larger than the sum of the width of shattered zone in faults and the separation of growth faults by vertical segmentation.

Keywords Fault edge · Fault zone structure · Segmentation growth · Micro-amplitude structure · Sealing · Safety distance

✉ Xiao-Fei Fu
  fuxiaofei2008@sohu.com

1 Laboratory of CNPC Fault Controlling Reservoir, Northeast Petroleum University, Daqing 163318, Heilongjiang, China

2 Science and Technology Innovation Team in Heilongjiang Province "Fault Deformation, Sealing and Fluid Migration", Daqing 163318, Heilongjiang, China

3 The State Key Laboratory Base of Unconventional Oil and Gas Accumulation and Exploitation, Earth Science College, Northeast Petroleum University, Daqing 163318, Heilongjiang, China

4 Earth Science College, Northeast Petroleum University, Daqing 163318, Heilongjiang, China

5 NO. 5 Production Plant, Huabei Oilfield Co., PetroChina, Xinji 052360, Hebei, China

Edited by Jie Hao

1 Introduction

About one-third of major onshore oilfields in East China are complexly faulted, and characterized by high water cut, high recovery percentage, high recovery rate, low reserve/

production ratio, and low oil recovery ratio (Yu 1998; Hu 2008; Jin et al. 2009). After long-term waterflooding, most of them are at the later stage of high waterflooding. However, the oil recovery is only 35 % or so, and approximately 50 % of reserves remain to be recovered. At the stage of high water cut and high recovery percentage of China National Petroleum Corporation (CNPC), annual oil production had accounted for 58 % of total oil production, and remaining recoverable reserves had reached 44 % of cumulative remaining recoverable reserves until the end of 2011 (Han 2010; Liu et al. 2008). The core of development and adjustment of old oilfields during the high water cut stage is "recognizing and recovering remaining oil," yet the remaining oil in fault oilfields concentrates at the edges of faults and poor sand bodies, as well as insufficiently injected and produced areas; therefore, the edges of faults are a major target for remaining oil enrichment and potential tapping (Jiang 2013). Currently, domestic and foreign research on remaining oil at fault edges of fault oilfields mainly focuses on reservoir heterogeneity, well pattern rationality, microscopic displacement experiments, potential tapping methods, etc. (Miall 1994; Peng et al. 2007; Yue et al. 2008; Hou et al. 2014; Feng et al. 2014), seriously neglecting the influence of fault zone structural features on remaining formation, enrichment, and development (Jin et al. 2012). This influence is manifest in two aspects. The first is the different sites of remaining oil owing to the configuration of faults and formation occurrence (Qi 2004), fault combination pattern, fault linkage status (hard linkage, soft linkage, cross linkage, superimposed deformation, etc.) (Kim and Sanderson 2005), and associated micro-amplitude structural differences. The foundation of research into remaining oil at the fault edges is to examine the microscopic combinations, linkages, and associated micro-amplitude structural features of faults. The second aspect is that the remaining oil at the fault edges is affected by fault sealing (Yielding 2002), apart from its relevance to the injection–recovery relationship, such as water channeling and oil leakage occurring frequently at footwall and hanging wall of faults. Thus, quantitative evaluation of fault sealing is the core of research into remaining oil at fault edges (Bretan et al. 2003; Fu et al. 2010; Liu et al. 2014). In this paper, the Putaohua oil layer of the Xingbei Oilfield in the Daqing Placanticline was selected for the study of faults and remaining oil. According to the dynamic variation of production from development wells controlled by fault edge injection–recovery relationships, the control of fault zone structural features on remaining oil at the fault edge was analyzed, and a systematic approach for predicting the remaining oil distribution at fault edges of fault oilfields during the high water cut stage was established.

2 Geological setting

The Songliao Basin is a typical rift basin (Gao 1980, 1983; Hu and Wang 1996; Zhang et al. 1996). There are three structural evolution stages of the basin including fault depression ($K_1h - K_1ych$), depression ($K_1d - K_2n$), and inversion ($K_2m - Q$) (Gao and Cai 1997; Yun et al. 2002). The Daqing Placanticline is a large anticlinal zone located in the central depression of the Songliao Basin (Fig. 1), composed of seven anticlines (Lamdian, Sartu, Xingshugang, Gaotaizi, Taipingtun, Putaohua, and Aobaota), which are mostly asymmetrical with gentle eastern flanks and steep western flanks. The Daqing Placanticline is located in the two oil generation areas of the Qijia-Gulong and Sanzhao Depressions, and formed a world-class oil field (Gao and Cai 1997). The first member of the Yaojia Formation is characterized as an interbedded sandstone–mudstone reservoir and is the major oil reservoir of the Daqing Placanticline (Fig. 2).

The Xingshugang Anticline is a secondary structure within the Daqing Placanticline, and the study area is located at the north of the anticline (Fig. 1). 42 tensor-shear faults developed here (Sun et al. 2013), dominantly NW–NNW striking and NE dipping, though a minority of them are NW dipping. Fault throw is usually less than 50 m, and extension is generally less than 4 km. More than 4000 development wells have been drilled in the area. At present, the water cut is 93 % and tapping the potential of remaining oil at the fault edges is essential to increase production.

3 Control of fault zone features on remaining oil distribution

Fault zone features primarily refer to the occurrence, combinations, linkage status, and associated micro-amplitude structural features (Fu et al. 2011) of fault zones, which affect fault edge injection–recovery relationships and thereby control remaining oil distribution.

3.1 Fault-reservoir configuration and different remaining oil distribution in the footwall and hanging wall

It is found by comparing development well productivity of parts of Fault F234 with different strikes that the development wells at edge of NNW-striking faults have higher productivity (Fig. 3) and richer remaining oil. At the first member of the Yaojia Formation in the Daqing Placanticline, the sedimentary system is fluvial-deltaic facies, source direction is from the north (Liu 2008; Liu 2010),

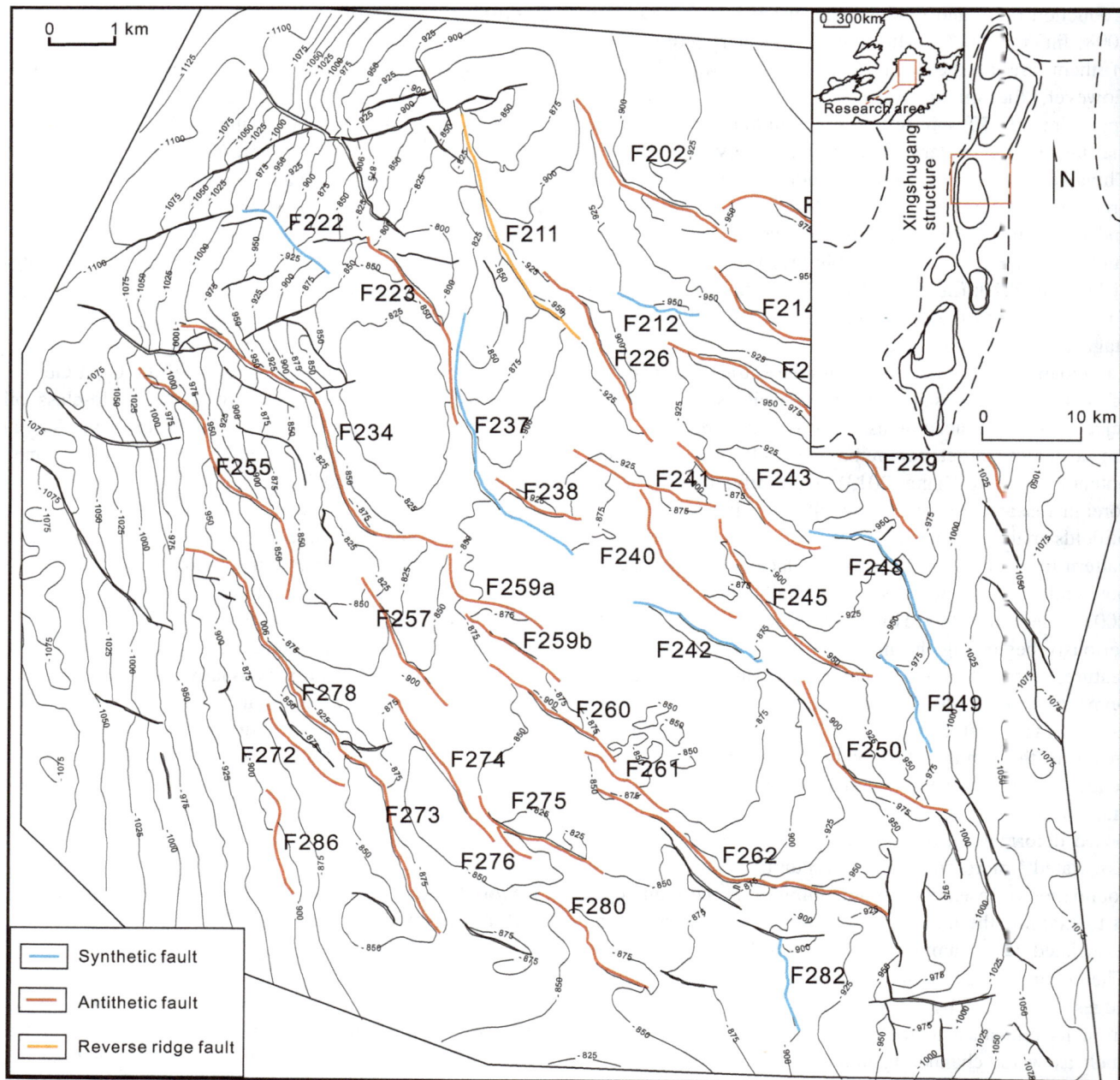

Fig. 1 Location and fault distribution of the study area

and channel sand is the major reservoir body. The channels have an approximately N-striking to NE-striking distribution and intersect NW-striking faults at high angles. Faults commonly form the enrichment boundary of remaining oil. Remaining oil tends to be found easily due to incomplete injection and recovery. At fault segments of the same trend, the larger the throw is, the higher the intensity of daily production gets. Remaining oil is the richest (Fig. 3) as the throw reaches the limit value. The limit value of throw of remaining oil enrichment at Fault F234 is between 56 and 68 m, while that of Fault F259–F262 is from 42 to 60 m.

The greater the fault throw, the richer the remaining oil, which is controlled by the two factors. One is that the oil layer in the first member of Yaojia Formation tends to be juxtaposed against an upper mudstone, so that the sealing capacity of faults becomes stronger with larger fault throw. The other one is that due to differential subsidence, the areas with greater throw of antithetic faulting and those that are closer to the structural high are more conducive to the enrichment of remaining oil, resulting in greater daily production. Faults could be classified into four types in light of fault-reservoir configurations (Cloos 1928; Hills

Period	Series	Fm.	Member	Stratigraphic column	Lithology	Depth	Seismic	Lake water depth (Deep / Shallow)	Source	Reservoir	Seal
Cretaceous	Upper	Nenjiang (N)	V		Grey-black, brown-red argillite mingled with sands	0–355	T_{03}				
			IV		Grey-green argillite mingled with sandstone and silt	0–300					
			III		Grey-black argillite interbedded with sandstone	0–130	T_{04}				
			II		Grey-black mudstone (shale) mingled with thin bed oil shale	0–150	T_{06}				
			I		Grey-black mudstone mingled with thin bed sandstone and mussel-shrimp debris limestone	0–220	T_{07}				
		Yaojia (Y)			Grey-green mudstone mingled with sands	0–150	T_1				
			I		Grey-green mudstone interbedded with sandstone, grey-black mudstone in the upper	0–80	T_{1-0}				
		Qingshankou (Qn)			Grey green mudstone mingled with sands	0–550	$T_{1'}$				
			I		Grey-black mudstone mingled with thin bed mussle-shrimp debris limestone and oil shale	0–160	T_1^3 / T_2				

Legend: Sandstone; Mud rock/thin bed sandstone; Mud/shale; Oil shale

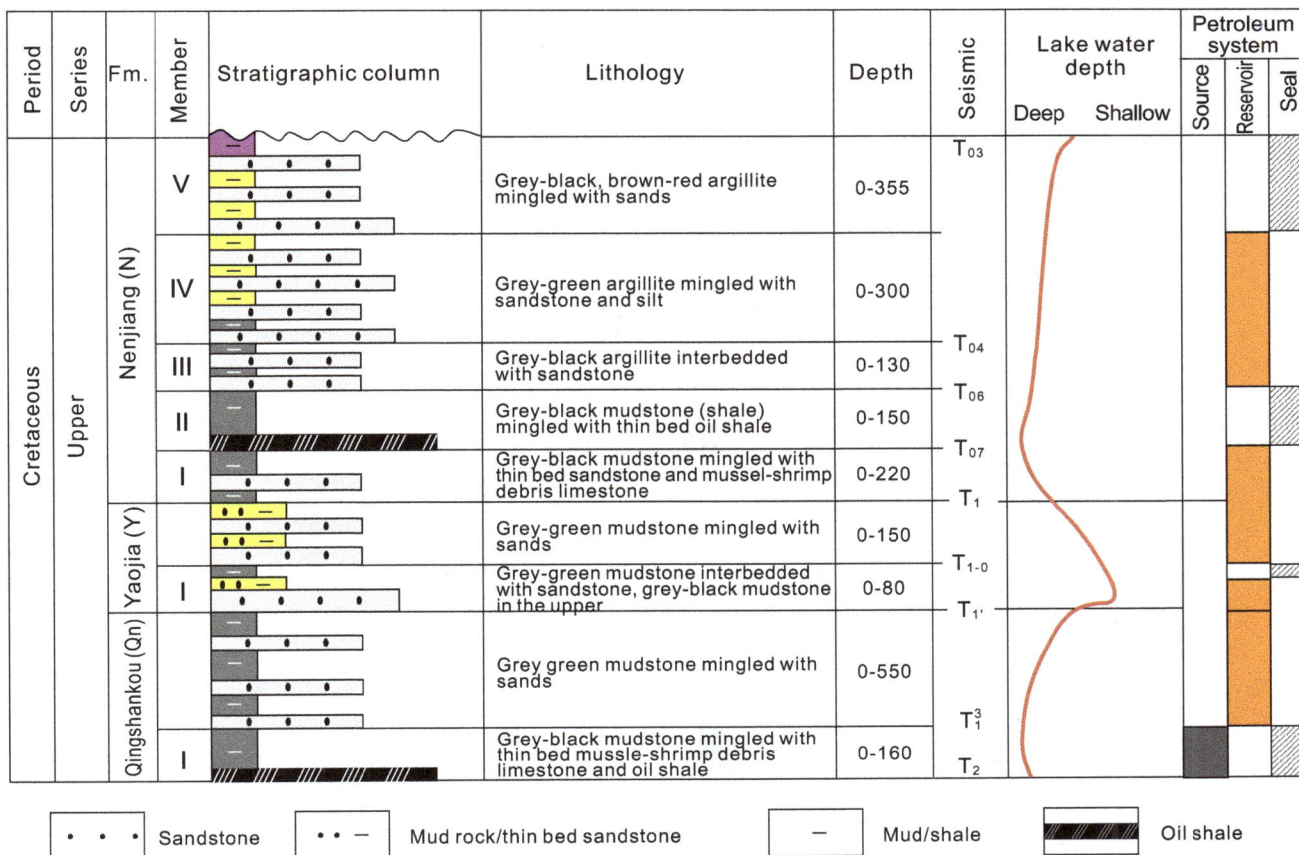

Fig. 2 Correlation map of strata hydrocarbon generation–storage–cover combinations in the Northern Songliao Basin

1953; Peacock et al. 2000; Qi 2004): antithetic faults, synthetic faults, ridge-like faults, and anti-ridge-like faults. In the Putaohua oil layer, antithetic faults (33) and synthetic faults (8) mainly developed. According to the dynamic data of fault edge development wells in the footwalls and hanging walls (Fig. 3), remaining oil is richer in the hanging walls of antithetic faults.

3.2 Fault linkage status identification and remaining oil distribution

It is universally observed that faults grow by segmentation in rift basins (Peacock 1991; Kim and Sanderson 2005; Wang et al. 2013), and the fault displacement–distance curve is one of the essential and most widely used approaches to identify segmented faulting. Displacement occurs in the low-value zone, i.e., the faults grow by segmentation (Peacock and Sanderson 1994, 1991; Fossen 2010; Fossen et al. 2007; Liu et al. 2012; Giba et al. 2012). Additionally, there are isolated faults. Displacement and throw are linearly related, so the throw-distance curve is applied to quantitatively define the faults that grow by segmentation and the isolated faults. Restricted by poor

seismic resolution, it is common that two segmented growth faults are interpreted as one. To effectively reduce uncertainty of fault interpretation, we use an approach derived from the research achievements of Soliva and Benedicto (2004), applying the "relay displacement (D, the aggregate displacement measured at the fault overlap center)/separation (S, the vertical distance of fault overlap)" to quantitatively affirm fault growth stages. According to relevant fault data that have been published at home and abroad (Soliva and Benedicto 2004; Soliva et al. 2008), on the basis of 3D seismic data from the Songliao Basin, relay displacement (D)/separation (S) data are tallied, completing the quantitative discrimination criterion for fault segment growth linkage. When D/S is smaller than 0.27, faults are at the overlap stage or at "soft-linkage" stage. When D/S is between 0.27 and 1, faults are at a stage of fault initiation, also at the "soft-linkage" stage. When D/S is larger than 1, faults are totally breached, or at the "hard-linkage" stage. Therefore, fault growth stages can be divided into the overlap, incipiently linked relays, and complete breaching (Wang et al. 2014) stages. For the 41 faults in the Xinbei development zone, comprehensive analysis and correction were conducted for the fault

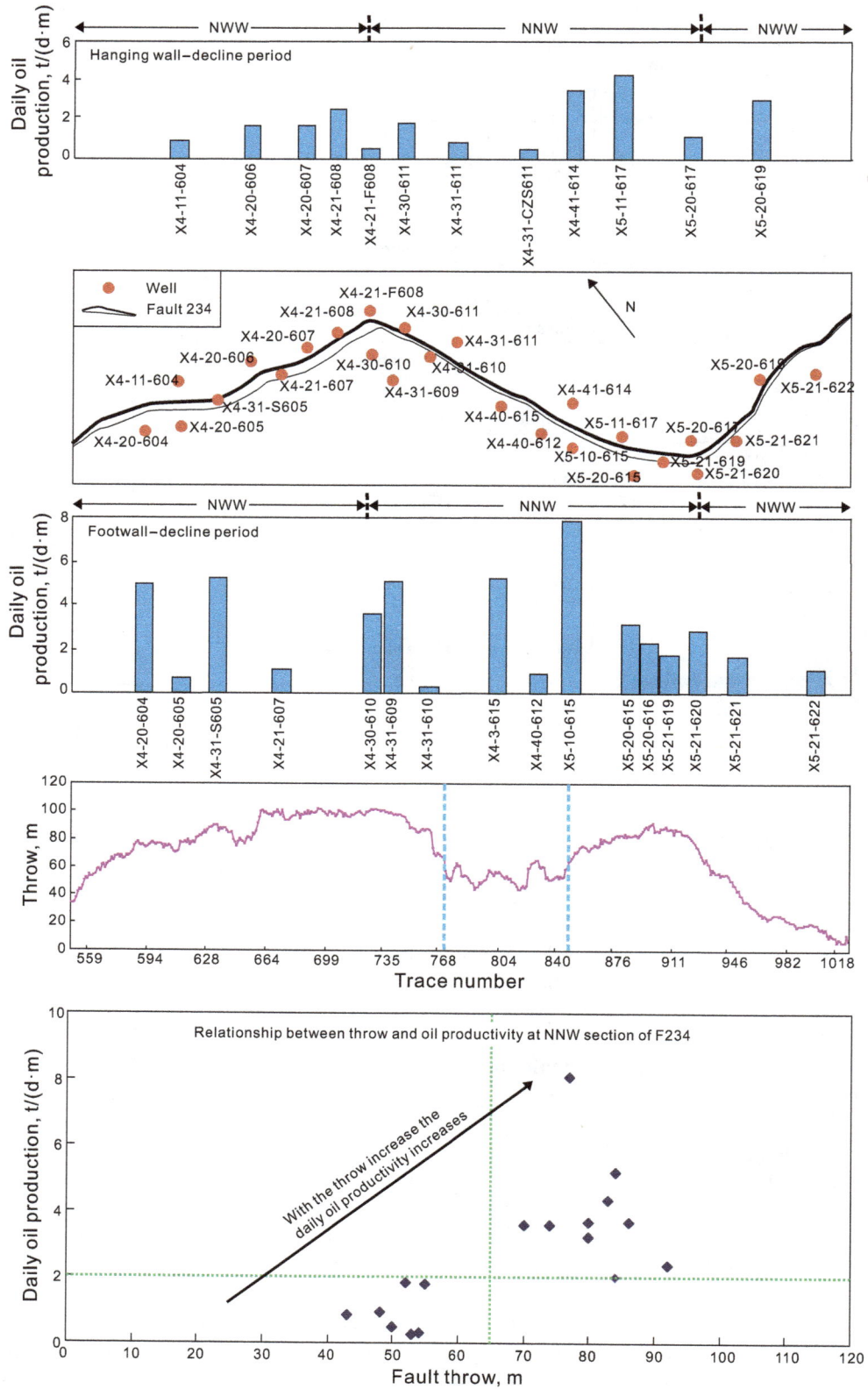

Fig. 3 Relationship between remaining oil distribution and oil productivity and fault throw. Oil production is the amount of oil production by one meter thickness and each day. t represents ton, d represents day, and m is the thickness of oil layer

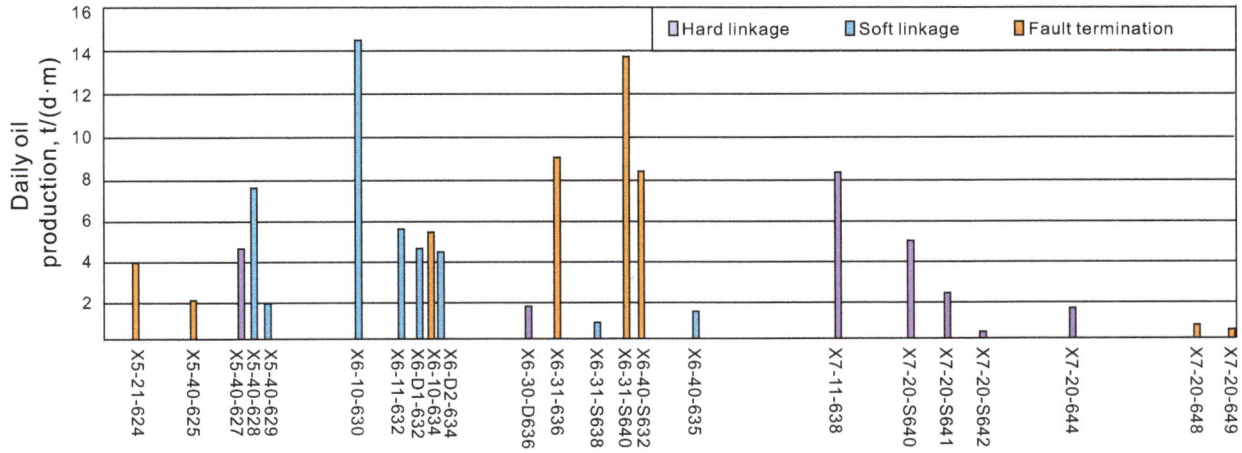

(a) Typical well daily oil production-initial

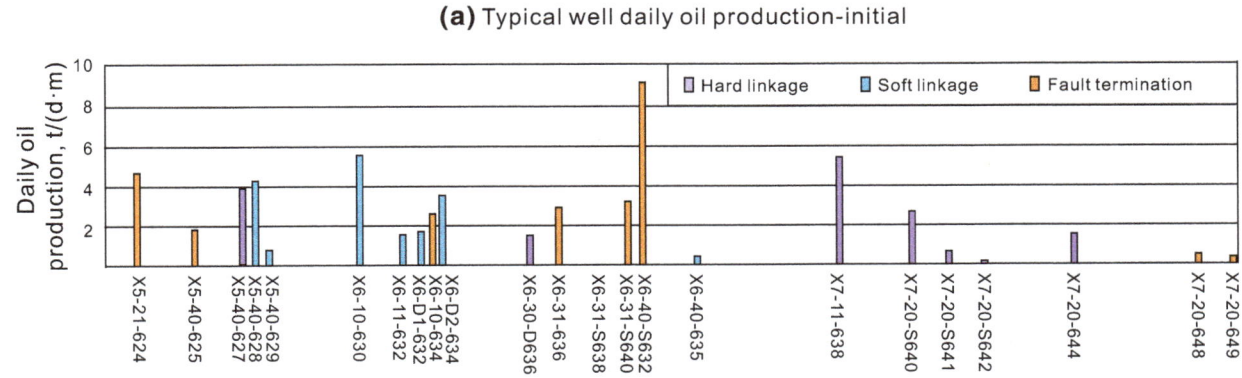

(b) Typical well daily oil production-decline

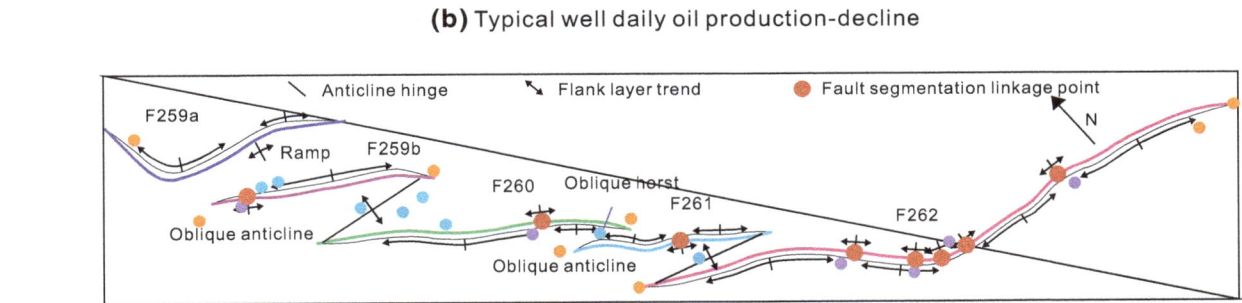

(c) Fault marginal positive micro-amplitude structure

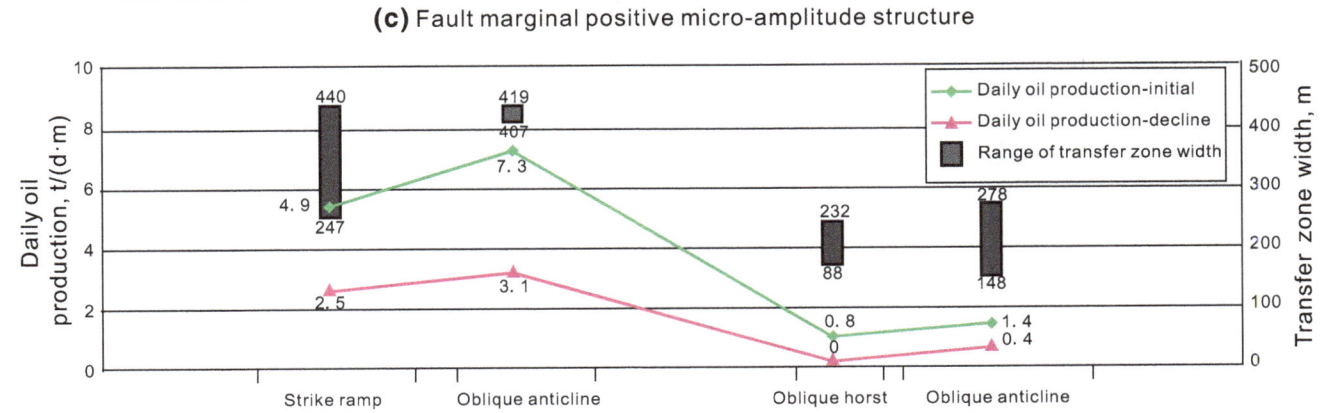

Fig. 4 Remaining oil distribution of different linkage status in the vicinity of faults from oil production data

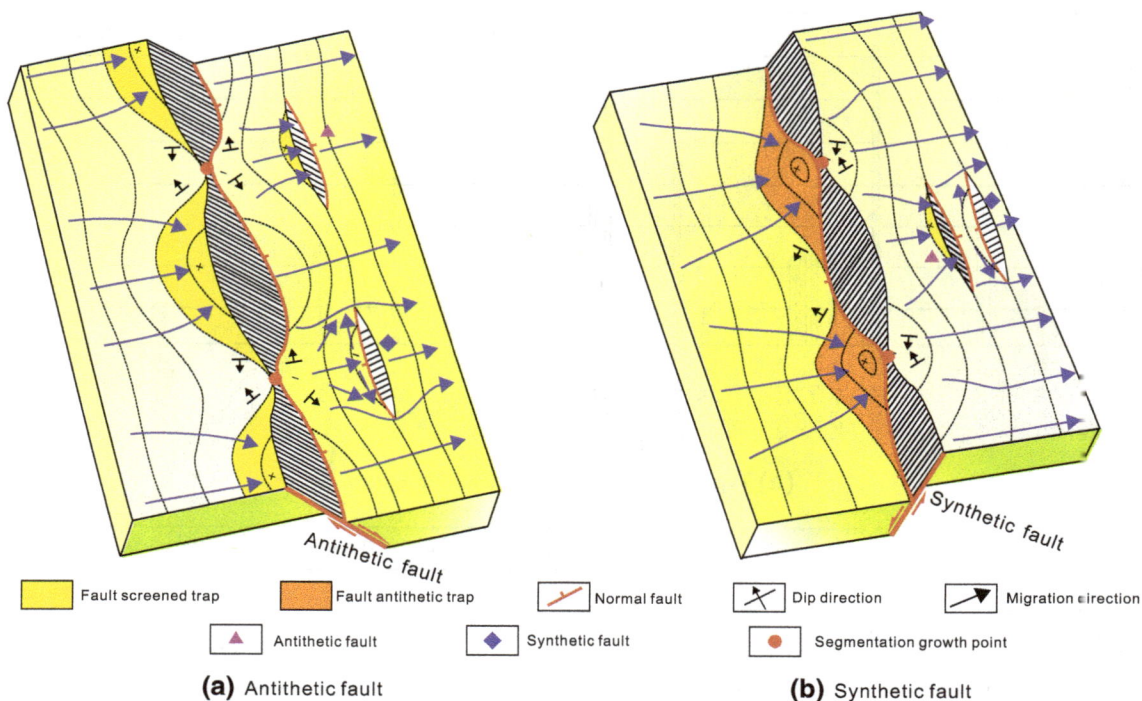

(a) Antithetic fault   (b) Synthetic fault

Fig. 5 Micro-amplitude structure of antithetic and synthetic faults during segmentation growth

combinations based on the quantitative discrimination criteria mentioned above. Near NNW-striking faults with remaining oil, remaining oil is rich near hard linkages and the end of faults, but it is less near soft linkages (Fig. 4). Structural configurations of soft linkages vary among different fault combinations, including relay ramps, oblique anticlines, and oblique grabens. Remaining oil enrichment in soft-linkage areas has a positive relationship with the width of the transfer zone (Fig. 4). Positive structures developed in the transfer zones of oblique anticlines are richer in remaining oil.

3.3 Types and remaining oil enrichment of micro-amplitude associated structures

There are two types of positive micro-amplitude structures developed at the edges of faults (Fig. 5). The first is tilted fault blocks controlled by footwalls of antithetic faults, and the second is lateral anticlines at hanging walls controlled by synthetic fault linkage points. Because of the control of the two types of micro-amplitude structures by the position of fault segmentation points, throw-distance curves can be used for effective identification of micro-amplitude structures. From the 278 antithetic fault throw-distance curves, 7 segmented growth points (Fig. 6) could be identified, where 8 micro-amplitude structures (Fig. 6) developed and some of them are interconnected. By comparing with daily oil production intensity, a large number of positive micro-

amplitude structures can be defined scattered throughout the hanging wall of Fault F234, and concentrated in the north and south. It is shown at the early production period that daily oil production of the wells at high positions in positive micro-amplitude structures is much higher than those in nonpositive ones (Fig. 7). Positive micro-amplitude structures are distributed wholly in footwalls, characterized by larger scale, and numerically fewer. Statistics show that daily oil production of the developing wells at high positions in positive micro-amplitude structures to the south is obviously higher than that far away from the positive micro-amplitude structures (Fig. 7). Comprehensive analysis reveals that remaining oil reservoirs in footwalls form more easily than those in hanging walls of antithetic faults. However, for the same fault block, remaining oil is richer in positive micro-amplitude structures developed at the fault edge, which indicates enrichment and distribution of remaining oil.

4 Internal structures of fault zones and their control on remaining oil distribution

4.1 Internal structures of fault zones and their relationship with remaining oil distribution

Conventional research on internal structure of fault zones is mainly based on outcrop area studies (Caine et al. 1996;

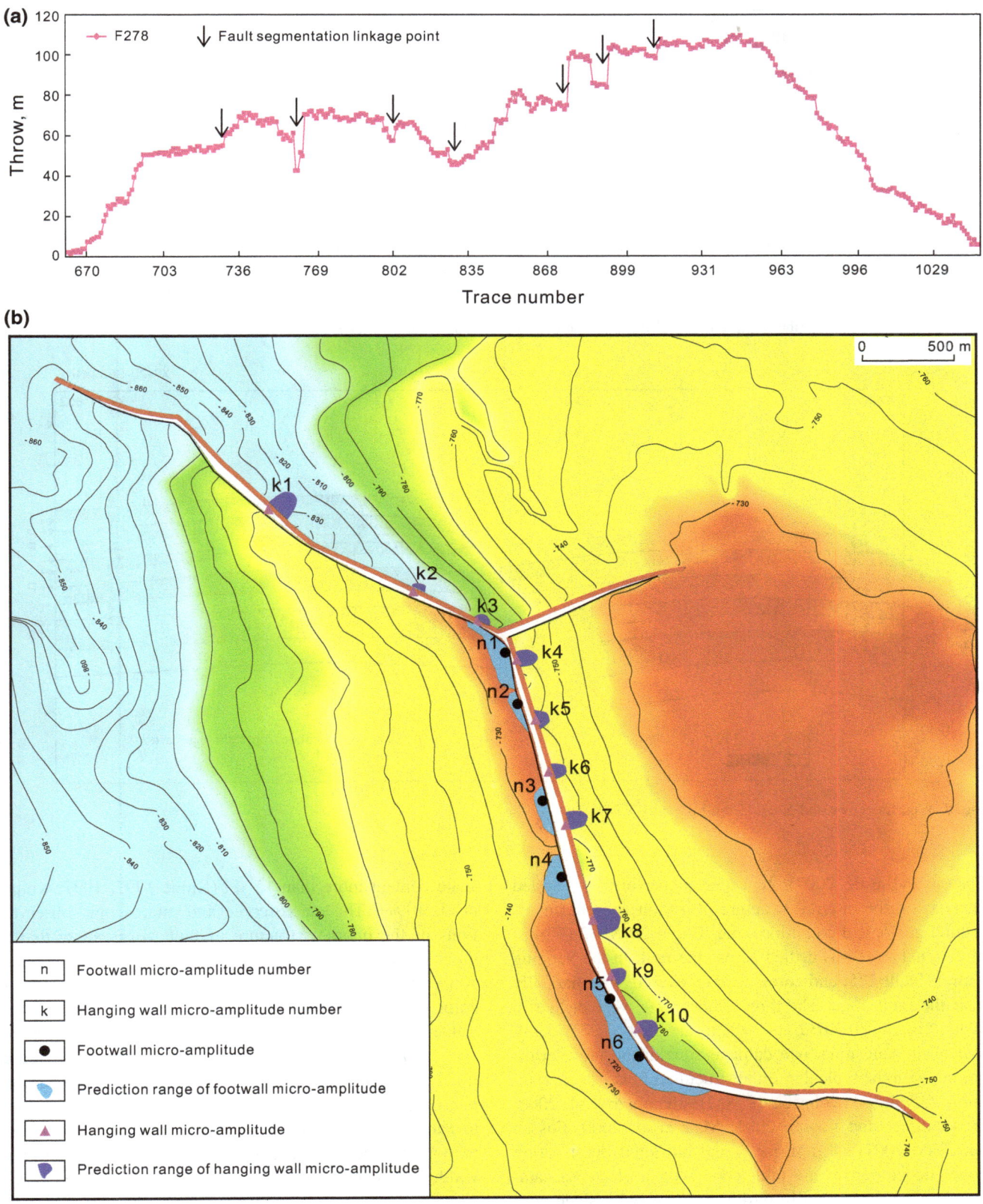

Fig. 6 Throw-distance curve and micro-amplitude structure distribution of No. 278 antithetic fault. **a** Throw-distance curve of No. 278 antithetic fault and **b** relation between fault segmentation and micro-amplitude structure distribution

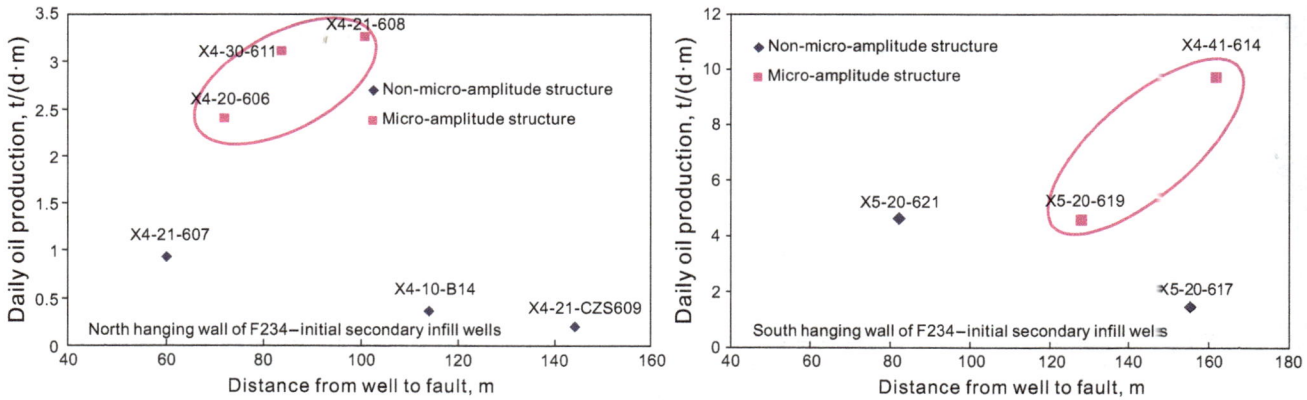

Fig. 7 Remaining oil distribution in and out of the micro-amplitude structure from oil production data

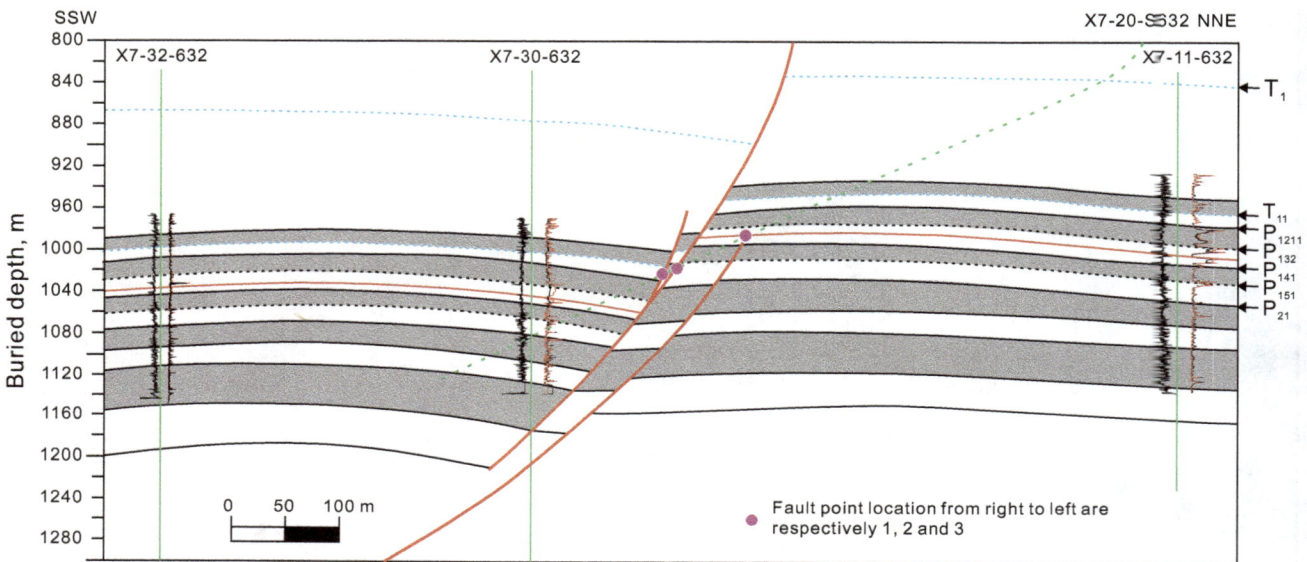

Fig. 8 Well trajectory and No. 275 fault point of X7-20-X632

Fossen and Bale 2007). In order to perform a detailed analysis on the internal structure of a fault zone in underground geological conditions, the Xing 7-20-X632 (Fig. 8) directional well was drilled across the fault in the hanging wall of Fault 275, and cores were recovered systematically near the fault zone. The three fault points were drilled in Well Xing 7-20-X632, of which the first (1272.15 m), located in sandstone, is a damage zone at the fault termination, with the typical internal structure of cataclastic fault zones (Aydin and Johnson 1978; Hesthammer et al. 2000; Hesthammer and Fossen 2001; Mair et al. 2000; Fossen and Bale 2007). Slip surfaces and clustered deformation zones developed in the fault core, with a clustered cataclastic zone developed in pure sandstone (with shale content less than 15 %), and clustered and layered silicate-frame fault rocks developed in impure sandstone (with

shale content more than 15 %) (Knipe 1992, 1997; Knipe et al. 1997). The macroscopic and microscopic observations of the internal structure of fault zones show that a great amount of deformation developed in the damage zone, which is characterized by a rib convex shape with its color darker than the parent rock under the microscope. Cataclastic zones are characterized by reduced particle size, poor sorting, and pores in the cataclastic zone that are filled by detrital material and appear argillaceous under the microscope (Fig. 9). As the distance from the slip surface to parent rock becomes smaller, the density of the deformation zone becomes higher and higher (Fig. 10), and the width of damage zone is approximately 4.5 m. The overall plane porosity ratio is about the sum of the effective pore area ratio and the residual oil area ratio. Research on the microstructural characteristics of damage zones indicates

Fig. 9 Macroscopic and microscopic characteristics of the internal structure of the fault zone

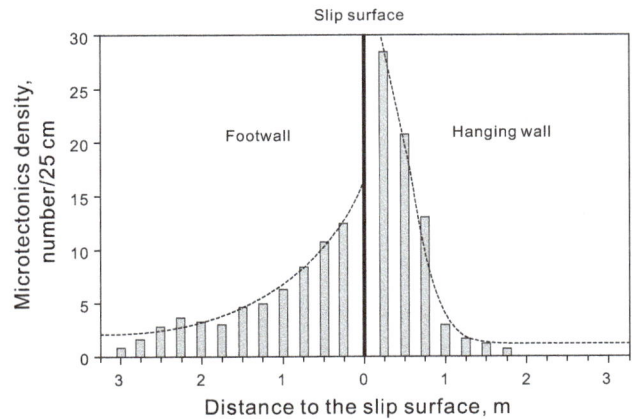

Fig. 10 Deformation band density change away from the slip surface of No. 1 fault point

(Fig. 11) that the total plane porosity ratio of the cataclastic deformation zone is 13.2 %, the effective plane porosity ratio is 1.5 %, and the oil area ratio is 11.7 %. The total plane porosity ratio of the surrounding (parent) rock is 23.3 %, the effective plane porosity ratio of the surrounding (parent) rock is 20.9 %, and the oil area ratio is 2.4 %. The results show that the deformation zone has lower porosity and permeability than the parent rocks.

4.2 Control of fault sealing properties on remaining oil distribution

The oil-bearing layer of the first member of the Yaojia Formation is a typical interbedded sand-mud reservoir. Fault sealing ability mainly relies on clay content filled in the fault zone. It is a typical fault rock seal (Fu et al. 2010). Based on the quantitative relationship between the shale gouge ratio (SGR) of fault rock and the height of hydrocarbon column [Eqs. (1) and (2)], we confirm that the critical fault sealing SGR value is 30 % (Fig. 12). Most of the reservoirs controlled by faults where the SGR is higher than 30 % are oil layers, i.e., faults are sealing ones. At the

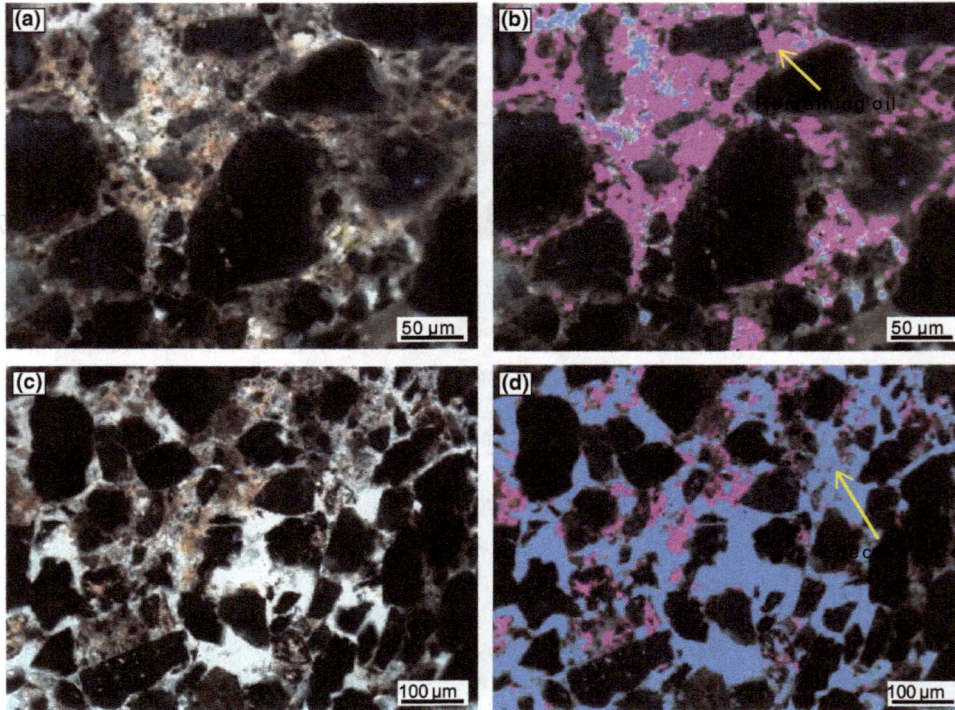

Fig. 11 Photomicrographs showing porosity and remaining oil distribution of the deformation zone and the surrounding rock. **a** Micro-scale characteristics of the deformation zone. **b** Micro-scale characteristics of the deformation zone by fluorometric analysis.

c Micro-scale characteristics of the surrounding rock. **d** Micro-scale characteristics of the surrounding rock by fluorometric analysis

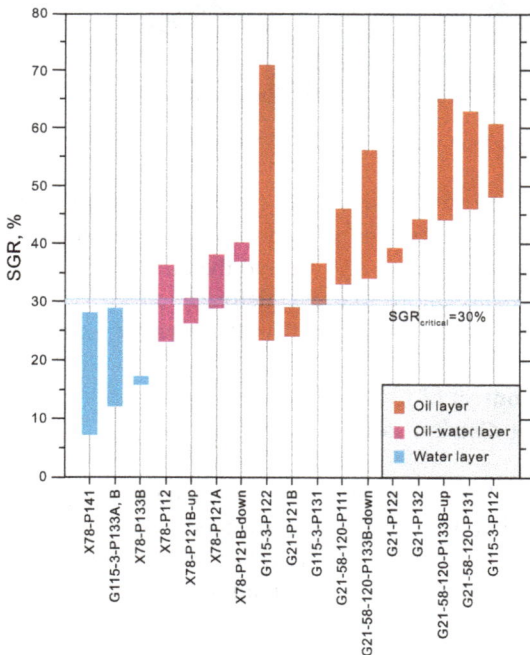

Fig. 12 Critical SGR (30 %) of fault sealing

position that SGR is lower than 30 %, most of the sand bodies controlled by faults are oil–water layers or water layers, i.e., faults have a higher leakage risk.

$$P_{\text{fault}} = (\rho_w - \rho_o)g = 10^{\left(\frac{\text{SGR}}{45}-0.5\right)} \tag{1}$$

$$H = \frac{10^{\left(\frac{\text{SGR}}{45}-0.5\right)}}{(\rho_w - \rho_o)g}, \tag{2}$$

where P_{fault} is the fault breakthrough pressure, MPa, ρ_w is the water density, kg/m^3, ρ_o is the oil density, kg/m^3, g is the acceleration of gravity, m/s^2, and H is the oil column height, m.

Taking Fault 234 as a detailed example (Fig. 13), the SGR of the fault surface, the breakthrough pressure, and the hydrocarbon column height are calculated. The minimum SGR occurs at the fault termination and the leakage risk is commensurately higher. The sealing ability of faults at the locations with larger throw is higher and remaining oil is richer there.

5 Prediction of remaining oil distribution at the fault edge

5.1 Types of remaining oil at the fault edge

Based on the injection–production relationship at the fault edge, the remaining oil is classified into three types. The first oil type is the oil remaining after imperfect injection

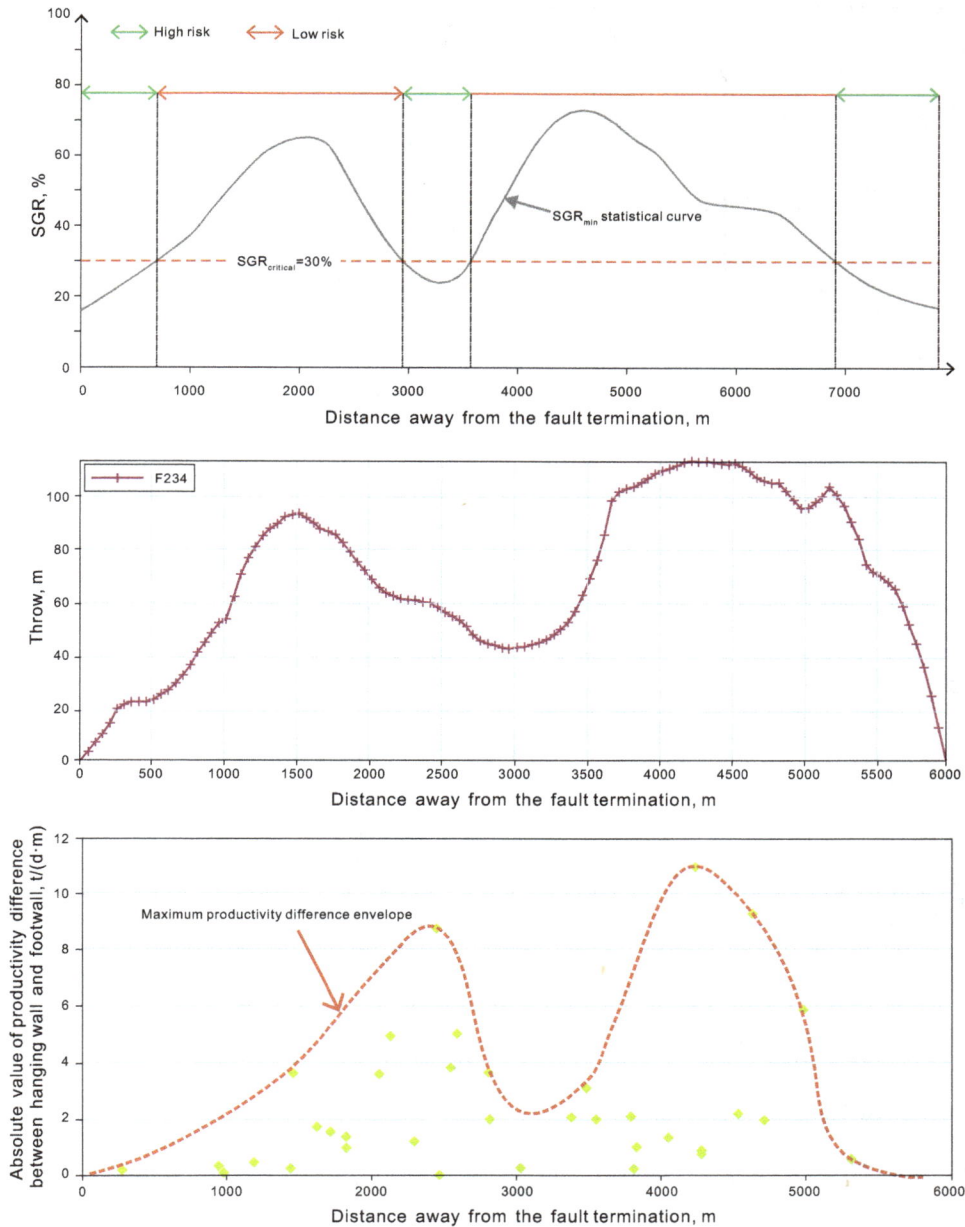

Fig. 13 Relation to fault seal and remaining oil distribution of No. 234 (productivity difference between hanging wall and footwall means the difference of oil production between hanging wall and footwall, which reflects the sealing capacity of fault)

and production. In some depositional time units, the matching of sand bodies and injection–production patterns is imperfect within some sand bodies. Some patterns in some localized areas have production without injection or have injection without production. As a result, remaining oil is due to imperfect injection and production. The effective thickness of sand bodies in PuI_2^2 unit formation to the south of Fault 234 is 1.92 m. Well Xing 5-2-21 and Well Xing 5-2-22 at the edge of Fault 234 have been perforated. However, oil wells Xing 5-21-619 and Xing 5-21-620 and water wells Xing 5-2-719 and Xing 5-2-122

outside the scope of well spacing have not been perforated, which belong to typical remaining oil types with injection but without production in individual sand bodies (Fig. 14a). The second type is oil remaining outside the controlling well pattern. Sand bodies in small underwater distributary channels are narrow and thin and have higher clay content and poorer physical properties and oiliness because of slightly lower river energy. However, narrow sand bodies lead to a low drilling rate, thus few wells or none have been drilled in the sand bodies within the range of hundreds of meters to several kilometers. However,

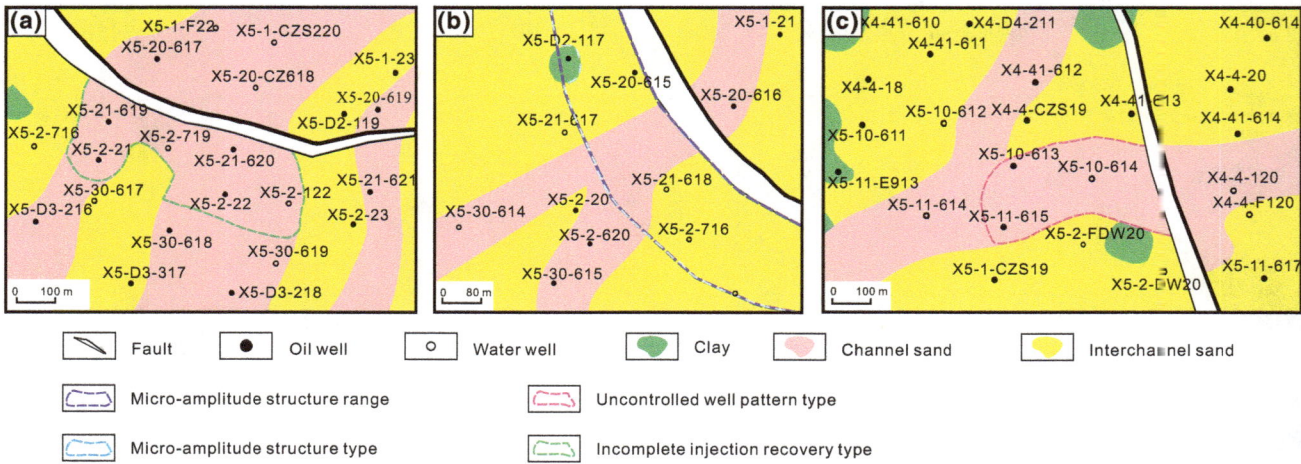

Fig. 14 Remaining oil type in the vicinity of faults

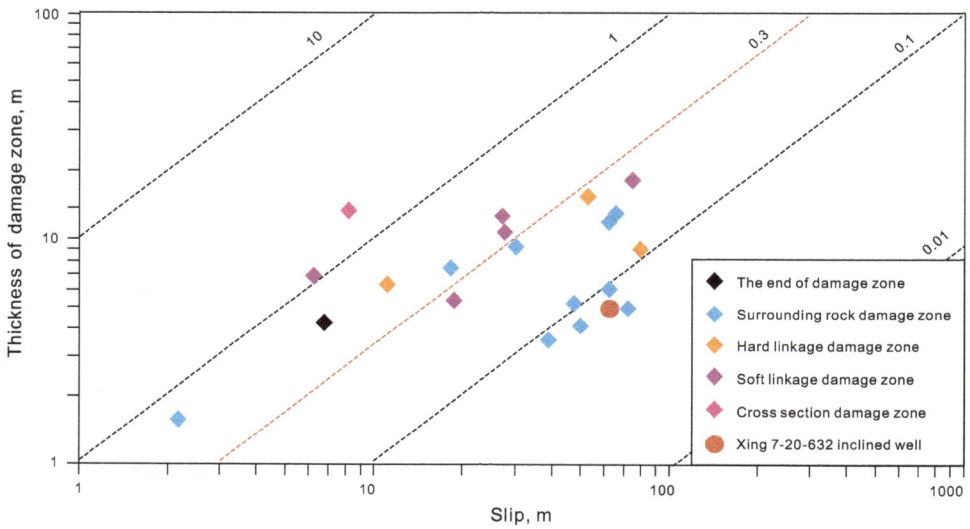

Fig. 15 Relation to throw and thickness of the damage zone

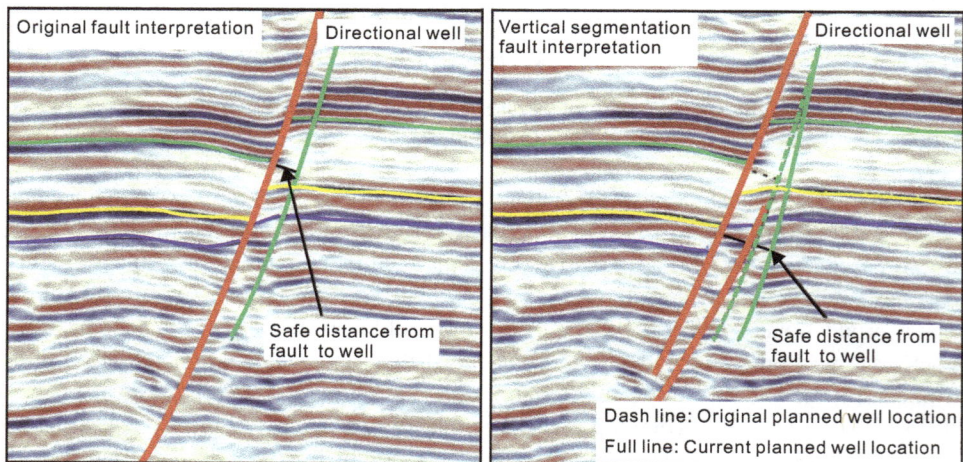

Fig. 16 Fault combination and safe distance from the high-angle well to the fault zone

there are few oil production wells without water injection or few water injection wells without oil production, so oil layers are not produced or produced with lower degree, seriously influencing the producing degree and productivity of oil layers. That amounts to no water injection in the oil layer, resulting in oil remaining outside the controlling well pattern. Sand bodies at the $SaII_5^1$ unit formation in central Fault 234 have an area of $0.116 \ km^2$ and an effective thickness of 0.8 m. There exists a small but thick channel sand at the fault edge. Because the three wells within the channel sand are not perforated, remaining oil is rich here (Fig. 14b). The third type is the oil remaining within micro-amplitude structures. At the fault edge, small anti-clines come from structural deformation. When sand bodies developed, oil remains, i.e., remaining oil of micro-amplitude structures. Sand bodies at the $SaII_8$ unit formation to the south of Fault 234 have an effective thickness of 1.04 m and are located in micro-amplitude structures. Well Xing 5-2-20 has been perforated outside the micro-amplitude structural trap, yet Well Xing 5-20-615, Well Xing 5-21-618, and Well Xing 5-2-716 at higher position in the structures have not been perforated. Therefore, it is confirmed that remaining oil is controlled by micro-amplitude structures in the trap (Fig. 14c).

5.2 Prediction of distribution and potential tapping methods of remaining oil at fault edges

On the basis of fault zone structures and their control on remaining oil distribution, and the types of remaining oil, a three-step approach for predicting distribution of remaining oil at fault edges was developed. First, we should determine faults and fault blocks with rich remaining oil and optimal NNW antithetic fault footwalls based on the matching relationship between faults and reservoirs. Second, we will determine the location of faults by superposing micro-amplitude structures, ranges of fault sealing oil, and the oblique anticlines formed by soft-linkage faults, and determining favorable locations for the distribution of remaining oil. Third, we should determine the horizon for tapping remaining oil potential in favorable settings on the basis of the injection–production relationships of all sublayers.

Remaining oil at fault edges is mainly tapped by multi-target directional wells. Rational well spacing (distance from fault to well) is determined, according to fault combination modes and widths of damage zones. Two issues have to be taken into account for safe remaining oil production. First is the width of damage zone. Owing to the strong heterogeneity of the reservoir in the damage zone, the porosity of the damage zone drops considerably, and the permeability declines by 2–3 orders of magnitude. This is unfavorable for efficient oil and gas development.

Meanwhile, casing damage or water injection leakage often occurs when drilling in the damage zone. So directional well deployment should keep away from the damage zone. Well coring and logging data show that the width of damage zone on the side of fault core is less than 15 m in general (Fig. 15), so the distance from well to fault should be larger than 15 m. The other issue is that safe remaining oil production at fault edges ignores an important geologic phenomenon—vertically segmented growth of faults (Rykkelid and Fossen 2002; Sperrevik et al. 2000; Færseth 2006). So a fault that is traditionally interpreted as a single through-going fault may be composed of two vertically *en echelon* overlying faults, and the separation between *en echelon* overlying faults has to be considered (Fig. 16). The safe distance of remaining oil well deployment at fault edges is the sum of the width of damage zone and the separation between vertically segmented growth faults.

6 Conclusions

(1) The greater the fault throw, the richer the remaining oil, which is controlled by two factors. One is that the oil layer at the first member of the Yaojia Formation tends to be juxtaposed against a top mudstone, so the sealing capacity of faults becomes stronger with increased fault throw. The other is that due to differential subsidence, areas with greater throw of antithetic faults and nearer to the structural high spot are more conducive to the enrichment of remaining oil.

(2) There are two major types of positive micro-amplitude structures developed at the fault edge: the first is tilted fault blocks controlled by the footwall of antithetic faults, and the second is the lateral anticlines at the hanging wall controlled by synthetic fault linkage points. Based on the principle of differential fault subsidence, a preliminary method is established to predict micro-amplitude structures using a throw-distance curve.

(3) The macroscopic and microscopic observation of the well cores of the first cross fault in the Songliao Basin confirmed that fault deformation is cataclastic, forming typical internal structures of cataclastic fault zones. Slip surfaces and clustered deformation zones are developed in the fault core. The density of the deformation zone becomes more and more closer to the slip plane. The deformation zone has lower porosity and permeability than the surrounding rock, and remaining oil in the damage zone concentrates in the deformation zone.

(4) Based on the quantitative relationship between the SGR of fault rocks and the height of the hydrocarbon

column, the critical SGR of the sealed fault in the first member of the Yaojia Formation of the Xingshugang structural zone is 30 %. In addition, the sealing ability of faults with larger fault throw is stronger, and remaining oil there is relatively richer.

(5) On the basis of the internal structure of fault zones, and its control on remaining oil and the type of remaining oil, a method is established that predicts the distribution of remaining oil. Two factors have to be taken into account for safe remaining oil well deployment distance at fault edges. These are the width of the damage zone and the separation of vertically segmented growth faults.

Acknowledgments The authors greatly appreciate the financial support from the Natural Science Foundation of China (Grant No. 41272151, 41472126), the Natural Science Foundation for Distinguished Young Scholars of Heilongjiang Province, China (Grant No. JC201304), the Joint Funds of the National Natural Science Foundation of China (Grant No. U1562214), and the Program for Huabei Oilfield (Grant No. HBYT-CY5-2015-JS-127).

References

Aydin A, Johnson AM. Development of faults as zones of deformation bands and as slip surfaces in sandstone. Pure appl Geophys. 1978;116(4-5):931–42.

Bretan P, Yielding G, Jones H. Using calibrated shale gouge ratio to estimate hydrocarbon column heights. AAPG Bull. 2003;87(3):397–413.

Caine JS, Evans JP, Forster CB. Fault zone architecture and permeability structure. Geology. 1996;24(11):1025–8.

Cloos H. Über antithetische Bewegungen. Geol Rundsch. 1928;19(3):246–51.

Færseth RB. Shale smear along large faults: continuity of smear and the fault seal capacity. J Geol Soc. 2006;163(5):741–51.

Feng CJ, Bao ZD, Yang L, et al. Reservoir architecture and remaining oil distribution of deltaic front underwater distributary channel. Pet Explor Dev. 2014;41(4):323–9 (in Chinese).

Fossen H. Structural geology. New York: Cambridge University Press; 2010. p. 152–85.

Fossen H, Bale A. Deformation bands and their influence on fluid flow. AAPG Bull. 2007;91(12):1685–700.

Fossen H, Schultz RA, Shipton ZK, et al. Deformation bands in sandstone: a review. J Geol Soc. 2007;164(4):755–69.

Fu XF, Li WL, Lv YF, et al. Quantitative estimation of lateral fault seal and application in hydrocarbon exploration. Geol Rev. 2011;57(3):387–97 (in Chinese).

Fu XF, Sha W, Yu D, et al. Lateral sealing of faults and gas reservoir formation in volcanic rocks in Xujiaweizi fault depression. Geol Rev. 2010;56(1):60–70 (in Chinese).

Gao MX. Tracing seismotectonic activity tendency—a possible approach for practical earthquake prediction. Recent Dev World Seismol. 1983;9:00 (in Chinese).

Gao RQ. Characteristics of the continental Cretaceous in the Songliao Basin. Acta Geol Sin. 1980;54(1):9–26 (in Chinese).

Gao RQ, Cai XY. Formation conditions and distribution of oil and gas fields in the Songliao Basin. Beijing: Petroleum Industry Press; 1997. p. 12–40.

Giba M, Walsh JJ, Nicol A. Segmentation and growth of an obliquely reactivated normal fault. J Struct Geol. 2012;39(2):253–67.

Han DK. Status and challenges for oil and gas field development in China and directions for the development of corresponding technologies. Eng Sci. 2010;12(5):51–7 (in Chinese).

Hesthammer J, Fossen H. Structural core analysis from the Gullfaks area, northern North Sea. Mar Pet Geol. 2001;18(3):411–39.

Hesthammer J, Johansen TES, Watts L. Spatial relationships within fault damage zones in sandstone. Mar Pet Geol. 2000;17(8):873–93.

Hills ES. Outlines of structural geology. London: Methuen; 1953. p. 22–36.

Hou J, Luo FQ, Li ZQ, et al. The critical description scale study on core microscopic and reservoir macroscopic remaining oil. Pet Geol Recovery Effic. 2014;21(6):95–8 (in Chinese).

Hu WR. Necessity and feasibility of PetroChina mature field redevelopment. Pet Explor Dev. 2008;35(1):1–5 (in Chinese).

Hu WS, Wang JL. Extensional tectonic evolution and petroleum accumulation in Songliao Basin. Pet Explor Dev. 1996;23(3):30–3 (in Chinese).

Jiang YL. Water-control and energy-saving & trapping potential test inside layer for La 9-232 well block. Energy Conserv Pet Petrochem Ind. 2013;7:9–11 (in Chinese).

Jin Q, Wang RP, He RW, et al. Identification and description of small faulted-block reservoirs. Acta Pet Sin. 2009;30(3):367–71 (in Chinese).

Jin Q, Zhou JF, Wang DP, et al. Identification of shattered fault zones and its application in development of fault-block oilfields. Acta Pet Sin. 2012;33(1):82–9 (in Chinese).

Kim YS, Sanderson DJ. The relation between displacement and length of faults. Earth Sci Rev. 2005;68(3):317–34.

Knipe RJ. Faulting processes and fault seal. Struct Tecton Model Appl Pet Geol. 1992;1:325–42.

Knipe RJ. Juxtaposition and seal diagrams to help analyze fault seals in hydrocarbon reservoirs. AAPG Bull. 1997;81(2):187–95.

Knipe RJ, Fisher QJ, Jones G, et al. Fault seal analysis: successful methodologies, application and future directions. Nor Pet Soc Spec Publ. 1997;7:15–38.

Liu Z, Lv YF, Sun YH, et al. Characteristics and significance of syngenetic fault segmentation in hydrocarbon accumulation: an example of Yuanyanggou fault in western sag, Liaohe Depression. J China Univ Min Technol. 2012;41(5):793–9 (in Chinese).

Liu ZB. High resolution sequence stratigraphy and hydrocarbon accumulation model of Fuyu reservoir in Sanzhao Depression. Ph.D. Thesis, Northeast Pet University. 2010 (in Chinese).

Liu ZB, Ma SZ, Sun Y, et al. High-resolution sequence stratigraphy division and depositional characteristics of Putaohua reservoir, Sanzhao Depression. Acta Sedimentol Sin. 2008;26(3):399–406 (in Chinese).

Liu ZB, Yan L, Gao F, et al. Enrichment rule and tapping methods of remaining oil on fault boundary at high water cut stage oil-fields: a case of Putaohua reservoir of the north fault block of Xingnan oilfield, Songliao Basin. J Northeast Pet Univ. 2014;38(4):52–8 (in Chinese).

Liu ZC. Improving oil recovery technology applications and development directions in Sinopec. Sinopec. 2008;4:5–8 (in Chinese).

Mair K, Main I, Elphick S. Sequential growth of deformation bands in the laboratory. J Struct Geol. 2000;22(1):25–42.

Miall AD. Reconstructing fluvial macroform architecture from two-dimensional outcrops: examples from the Castlegate Sandstone, Book Cliffs, Utah. J Sediment Res. 1994;64(2):145–58.

Peacock DCP. Displacement and segment linkage in strike-slip fault zones. J Struct Geol. 1991;13(9):1025–35.

Peacock DCP, Sanderson DJ. Displacements, segment linkage and relay ramps in normal fault zones. J Struct Geol. 1991;13(6):721–33.

Peacock DCP, Sanderson DJ. Geometry and development of relay ramps in normal fault systems. AAPG Bull. 1994;78(2):147–65.

Peacock DCP, Knipe RJ, Sanderson DJ. Glossary of normal faults. J Struct Geol. 2000;22(3):291–305.

Peng SM, Zhou HT, Li HY, et al. Phased flow unit model establishment and remaining oil prediction: An example from Jing 11 block in Bieguzhuang Oilfield. Pet Explor Dev. 2007;34(2):216–21 (in Chinese).

Qi JF. Two tectonic systems in the Cenozoic Bohai Bay Basin and their genetic interpretation. Geol China. 2004;31(1):15–22 (in Chinese).

Rykkelid E, Fossen H. Layer rotation around vertical fault overlap zones: observations from seismic data, field examples, and physical experiments. Mar Pet Geol. 2002;19(2):181–92.

Soliva R, Benedicto A. A linkage criterion for segmented normal faults. J Struct Geol. 2004;26(12):2251–67.

Soliva R, Benedicto A, Schultz RA, et al. Displacement and interaction of normal fault segments branched at depth: Implications for fault growth and potential earthquake rupture size. J Struct Geol. 2008;30(10):1288–99.

Sperrevik S, Færseth RB, Gabrielsen RH. Experiments on clay smear formation along faults. Pet Geosci. 2000;6(2):113–23.

Sun YH, Chen YB, Sun JG, et al. Evolutionary sequence of faults and the formation of inversion structural belts in the northern Songliao Basin. Pet Explor Dev. 2013;40(3):275–83 (in Chinese)

Wang HX, Li MH, Shen ZS, et al. The establishment and geological significance of quantitative discrimination criterion of fault segmentation growth: an example from Saertu reservoir in Xingbei development area of Songliao Basin. Geol Rev. 2014;60(6):1259–64 (in Chinese).

Wang HX, Lv YF, Fu XF, et al. Formation, evolution and reservoir-controlling mechanism of relay zone in rift basin. Geol Sci Technol Inf. 2013;34(4):102–10 (in Chinese).

Yielding G. Shale gouge ratio: calibration by geohistory. Norwegian Pet Soc Spec Publ. 2002;11(2):1–15.

Yu SD. The development models of complicated fault-block sandstone oilfield. Beijing: Petroleum Industry Press; 1998. p. 1–35 (in Chinese).

Yue DL, Wu SH, Cheng HM, et al. Numerical reservoir simulation and remaining oil distribution patterns based on 3D reservoir architecture model. J China Univ Pet. 2008;32(2):21–7 (in Chinese).

Yun JB, Jin ZJ, Yin JY. Characteristics of inherited fault belts and their effect on hydrocarbon accumulation. Geotecton Metallog. 2002;26(4):379–85 (in Chinese).

Zhang GC, Xu H, Liu HF, et al. Inversion structures in relation to oil and gas field distribution in Songliao Basin. Acta Pet Sin. 1996;17(2):9–14 (in Chinese).

14

Where is the future of China's biogas? Review, forecast, and policy implications

Lei Gu[1] · Yi-Xin Zhang[2] · Jian-Zhou Wang[3] · Gina Chen[4] · Hugh Battye[5,6]

Abstract This paper discusses the history and present status of different categories of biogas production in China, most of which are classified into rural household production, agriculture-based engineering production, and industry-based engineering production. To evaluate the future biogas production of China, five models including the Hubbert model, the Weibull model, the generalized Weng model, the H–C–Z model, and the Grey model are applied to analyze and forecast the biogas production of each province and the entire country. It is proved that those models which originated from oil research can also be applied to other energy sources. The simulation results reveal that China's total biogas production is unlikely to keep on a fast-growing trend in the next few years, mainly due to a recent decrease in rural household production, and this greatly differs from the previous goal set by the official

✉ Yi-Xin Zhang
zhangyx13@lzu.edu.cn

✉ Jian-Zhou Wang
wangjz@dufe.edu.cn

[1] College of Earth and Environmental Sciences, Lanzhou University, Lanzhou 730000, China

[2] School of Mathematics and Statistics, Lanzhou University, Lanzhou 730000, China

[3] School of Statistics, Dongbei University of Finance and Economics, Dalian 116025, China

[4] School of Law, Northwestern University, Chicago, IL 60611, USA

[5] School of History and Culture, Lanzhou University, Lanzhou 730000, China

[6] Department of Economics, The London School of Economics and Political Science, London WC2A 2AE, UK

Edited by Xiu-Qin Zhu

department. In addition, China's biogas production will present a more uneven pattern among regions in the future. This paper will give preliminary explanation for the regional difference of the three biogas sectors and propose some recommendations for instituting corresponding policies and strategies to promote the development of the biogas industry in China.

Keywords Biogas production · China · Temporal–spatial · Forecast · Policy

1 Introduction

When faced with a global energy crisis, we have seen considerable efforts and progress in exploring effective and sustainable energy during recent years. Among different types of energy, biogas is regarded as an important and versatile one that not only is a sustainable fuel source for producing electricity, heat, or driving power, but it also has multiple advantages in environment improvement, greenhouse gas reduction, carbon capture, and fertilizer production, even social benefits like cost and labor saving in some cases (Cheng et al. 2011; Barnhart 2013; Gosens et al. 2013; Zhang et al. 2013). Therefore, biogas is widely used in both developed and developing countries nowadays.

Due to different natural resource endowments, climate conditions, technologies and industrialization development levels, and socioeconomic status, the production of biogas varies among different cases. According to the statistical data by the International Energy Agency (IEA), the entire biogas production of the world was about 42 Giga cubic meters in 2008 (IEA 2010). In detail, the countries of the Organization for Economic Co-operation and Development

(OECD) have an annual production of 27 Giga cubic meters; non-OECD countries produce approximately 15 Giga cubic meters biogas a year. Some European countries like Germany represent the advanced level of biogas production in the world, not only in production amount, but also in technology and policy frameworks (Poeschl et al. 2010). The utilization of biogas shows a number of diverged functions and has achieved a high level of industrialization and commercialization in OECD countries (GARR 2012; U.S.EPA 2015; Couture and Gagnon 2010; FIT 2012), such as electricity production, biogas vehicles, and other transport systems (SEA 2012; Fallde and Eklund 2015; Olsen et al. 2013; NIFU 2015; Gildas 2010; Bojesen et al. 2014). Recently, some crucial changes have happened in biogas production among those countries: some Central or East European countries including Poland, Hungary, Slovakia have made fast progress in biogas development; the number of biogas power stations continues to grow in a steady rate in Sweden, UK, and Denmark, but the biogas industries in Germany, Austria, and Italy have gradually stagnated in the past few years. As the biggest contributor in Europe, the stagnation of Germany's biogas industry may have a strong influence on Europe as a whole. Thus, the European Biogas Association believes that the future is not bright for biogas development in Europe (EBA 2014).

Biogas production in developing countries varies greatly. The data from the IEA show that only 36 % of the total biogas is produced by non-OECD countries. China produces 98.4 % of biogas among the non-OECD countries in 2008 (IEA 2010). Most of developing countries have just accelerated their pace to develop their biogas industries in recent years. Factors such as industry structure and socioeconomic status have direct influence on biogas production and its dissemination. As a result, in such countries as China, India, Nepal, and Bangladesh, small-scale and economic domestic biogas plants have become the mainstream of biogas technology and are widely used by people in rural areas in the household mode (Tomar 1995; Bhat et al. 2001; Singh and Maharjan 2003; Alam 2008; Gautam et al. 2009; Chen et al. 2010; Dimpl 2010). The main materials for household biogas plants are livestock manure, human excreta and agriculture residues, among which livestock manure (pig or cow) plays a crucial part in biogas generation (Bond and Templeton 2011; Kabir et al. 2013). Some research indicates that there are many potential resources for creating biogas in developing nations, but with hurdles such as socioeconomic factors, climate conditions, technology, and institutional frameworks (Jiang et al. 2011; Mwirigi et al. 2014). A traditional domestic biogas plant primarily creates cooking fuel and fertilizers for farmers, but seldom generates electricity or fuel for market use. However, in recent years, medium- or large-scale multi-purposes biogas plants have been introduced and developed in countries such as China, Brazil, and India (Jiang et al. 2011; Coimbra-Araújo et al. 2014; Singh and Jash 2015), and have gradually closed the gap with developed countries in biogas productivity in the past few years.

2 Overview of biogas development in China

2.1 History of China's biogas development

China is one of the earliest countries that has developed and utilized biogas for almost a century. By the end of the nineteenth century, simple biogas digesters had appeared in the coastal areas of Southern China. In the 1920s, Luo Guorui in Taiwan has created the first water-pressure biogas digester (He et al. 2013). It was not until the 1970s that millions of household biogas tanks were installed in rural areas, when China's government first set up measures to extend biogas use to overcome the energy shortage in rural areas. Since the 1980s, the development of biogas infrastructure had been included into the national long-term development program by the central government of China (Jiang et al. 2011).

After the 1990s, due to the deep implementation of open and reform policy, China has experienced fast acceleration in industrialization and urbanization, accompanied by environmental deterioration and sharply rising energy demand. As a consequence, China has become one of the world's top energy-consuming countries, together with a rapidly degrading eco-environment (China's National Energy Administration, NEA 2012).

From 2003, the national debt aid program has supported a new round of biogas promotion, especially for rural biogas implementation by the Ministry of Agriculture (MOA) of China (MOA 2002, 2003). The central government of China paid much more attention to biogas development by offering direct financial support, from 1000 million CNY (China Yuan) in 2003–2005 to 2500 million CNY in 2006–2007 with increasing focus on biogas engineering projects after the "Renewal Energy Law" implemented in 2006 (NPC 2005; MOA 2007). The support and aid reached to 5000 million CNY in 2010, leading to a drastic rise of biogas users in the first decade of twenty-first century: from 11 million family users in 2003 to 43 million in 2013, and from 2300 biogas engineering projects in 2003 to nearly 10,000 projects in 2013 (calculated from data reported by MOA; several years used). Production capacity by household biogas digesters increased from only 4.5 Giga cubic meters in 2003 to 16 Giga cubic meters in 2013, amounting to 8000 Ktoe oil equivalent; total biogas production of large-scale biogas projects increased from 0.2 Giga to 2 Giga cubic meters during the 10 years, accounting for 12.4 % of natural gas consumption and 1000 Ktoe oil equivalent in 2013 (calculated from data reported by MOA 2013). After 2009, China has enhanced its support for biogas engineering projects by offering subsidies from 25 % to 45 % of the whole cost of projects, especially allocating more aid to Mid-Western areas and innovative projects, setting up policies similar to feed-in tariffs to promote power generation through biogas plants. Meanwhile, local biogas service systems were established by for improving the efficiency of biogas production and utilization. In 2014, central and local government still invested more than 2500 million CNY, of which the total biogas engineering construction accounting for 40 %, higher than the 18.4 % in 2008.

A long-term goal planned by China's National Development and Reform Commission (NDRC) claims that the available biogas production in 2020 will be 44 Giga cubic meters, from which 30 Giga cubic meters biogas will be produced by household users and 14 Giga cubic meters biogas will be generated from biogas engineering projects (NDRC 2007). Another short-term goal on biogas development has been set up by China's National Energy Administration (NEA): By 2015, the number of the biogas household users will reach 50 million with total biogas production of 19 Giga cubic meters. Large-scale biogas projects focused on agricultural and industrial (including municipal) wastes will produce 2.5 Giga and 0.5 Giga cubic meters of biogas, respectively, with more biogas being used to produce electricity production (NEA 2012).

As shown in Fig. 1, in general, biogas from rural household biogas digesters comprises the major proportion of total biogas production and shows a long and steady growth. However, according the updated data by the Ministry of Agriculture of China, in recent years, household biogas generation seems to have slowed down to a pre-2011 level,

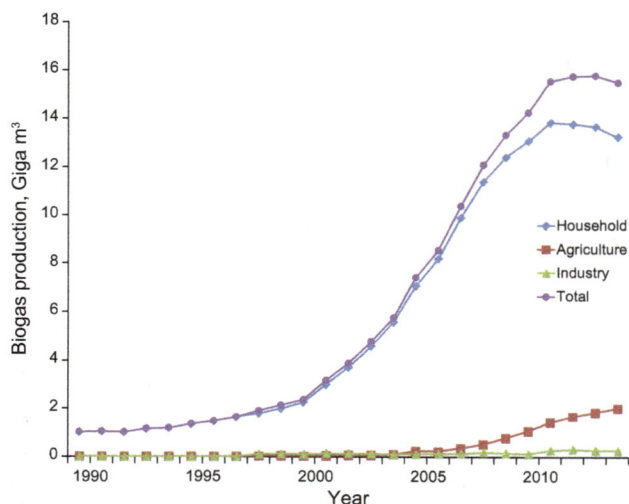

Fig. 1 Biogas production of different sectors from 1989 to 2014

and probably has reached a peak in many provinces (calculated from MOA 1989–2014). In contrast, biogas from agriculture-based biogas engineering projects shows a dramatic growth trend with increasing proportion, and starts to play a more important role in the biogas production of China. Furthermore, biogas from industry-based biogas engineering projects rises gradually with some slight fluctuations, exceeded by agriculture sector after 2004 (calculated from MOA 1989–2014). In the past few decades, China has accelerated its reform pace in urbanization, industrialization, and even energy and agriculture transformation. Therefore, China's biogas production is undertaking a dynamic and complicated transition, quite different from those of developed countries, and also different from those countries which rely on traditional agriculture or livestock husbandry.

2.2 Details of biogas sectors in China

During past years, Chinese people have introduced, modified, or created various models for biogas of different sectors according to local circumstances.

Firstly, rural household biogas plants are the largest contributor to biogas production in China. As the earliest form of biogas digester, the round-shaped water-pressure biogas plant was once popular among farmers in China (He et al. 2013), and also welcomed by many developing countries with a name of "China-mode biogas" digester (Zeng et al. 2007). Recently other types of technology have been developed in various areas of China. Strong back-flow biogas digesters have been introduced in Jiangxi province. The meandering-distribution biogas digester is a major type of digester in Yunnan province. The biogas digesters in Hunan province have separated gas storage with a floating cover. In the northwest region, biogas digesters with circumflex technology are applied to solve problems in resource updating and production efficiency (LSA 2010). Household biogas-based systems in different areas also vary from each other, such as newly developed biogas-kitchen-farm-toilet model (MOA 2008), pig-biogas-fruit model in South China (Chen 1997; MOA 2001a, b), 4-in-1 biogas system in Northern China with a greenhouse (MOA 2001a, b), and 5-in-1 model in Northwest China with water and solar infrastructures (Yue 1997; MOA 2006, 2013).

Secondly, agriculture-based biogas engineering projects have been promoted in China since late 1990s by local and central governments, and the basic technology was introduced and fixed. As Fig. 2 presents, the system of agriculture-based biogas engineering consists of a fermentation tank, gas storage room, bioliquid storage room, gas purification system, biogas electricity generator, biogas transportation system and other related parts (MOA 2006). Most of these biogas engineering projects use either the upflow solid reactor (USR) or the continuous stirred tank reactor (CSTR) as reactor

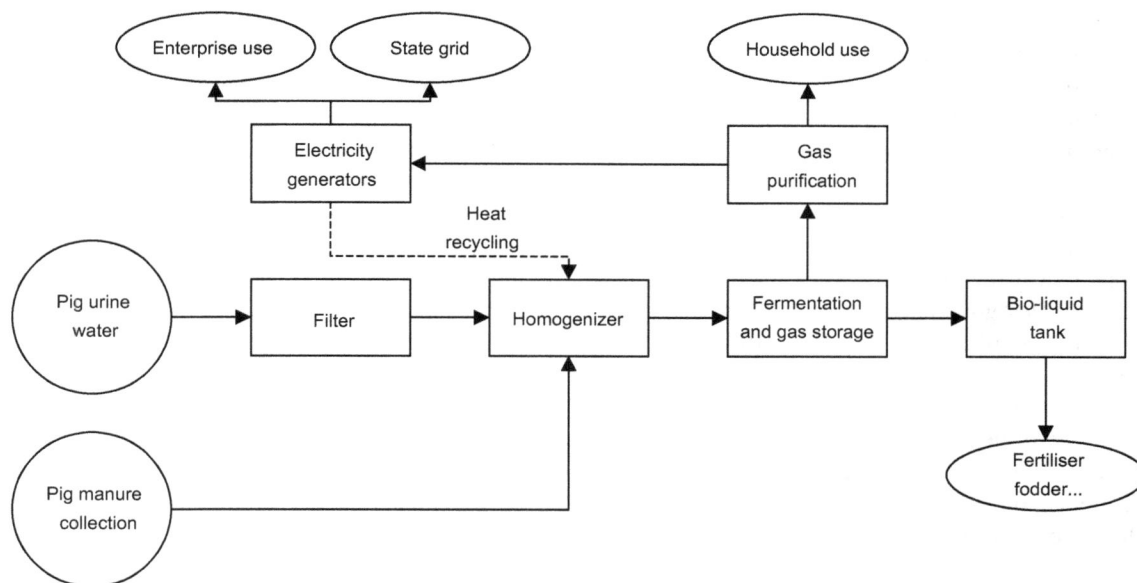

Fig. 2 Entire system of biogas engineering project with a pig farm

technology, and these technologies have been used for more than 20 years (HWEEC 2011). For these engineering projects, though climate condition is no longer the most important, enough raw materials should be guaranteed, and the project should be well maintained both technologically and financially. Biogas production involves many different stakeholders in this market-oriented model, far different from the model of the household biogas plant for families' use.

Thirdly, China's industrial biogas sector ranks the third largest biogas contributor. China's industry-based biogas project started in Henan province in the 1960s, dealing with the wastewater of an alcohol factory (Zhang et al. 2008). There are increasing numbers of industry-based biogas projects in China that tackle rising industrial liquid and solid waste in paper making, food or liquor manufacturing, printing and dyeing, among other challenges. With regard to core regeneration technology of biogas production, the upflow anaerobic sludge blanket (UASB) technology is used in more than half of all biogas plants, followed by CSTR and USR (HWEEC 2011). In addition, municipal sewage-cleaning biogas projects, landfill gas projects, and other biogas plants are small but non-neglected parts of the biogas industry in China, and part of biogas they produce is often reported together with the industry-based engineering projects (NEA 2012).

3 Literature review

Undoubtedly, biogas production is a comprehensive process with complex relationships between nature and humans. So with regard to the issues of biogas in China,

there are various perspectives: not only about resource, climate and technology, but also with social, economic, cultural, and political considerations.

Many studies on China's biogas are from just one micro-perspective. Some focus on the introduction of new biogas technology. During its long history, China has developed a variety of household-based biogas technologies (Cheng et al. 2013): glass fiber-reinforced plastic digesters, plastic soft/hard digesters, and solar fiberglass reinforced plastic (FRP) biogas digester (Shi et al. 2008), as well as different comprehensive biogas systems as mentioned above. However, studies related to technological innovations of large-scale biogas engineering projects are rarely reported in China. In fact, most of the high-level biogas engineering technology in China is introduced from developed countries.

It is evident that the biogas systems mentioned above have benefits for energy consumption, environment improvement, labor and cost saving, fertilizer production, etc., compared to many non-renewable energy forms, so many studies analyze efficiency, sustainability or performance of biogas plants. Some studies are focused on the carbon emission reduction by using the life cycle analysis method and energy analysis, demonstrating that biogas systems are effective for reducing the emission of greenhouse gases and other waste gases (Wang et al. 2014; Zhang and Wang 2005; Yang and Chen 2014). Some studies evaluate the performance of biogas plants in field studies including the interview and questionnaire method, which is more realistic and helpful according to the local situation. He and coworkers has made a survey in

Shandong province and found that bioenergy systems should be selected based on the local circumstances (He et al. 2013). Although biogas has many advantages, it is still controversial whether these advantages can be fully explored due to different environments.

Recently, more researchers have expressed interests in comprehensive factors that influence biogas development, especially social and political reasons. According to a questionnaire survey in five provinces, Qu et al. (2013) believes that besides government promotion, other social factors also have significant effects on households' decision of biogas use. Another study suggests that the net effect of the current subsidy policy on rural household biogas use is nearly negligible (Sun et al. 2014). Frankly speaking, these studies have already posed challenges to the current policy on biogas in China. Is it necessary to change China's current biogas policy and plan?

From a macroperspective, most of the studies related to China's biogas are of qualitative, with consideration to introduction, overview and experiences (Wang et al. 2012; Li and Xue 2010; Jiang et al. 2011; Chen et al. 2012). However, only a few of the studies are quantitatively based with a macroperspective. According to some studies, China has abundant natural resources to produce biogas, but with low utilization rate and significant regional differences (Yang et al. 2012, Chang et al. 2014); Some other research paid attention not only to the regional differences, but also to the regional policy and strategy level (Deng et al. 2014). These studies can provide guidance to decision-making in a more accurate and thorough way. With a new temporal–spatial perspective, this paper aims to investigate the status of China's biogas production with respect to all sectors, and to predict its development in the foreseeable future through mathematical models. Then, according to the results associated with the current background and conclusions from the literature, this paper will discuss the possible reasons and explanations for temporal–spatial transformation in biogas production, and also give corresponding policy implications and strategies in the final part of the paper. Hopefully, this study could provide valuable insight into the future of China's biogas industry, and also give guidance to those developing countries with similar circumstances.

4 Methodologies

4.1 Time-dependent approach and technology diffusion

In this study, a time-dependent approach and time series methods were employed to investigate and forecast the trend of biogas production. On one hand, biogas production tends to be a grey box system that is not entirely and accurately understood by scientists, because it is difficult to explore all natural and social factors that influence the transformation of biogas production; in addition, data on influencing factors are either qualitative or quantitative, and sometimes insufficient too. On the other hand, based on observed data, time-dependent approaches and time series methods have their own advantages: focusing on the inter-relation and consistency of existing data series, and taking time as its only independent variable without consideration of other factors temporarily. The time-dependent approach does not provide long-term forecasts, and is proved to be successful only for short period prediction, when conditions do not change dramatically.

The time-dependent approach to technology diffusion was first introduced and applied by Mansfield (1961) in industrial and high technology fields with a logistic model. However, the real situation in the society is actually more complex than the original version of the logistic model. Due to a variety of situations, the facts can vary sometimes, for example, the trend may decline after reaching a peak, or the curve has multiple cycles. Later, to cope with different situations of transformation, researchers introduced some derivative logistic models with more complex features (including flexible growth or decline, symmetric bell-shaped, asymmetric bell-shaped and multiple cycled): Floyd model, Gompertz curve, Sharif–Kabir model, nonsymmetric responding logistic model, nonuniform influence model, Stanford Research Institute model (Mahajan et al. 1990), Hubbert model (Hubbert 1982; Bartlett 2000; Tao and Li 2007; Maggio and Cacciola 2009; Mohr and Evans 2009), Weibull model, generalized Weng model (Weng 1991 Chen and Hu 1996), H–C–Z model (Hu et al. 1995), Boltzmann model, GHB model, and multi-cycle Hubbert model (Lynch 2002; Guseo et al. 2007; Reynolds and Kolodziej 2008; Mohr and Evans 2010). It is proved to be credible that these modified logistic-like models have been adopted by researchers in many fields to simulate and forecast not only the process of innovation diffusion, but also resource transformation, such as water, fish, population, and energy sources including oil, gas and coal (Mahajan et al. 1990; Laherrère 1997, 2000; Bardi and Yaxley 2005; Palaniappan and Gleick 2008; Brandt 2010; Sorrell 2010; Hook et al. 2011).

With regards to biogas production, it can be seen as a type of resource production through a process of technology diffusion. A general process of technology diffusion is illustrated in Fig. 3. Some studies also indicate that technology diffusion has its own life cycle, including ascent phase, maturity phase, and descent phase, showing a bell-shaped curve that looks similar to some logistic-like models (Mansfield 1961; Mahajan et al. 1990). Similar to a bell-shaped curve, some research on technology adoption life cycle draws a conclusion of five stages: innovators, early adopters, early majority, late majority, and laggards

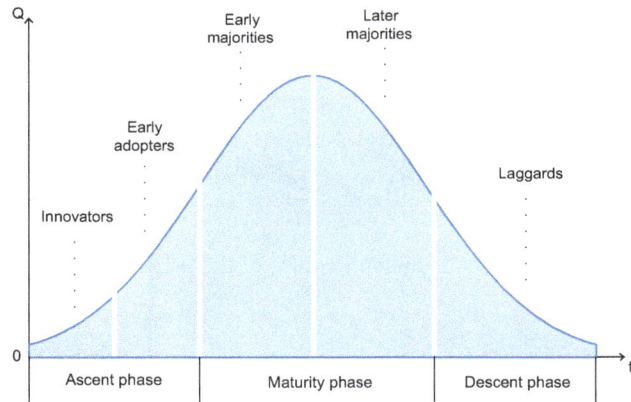

Fig. 3 Process and stages of technology diffusion (modified from Rogers 2003). *Notes t* represents time and *Q* represents the degree of technology evolution

(Rogers 2003). With regards to China's biogas production, according to the data reported by MOA (2013, 2014), the total biogas production has achieved the first peak in 2013 (15.8 Giga cubic meters), then decreased in 2014 (Fig. 1). In terms of household biogas production, the first peak value appeared in 2012 (13.8 Giga cubic meters) and then production declined. In detail, 22 out of 31 provinces directly revealed a bell-shaped curve, showing a declining trend after reaching the peak from 2011 to 2013; 6 of 31 provinces slowed down their growth rate and were about to achieve their peak value, probably located at the early maturity phase. For agriculture-based engineering projects, although they have shown a growth trend during the past decades, the annual growth rate actually became smaller than ever in recent years (from 45 % in 2008 to 9 % in 2014), which means it is probably reaching the early stage of the maturity phase of a bell-shaped curve (Fig. 3). Moreover, the industry-based projects present a form of multiple bell-shaped curves. As historical data demonstrates, in many cases, the trend of biogas production is related to bell-shaped curve(s) or part of a bell-shaped curve, so it is reasonable to consider the temporal change of biogas production as a technology diffusion process that fits some logistic-like curve in a certain period. Thus, some of modified logistic-like models will be applied to forecast the biogas production of China in the short term.

4.2 Forecast models

We take N_p as the accumulative production from the first year to the tth year and $Q(t)$ as the production at t year, which satisfies:

$$N_p(t) = \int_0^t Q(t)\,dt. \tag{1}$$

On the case that the production of biogas is finite, when $t \to +\infty$, that is, $Q(t) \to 0$, the total production of biogas is defined as:

$$N_R = N_p(t)_{t\to+\infty} = \int_0^{+\infty} Q(t)\,dt. \tag{2}$$

If $f(t)$ and $F(t)$ are the probability density function (PDF) and accumulative density function (CDF) of the sample data respectively, then

$$F(t)_{t\to+\infty} = \frac{N_p(t)_{t\to+\infty}}{N_R} = \int_0^{+\infty} \frac{Q(t)}{N_R}\,dt = 1 \tag{3}$$

and

$$f(t) = \frac{Q(t)}{N_R}. \tag{4}$$

Here we will discuss four different modified models: the Hubbert model, the generalized Weng model, the H–C–Z model and the Weibull model, showing different types of bell-shaped transformation. The specific description of each model is displayed in Table 1.

To determine the value of a, b and c, conduct the division of annual production to accumulative production, then we can get:

$$\frac{Q}{N_p} = \frac{abN_R e^{-bt}}{N_R(1+ae^{-bt})} = \frac{abN_p e^{-bt}}{N_R} = \frac{bN_p(1+ae^{-bt})}{N_R} - \frac{bN_p}{N_R} \tag{5}$$

and then we can get: $N_R = \alpha/\beta$ and $b = \beta N_R$.

Besides, with $N_R = \alpha/\beta$, $\log((N_R - N_p)/N_p) = A - Bt$.

Therefore, the parameter a and b can be given by $a = 10^A$ and $b = 2.303B$.

The parameters in the Weibull model, generalized Weng model, and H–C–Z model can be calculated in a similar way.

Considering the uncertainty of some cases, we use the Grey model as a supplementary. Grey theory is a study containing the known and unknown information at the same time, constructing a Grey model to forecast from a limited, discrete data. The modeling process is as follows:

The original sequence, $\{x_t,\ t = 1, 2, \ldots, n\}$, is done with the accumulative generation operation (AGO). As a result, we get $\{x_t^{(1)},\ t = 1, 2, \ldots, n\}$, and $x_t^{(1)} = \sum_{i=1}^t x_i, t = 1, 2, \ldots, n.$

$$\frac{dx^{(1)}}{dt} + ax^{(1)} = b. \tag{6}$$

Equation (6) presents the differential equation based on the AGO sequence. Correspondingly, the discrete form of GM (1, 1) is given by Eq. (7)

Table 1 Description of different models

Models	Hubbert	Weibull	Generalized Weng	H-C-Z	GM (1, 1)
Annual production	$Q = \dfrac{abN_R e^{-bt}}{(1+ae^{-bt})^2}$	$Q = at^b e^{-(t^{b+1}/c)}$	$Q(t) = at^b e^{-(t/c)}$	$Q = aN_R \exp[-(a/b)\exp(-bt) - bt]$	$\hat{x}^{(0)}(t) = \left(x^{(0)}(0) - \dfrac{b}{a}\right)(1 - e^{-at})e^{-at}$
Maximum annual production	$Q_{max} = 0.25bN_R$	$Q_{max} = a\left[\dfrac{bc}{2.718(b+1)}\right]^{b/(b+1)}$	$Q_{max} = a\left(\dfrac{bc}{2.718}\right)^b$	$Q_{max} = 0.3679bN_R$	—
Accumulative production	$N_p = \dfrac{N_R}{1+ae^{-bt}}$	$N_p = N_R\left[1 - e^{-(t^{b+1}/c)}\right]$	$N_R = ac^{b+1}\Gamma(b+1)$	$N_p = N_R \exp[-(a/b)\exp(-bt)]$	$\hat{x}^{(1)}(t) = \left(x^{(0)}(0) - \dfrac{b}{a}\right)e^{-at} + \dfrac{b}{a}$
Accumulative production at peak time	$N_{pm} = 0.5N_R$	$N_{pm} = 0.3679N_R$	—	$N_{pm} = 0.3679N_R$	—
Peak time	$t_m = \dfrac{1}{b}\ln a$	$t_m = \left(\dfrac{bc}{b+1}\right)^{1/(b+1)}$	$t_m = bc$	$t_m = \dfrac{\ln(a/b)}{b}$	—

– means the item of the model is nonexistent

$$x^{(1)}(t) + az^{(1)}(t) = b \tag{7}$$

where $z^{(1)}(t) = \frac{x^{(1)}(t)+x^{(1)}(t-1)}{2}$.

With least square method (LSM), the solution of the difference Eq. (6) is given by:

$$\hat{x}^{(1)}(t) = \left(x^{(0)}(0) - \frac{b}{a}\right)e^{-at} + \frac{b}{a} \tag{8}$$

Therefore, the simulated values of the original sequence are:

$$\hat{x}^{(0)}(t) = \left(x^{(0)}(0) - \frac{b}{a}\right)(1 - e^{-at})e^{-at} \tag{9}$$

In the above process, GM (1, 1) contains three mainly basic parts: accumulative generation operation (AGO), inverse accumulative generation operation (IAGO) and forecast model GM (1, 1). AGO can deal with the stochastic and chaos of the data, which can transfer the discrete and irregular time series data to the strict monotone increasing smoothed time series. Then the difference equation is established with the AGO series, whose results will be conducted with the IAGO. Based on the above two steps, the GM (1, 1) is ultimately constructed.

4.3 Optimization algorithm

The cuckoo search algorithm is inspired by the breeding behavior of cuckoos in combination with Lévy flight distribution. A Lévy flight is a random walk with step-lengths drawn from the Lévy distribution with a heavy-tailed probability, which was featured by Yang and Deb who worked to improve the search efficiency of the cuckoos (Yang and Deb 2010).

According to Yang and Deb (2010), for multi-objective optimization, three idealized rules in the Cuckoo Search Algorithm are declared below: (1) N eggs (the solutions of n objectives) are laid by each cuckoo at a time and placed in a randomly chosen nest; (2) the best nests that carried high-quality eggs (solutions) from each generation are kept for the next generation; (3) the number of available host nests is fixed and those nests in which a host can discover alien eggs with probability $p_\alpha \in [0, 1]$ are discarded and removed from further calculations.

Therefore, in the searching process, specifically, when generating new solutions $\left(x_{t+1}^{(i)}\right)$, the local random walk can be expressed as:

$$x_{t+1}^{(i)} = x_t^{(i)} + \beta s \oplus H(p_\alpha - \varepsilon) \oplus \left(x_t^{(j)} - x_t^{(k)}\right) \tag{10}$$

where $x_{t+1}^{(i)}$ and $x_t^{(i)}$ represent the location of the ith host nest at tth generation, while $x_t^{(j)}$ and $x_t^{(k)}$ are two random different solutions at tth generation. $\beta > 0$ is the size scaling factor, $H(\cdot)$ is a Heaviside function, and s is the

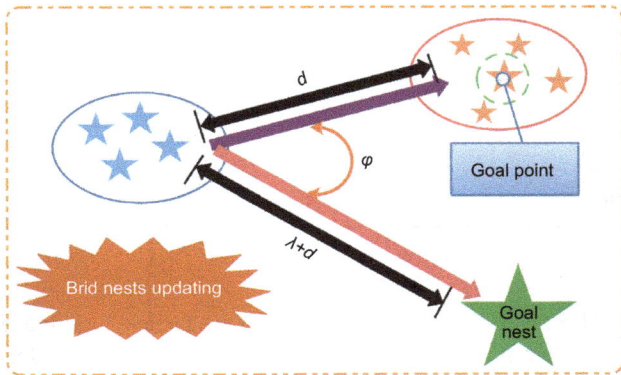

Fig. 4 Searching mechanism of the cuckoo search algorithm

step size. The product $\oplus$ indicates entry-wise multiplications.

The global random walk is performed by Lévy flights and the searching mechanism is illustrated in Fig. 4.

$$x_{t+1}^{(i)} = x_t^{(i)} + \beta \oplus \text{Lévy}(\lambda) \tag{11}$$

where the value of Lévy (λ) is calculated from the following Lévy distribution.

$$\text{Lévy}(\lambda) = \frac{\lambda \Gamma(\lambda) \sin(\pi\lambda/2)}{\pi s^{1+\lambda}} \tag{12}$$

with the Gamma function $\Gamma(z) = \int_0^{+\infty} t^{z-1} e^{-t} \mathrm{d}t$.

With the cuckoo search algorithm, the parameter b and parameter a in the generalized Weng model and Weibull model, respectively, can be calculated more efficiently and accurately, but originally calculated through trial and error.

4.4 Grey relational analysis

Grey relational analysis is widely used for testing the closeness of different sequences of number by comparing with their curve shapes. In this study, it will be used to analyze the relationship between local biogas production and its factors. At first, the original data are processed via normalization.

The reference sequence is expressed as $X_0 = \{x_0(k) | k = 1, 2, \ldots, n\}$, and the factor sequence is expressed as $X_i = \{x_i(k) | k = 1, 2, \ldots, n\}$, $i = 1, 2, \ldots, N$.

The Grey relational degree between X_i and X_0 is defined as:

$$r_i = \frac{1}{n} \sum_{k=1}^{n} \xi_i(k), k = 1, 2, \ldots, n \tag{13}$$

where

$$\xi_i(k) = \frac{\min\limits_{i} \min\limits_{k} |x_0(k) - x_i(k)| + \rho \max\limits_{i} \max\limits_{k} |x_0(k) - x_i(k)|}{|x_0(k) - x_i(k)| + \rho \max\limits_{i} \max\limits_{k} |x_0(k) - x_i(k)|} \tag{14}$$

Finally, a comparison is made among the values of r_i $(i = 1, 2, \ldots, N)$, in order to identify the importance of each factor.

5 The simulation analysis and forecast of biogas production in China

China's biogas production is currently undergoing a process of drastic transformation, with significant regional differences and sector disparities. Therefore, it is essential to make a thorough assessment and prediction on the biogas production in China from both temporal and spatial perspectives.

5.1 Data sets and preliminary analysis

The data sets contain three sectors, including biogas capacity from household biogas digesters (1989–2014), agriculture-based biogas projects (1996–2014) and industry-based biogas projects (1996–2014) in 31 provinces in China. The data are reported officially by the MOA of China. The missing part, the value of the biogas production in the year of 2000, is addressed using the piecewise cubic hermite interpolating polynomial (PCHIP), which is also called contact difference. In order to keep the derived function as close to the original function as possible, it requires the derived function not only to have the same values, but also the same derivative values on the inserted nodes.

In cubic polynomial $P_3(x) = y_0 \varphi_0(x) + y_1 \varphi_1(x) + h y_0' \phi_0(x) + h y_1' \phi_1(x)$, $\varphi_0, \varphi_1, \phi_0, \phi_1$ are the cubic polynomials needed to be determined which satisfy:

$$\begin{aligned}
&\varphi_0(0) = 1, \varphi_0(1) = \varphi_0'(0) = \varphi_0'(1) = 0 \\
&\varphi_1(1) = 1, \; \varphi_1(0) = \varphi_1'(0) = \varphi_1'(1) = 0 \\
&\phi_0'(0) = 1, \; \phi_0(0) = \phi_0(1) = \phi_0'(1) = 0 \\
&\phi_1'(1) = 1, \; \phi_1(0) = \phi_1(1) = \phi_1'(0) = 0
\end{aligned} \tag{15}$$

Then the cubic polynomial can be written as:

$$\begin{aligned}
P_3(x) = {} & y_0 \varphi_0\left(\frac{x - x_0}{h}\right) + y_1 \varphi_1\left(\frac{x - x_0}{h}\right) + h y_0' \phi_0\left(\frac{x - x_0}{h}\right) \\
& + h y_1' \phi_1\left(\frac{x - x_0}{h}\right)
\end{aligned} \tag{16}$$

where $h = x_1 - x_0$, x_0, x_1 are two random numbers.

From the perspectives of regional data trends, the value of household biogas digesters of different provinces shows great diversity in transformation dynamics. Though fluctuation sometimes occurred, it is generally divided into three forms: upward, bell-shaped, and downward. In fact, according to technology diffusion laws, the upward trend can be considered as the ascent phase of the technological life cycle, and the downward ones are actually the descent phase of technological cycle. With regard to engineering projects, biogas production of agriculture-based projects appears with an analogous upward trend in most provinces; while biogas production of industry-based project varies among different provinces, showing unsmooth growth or decline trends with fluctuations in most places, probably because of uncertain life cycle of industry-based projects.

5.2 Evaluation criteria

Since the forecast results affect the model itself directly, it is necessary to evaluate the performance of each model. In this paper, three indicators: mean absolute error (MAE), mean absolute percentage error (MAPE) and root-mean square error (RMSE) are adopted to assess the forecast accuracy of each model, respectively.

$$\text{MAE} = \frac{1}{n} \sum_{i=1}^{n} |\hat{y}_i - y_i| \tag{17}$$

$$\text{MAPE} = \frac{1}{n} \sum_{i=1}^{n} \left| \frac{\hat{y}_i - y_i}{y_i} \times 100 \right| \tag{18}$$

$$\text{RMSE} = \sqrt{\frac{1}{n} \sum_{i=1}^{n} (\hat{y}_i - y_i)^2} \tag{19}$$

where y_i and $\hat{y}_i$ represent the observed value and simulated value of the ith data point for the performance evaluation, and n is the total number of data points used for the performance evaluation and comparison.

Besides, to evaluate the performance of each model consistently, the criteria of MAPE are listed in Table 2.

5.3 Result and analysis

5.3.1 Household biogas production with temporal–spatial analyses

Since the production of biogas per household in China and each province almost demonstrates the characteristics that the shape of the production curve is the left side of a bell-shaped curve, five models, namely the Hubbert model, the generalized Weng model, the Weibull model, the H–C–Z model, and the Grey model, are adopted to study the biogas production. The comparison of five models is demonstrated in Fig. 5, where Fig. 5a illustrates the MAE, MAPE, and RMSE of five models with the radar map and Fig. 5b shows the simulated values of the generalized Weng model which has the minimum error and the errors between them. For simulating national biogas production, this model fits the data series with an R-squared value of 0.9942.

Table 2 Criteria of MAPE

MAPE, %	Forecast performance
<20	Excellent
20–30	Good
30–50	Reasonable
>50	Incorrect

The results of five models listed in Tables 3, 4 and 5 reveal that for most of provinces in China, the four models (the Hubbert model, the generalized Weng model, the Weibull model, and the H–C–Z model) with smaller errors are much plausible to the present status of China, compared to GM (1, 1). For the exceptional cases, such as Guangdong, Sichuan, and Xinjiang provinces, the simulated accuracy of the Grey model is much higher than other four models, which mainly results from the fluctuation and the current growth of the production data.

For most provinces in China, especially in the household biogas production of the whole nation shown in Fig. 5b, the generalized Weng model can portray the trend more precise than other four models in terms of the MAPEs listed in Table 4. According to Table 2, the MAPEs of different provinces indicate that the generalized Weng model offers excellent forecast results for nearly half the cases and its results are also authentic overall. The H–C–Z Model also has some excellent performances and good performance for most cases. The Hubbert model and Weibull model conduct good and reasonable results in about half the different provinces, while GM (1, 1) is inappropriate for some provinces due to the fact that the MAPE is over 50 %. Therefore, we choose the prediction value calculated by generalized Weng model, and the forecast of household biogas production is shown in Fig. 5b, which indicates that the total household biogas production will be about 10 Giga cubic meters.

When considering both time dynamics and regional difference, Fig. 6 shows that the spatial diffusion process can be described by four cross sections which have some correlations with spatial factors. In Fig. 6, the relative productivity, referring to the ratio of local value to the maximum value, is used as a mapping index. In all the four periods, Sichuan (the darkest brown color in Fig. 6) has always been the most productive province of the whole nation. In the early 1990s, Jiangsu from the coastal area, as well as Hunan and Hubei from central China also had plenty of household biogas production. It was not until 2000 that household biogas production diffused toward adjacent southwestern and central areas, such as Yunnan, Guangxi, and Jiangxi (relatively dark brown color in Fig. 6b), which probably have the most suitable climate

(a)

(b) The household biogas production of China, m³

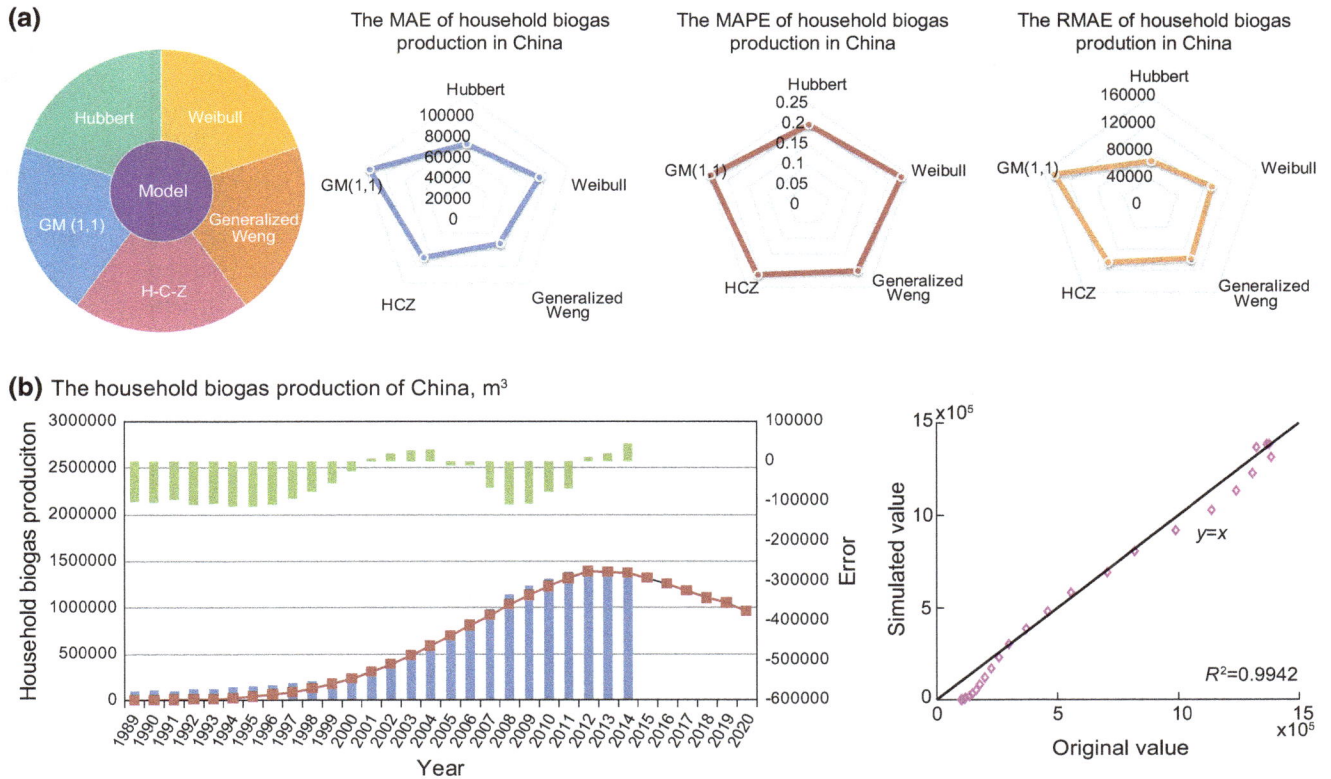

Fig. 5 Results of five models on the household biogas production in China. **a** The MAE, MAPE, RMSE of five models. **b** The actual values are displayed by the *blue histogram*, and simulated values of generalized Weng model which has the minimum error are displayed by the *red line* and the errors between them are displayed by the *green histogram*

conditions for household biogas production. On the other hand, people who use household biogas digesters in the east coast regions make up a lower proportion than before. After 10 years of promotion by government and the progress of technology, a considerable number of household biogas plants have been built and used for biogas production. It has spread north to more places, such as Henan, Hebei, and Shandong (medium brown color in the central China of Fig. 6c) and these have become important regions of biogas production; even in Tibet, there are also a number of family-sized biogas plants, however, household biogas is no longer produced in Shanghai. In contrast, Sichuan, Guangxi, and Yunnan (relatively dark brown color in Fig. 6c) were still the regions where household biogas production was the highest in the early years of 2010s. But in around 2012, most provinces started to decelerate their speed of increase of biogas production, and some of them had already reached their peaks. According to the result of forecasts, household biogas will decrease in most provinces in the early 2020s, and be not as widely used as before, and finally show an agglomeration effect in southwest regions (in Fig. 6d) that have more suitable climate condition for all year biogas use, and more small-sized or family-sized livestock raising than any other region in China.

In the Chinese context, there might be diverse reasons related to these phenomena which have occurred over the last few years: the decrease in small-scale agriculture/livestock production (large-scale agriculture/livestock breeding rising), the rural population reduction, the more centralized rural residential patterns, the increased choices for other types of energy, etc. (Wang et al. 2012). It is obvious that these processes seem too hard to reverse in contemporary China under the fast pace of urbanization, industrialization as well as other reforms. Meanwhile, regional differences in agro-climate, resources, socioeconomic status and policy also make considerable diversity in household biogas production (Yang et al. 2012).

In order to analyze the reasons for regional household biogas development during recent years, further analysis is made by using various natural and socioeconomic indexes: climate condition (months above 10 °C in a year), the scale of livestock breeding fit for household (number of existing pigs in less than 50 scale farms), rural household energy demand (the proportion of alternative energy sources: coal, oil, wood, electricity, etc.), rural population level (the proportion of local rural population), local rural economic condition (rural income per person), government support (the investment from central to local government for

Table 3 MAE of five models of household biogas production

Province	Hubbert	Weibull	Generalized Weng	H–C–Z	GM (1, 1)
China	5.43E+04	7.12E+04	5.23E+04	6.92E+04	9.53E+04
Anhui	1.11E+03	1.17E+03	9.45E+02	1.01E+03	1.30E+03
Chongqing	3.68E+03	3.98E+03	2.83E+03	2.86E+03	7.73E+03
Fujian	1.60E+03	1.95E+03	1.46E+03	1.49E+03	2.42E+03
Gansu	9.81E+02	1.09E+03	8.91E+02	4.23E+03	4.23E+03
Guangdong	2.34E+03	1.76E+03	9.14E+02	1.14E+03	8.99E+02
Guangxi	1.65E+04	1.97E+04	1.65E+04	1.65E+04	2.92E+03
Guizhou	1.97E+03	2.07E+03	1.79E+03	1.82E+03	5.05E+03
Hainan	6.50E+02	9.14E+02	6.31E+02	6.75E+02	1.84E+03
Hebei	4.14E+03	5.07E+03	4.14E+03	4.14E+03	1.05E+04
Heilongjiang	7.21E+02	1.10E+03	7.21E+02	7.21E+02	2.62E+03
Henan	4.72E+03	5.14E+03	4.72E+03	4.72E+03	9.77E+03
Hubei	6.46E+03	7.48E+03	5.33E+03	6.46E+03	7.08E+03
Hunan	4.94E+03	4.94E+03	4.94E+03	4.94E+03	1.14E+04
Jiangsu	2.26E+03	3.16E+03	1.81E+03	1.52E+03	2.45E+03
Jiangxi	4.76E+03	7.40E+03	3.29E+03	3.68E+03	5.29E+03
Jilin	1.93E+02	4.94E+03	2.03E+02	2.88E+02	2.88E+02
Liaoning	9.39E+02	1.12E+03	7.68E+02	6.31E+02	5.29E+03
Inner Mongolia	5.24E+02	5.62E+02	5.16E+02	5.39E+02	2.28E+03
Ningxia	5.20E+02	6.10E+02	4.50E+02	3.97E+02	5.55E+02
Qinghai	2.74E+02	5.22E+02	2.74E+02	2.74E+02	6.45E+02
Shandong	7.12E+03	7.95E+03	6.59E+03	5.82E+03	7.14E+03
Shanxi1	1.83E+03	2.00E+03	1.80E+03	1.65E+03	3.35E+03
Shanxi2	1.63E+03	1.88E+03	1.39E+03	1.51E+03	3.34E+03
Sichuan	1.77E+04	1.78E+04	1.74E+04	1.12E+04	9.89E+03
Tianjin	1.67E+02	1.75E+02	1.72E+02	8.19E+01	1.65E+02
Tibet	8.95E+02	1.15E+03	7.43E+02	6.45E+02	1.76E+03
Xinjiang	1.07E+03	1.42E+03	1.03E+03	7.64E+02	3.37E+02
Yunnan	4.87E+03	1.11E+04	3.17E+03	5.02E+03	5.88E+03
Zhejiang	1.34E+03	1.09E+03	5.08E+02	9.07E+02	1.07E+03

building biogas digesters). By calculating the Grey relational degree to local household biogas production, the importance of factors are as below: government support (0.781) > rural population level (0.763) > energy demand (0.757) > climate condition (0.752) > local rural economic condition (0.750) > scale of livestock breeding (0.741). Hence, government support is the major factor that has impacted on rural household biogas development, which indicates that rural household biogas development is mainly from government promotion. Policy promotion is from both central and local governments. After 2003, the central government enhanced the support for rural biogas construction by offering subsidies: from 800/1000/1200 CNY per household before 2008 and then up to 1000/1200/ 1500 CNY per household after 2008 (for east/middle/west region, respectively). The financial support from central government in a certain period can be calculated by the number of newly built biogas units of every province and the local subsidy standard. Also improvement of biogas service systems is another way of policy support: after 2009, the number of local service centers increased and the central government provided 20,000/35,000/45,000 CNY per year for every service center (in east/middle/west region) (MOA 2008). Moreover, some local policies also take many solutions including providing direct investment and encouraging international cooperation: in some cases in Sichuan province, household biogas production is introduced into the international carbon trade system due to local policy support. However, new policies from central government have re-oriented its priority to large-scale biogas engineering projects (MOA 2015), so household biogas production is unlikely to grow as fast as before, but could be still useful in some part of China. Besides, the rural population level and energy demand are important reasons for rural household biogas usage. Climate conditions, local rural economic conditions, and scale of

Table 4 MAPE of five models of household biogas production

Province	Hubbert	Weibull	Generalized Weng	H–C–Z	GM (1, 1)
China	0.2013	0.2312	0.1986	0.2084	0.2466
Anhui	0.2423	0.2932	0.2234	0.2303	0.2536
Chongqing	0.2092	0.2312	0.1712	0.2021	0.3686
Fujian	0.2259	0.2592	0.1995	0.2033	0.4236
Gansu	0.2865	0.3312	0.2712	0.2864	0.4386
Guangdong	0.1249	0.1249	0.1249	0.1249	0.1249
Guangxi	0.2524	0.2924	0.2343	0.2245	0.6271
Guizhou	0.3532	0.3753	0.2932	0.3227	0.4803
Hainan	0.2814	0.3325	0.2214	0.2799	0.4737
Hebei	0.2604	0.3262	0.2044	0.2488	0.4921
Heilongjiang	0.3315	0.3908	0.2853	0.2915	0.5389
Henan	0.2711	0.3215	0.2137	0.2517	0.6064
Hubei	0.2055	0.2380	0.1744	0.2055	0.2114
Hunan	0.1982	0.2928	0.1852	0.1982	0.5145
Jiangsu	0.2141	0.2687	0.1868	0.1579	0.2291
Jiangxi	0.2229	0.2509	0.1940	0.1821	0.5252
Jilin	0.3238	0.3382	0.2329	0.3382	0.3174
Liaoning	0.2401	0.2600	0.2303	0.2212	0.5252
Inner Mongolia	0.4463	0.3634	0.3446	0.4036	0.8100
Ningxia	0.2576	0.3184	0.2067	0.2257	0.3562
Qinghai	0.2918	0.3443	0.2918	0.2918	0.4423
Shandong	0.3241	0.3546	0.2752	0.2572	0.3286
Shanxi1	0.3594	0.3806	0.3413	0.3204	0.3167
Shanxi2	0.3177	0.3504	0.2941	0.2791	0.3825
Sichuan	0.2680	0.2807	0.2402	0.1683	0.1368
Tianjin	0.2753	0.2909	0.2827	0.2340	0.3572
Tibet	0.2159	0.2521	0.2131	0.2251	0.3931
Xinjiang	0.2345	0.2734	0.2329	0.2068	0.1204
Yunnan	0.2063	0.1947	0.1802	0.1723	0.2939
Zhejiang	0.2713	0.2534	0.1712	0.2704	0.3732

livestock breeding are also relevant to rural household biogas production.

5.3.2 Biogas production of engineering projects with temporal–spatial analyses

Data for biogas engineering projects have been reported since the late 1990s by MOA (1996–2004); and the feature of these data series in different regions shows different patterns in comparison with household ones: Some are monotonously changed, while others have obvious fluctuations, thus forecast models should be more diverse according to circumstances. Considering the fluctuation of the some biogas engineering projects, it is essential to apply a multi-curve model to forecast the production of industry-based biogas. The biogas production of agriculture- and industry-based engineering projects of China is illustrated in Fig. 7a, b. By using the same approach in simulating household biogas production, we find that generalized Weng model is still a good choice for simulating agriculture-based engineering biogas production, with the best value of MAE (7.04E+03), MAPE (0.1913), RMSE (1.26E+04), and R-squared (0.9922) when calculating total biogas production; likewise, the (multi-cycle) Hubbert model is most suitable to simulate industry-based engineering biogas production, with the most satisfactory values of MAE (6.15E+03), MAPE (0.3137), RMSE (8.65E+03), and R-squared (0.7516) for evaluating national biogas production. For some provinces with unstable fluctuations which the five models cannot fit well, the least square method is used to estimate the future value. The regional difference is shown in Fig. 8, round shape of colors represents the biogas relative productivity of agriculture-based engineering projects, while the blue color from light to dark refer to relative biogas productivity of industry-based engineering projects.

Table 5 RMSE of five models of household biogas production

Province	Hubbert	Weibull	Generalized Weng	H–C–Z	GM (1, 1)
China	6.60E+04	9.18E+04	9.76E+04	1.04E+05	1.46E+05
Anhui	1.38E+03	1.38E+03	1.38E+03	1.38E+03	1.80E+03
Chongqing	4.31E+03	4.31E+03	4.31E+03	4.31E+03	1.16E+04
Fujian	2.14E+03	2.37E+03	2.14E+03	2.14E+03	3.43E+03
Gansu	1.45E+03	2.14E+03	1.02E+03	8.76E+03	8.76E+03
Guangdong	1.68E+03	1.68E+03	1.68E+03	1.68E+03	1.68E+03
Guangxi	2.94E+04	3.30E+04	2.94E+04	2.94E+04	3.87E+04
Guizhou	2.99E+03	3.19E+03	2.99E+03	2.99E+03	7.07E+03
Hainan	9.58E+02	1.41E+03	9.26E+02	1.10E+03	2.31E+03
Hebei	6.33E+03	8.44E+03	6.33E+03	6.33E+03	2.80E+05
Heilongjiang	1.04E+03	1.35E+03	1.04E+03	1.04E+03	3.97E+03
Henan	6.79E+03	7.31E+03	6.27E+03	6.39E+03	1.43E+04
Hubei	7.75E+03	8.32E+03	6.42E+03	7.75E+03	1.10E+04
Hunan	6.50E+03	6.50E+03	6.50E+03	6.50E+03	1.38E+04
Jiangsu	2.61E+03	3.27E+03	2.48E+03	2.19E+03	2.94E+03
Jiangxi	4.44E+06	1.10E+04	4.68E+03	5.00E+03	6.92E+03
Jilin	2.46E+02	6.50E+03	3.14E+02	5.93E+02	4.51E+02
Liaoning	1.33E+03	1.56E+03	1.14E+03	1.02E+03	6.92E+03
Inner Mongolia	7.62E+02	7.62E+02	7.62E+02	7.62E+02	2.72E+03
Ningxia	6.36E+02	7.83E+02	5.24E+02	5.17E+02	6.82E+02
Qinghai	4.52E+02	7.43E+02	4.52E+02	4.52E+02	7.80E+02
Shandong	9.18E+03	1.04E+04	7.88E+03	6.34E+03	8.66E+03
Shanxi1	2.51E+03	2.57E+03	2.55E+03	2.00E+03	4.12E+03
Shanxi2	2.20E+03	2.55E+03	2.29E+03	2.09E+03	3.81E+03
Sichuan	2.01E+04	1.99E+04	2.12E+04	1.38E+04	1.19E+04
Tianjin	2.10E+02	2.52E+02	2.34E+02	1.06E+02	2.12E+02
Tibet	1.41E+03	1.81E+03	9.58E+02	6.96E+02	2.33E+03
Xinjiang	1.73E+03	2.26E+03	1.61E+03	1.14E+03	4.45E+02
Yunnan	6.23E+03	2.01E+04	4.54E+03	6.20E+03	7.40E+03
Zhejiang	1.85E+03	1.35E+03	6.06E+02	1.01E+03	1.21E+03

As shown in Fig. 7a, in the next few years, the total production of agriculture-based biogas projects will probably increase steadily with an annual growth rate of 8 %, reaching a peak of almost three Giga cubic meters in around 2020. From a spatial perspective (Fig. 8), agriculture-based engineering projects have diffused from southeast coastal regions inland to provinces such as Henan and Sichuan, and become increasingly widely used on a national scale during the last decades. In early 2000s, Fujian (red spot in Fig. 8a) played the leading role in agriculture-based biogas projects among regions; however, in early 2010s, Sichuan (red spot in Fig. 8b) generated the most abundant biogas, followed by those regions located in the eastern coast. Based on this prediction (under current policy support and situation), in 2020, the spatial pattern will not change a lot from 2010, and the current top provinces will hold the lead positions and the productivity will

begin to reach a peak value, while some provinces like Heilongjiang with rich resources likely to become potential growth points.

From Fig. 7b, we can see many fluctuations with a gradual growing trend happen in industry-based biogas production. By 2020, according to forecast results, based on current status and policy, the amount of industry-based biogas production will reach about 0.3 Giga cubic meters. From a spatial perspective (Fig. 8), in the early 2000s, engineering projects of this type were built in mid and eastern areas like North China Plain, where Shandong province (the darkest blue color in Fig. 8a) ranked the first place. Recently the development of industry-based project presents an agglomeration effect in certain locations, for instance, Henan province (the darkest blue color in Fig. 8b) showed a rapid growth and became the core area of industry-based biogas production in the early 2010s, yet

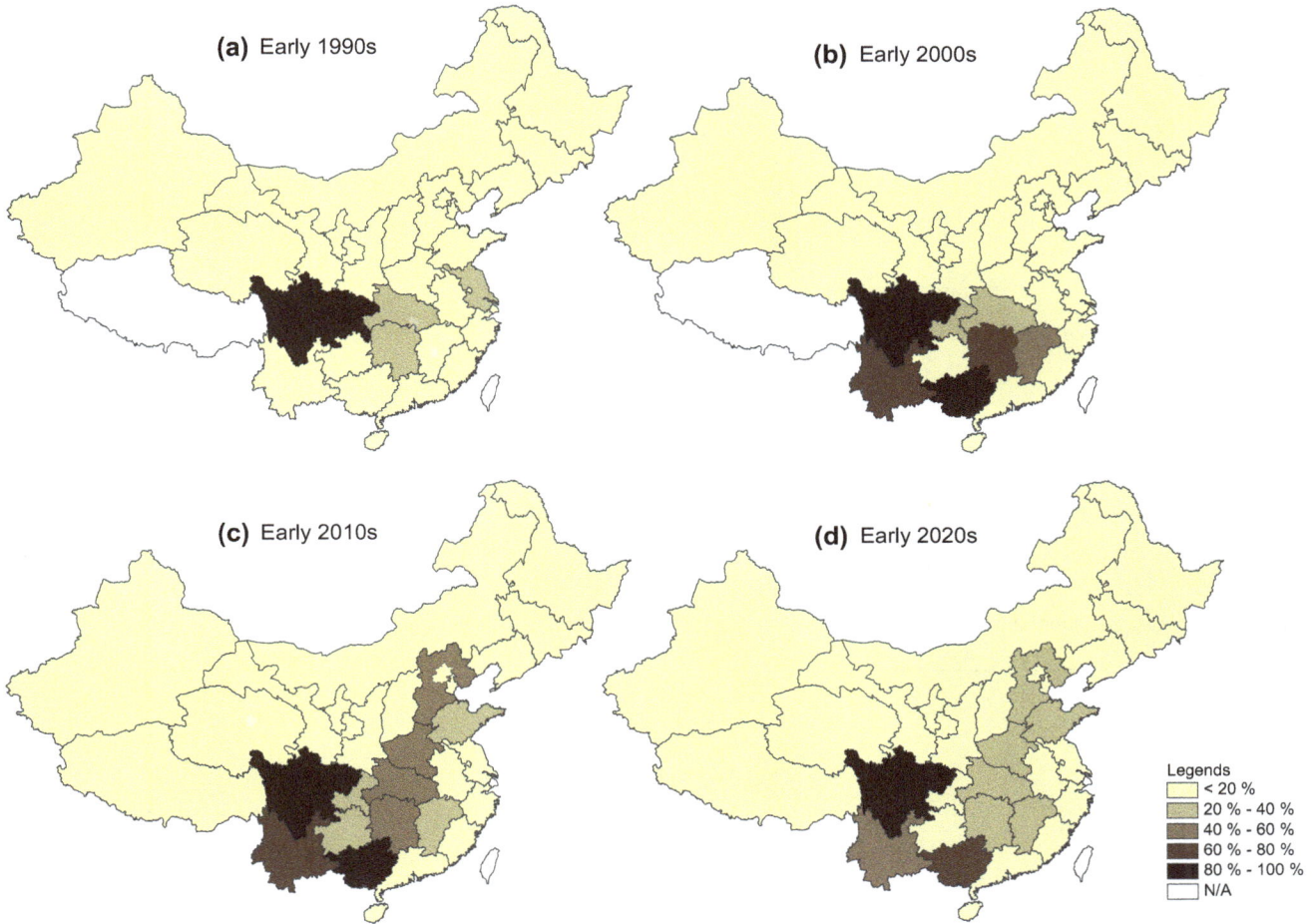

Fig. 6 Regional difference of household biogas production from 1990 to 2020

nearly half of provinces (white color in Fig. 8b) have produced little industrial-based biogas in past few years. These include Guangdong, Fujian, and Zhejiang which used to have a number of industry-based biogas plants, and this phenomenon is likely to have been caused by manufacturing industry transferring process from coastal areas to more inland areas in recent years. In the 2020s, the forecasts illustrate that the spatial pattern will not change much from 2010, which could be caused by path dependence, but there exists a slight trend from coastal area to the interior for industry-based biogas production.

Comparing between agriculture- and industry-based biogas engineering projects, we observe that agriculture-based engineering biogas production nearly follows a generalized Weng model which is an asymmetric bell shape, with a faster growing trend than declining trend (probably due to strong promotion by government), while industry-based engineering biogas production follow a (multi-cycle) symmetric bell-shaped model (Hubbert model) with almost the same rate of climbing and decreasing (possibly similar to normal technology diffusion processes). However, due to the characteristics of the two sectors, the industrial production has fluctuated more than that from the agriculture sector and thus is hard to simulate and predict, with only about an R-squared value of 0.75 compared to 0.99 for the agriculture sector.

In term of factors that drive the development of engineering projects, some studies of agriculture-based biogas projects (Li and Xue 2010; Jiang et al. 2011; Wang et al. 2012) show that the main influencing factors are slightly different from the household sector, such as raw material resources, economic conditions, technological level, and government support. The indexes selected for those factors are as follows: raw material resources (the number of livestock from large farms), economic conditions (local income per person), marketization of technology (the number of local biogas-related companies) and government support (the investment from central to local government for building engineering-scale biogas plants). By calculating the Grey relational degree to local biogas production of agriculture-based engineering projects, the factors are sorted as below: marketization of technology

(a) The biogas production of agriculture-based engineering projects, m³

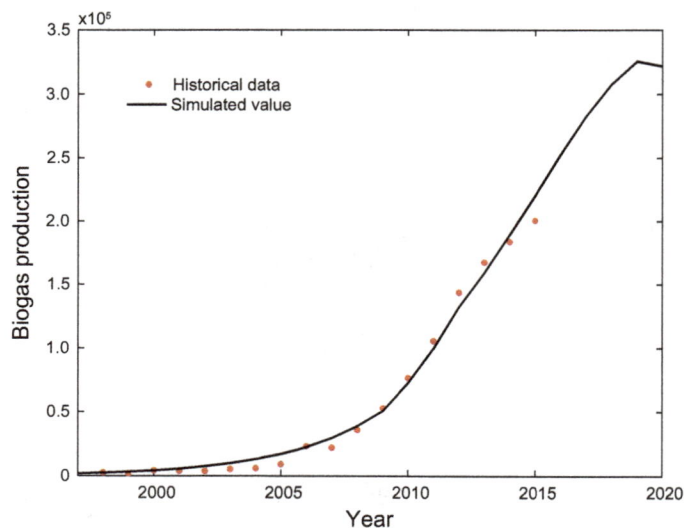

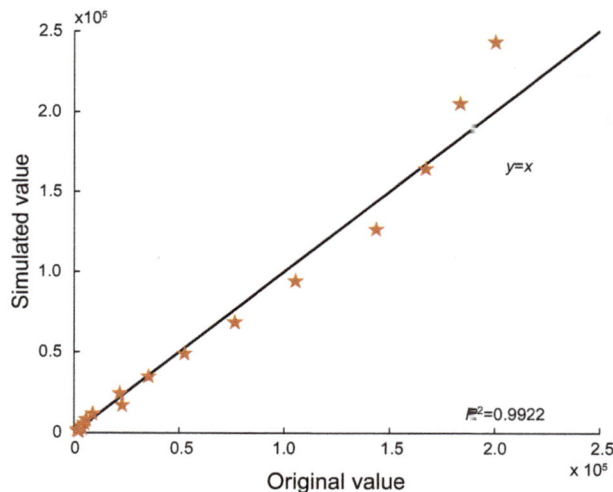

(b) The biogas production of industry-based engineering projects, m³

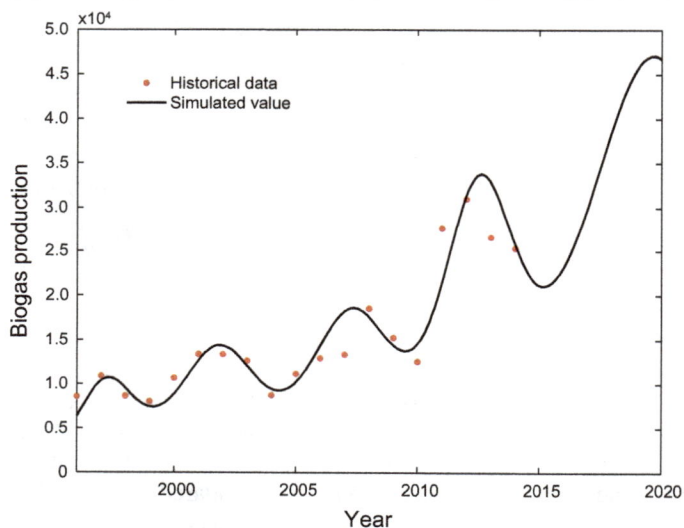

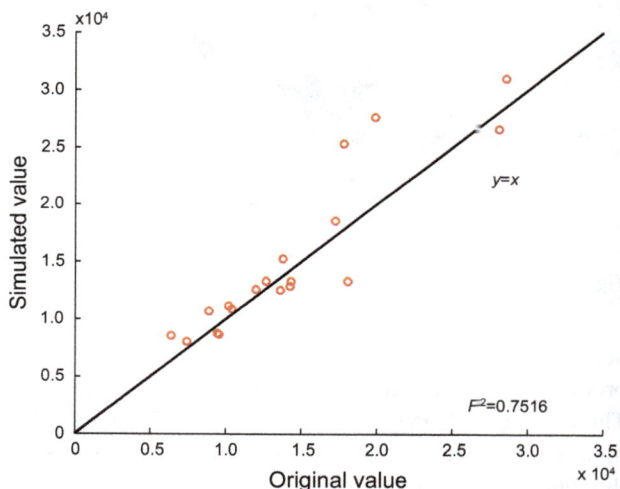

Fig. 7 Biogas production forecast of engineering projects in China. **a** The biogas production of agriculture-based engineering projects of China, **b** The biogas production of industry-based engineering projects of China

(0.758) > government support (0.736) > economic condition (0.728) > raw material resources (0.715). Therefore, marketization of technology is the most significant factor to building agriculture-based projects, for example, with the largest number of biogas companies, Sichuan province ranks the first place among all regions in engineering biogas production recently. Government support and economic conditions are also important factors, followed by raw material resources. With regard to industry-based engineering projects, this sector received much less support from government than the agricultural sector, so the productivity is relatively limited nowadays. According to data

in Fig. 8, regions with a long tradition in industry-based biogas production still keep their advantage; in contrast, some provinces with low past biogas production have stopped producing biogas recently. Therefore, an obvious territorial division will gradually form due to historical path dependence. In conclusion, biogas engineering projects are much more market-oriented and technology-dependent than the household sector, and policy support from government is still a necessary factor.

From a policy perspective, for engineering-scale biogas projects, the central government offers 25 %/35 %/45 % out of total investment (up to 150/200/250 million CNY for

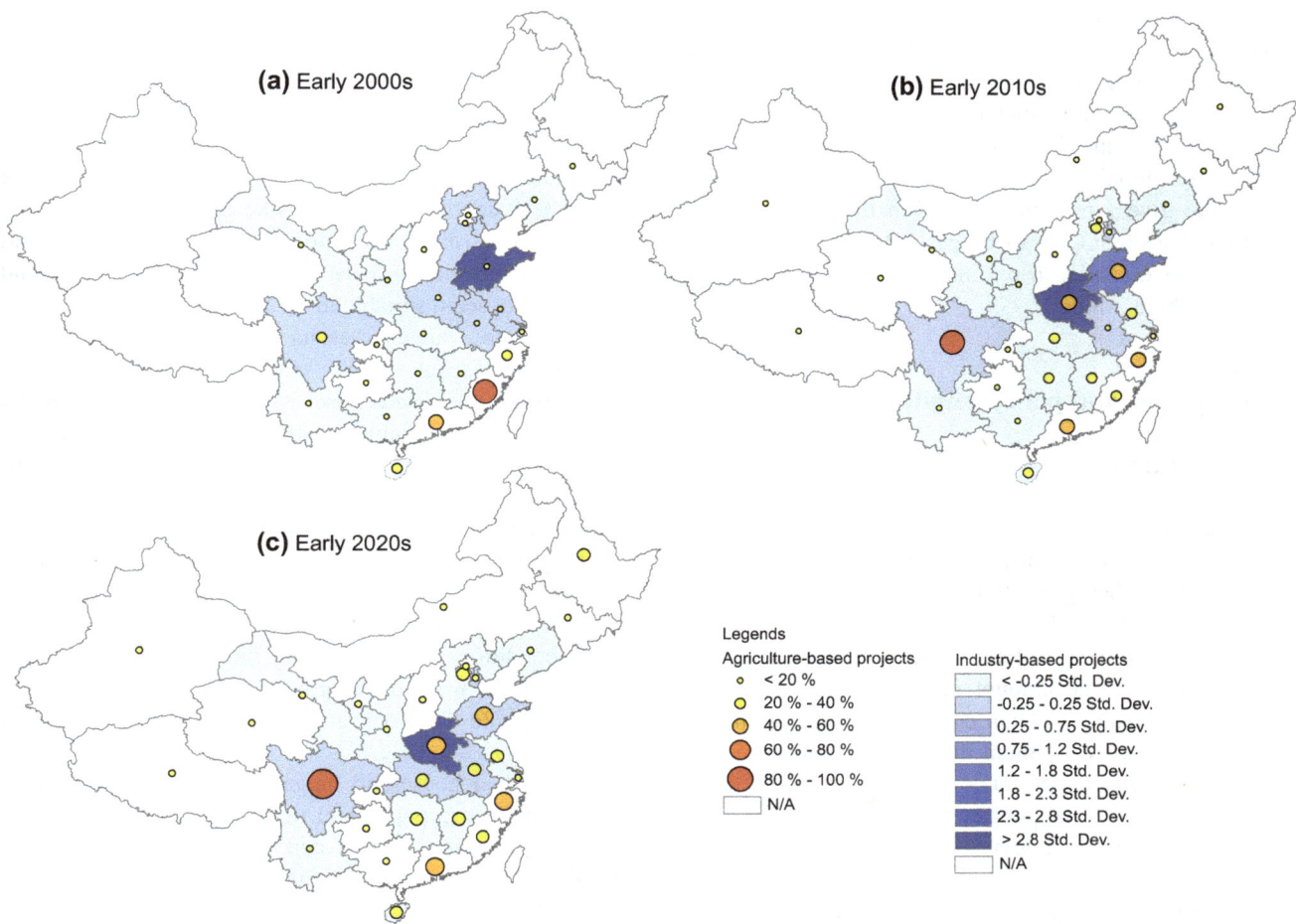

Fig. 8 Regional difference of biogas engineering projects production from 2000 to 2020

east/middle/west regions) for one newly built project since 2009 (MOA 2008). After 2015, new policies concentrate on updating existing biogas industry structure, and give priority to the development of large-scale biogas and bio-based gas projects (MOA 2015). In this context, the productivity of large-scale biogas engineering projects will probably increase to a new level; bio-based gas projects will become a new source for biogas production; yet the new policy might affect motivation of building small- and medium-scale biogas projects. In general, the biogas production from agriculture-based engineering projects will probably grow steadily under current policy promotion. Considering the expensive cost, high technology dependence and various management difficulties for building and maintaining large-scale biogas projects, future policy should not only ensure enough subsidies from government, but also pay attention to how to encourage the entrance of private capital and how to cooperate with enterprises and technical institutions, because the development of engineering projects is closely related to market and technology.

5.3.3 Uncertainty notes

The simulation and forecast results by these curve-fitting methods are on the basis of existing data under relatively stable current status, in other words, those curves can successfully be used for extrapolation of historical trends and to predict future outcomes as long as there is no significant change in underlying economic and technological conditions (Hook et al. 2011). According to this, forecasts in this study are just for the short term (until 2020), and they are not recommended for long period prediction, because of long run uncertainty in economic, technical and political aspects. For example, although the total rural household biogas production is decreasing during the past few years, it is still likely to grow and reach to another peak in the future due to technological updates and other reasons. In the industry-based engineering sector, due to complicated industry shift process, more fluctuations are likely to occur, as a result, the predicted values tend to be uncertain compared to the agricultural and household sectors, however, biogas production in the industry sector is one order of magnitude lower than the agriculture sector, between 0.1 and 0.4 Giga cubic meters and hence only a small proportion of the nation's total production (between 10 and 20 Giga cubic meters), thus uncertainty in the industry sector cannot influence the total biogas production significantly. Finally, the results in this study are just intended to give a reference based on the existing trend, which indicates that China's biogas industry is at a turning point, so it is essential to rethink the policy and planning related to China's biogas.

6 Conclusions and policy implications

6.1 General trend

As is shown with the previous data, under the current policy and status, the total amount of China's biogas production will probably continue to decrease in the next decade, which is largely a result of the declining use of rural household biogas digesters. This research implies that by the year 2020 the amount of biogas production will probably decrease to 10 Giga cubic meters for the household sector, yet grow to about three Giga cubic meters for engineering projects (in total less than 20 Giga cubic meters). Thus, the total biogas production is unlikely to grow as fast as before in next few years, even with a slight decrease in 2014. The results show a large discrepancy with the National Development and Reform Commission of China's 2007 projected long-term goal of a total of 44 Giga cubic meters biogas production in 2020 (NDRC 2007), as well as with the National Energy Administration's 2012 short-term goal of an estimated 19 Giga cubic meters of biogas produced by households in 2015 (NEA 2012), but the newest data in 2014 is 13.2 Giga cubic meters and the actual trend is still downward under the current context. Thus, the analyses detailed above demonstrate that these projections seem inaccurate.

It appears that the former planners and policy makers overestimated the growth rate of biogas production of China. This could be attributed to neglecting consideration of the technological evolution of biogas, with its cycles and peaks, as well as other factors such as rural population loss, and the decline of small livestock breeding in rural areas. Overlooking the above-mentioned factors thus resulted in far more optimistic predictions of China's biogas production, especially with regards to the household biogas sector. This type of overestimation could well cause overinvestment or miss-investment by the government, families or other agencies. Therefore, a scientific prediction of biogas production is of great significance for planners and policy makers, enabling them to set up suitable goals and regulations on biogas development that reflect reality.

6.2 Regional difference

China is a vast country with as well great variations in altitude, resulting in huge climate variations. This is one of many factors that have a profound impact on the anaerobic fermentation of rural household biogas use. This is also the reason why southwestern areas like Sichuan and Guangxi play a leading role in the production and use of household biogas. In fact, it has been demonstrated that household plants do not seem suitable in some arid and cold areas; for example many failed cases have been reported in Inner Mongolia and Heilongjiang (Cui and Ma 2009), where alternative energy sources like wind or solar power may be more effective for family use. Therefore, future policy concerning household biogas should prioritize the southwest and those regions fit for biogas development, instead of giving relatively equal support to most regions, thus avoiding unnecessary finance and resource waste. Furthermore, even within a province, there still exist large regional differences. For instance, Southeastern Gansu has better conditions for developing household biogas compared with the west of the province. Hence, local decision makers should consider a more localized and detailed policy.

Rather than household biogas production engineering projects are more popular in those advanced regions that have better socioeconomic and technological conditions; abundant raw material is another significant factor (Li and Xue 2010). Thus, Sichuan province, a relatively developed inland region with competitive advantages in agriculture, leads in both household and engineering biogas production. As for those less developed areas with good material conditions for biogas production, local policy makers can make efforts to introduce outside capital and technology into the local biogas industry by offering subsides and other preferential treatment. The national government can also offer more political and financial support to these areas in order to help mitigate the gap in production between some regions, and avoid unnecessary waste on those regions that are not fit for certain types of biogas industry.

With regards to zoning policies on biogas in the past policymakers set a very rough division of the "Eastern region," the "Middle–Northeastern region" and the "Western region," mainly from a geographic neighboring concept (MOA 2005, 2007). However, as biogas is a mixed sphere with natural and human factors, past zoning policies seem too simple, inflexible, and thus questionable. For example, Sichuan province should not be considered with many other western provinces when making regional policies. Thus, detailed zoning plans should be considered based on the natural environment, resources, the local economy, technology, and other criteria.

6.3 Sector disparity

In the past few decades, with strong promotion by the government, biogas from the agricultural sector including rural household biogas and agricultural engineering projects occupies the major part of China's biogas production. The policy of the first decade of the twenty-first century focused much attention on rural household biogas before shifting its focus to agricultural engineering projects, and those policies supporting biogas development were mainly set up by the Ministry of Agriculture (MOA 2005, 2007; NDRC/MOA 2008). As a result of that, biogas produced by agricultural engineering projects has steadily increased throughout the nation. According to forecasts in this study, the average annual growth rate will probably be as high as 8 % until 2020, and probably reach its first peak around that time, with even faster growth in some regions. In contrast, it has been challenging for the industry sector to develop biogas projects over the last few decades, because the Ministry of Industry and Information Technology has provided less policy support. However, as the "factory of the world," China has great potential for biogas generation within its industrial sector. In addition, biogas engineering projects also need the support of market and technology. Admittedly, in these fields, China has a long way to go to catch up with developed countries.

Consequently, on one hand, continued policy support should be given to agricultural projects in the next decade, especially for those mid-west regions which have a potential in resources; on the other hand, support for rural household construction should be given only in suitable areas. Meanwhile, more preferential policies should be given to the industry sector (prior to those advanced regions in manufacturing industry like Henan). Hence, policy makers of different sectors in China should rethink and redirect the policies according to the sector's own feature, the spatial pattern, as well as scientific forecasts. Last but not least, communication and cooperation among different official departments including energy, agriculture, industry and municipal works should be valued and encouraged in the future; also collaboration among government, enterprises and technical institutions are of great importance for future development of biogas, especially for biogas engineering projects.

6.4 Outlook for the future

Based on the predicted results, although the total productivity of biogas in China will probably become static or decrease for a period and rural household biogas plants will not be as widely used as they were in the past, it is also notable that China is making great progress in biogas production, especially through agriculture-based engineering projects. Therefore, China's biogas industry is and will continue to be in a journey of transformation. In recent years, China's government has paid more attention to the eco-environment and renewable energy issues; in recent years, it promulgated a series of new policies related to ecological civilization: the newly revised 'Environmental Protection Law' (NPC 2014), the 'Implementation plan for green industry development action,' and the 'Plan for water pollution prevention and governance'; further, the development of bio-based energy is suggested to become a priority in China's 13th Five-year Plan (Liu et al. 2015; NDRC 2015). All of the above should result in good prospects for China's biogas industry, especially for large-scale biogas engineering projects. In turn a thriving biogas industry will hopefully be beneficial to China's energy structure and environmental protection efforts.

However, there are still some disadvantages and drawbacks that hinder the development of China's biogas sector, as mentioned in previous literature (Chen et al. 2010; Jiang et al. 2011). Firstly, the technology, especially in biogas engineering projects, is still at a rudimentary level compared with the technologies of developed countries (Chen et al. 2010). In addition, a large number of biogas plants have quality problems and lack maintenance services. Therefore, it is imperative for governments and enterprises to invest more in research and development in biogas technology and develop a more educated and skilled pool of technicians and administrators. It is also important to enhance international cooperation and exchange in the sharing of best practices. Secondly, the utilization of biogas is relatively basic, and neither very commercialized nor industrialized. Thus, it is essential to provide more favorable policies that encourage the multi-utilization of biogas, for example in power or heat generation, vehicle fuel and so on; in addition, connecting biogas to the international carbon market will also result in increased profit. Thirdly, the institutional framework building should be focused on for China's biogas sector. Over the past several years, China's biogas development has excessively depended on government subsidies, and the entrepreneurial spirit of enterprises and other non-public sector actors has not been tapped into for building high-quality biogas plants. Government subsidies account for roughly 50 % of the financing for a typical biogas engineering project (NDRC/ MOA 2011), often leaving room for corruption and waste, and leading to the low quality of many biogas projects. As a result, better policies such as encouraging a shareholding system, third-party monitoring, and specific legislation with strict enforcement are needed to create the appropriate institutional environment, good public–private relationships, and a tight collaborative network among governments, capitalists, enterprises, consumers, NGOs, and other stakeholders. In conclusion, future development of China's biogas is not only partly due to the changes in quantity, but also depends on the quality improvement and institutional innovation, in order to extend technological life cycle or create opportunities for the next peak. Further research is needed for detailed explanations and solutions to China's biogas development.

Acknowledgments This research is supported by the National Natural Science Foundation of China (Grant No. 71171102). The authors would like to thank all the people who have given kind support to this study.

References

Alam J. Biogas energy for rural development: opportunities, challenges and lacuna of implementation, Nepal. SESAM/ARTES South Asian Regional Workshop. 2008. http://www.iim.uni-flensburg.de/sesam/.

Bardi U, Yaxley L. How general is the Hubbert curve? The case of fisheries. In: Proceedings of ASPO workshop Lisbon. 2005. http://europe.theoildrum.com/files/bardiyaxleyaspo2005.pdf.

Barnhart S. From household decisions to global networks: biogas and the allure of carbon trading in Nepal. Prof Geogr. 2013;3:345–53.

Bartlett AA. An analysis of U.S. and world oil production patterns using Hubbert-style curves. Math Geol. 2000;32(1):1–17.

Bhat PR, Chanakya HN, Ravindranath NH. Biogas plant dissemination: success story of Sirsi, India. Energy Sustain Dev. 2001;5(1):39–46.

Bojesen M, Birkin M, Clarke G. Spatial competition for biogas production using insights from retail location models. Energy. 2014;68:617–28.

Bond T, Templeton MR. History and future of domestic biogas plants in the developing world. Energy Sustain Dev. 2011;15(4):347–54.

Brandt AR. Review of mathematical models of future oil supply: historical overview and synthesizing critique. Energy. 2010;35(9):3958–74.

Chang IS, Wu J, Zhou C, et al. A time-geographical approach to biogas potential analysis of China. Renew Sustain Energy Rev. 2014;37:318–33.

Chen L, Zhao L, Ren C, et al. The progress and prospects of rural biogas production in China. Energy Policy. 2012;51(4):58–63.

Chen R. Livestock-biogas-fruit systems in South China. Ecol Eng. 1997;8(1):19–29.

Chen Y, Hu J. Review and derivation of the Weng Model. China Offshore Oil Gas (Geol). 1996;10:317–24 (in Chinese).

Chen Y, Yang G, Sweeney S, et al. Household biogas use in rural China: a study of opportunities and constraints. Renew Sustain Energy Rev. 2010;14(1):545–9.

Cheng S, Li Z, Shih J, et al. A field study on acceptability of 4-in-1 biogas systems in Liaoning province, China. Energy Proc. 2011;5(1):1382–7.

Cheng S, Li Z, Mang H, et al. A review of prefabricated biogas digesters in China. Renew Sustain Energy Rev. 2013;28(8):738–48.

Coimbra-Araújo CH, Mariane L, Bley Júnior CB, et al. Brazilian case study for biogas energy: production of electric power, heat and automotive energy in condominiums of agroenergy. Renew Sustain Energy Rev. 2014;40:826–39.

Couture T, Gagnon Y. An analysis of feed-in tariff remuneration models: implications for renewable energy investment. Energy Policy. 2010;38(2):955–65.

Cui X, Ma Y. Unusual scrappage of thousands of biogas digesters. China Econ Wkly. 2009;34:1–4 (in Chinese).

Dimpl E. Small-scale electricity generation from biomass. Part II: Biogas. 2010. http://www.gvepinternational.org/sites/default/files/resources/gtz2010-en-small-scale-electricity-generation-from-biomass-part-i.pdf.

Deng Y, Xu J, Liu Y, et al. Biogas as a sustainable energy source in China: regional development strategy application and decision making. Renew Sustain Energy Rev. 2014;35:294–303.

European Biogas Association (EBA). New German Renewable Energy Act (EEG) shoots biogas in Germany. 2014. http://european-biogas.eu/policies/position-papers/:2014.

Fallde M, Eklund M. Towards a sustainable socio-technical system of biogas for transport: the case of the city of Linköping in Sweden. J Clean Prod. 2015;98:17–28.

Feed-In Tariffs Ltd (FIT). Listing of all generation tariff levels for the current period [EB/OL]. 2012. http://www.fitariffs.co.uk/eligible/levels/.

Gautam R, Baral S, Herat S. Biogas as a sustainable energy source in Nepal: present status and future challenges. Renew Sustain Energy Rev. 2009;13(1):248–52.

German Agency for Renewable Resources (GARR). Bioenergy in Germany: facts and figures. 2012. http://www.biodeutschland.org.

Gildas LS. Perspectives for a European standard on biomethane: a biogasmax proposal. 2010. http://www.biogasmax.eu/.

Gosens J, Lu Y, He G, Bluemling B, et al. Sustainability effects of household-scale biogas in rural China. Energy Policy. 2013;54(3):273–87.

Guseo R, Valle AD, Guidolin M. World oil depletion models: price effects compared with strategic or technological interventions. Technol Forecast Soc Change. 2007;74(4):452–69.

Hangzhou Water Environment Engineering Co., Ltd (HWEEC). Report on the biogas project of the Jinzhu pig farm in Taicang. 2011;1–5 (in Chinese).

He G, Bluemling B, Mol APJ, et al. Comparing centralized and decentralized bio-energy systems in rural China. Energy Policy. 2013;63(6):34–43.

Hu J, Chen Y, Zhang S. A new model for predicting production and reserves of oil and gas fields. Acta Pet Sin. 1995;16:79–86 (in Chinese).

Hubbert, MK. Techniques of prediction as applied to production of oil and gas. NBS Special Publication, US Department of Commerce. 1982;1–121.

Hook M, Li J, Oba N, Snowden S. Descriptive and predictive growth curves in energy system analysis. Nat Resour Res. 2011;20(2):103–16.

International Energy Agency (IEA). IEA Bioenergy Task 37, Biogas country overview. 2010. http://www.iea-biogas.net/.

Jiang X, Sommer SG, Christensen KV. A review of the biogas industry in China. Energy Policy. 2011;39(10):6073–81.

Kabir H, Yegbemey RN, Bauer S. Factors determinant of biogas adoption in Bangladesh. Renew Sustain Energy Rev. 2013;28:881–9.

Laherrère J. Multi-Hubbert modeling. 1997. http://www.hubbertpeak.com/laherrere/multihub.htm.

Laherrère J. Distribution of field sizes in a petroleum system; parabolic fractal, lognormal or stretched exponential. Mar Pet Geol. 2000;17:539–46.

Lynch CM. Forecasting oil supply: theory and practice. Q Rev Econ Finance. 2002;42(2):373–89.

Li J, Xue M. Review and prospects of biogas development in China. Renew Energy Resour. 2010;28(3):1–5 (in Chinese).

Liaoning Science Association (LSA). Household biogas types, biogas and eco-agriculture technology. Shenyang: Liaoning Science & Technology Press; 2010. p. 17–30 (in Chinese).

Liu H, Cui J, Chen Y. Open for advice, China plans for environmental development in the earlier stage of "13th Five-year" period. 2015. http://news.xinhuanet.com/ (in Chinese).

Maggio G, Cacciola G. A variant of the Hubbert curve for world oil production forecasts. Energy Policy. 2009;37:4761–70.

Mahajan V, Muller E, Bass FM. New product diffusion models in marketing: a review and directions for research. J Mark. 1990;54(1):125–77.

Mansfield E. Technical change and the rate of imitation. Econometrica. 1961;29:741–66.

Ministry of Agriculture (MOA). China agriculture statistical report (1989–2014). Beijing China Agriculture Press. Several Years Used (in Chinese).

Ministry of Agriculture (MOA). National rural biogas construction plan, 2003–2010. 2002. http://www.biogas.cn/CN/B_Policy.aspx (in Chinese).

Ministry of Agriculture (MOA). Measures for the administration of national bonds for construction of rural biogas (Trial). 2003. http://www.biogas.cn/CN/B_Policy.aspx (in Chinese).

Ministry of Agriculture (MOA). Development plan for the agricultural bioenergy industry 2007–2015. 2007. http://www.biogas.cn/CN/B_Policy.aspx (in Chinese).

Ministry of Agriculture (MOA). Household-scale biogas & integrated farming system: specification of design, construction and use for the southern model. 2001. http://www.biogas.cn/CN/B_Technology.aspx (in Chinese).

Ministry of Agriculture (MOA). Household-scale biogas & integrated farming system: specification on design, construction and use for the northern model. 2001. http://www.biogas.cn/CN/B_Technology.aspx (in Chinese).

Ministry of Agriculture (MOA). Household-scale biogas & integrated farming system: specification on design, construction and use for the northwest model. 2013. http://www.biogas.cn/CN/B_Technology.aspx/ (in Chinese).

Ministry of Agriculture (MOA). National rural biogas construction plan, 2006–2010. 2005. http://www.biogas.cn/CN/B_Policy.aspx (in Chinese).

Ministry of Agriculture (MOA). Technology criterion on rural biogas digester and three renovations. 2008. http://www.biogas.cn/CN/B_Technology.aspx/ (in Chinese).

Ministry of Agriculture (MOA). Technical code for biogas engineering. 2006. http://www.biogas.cn/CN/B_Technology.aspx/ (in Chinese).

Ministry of Agriculture (MOA). The updating and transformation program for rural biogas engineering projects. 2015. http://www.biogas.cn/CN/B_Policy.aspx/ (in Chinese).

Mohr SH, Evans GM. Combined generalized Hubbert–Bass Model approach to include disruptions when predicting future oil production. Nat Resour. 2010;1:28–33.

Mohr SH, Evans GM. Forecasting coal production until 2100. Fuel. 2009;88(11):2059–67.

Mwirigi J, Balana BB, Mugisha J, et al. Socio-economic hurdles to widespread adoption of small-scale biogas digesters in Sub-Saharan Africa: a review. Biomass Bioenergy. 2014;70:17–25.

National Development and Reform Commission (NDRC). Medium and long-term development plan for renewable energy in China. 2007. http://www.biogas.cn/CN/B_Policy.aspx (in Chinese).

National Development and Reform Commission/Ministry of Agriculture (NDRC/MOA). Announcement on the strengthening of

rural biogas construction work in 2008. 2008. http://moa.gov.cn/ (in Chinese).

National Development and Reform Commission (NDRC). Lists of important research projects in the earlier stage of "13th Five-year Plan". 2015. http://www.sdpc.gov.cn/ (in Chinese).

National Development and Reform Commission/Ministry of Agriculture (NDRC/MOA). Central budget reserved for rural biogas construction. 2011. http://moa.gov.cn (in Chinese).

National Energy Administration (NEA). The "12th Five-year Plan" on bio-energy development. 2012. http://www.biogas.cn/CN/B_Policy.aspx (in Chinese).

NIFU. Nordic pathways for sustainable transport and energy. 2015. http://www.nifu.no/en/projects/topnest/.

National People's Congress (NPC). Renewable energy law of the People's Republic of China. 2005. http://www.npc.gov.cn/ (in Chinese).

National People's Congress (NPC). Environmental protection law of the People's Republic of China (Revised). 2014. http://www.npc.gov.cn/ (in Chinese).

Olsen DS, Klitkou A, Eerola A, et al. Analysis of biofuels policy in the Nordic countries. 2013. http://www.topnest.no/attachments/article/.

Palaniappan M, Gleick PH. The world's water 2008–2009, Ch 1. Pacific Institute. 2008. http://www.worldwater.org/data20082009/.

Poeschl M, Ward S, Owende P. Prospects for expanded utilization of biogas in Germany. Renew Sustain Energy Rev. 2010;14(7):1782–97.

Qu W, Tu Q, Bluemling B. Which factors are effective for farmers' biogas use? Evidence from a large-scale survey in China. Energy Policy. 2013;63:26–33.

Reynolds D, Kolodziej M. Former Soviet Union oil production and GDP decline: Granger causality and the multi-cycle Hubbert curve. Energy Econ. 2008;30(2):271–89.

Rogers EM. Diffusion of innovations. 5th ed. New York: Free Press; 2003. p. 280–5.

Singh K, Jash T. Performance analysis of micro turbine-based grid-connected biogas power plant in Purulia in West Bengal, India. Clean Technol Environ Policy. 2015;17(3):789–95.

Singh M, Maharjan KL. Contribution of biogas technology in well-being of rural hill areas of Nepal: a comparative study between biogas users and non-users. J Int Dev Coop. 2003;9:43–63.

Shi L, Zheng Y, Deng C, et al. Design of new type solar FRP biogas digester. J Hebei Norm Univ Sci Technol. 2008;22:54–7 (in Chinese).

Sorrell S. Hubbert's legacy: a review of curve-fitting methods to estimate ultimately recoverable resources. Nat Resour Res. 2010;19(3):209–30.

Sun D, Bai J, Qiu H, Cai Y. Impact of government subsidies on household biogas use in rural China. Energy Policy. 2014;73:748–56.

Swedish Energy Agency (SEA). Energy in Sweden. 2012. http://www.energimyndigheten.se.

Tao Z, Li M. What is the limit of Chinese coal supplies—a STELLA model of Hubbert peak. Energy Policy. 2007;35(6):3145–54.

Tomar SS. Status of biogas plants in India—an overview. Energy Sustain Dev. 1995;1(5):53–6.

U.S. Environmental Protection Agency (U.S. EPA). Landfill methane outreach program. 2015. http://www.epa.gov/lmop/.

Wang F, Cai Y, Qiu H. Current status, incentives and constraints for future development of biogas industry in China. Trans CSAE. 2012;28(1):184–9 (in Chinese).

Wang X, Chen Y, Sui P, et al. Efficiency and sustainability analysis of biogas and electricity production from a large-scale biogas project in China: an energy evaluation based on LCA. J Clean Prod. 2014;65(4):234–45.

Weng W. Theory of forecasting. Beijing: International Academy Publishers; 1991. p. 1–20.

Yang J, Chen B. Energy analysis of a biogas-linked agricultural system in rural China—a case study in Gongcheng Yao Autonomous County. Appl Energy. 2014;118(1):173–82.

Yang X, Deb S. Engineering optimisation by Cuckoo search. Int J Math Model Numer Optim. 2010;1(4):330–43.

Yang Y, Zhang P, Li G. Regional differentiation of biogas industrial development in China. Renew Sustain Energy Rev. 2012;16(9):6686–93.

Yue Z. The technical content and benefit of the 5-in-1 biogas model in Weibei orchard, Shaanxi. China Biogas. 1997;15(3):43–4 (in Chinese).

Zeng X, Ma Y, Ma L. Utilization of straw in biomass energy in China. Renew Sustain Energy Rev. 2007;11(5):976–87.

Zhang L, Wang C, Song B. Carbon emission reduction potential of a typical household biogas system in rural China. J Clean Prod. 2013;47(5):415–21.

Zhang P, Li X, Yang Y, et al. Greenhouse gas mitigation benefits of large and middle-scale biogas project in China. Trans CSAE. 2008;24(9):239–43 (in Chinese).

Zhang P, Wang G. Contribution to emission reduction of CO_2 and SO_2 by household biogas construction in rural China: analysis and prediction. Trans CSAE. 2005;21(12):147–51 (in Chinese).

15

Virtual sensing for gearbox condition monitoring based on kernel factor analysis

Jin-Jiang Wang[1] · **Ying-Hao Zheng**[1] · **Lai-Bin Zhang**[1] · **Li-Xiang Duan**[1] · **Rui Zhao**[2]

Handling editor: Jian Shuai

Abstract Vibration and oil debris analysis are widely used in gearbox condition monitoring as the typical indirect and direct sensing techniques. However, they have their own advantages and disadvantages. To better utilize the sensing information and overcome its shortcomings, this paper presents a virtual sensing technique based on artificial intelligence by fusing low-cost online vibration measurements to derive a gearbox condition indictor, and its performance is comparable to the costly offline oil debris measurements. Firstly, the representative features are extracted from the noisy vibration measurements to characterize the gearbox degradation conditions. However, the extracted features of high dimensionality present nonlinearity and uncertainty in the machinery degradation process. A new nonlinear feature selection and fusion method, named kernel factor analysis, is proposed to mitigate the aforementioned challenge. Then the virtual sensing model is constructed by incorporating the fused vibration features and offline oil debris measurements based on support vector regression. The developed virtual sensing technique is experimentally evaluated in spiral bevel gear wear tests, and the results show that the developed kernel factor analysis method outperforms the state-of-the-art feature selection techniques in terms of virtual sensing model accuracy.

Keywords Gearbox condition monitoring · Virtual sensing · Feature selection and fusion

✉ Jin-Jiang Wang
  jiw09005@foxmail.com; jwang@cup.edu.cn

✉ Lai-Bin Zhang
  zhanglb@cup.edu.cn

1 School of Mechanical and Transportation Engineering, China University of Petroleum, Beijing 102249, China

2 School of Electrical and Electronic Engineering, Nanyang Technological University, Singapore 639798, Singapore

Edited by Yan-Hua Sun

1 Introduction

In the oil and gas industry, the gearbox is a very important element and its health and safety are critical to the smooth operation and efficiency of relevant facilities. However, gearboxes generally work under complex conditions, which may accelerate degradation and further induce different defects such as fatigue crack, pitting. These gearbox defects may even cause the breakdown of the whole system, leading to significant economic losses, costly downtime, and catastrophic damage. Therefore, gearbox condition monitoring is of great significance to its safe operation and maintenance schedule.

Increasing need for gearbox reliability has accelerated the integration of sensing techniques for condition monitoring. These sensing techniques could be roughly categorized into indirect sensing and direct sensing techniques, of which vibration and oil debris analysis are two typical sensing techniques, respectively. Various vibration analysis techniques have been widely investigated in gearbox fault diagnosis and prognosis. For instance, a fault characteristic order (FCO) analysis method was proposed to extract the vibration signal components related to rotational speed from the time–frequency representation (TFR) for gear fault detection under time-varying rotational speed (Wang et al. 2014). A modified cantilever beam model was investigated to analytically evaluate the time-varying mesh stiffness of a planetary gear set for the detection of crack severity and location via vibration analysis (Liang et al. 2014). The vibration signal properties of a planetary gearbox were investigated to differentiate healthy and cracked tooth conditions (Liang et al. 2015). A neuro-fuzzy approach was investigated by Samanta and Nataraj (2008) for modeling and prediction of gearbox dynamics utilizing various health indices. Defect prognostics were performed to estimate the temporal evolution of features (Wang 2007). The degradation condition prediction of a planetary gearbox was also investigated by Hussain and Gabbar (2013) based on acoustic phenomena along with neural networks and neuro-fuzzy approaches.

Vibration sensing is the most commonly used online monitoring technique because of its rugged cost-effective, but the sensing quantity is an indirect indicator of gearbox condition due to the low signal-to-noise ratio (SNR) of the sensing measurement. On the other hand, oil debris analysis, as a direct sensing method, provides an alternative solution by offline inspection of the gearbox condition. The oil debris measurements can directly indicate the machinery condition with high accuracy. An oil debris analysis was performed by Bolander et al. (2009) to estimate the spall size as the damage progressed in aircraft engine prognosis. It can be found that oil debris monitoring is suitable to provide an early indication and quantification of internal damage of a gearbox, but it is far from convenient or cost-effective because of high cost and human intervention requirements during normal operations of gearboxes.

To bridge the gap between direct sensing and indirect sensing, virtual sensing has emerged as a viable, noninvasive, and cost-effective method to infer difficult-to-measure or expensive-to-measure parameters based on computational models (Tham et al. 1991). It has been investigated for active noise and vibration control (Petersen et al. 2008), industrial process control (Cheng et al. 2004), building operation optimization (Ploennigs et al. 2011), lead-through robot programming (Ragaglia et al. 2016), product quality of hydrodesulfurization (HDS) (Shokri et al. 2015), and tool condition monitoring (Bustillo et al. 2011; Li and Tzeng 2000). Data-driven virtual sensing techniques are favorable by fusing the extracted features from noisy online measurements to infer the difficult-to-measure parameters based on artificial intelligence models (Gelman et al. 2013). A good feature representation method should be able to remove the irrelevant and redundant features, while preserving important (geometric or statistical) properties of the original data (Bolón et al. 2015). Thus, it is critical to devise a systematic feature selection and representation scheme to extract and select the most representative features. Moreover, it may reduce the computational complexity and storage, improve the

efficiency of virtual sensing models, and provide insight for knowledge discovery.

Different feature selection and fusion techniques have been developed including principal component analysis (PCA) and its kernel version (He et al. 2007; Schölkopf et al. 1998), factor analysis (FA) (Bishop 2006), dominant feature identification method (Zhou et al. 2011), minimum redundancy maximum relevance technique (Peng et al. 2005), and locally linear embedding (LLE) (Roweis and Saul 2000). In the above methods, PCA, FA, and kernel PCA are widely used in machine learning and data mining. Kernel PCA, as a nonlinear extension of PCA, was developed to explore the nonlinear relationship among variables by the use of a kernel function. Owing to high computational efficiency and nonlinear projection ability of using kernel functions, kernel methods are applied to extend the traditional linear model for feature selection and fusion. FA, as a linear model based on second-order statistics, generally has difficulty processing real-world data which has non-Gaussian distributions. The kernel methods provide the inspiration to develop a new feature selection and fusion method based on FA for feature fusion in gearbox condition monitoring.

To better utilize the sensing measurements for gearbox condition monitoring, this paper presents a virtual sensing technique based on artificial intelligence by fusing the low-cost online vibration measurements to infer the gearbox condition, and its performance can be comparable to the costly offline oil debris measurements. Firstly, the representative features are extracted from the noisy vibration measurements to characterize the gearbox degradation conditions. Next, a new nonlinear feature selection and fusion method, named kernel factor analysis (KFA), is proposed to reduce the feature dimensionality. Then the virtual sensing model is constructed by incorporating the fused vibration features and offline oil debris measurements based on support vector regression. The developed virtual sensing technique is experimentally evaluated in the spiral bevel gear wear tests, and the results show that the developed kernel factor analysis method outperforms the state-of-the-art feature selection techniques in terms of virtual sensing model accuracy.

The main contributions of this study rest on: (1) A virtual gearbox condition sensing framework is proposed to bridge the gap between vibration analysis and oil debris monitoring methods, and (2) a new feature selection and representation method (KFA) is presented to exploit the nonlinear representative features with non-Gaussian distributions, and the effectiveness of the KFA method is experimentally validated by the gear wear study. The rest of this paper is constructed as follows. After introducing the theoretical background of conventional feature representation techniques in Sect. 2, the details of the kernel factor analysis-based virtual sensing model are then discussed in Sect. 3. The effectiveness of the presented technique is experimentally demonstrated in Sect. 4 based on direct and indirect sensing data acquired from a spiral bevel gear case study. Finally, conclusions are drawn in Sect. 5.

2 Theoretical framework

2.1 Principal component analysis and its kernel variant

Principal component analysis (PCA) has been widely investigated for dimensionality reduction of feature space. It transforms a set of observations of possible correlated variables into a set of uncorrelated variables called principal components, where the first principal component has the largest variance, and each succeeding principal component has comparative lower variance orthogonal to the preceding principal components. However, if the sample data has more complicated structures which cannot be well represented in a linear subspace, PCA may be not applicable. Kernel principal component analysis (KPCA) generalizes the traditional PCA to the nonlinear dimensionality reduction method by incorporating kernel techniques. The key idea of KPCA is to define a nonlinear transformation $\phi(\bullet)$ which transforms the sample data into a high-dimensional data space, where each data point X_i is projected to a point $\phi(X_i)$. Then, the traditional PCA is performed in the new feature space (He et al. 2007). The first several principal components can well represent the original data with minimal mean squared approximation error, and thus KPCA has been widely used in the dimensionality reduction applications (He et al. 2007).

2.2 ISOMAP method

The ISOMAP algorithm extends the metric multidimensional scaling (MDS) method by integrating the geodesic distances instead of pairwise Euclidean distances to compute the graph shortest path distances (Tenenbaum et al. 2000). The key idea of ISOMAP is to find a low-dimensional embedding of data points, which is characterized as a nonlinear and global optimal method since only one free parameter (e.g., ε or K) needs to be optimized. The implementation of this algorithm mainly includes the following steps. Firstly, a neighborhood graph G is constructed by determining which points are neighbors on the manifold M based on the distances $d_X(i, j)$ between pairs of points i, j in the input space X, where ε-ISOMAP and K-ISOMAP methods can be used to determine the neighborhood points. Secondly, the geodesic distances $d_M(i, j)$

between all pairs of points on the manifold M are estimated by computing the shortest path distances $d_G(i, j)$ in the graph G. Finally, the classical MDS is applied to the matrix D_G, constructing an embedding of the data in a d-dimensional space Y which best preserves the manifold's estimated intrinsic geometry. However, a typical shortcoming of ISOMAP method is the high computational complexity, characterized by the full matrix eigenvector decomposition (Tenenbaum et al. 2000).

2.3 Locally linear embedding algorithm

Locally linear embedding (LLE), as a representative manifold learning technique, is a nonlinear dimension reduction method by mapping the high-dimensional data to a lower dimensional space while preserving the essential properties of the raw data. It attempts to discover the underlying nonlinear structure (nonlinear manifold) in high-dimensional data by exploiting the local symmetries of linear reconstructions (Roweis and Saul 2000). Like the ISOMAP algorithm, the implementation of the LLE method also requires several steps. First of all, the neighbors of each data point x_i are obtained by calculating the Euclidean distances between neighbor points and the data point of interest. Next, the weights matrix W is computed by minimizing the reconstruction error of the data point from its neighbors. Finally, each high-dimensional observation X is mapped to a low-dimensional vector Y representing the global internal coordinates on the manifold. When implementing the LLE algorithm, only one free parameter needs to be optimized, which is quite straightforward. Therefore, once the number of neighbors per data point K is chosen, the optimal weights W_{ij} and coordinates Y_i are computed by standard methods in linear algebra. However, this method is sensitive to noise and prone to ill-conditioned eigen issues, which may lead to unsatisfactory performance of feature selection and fusion.

2.4 Factor analysis

Factor analysis (FA), as a typical variance-based feature selection and representation technique, is different from traditional PCA which is formulated on matrix decomposition. FA is a linear Gaussian latent variable model and releases constraint by forming a diagonal covariance (Bishop 2006). The constructed factor model represents the original variable by a linear combination of latent and measured variables. Thus FA is able to deal with the uncertainty of extracted features in gearbox condition monitoring, since the gearbox degradation process under complex operating conditions (such as accumulation of fatigue, crack propagation, wear) may be subject to uncertainty and changeable operations. Moreover, FA is invariant to the component-wise rescaling of the feature space for input data by preserving the intrinsic data structures (Bishop 2006). Unfortunately, FA is a linear model based on the second-order statistics, which means that the processed data need to obey Gaussian distributions. However, the vibration signals in practice contain much noise and a variety of frequency components obeying non-Gaussian distributions. By taking into account the high computational efficiency and nonlinear projection ability of kernel functions, a kernel factor analysis is investigated for feature selection and representation in gearbox condition monitoring.

3 Proposed virtual sensing framework

During the normal operation process, online sensing techniques such as accelerometer and tachometer signals are continuously recorded to reflect gearbox conditions, but they are indirect indicators of gearbox conditions. On the other hand, oil debris is usually measured offline by experienced engineers to inspect the gearbox conditions, but it can directly reflect the gearbox condition. The proposed virtual sensing model for gearbox condition monitoring takes advantage of online measurements to estimate the gearbox conditions which are comparable to the oil debris measurements based on artificial intelligence as illustrated in Fig. 1. The virtual sensing framework mainly consists of four modules: (1) a data acquisition system capable of measuring vibration measurements during gearbox operation, (2) a feature extraction module to extract the representative gearbox condition indicators (CIs) by preprocessing the raw noisy measurements, (3) a kernel factor analysis-based feature fusion module to select

and fuse the extracted features for dimension reduction, and (4) a support vector regression-based artificial intelligence model to infer gearbox conditions from the fused features. The developed virtual sensing method is a complement to direct sensing or indirect sensing and provides a more effective tool for gearbox condition monitoring. The details of each module are discussed below.

3.1 Data acquisition and feature extraction

The vibration signal measurements are usually collected continuously to characterize the gearbox condition. Due to the poor signal-to-noise ratio (SNR) and multi-component interaction in a gearbox, vibration signal processing is required to de-noise the signal and extract defective signatures. In this study, a total of 21 features or condition indictors (CIs) from time, frequency, and time–frequency domains are investigated including (1) time synchronous averages (TSA): root mean square (RMS), kurtosis (KT), peak-to-peak (P2P), crest factor (CF); (2) residual RMS, KT, P2P, CF; (3) energy operator RMS, KT; (4) energy ratio; (5) FM0; (6) sideband level factor; (7) narrowband (NB) RMS, KT, CF; (8) amplitude modulation (AM) RMS, KT; (9) derivative AM KT; and (10) frequency modulation (FM) RMS, KT. The detailed formulation of these condition indicators have been published (Zakrajsek et al. 1993; Wemhoff et al. 2007).

3.2 Feature selection and fusion

The extracted features are formulated as feature vectors and further constructed as feature space of high dimensionality. To remove irrelevant and redundant features, and to improve model computational efficiency, a proper feature selection and fusion strategy is needed to lower the dimension of feature space. In the factor analysis, the

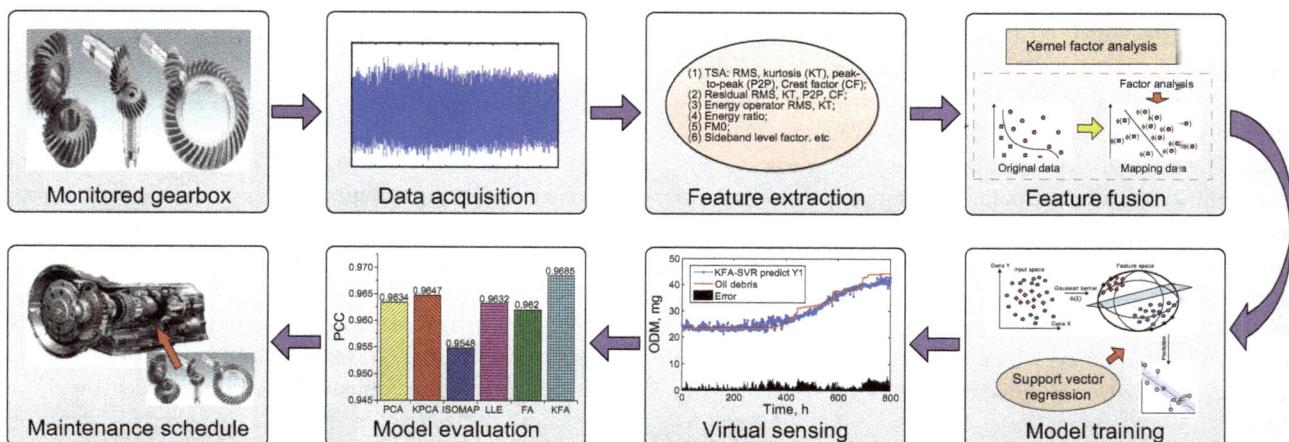

Fig. 1 Diagram of the developed virtual sensing model

feature set X is defined as a linear combination of latent variable set Z plus a noise term as follows:

$$X = WZ + \mu + \varepsilon \tag{1}$$

where W is a $D \times M$ factor loading matrix capturing the correlations behind the extracted feature variables; $\mu \in \mathbb{R}^D$ is the mean vector for feature set X; and ε denotes a D-dimensional Gaussian noise with zero mean and a diagonal covariance, i.e., $\varepsilon \sim \mathbb{N}(0, \Psi)$, where Ψ is a $D \times D$ diagonal matrix modeling the independent noise variance for each original dimension. For conciseness, the mean vector μ is ignored in the following derivation, since the data are easily assumed to be zero-centered after preprocessing.

With the proper transformation, the original features could be well represented by a low-dimensional latent variable space. However, an underlying constraint in factor analysis is that the variables follow Gaussian distributions which are difficult to meet in real-world gearbox condition monitoring. Thus, a kernel version of the factor analysis method is formulated to tackle this issue. The original features are projected into a new feature space $\mathcal{B}$ with a mapping function ϕ, and the new feature matrix is generated and written by:

$$F = \begin{pmatrix} \Phi(x_1) \\ \Phi(x_2) \\ \vdots \\ \Phi(x_n) \end{pmatrix} \tag{2}$$

Then, the FA is introduced in the new feature space $\mathcal{B}$, which can be treated as performing a nonlinear FA in the original space. Similar to the FA method, the data in the new space $\mathcal{B}$ can also be represented as follows:

$$F = WT + E \tag{3}$$

where $W \in \mathbb{R}^{N \times M}$ that is the projected latent data matrix, $T \in \mathbb{R}^{M \times P}$, and $E \sim \mathbb{N}_{N,P}(0, \Psi \otimes I_M)$ that is the noise variance matrix following the independent and identical distribution. To estimate model parameter set θ containing $\{W, \Psi\}$, the expectation–maximization (EM) algorithm is used, which is an iterative method proposed for maximum likelihood of a latent probabilistic model (Dempster et al. 1977, Moon 1996). Considering that the new data matrix F is centered, the parameter estimation in E-steps and M-steps can be obtained as (Wang et al. 2016):

$$W_{q+1} = K_{\text{norm}} \Psi_q^{-1} W_q E[T]^{\text{T}} (I + G_q W_q^{\text{T}} \Psi_q^{-1} K_{\text{norm}} \Psi_q^{-1} W_q)^{-1} \tag{4}$$

$$\Psi_{q+1} = \frac{1}{N} \text{diag}\{K_{\text{norm}} - W_{q+1} G_q W_q^{\text{T}} \Psi_q^{-1} K_{\text{norm}}\} \tag{5}$$

$$G_q = (I + W_q^{\text{T}} \Psi_q^{-1} W_q)^{-1} \tag{6}$$

where q and $q + 1$ represent two successive iteration steps. The kernel factor analysis only needs to address the kernel

matrix K, which is different from the traditional FA and has a more efficient learning process.

3.3 Virtual sensing model construction

The selected features are fed into the artificial intelligence model to construct the virtual sensing technique. Different artificial intelligence techniques could fit the purpose including artificial neural network (Dong et al. 2010), support vector regression (Widodo and Yang 2007), and fuzzy logic (Gokulachandran and Mohandas 2015). The artificial neural network technique has been widely investigated, but it requires a large amount of historical data for model training and suffers from local optima and overfitting issues. Support vector regression raises much attention because of high generalization capability and lower training sample requirements (Widodo and Yang 2007). Considering only a limited number of labeled experimental data sets are available, the support vector regression is selected to build the virtual sensing model in this study. During the model construction process, the selected features from vibration measurements are taken as the inputs while the oil debris measurements are treated as the outputs. The selection of parameters and kernel functions in the support vector regression model is determined using a grid search algorithm following a leave-one-out cross-validation method. Then the built support vector regression model fuses the selected features from vibration measurements to infer the gearbox condition indicator which is comparable with the oil debris measurements for gearbox condition monitoring.

4 Experimental studies

4.1 Data preparation

Experimental data obtained from a spiral bevel gear case study (Dempsey et al. 2002) is used to evaluate the

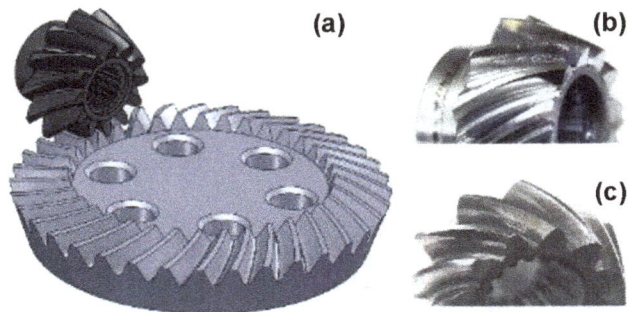

Fig. 2 Illustration of gearbox test. a Bevel gear test rig. b Damaged spiral bevel gear in experiment Y1. c Damaged spiral bevel gear in experiment Y3 (Dempsey et al. 2002)

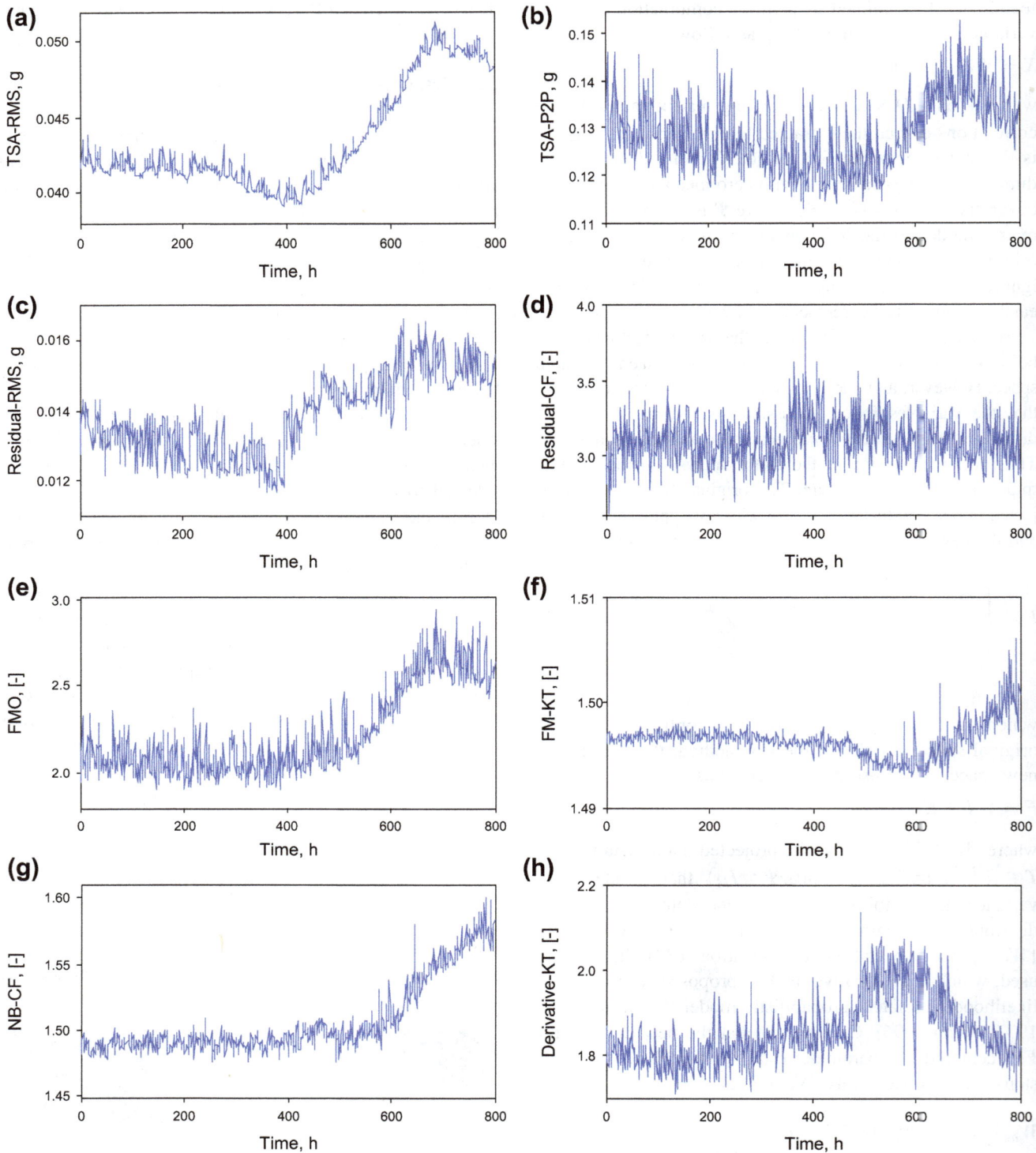

Fig. 3 Exemplified feature sets extracted from vibration measurements

presented virtual sensing method. The schematic diagram of the bevel gear test rig is shown in Fig. 2. A number of gear wear tests were performed until surface fatigue occurred, during which the vibration and oil debris measurements were collected to characterize the gearbox conditions. Vibration data was measured by two accelerometers located on the left and right pinion shaft bearing housing. They were collected once per minute

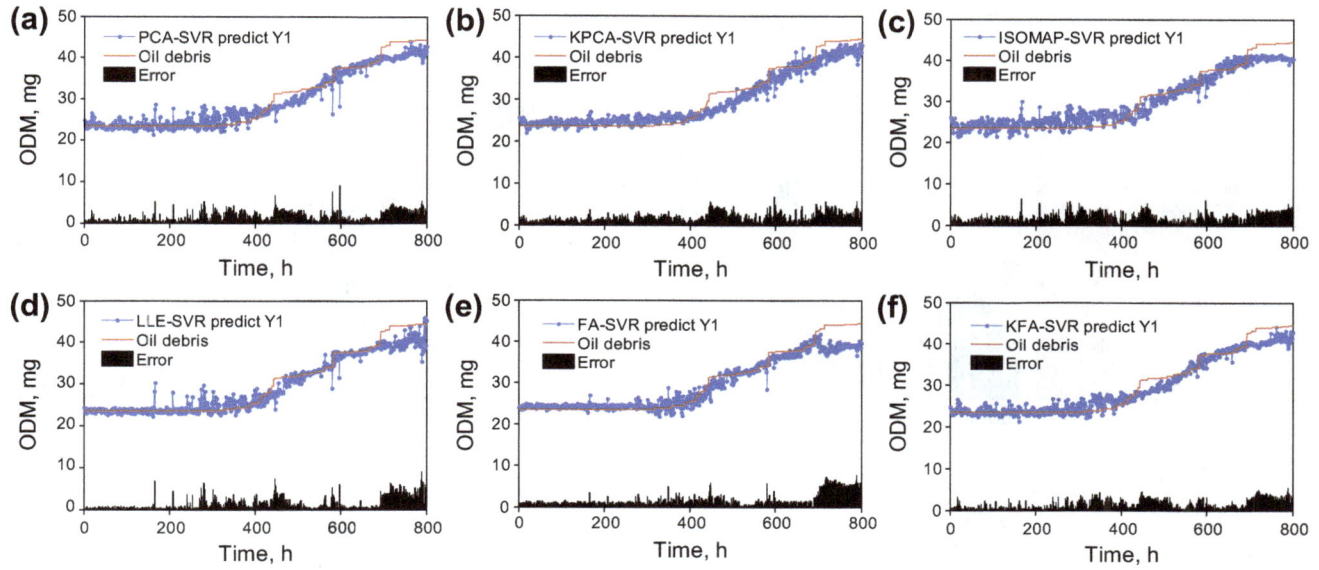

Fig. 4 Performance comparison of different virtual sensing schemes using dataset Y1

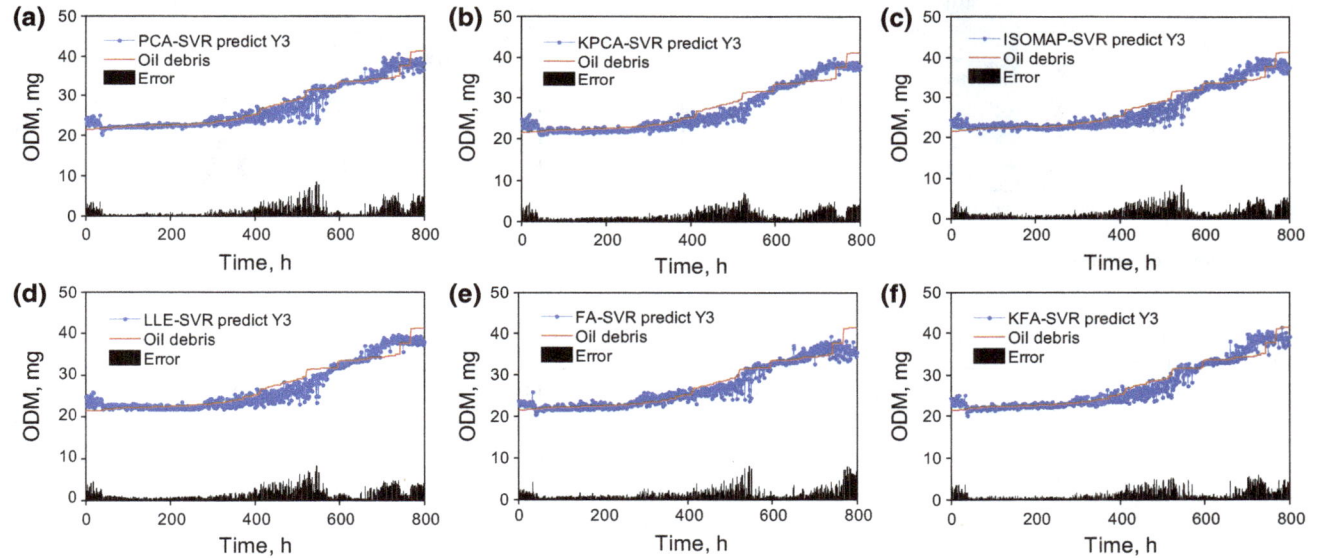

Fig. 5 Performance comparison of different virtual sensing schemes using dataset Y3

Table 1 Quantitative evaluation criteria for performance comparison	Metrics	Mathematic formulation
	Pearson correlation coefficient (PCC)	$PCC = \dfrac{\sum_i (y_i - \bar{y})(\widehat{y}_i - \bar{\widehat{y}})}{\sqrt{\sum_i (y_i - \bar{y})^2 \sum_i (\widehat{y}_i - \bar{\widehat{y}})^2}}$
	Root-mean-square error (RMSE)	$RMSE = \sqrt{\dfrac{1}{N} \sum_{i=1}^{N} (\widehat{y}_i - y_i)^2}$
	Mean absolute error (MAE)	$MAE = \dfrac{1}{N} \sum_{i=1}^{N} |y_i - \widehat{y}_i|$
	Mean absolute percentage error (MAPE)	$MAPE = \dfrac{1}{N} \sum_{i=1}^{N} \dfrac{|y_i - \widehat{y}_i|}{y_i}$

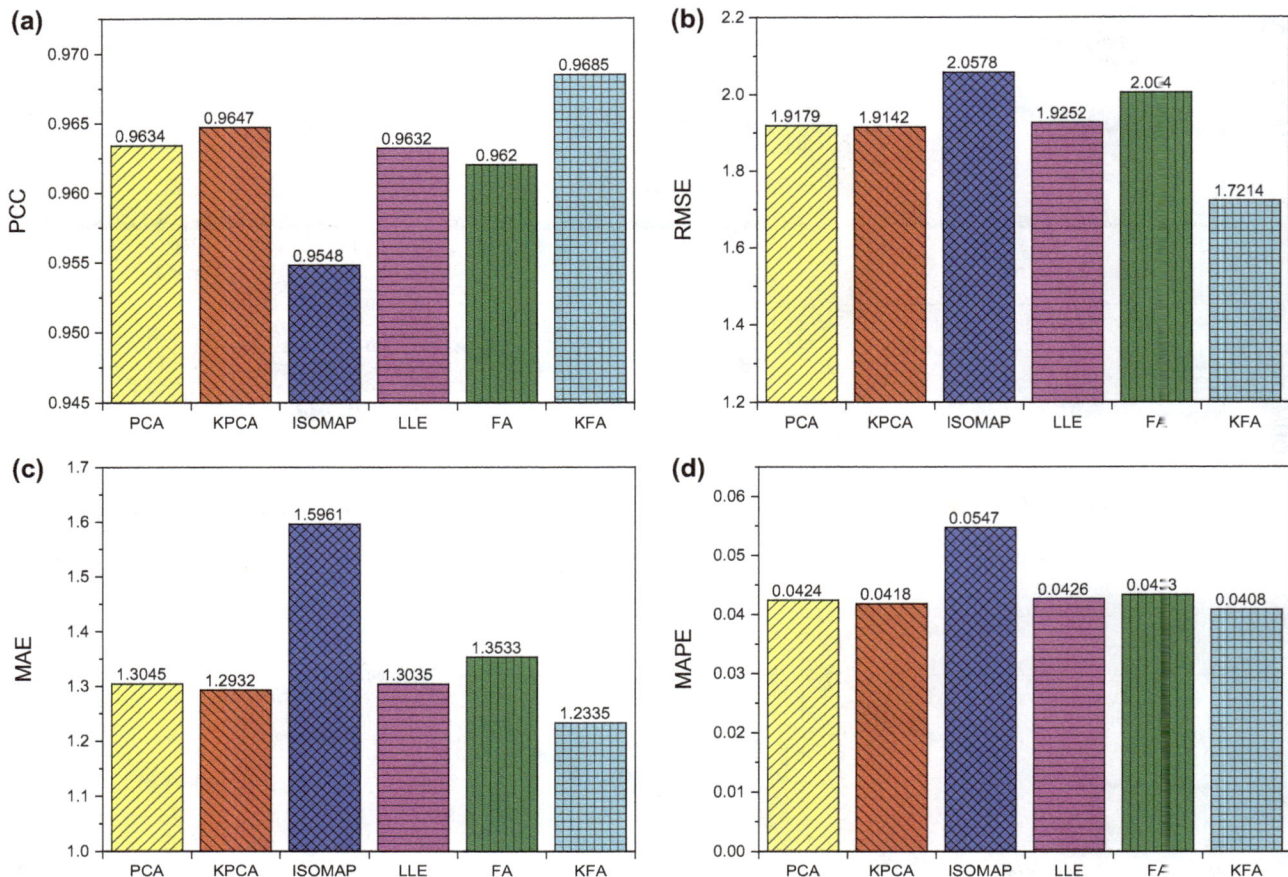

Fig. 6 Performance comparison of different virtual sensing schemes using different criteria. **a** PCC. **b** RMSE. **c** MAE. **d** MAPE

Table 2 Performance comparison of different virtual sensing schemes

Methods	PCC	RMSE	MAE	MAPE
PCA based	0.9634 ± 0.0089	1.9179 ± 0.0240	1.3045 ± 0.1370	0.0424 ± 0.0042
KPCA based	0.9647 ± 0.0121	1.9142 ± 0.0513	1.2932 ± 0.0318	0.0418 ± 0.0024
ISOMAP based	0.9548 ± 0.0116	2.0578 ± 0.0995	1.5961 ± 0.1806	0.0547 ± 0.0055
LLE based	0.9632 ± 0.0115	1.9252 ± 0.0095	1.3035 ± 0.0984	0.0426 ± 0.0054
FA based	0.962 ± 0.0103	2.004 ± 0.2077	1.3533 ± 0.1632	0.0433 ± 0.0028
KFA based	0.9685 ± 0.0081	1.7214 ± 0.0850	1.2335 ± 0.1246	0.0408 ± 0.0027

using a sampling rate of 100 kHz for 2 s duration. The shaft speed was measured by an optical sensor once per each gear shaft revolution, generating time synchronous averages (TSA). Oil debris data were collected using a commercially available oil debris sensor to detect the pitting damage on spiral bevel gears (Howe and Muir 1998).

A total of 21 representative features (as discussed in Sect. 3.1) are extracted from time, frequency and time–frequency domains by preprocessing the time synchronous averaging signal. The exemplified features are shown in Fig. 3. Whitening and eigenvalue decomposition (EVD) are firstly performed to select six dominant features by

preserving almost 95% of the cumulative variances. Next, kernel factor analysis is performed for dimension reduction to remove the irrelevant and redundant features.

4.2 Performance evaluation

The presented virtual sensing model is used to exploit the complex relationship between the vibration and oil debris analysis methods. A total of three sets of gearbox life test data (e.g., Y1, Y2, and Y3, etc.) are available. The leave-one-out strategy is followed to cross validate the performance of the virtual sensing model. More specifically, two data sets are chosen for model training, and the remaining one is for model testing. Firstly, the SVR model is built by optimizing the cost parameter C and Gaussian kernel parameter γ using the grid search method to prevent overfitting. Next, the selected features obtained by KFA are fed into the constructed SVR model to infer gearbox conditions. To compare the performance of KFA, several state-of-the-art dimension reduction techniques are also investigated including PCA, KPCA, LLE, ISOMAP, and FA. The virtual sensing results of these different feature selection schemes are shown in Figs. 4 and 5 using different sets of experimental data. It is found that the predicted gear conditions by these virtual sensing models generally follow the trend of the actual oil debris measurement.

To quantitatively compare the performance of different virtual sensing schemes, different criteria are investigated including the Pearson correlation coefficient (PCC), root-mean-square error (RMSE), mean absolute error (MAE), and mean absolute percentage error (MAPE) as illustrated in Table 1. In the evaluation indexes, y represents the actual oil debris measurement and $\hat{y}$ is the estimated oil debris measurement using the virtual sensing model. Generally, the larger the PCC value, the better the model performance, while the lower the RMSE/MAE/MAPE value, the better the model performance.

According to the above evaluation criteria, the performance of these virtual sensing schemes is compared and the results are shown in Fig. 6 and Table 2. It can be found that different feature selection techniques play important roles in the performance of virtual sensing models. The KFA-based virtual sensing model outperforms the conventional feature selection based virtual sensing models. By incorporating the kernel techniques into factor analysis, the superiority of KFA method is demonstrated to tackle the uncertainty and non-Gaussian features in the vibration measurements for feature selection and fusion.

5 Conclusions

Virtual sensing, as a complement to direct sensing or indirect sensing, provides a new perspective for machinery condition monitoring. According to the results obtained in this study, the conclusions can be drawn as follows.

(1) A new virtual sensing technique is presented for gearbox condition monitoring by taking the merits of vibration analysis and oil debris analysis methods.
(2) By incorporating the kernel technique into the factor analysis, a new feature selection method (KFA) is presented to exploit the nonlinear representative features with non-Gaussian distributions.

(3) The effectiveness of the presented virtual sensing method is validated in the experimental studies of gear wear, and the comparison results show that the presented KFA scheme outperforms the conventional feature selection techniques in terms of virtual sensing model accuracy.

A variety of experimental tests will be performed to evaluate the robustness of the proposed method in our next-step research.

Acknowledgements This research acknowledges the financial support from the National Science Foundation of China (No. 51504274 and No. 51674277), the National Key Research and Development Program of China (No. 2016YFC0802103), and the Science Foundation of China University of Petroleum, Beijing (No. 2462014YJRC039 and 2462015YQ0403).

References

Bishop CM. Pattern recognition and machine learning. New York: Springer Science; 2006.

Bolander N, Qiu H, Eklund N, et al. Physics-based remaining useful life prediction for aircraft engine bearing prognosis. In: Annual conference of the prognostics and health management society. 2009.

Bolón-Canedo V, Sánchez-Maroño N, Alonso-Betanzos A. Recent advances and emerging challenges of feature selection in the context of big data. Knowl Based Syst. 2015;86:33–45. doi:10.1016/j.knosys.2015.05.014.

Bustillo A, Correa M, Renones A. A virtual sensor for online fault detection of multitooth-tools. Sensors. 2011;11(3):2773–95. doi:10.3390/s110302773.

Cheng JW-J, Chao TC, Chang LH, et al. A model-based virtual sensing approach for the injection molding process. Polym Eng Sci. 2004;44(9):1605–14. doi:10.1002/pen.20158.

Dempsey PJ, Handschuh RF, Afjeh AA. Spiral bevel gear damage detection using decision fusion analysis. In: The 5th international conference on information fusion, Cleveland, OH, USA; July 8–11, 2002. doi:10.1109/icif.2002.1021136.

Dempster AP, Laird NM, Rubin DB. Maximum likelihood from incomplete data via the EM algorithm. J R Stat Soc Ser B Methodol. 1977;39(1):1–38. doi:10.2307/2984875.

Dong X, Wang S, Sun R, et al. Design of artificial neural networks using a genetic algorithm to predict saturates of vacuum gas oil. Pet Sci. 2010;7(1):118–22. doi:10.1007/s12182-010-0015-y.

Gelman L, Petrunin I, Jennions IK, et al. Diagnostics of local tooth damage in gears by the wavelet technology. Int J Progn Health Manag. 2013;3:52–8.

Gokulachandran J, Mohandas K. Comparative study of two soft computing techniques for the prediction of remaining useful life of cutting tools. J Intell Manuf. 2015;26(2):255–68. doi:10.1007/s10845-013-0778-2.

He Q, Kong F, Yan R. Subspace-based gearbox condition monitoring by kernel principal component analysis. Mech Syst Signal Process. 2007;21(4):1755–72. doi:10.1016/j.ymssp.2006.07.014.

Howe B, Muir D. In-line oil debris monitor (ODM) for helicopter gearbox condition assessment. In: Defense Technical Information Center, VA; 1998.

Hussain S, Gabbar HA. Vibration analysis and time series prediction for wind turbine gearbox prognostics. IJPHM Special Issue on Wind Turbine PHM (Color). 2013:69.

Li CJ, Tzeng TC. Multimilling-insert wear assessment using non-linear virtual sensor, time-frequency distribution and neural networks. Mech Syst Signal Process. 2000;14(6):945–57. doi:10.1006/mssp.1999.1282.

Liang X, Zuo MJ, Hoseini MR. Vibration signal modeling of a planetary gear set for tooth crack detection. Eng Fail Anal. 2015;48:185–200. doi:10.1016/j.engfailanal.2014.11.015.

Liang X, Zuo MJ, Pandey M. Analytically evaluating the influence of crack on the mesh stiffness of a planetary gear set. Mech Mach Theory. 2014;76:20–38. doi:10.1016/j.mechmachtheory.2014.02.001.

Moon TK. The expectation-maximization algorithm. IEEE Signal Process Mag. 1996;13(6):47–60.

Peng H, Long F, Ding C. Feature selection based on mutual information criteria of max-dependency, max-relevance, and min-redundancy. IEEE Trans Pattern Anal Mach Intell. 2005;27(8):1226–38. doi:10.1109/tpami.2005.159.

Petersen CD, Fraanje R, Cazzolato BS, et al. A Kalman filter approach to virtual sensing for active noise control. Mech Syst Signal Process. 2008;22(2):490–508. doi:10.1016/j.ymssp.2007.06.007.

Ploennigs J, Ahmed A, Hensel B, et al. Virtual sensors for estimation of energy consumption and thermal comfort in buildings with underfloor heating. Adv Eng Inform. 2011;25(4):688–98. doi:10.1016/j.aei.2011.07.004.

Ragaglia M, Zanchettin AM, Bascetta L, et al. Accurate sensorless lead-through programming for lightweight robots in structured environments. Robot Comput Integr Manuf. 2016;39:9–21. doi:10.1016/j.rcim.2015.11.002.

Roweis ST, Saul LK. Nonlinear dimensionality reduction by locally linear embedding. Science. 2000;290(5500):2323–6. doi:10.1126/science.290.5500.2323.

Samanta B, Nataraj C. Prognostics of machine condition using soft computing. Robot Comput Integr Manuf. 2008;24(6):816–23. doi:10.1016/j.rcim.2008.03.011.

Schölkopf B, Smola A, Müller KR. Nonlinear component analysis as a kernel eigenvalue problem. Neural Comput. 1998;10(5):1299–319. doi:10.1162/089976698300017467

Shokri S, Sadeghi MT, Marvast MA, et al. Improvement of the prediction performance of a soft sensor model based on support vector regression for production of ultra-low sulfur diesel. Pet Sci. 2015;12(1):177–88. doi:10.1007/s12182-014-0010-9.

Tenenbaum JB, De Silva V, Langford JC. A global geometric framework for nonlinear dimensionality reduction. Science. 2000;290(5500):2319–23. doi:10.1126/science.290.5500.2319.

Tham MT, Montague GA, Morris AJ, et al. Soft-sensors for process estimation and inferential control. J Process Control. 1991;1(1):3–14. doi:10.1016/0959-1524(91)87002-F.

Wang J, Xie J, Zhao R, et al. A new probabilistic kernel factor analysis for multisensory data fusion: application to tool condition monitoring. IEEE Trans Instrum Meas. 2016;65(11):2527. doi:10.1109/tim.2016.2584238.

Wang T, Liang M, Li J, et al. Rolling element bearing fault diagnosis via fault characteristic order (FCO) analysis. Mech Syst Signal Process. 2014;45(1):139–53. doi:10.1016/j.ymssp.2013.11.011.

Wang W. An adaptive predictor for dynamic system forecasting. Mech Syst Signal Process. 2007;21(2):809–23. doi:10.1016/j.ymssp.2005.12.008.

Wemhoff E, Chin H, Begin M. Gearbox diagnostics development using dynamic modeling. In: AHS 63rd annual forum, Virginia Beach, VA; 2007.

Widodo A, Yang BS. Support vector machine in machine condition monitoring and fault diagnosis. Mech Syst Signal Process. 2007;21(6):2560–74. doi:10.1016/j.ymssp.2006.12.007.

Zakrajsek JJ, Townsend DP, Decker HJ. An analysis of gear fault detection methods as applied to pitting fatigue failure data. In: 47th Mechanical Failure Prevention Group Virginia Beach, Virginia; April 13–15, 1993.

Zhou JH, Pang CK, Zhong ZW, et al. Tool wear monitoring using acoustic emissions by dominant-feature identification. IEEE Trans Instrum Meas. 2011;60(2):547–59. doi:10.1109/tim.2010.2050974.

16

Waxes in asphaltenes of crude oils and wax deposits

Yulia M. Ganeeva[1,2] · Tatiana N. Yusupova[1,2] · Gennady V. Romanov[1,2]

Abstract Composition and molecular mass distribution of *n*-alkanes in asphaltenes of crude oils of different ages and in wax deposits formed in the borehole equipment were studied. In asphaltenes, *n*-alkanes from C_{12} to C_{60} were detected. The high molecular weight paraffins in asphaltenes would form a crystalline phase with a melting point of 80–90 °C. The peculiarities of the redistribution of high molecular paraffin hydrocarbons between oil and the corresponding wax deposit were detected. In the oils, the high molecular weight paraffinic hydrocarbons C_{50}–C_{60} were found, which were not practically detected in the corresponding wax deposits.

Keywords Waxes · Asphaltenes · Hydrocarbon composition · Crude oil · Wax deposit

1 Introduction

Waxes and asphaltenes are the most important constituents of crude oils because they have a great influence on crude oil properties. Waxes are a complex mixture of solid (at ambient temperature) hydrocarbons which consist mainly of paraffin hydrocarbons with a small amount of naphthenic and aromatic hydrocarbons as well as polar compounds. There are two general classes of petroleum waxes. Waxes that are composed primarily of normal alkanes (*n*-alkanes) with a chain length of C_{18}–C_{30} and crystallize in large flat plates (macrocrystalline structures) with a melting point of 40–60 °C are referred to as paraffin waxes (Jowett 1984). Waxes, which consist primarily of the higher molecular weight *n*-alkanes within C_{30}–C_{60}, with a high share of iso-paraffins and naphthenes, and have much higher melting point ranges (above 60–90 °C), are referred to as microcrystalline waxes. According to Musser and Kilpatrick (1998), paraffinic and microcrystalline waxes have molecular weight ranges of 350–600 and 300–2500, respectively.

The ratio of paraffinic to microcrystalline waxes in oils depends on origin of oils (Tissot and Welte 1978; Musser and Kilpatrick 1998; Philp 2004; Fazeelat and Saleem 2007). However, in most cases, the higher molecular weight paraffin hydrocarbons (microcrystalline type) are present in very low concentrations or are absent in produced oils. This is due to their low mobility and limited solubility in crude oil. As a result, these components remain in the reservoir rocks or precipitate out in borehole equipment when the temperature and pressure change. It was demonstrated repeatedly that high molecular weight *n*-alkanes were concentrated in solid deposits formed in well equipment and pipelines (Philp et al. 1995; Chouparova et al. 2004). Moreover, the concentrations of high molecular weight paraffin hydrocarbons in oils, which are determined by high temperature gas chromatography (HTGC, one of the main methods for investigating hydrocarbon composition of oils) (Lipsky and Duffy 1986; Hawthorne and Miller 1987), are underestimated because of the poor solubility of the high molecular weight hydrocarbons in the solvent and the difficulty in eluting of

✉ Yulia M. Ganeeva
yuyand@yandex.ru

[1] A.E. Arbuzov Institute of Organic and Physical Chemistry, Kazan Scientific Center, Russian Academy of Sciences, 8 Arbuzov St., Kazan, Russian Federation 420088

[2] Institute of Geology and Petroleum Technologies, Kazan (Volga region) Federal University, 18 Kremlyovskaya St., Kazan, Russian Federation 420008

Edited by Yan-Hua Sun

such high boiling components even at 440 °C, resulting in the bias of response toward the shorter-chain homologs (Philp et al. 1995).

Asphaltenes are the heaviest and most polar compounds in crude oils with the highest aromaticity and polarity. They consist of fused polynuclear aromatic systems, heteroatoms (S, N, and O), alkyl chains, and heavy metals (e.g. Ni, V). Asphaltenes are defined according to their solubility, i.e., as the fraction of the crude oil which is soluble in toluene and precipitates in low boiling point alkanes (n-pentane, n-heptane). Two of the representative structures for the asphaltene molecules belong to the "island" and "archipelago" structures (Groenzin and Mullins 2000; Acevedo et al. 2007). According to "island" structure, the most dominant part of the molecule is a planar polycyclic aromatic system containing on average 7 aromatic rings, and this aromatic system has poor alkyl substitution. The mean molecular weight for "island" asphaltenes is 750 with a distribution 500–1000, and the mean size is 2 nm (Mullins et al. 2012). The "archipelago" model (Acevedo et al. 2007; Strausz et al. 2008) represents several fused ring systems that are interconnected by alkyl chains. These two molecular structures probably coexist, and the debate is now centered on their proportions and specific roles in an asphaltene mixture.

As can be seen, asphaltenes and waxes differ greatly in chemical properties from each other and from other oil fractions. Asphaltenes are the most aromatic and polar fraction, waxes—most aliphatic and nonpolar. However, despite these differences, these two oil fractions are similar in that they have a high molecular weight, aggregate or associate in solutions, and have a limited solubility in crude oil. The waxes and asphaltenes have a dramatic influence on the properties of crude oil and can cause problems related to crystallization and aggregation, respectively. Changes in pressure, oil composition, and temperature often induce paraffin crystallization and/or asphaltene flocculation and precipitation. Wax- and asphaltene-related problems include solid deposits, stabilization of a water–oil emulsion, and sludge production. Determination of wax and asphaltene contents is a routine analytical requirement for crude oil quality control. In order to study the composition and properties of waxes and asphaltenes, they are isolated from oil by specially developed procedures. In general, the asphaltenes are isolated from oils first; waxes are isolated from deasphalted oil. In other words, the waxes are isolated from the oil system from which the asphaltenes have been removed beforehand [from maltenes (Branthaver et al. 1983; Elsharkawy et al. 2000; Fazeelat 2006; Alcazar-Vara and Buenrostro-Gonzalez 2011) or saturates (Yang and Kilpatrick 2005; Lu et al. 2008)]. However, the traditional methods used to isolate asphaltenes by adding 40 volumes in excess of low boiling point paraffin

hydrocarbons can produce a fraction which is contaminated with a significant amount of waxes (Chouparova and Philp 1998; Thanh et al. 1999; Liao et al. 2006; Acevedo et al. 2009; Coto et al. 2011). The presence of paraffin hydrocarbons in asphaltenes can be explained by their coprecipitation as a result of low solubility in low molecular weight n-alkanes (Thanh et al. 1999; Coto et al. 2011), by interaction between the alkyl lateral chains (C_7–C_{10}) of asphaltenes with high molecular weight n-alkanes (Garcia and Carbognani 2001; Stachowiak et al. 2005; Ganeeva et al. 2014), and by their adsorption and/or occlusion inside asphaltene aggregates (Liao et al. 2005, 2006; Acevedo et al. 2009; Gray et al. 2011). Now, it is well-known that the asphaltenes exist as three-dimensional aggregates due to the strong intermolecular forces such as hydrogen bonding and π-bonding (Murgich 2002) and even at very low concentration as 50–150 mg/L in a good solvent such as toluene (Mullins et al. 2012). The wax coprecipitation increases the apparent yield of asphaltenes, decreases the yield of waxes, and leads to incorrect determination of the composition and properties of asphaltenes and waxes, to provide misleading and ambiguous results in modeling and treatment programs. Methods for the separation of waxes from asphaltenes have been developed (Thanh et al. 1999; Schabron et al. 2012).

It is impossible to avoid the coprecipitation of high molecular paraffin hydrocarbons with asphaltenes under the standard procedure of oil deasphalting. This phenomenon can be used to concentrate the most high molecular paraffin hydrocarbons to study their composition and properties in some detail. Hence the goal of this paper is focused on an examination of the waxes in asphaltenes of different oils and corresponding solid deposits.

2 Experimental

Five crude oils (designated as oil 1–5) and three solid deposits (designated as deposit 3–5) were selected from oil fields of Russia and Kyrgyzstan (Table 1). The crude oils 1–3 were produced from oil reservoirs of different ages in the Samara region (Russia): Carboniferous (Bashkirian and Tournaisian stages) and Devonian periods respectively. The crude oils 4 and 5 were produced from the giant Romashkinskoe oil field of Tatarstan (Russia) and the Maily-Su oil field (Kyrgyzstan), respectively. The solid deposits 3–5 formed in the borehole equipment during production of crude oils 3–5, respectively. Chemical composition, some properties of these crude oils and solid deposits, as well as the age of the oil reservoirs, from which the crude oils were produced, are shown in Table 1.

Asphaltenes were precipitated by a 40-fold excess dilution of the crude oil with petroleum ether (boiling point

Table 1 Properties of crude oils and solid deposits

Symbol	Resource	Waxes, wt%	Asphaltenes, wt%	Density, g/cm^3	Viscosity, cSt	T_{wa} for crude oil/T_{melt} for solid deposit, °C[a]
Crude oil						
Crude oil 1	Russia, Samara region, Bashkirian stage (Carboniferous period)	3.6	0.3	0.7861	2.2	
Crude oil 2	Russia, Samara region, Tournaisian stage (Carboniferous period)	7.7	3.8	0.8551	16.2	
Crude oil 3	Russia, Samara region, Devonian period	25.0	1.4	0.9100	72.2	46.1
Crude oil 4	Russia, Tatarstan, Devonian period	6.0	4.7	0.9180	55.7	50.1[b]
Crude oil 5	Kyrgyzstan	6.1	6.6	0.9100	48.9	30.4
Deposit						
Deposit 3	Russia, Samara region, Devonian period	72.1	0.8			80.3
Deposit 4	Russia, Tatarstan, Devonian period	65.9	3.1			75.9
Deposit 5	Kyrgyzstan	49.4	6.6			72.4

[a] T_{wa} is the wax appearance temperature and T_{melt} is the melting temperature

[b] The content of water in this crude oil is 19.8 wt%. Other crude oils contain no water

40–70 °C). The precipitated asphaltenes were washed in a Soxhlet apparatus with petroleum ether until the filtrate was colorless. Then the asphaltenes in the filter were washed out with benzene, which was then evaporated. Waxes were isolated from the deasphalted resin-free crude oil according to the European standard method EN 12606-2. The deasphalted oil (maltenes) was adsorbed onto silica gels, previously activated at 120 °C overnight. After adsorption the silica gel was washed with a mixture of 3:1 petroleum ether/carbon tetrachloride, the obtained extract was then concentrated to yield resin-free oil. The resin-free oil was dissolved in a mixture of 1:1 acetone/benzene in the ratio of 1 g sample per 10 mL solvent and thereafter cooled to the temperature of -20 ± 1 °C. The waxes present in the oil precipitated, and then they were washed by hot benzene and finally weighed after solvent removal. Resins with silica gels were washed off with a mixture of 1:1 benzene/isopropanol.

Asphaltenes and waxes from solid deposits were isolated similarly. It should be noted that before the study the solid deposits were purified from mechanical impurities.

Hydrocarbon composition of waxes was analyzed by gas chromatography (GC) on a Perkin-Elmer chromatograph with a flame-ionization detector in the mode of temperature programming from 20 to 400 °C. PE-5ht capillary column (30 m long, 0.25 mm in internal diameter) and 5 %-phenil-95 %-methylpolysiloxane with a film thickness of 0.1 μm were used as the stationary phase. The flow rate of the helium carrier gas was set at 2 cm^3/min. 1 % solutions of the samples in carbon tetrachloride were prepared. The calibration was carried out using n-C_{20}.

As was shown by Ganeeva et al. (2014), the n-alkanes dominate the chromatograms of oils, waxes, and asphaltenes; therefore, further we discuss only the

molecular mass distribution (MMD) of n-alkanes in waxes, estimated by the method of internal normalization.

Thermal characterization was carried out with a C80 calorimeter (Setaram, France). The temperature calibration was performed using indium. The energy calibration was performed using the Joule effect method in the factory and checked by measuring the heat of fusion of naphthalene. Each sample (5–10 mg) was first heated to 100 °C, held isothermally for 1 min, and then cooled to ambient temperature at a rate of 1 °C/min. In order to delete any thermal history effects, two heating/cooling cycles were recorded, so that crystallization and melting properties were obtained from the second cycle. Measured thermal characteristics include temperatures and enthalpies of melting and crystallization of waxes. The temperature of melting (T_{melt}) was determined as the extreme point (minima) on heat flow curves. The enthalpy of melting was calculated as the area defined by the heat flow curve and base line between the points of onset and end of the endothermic effect on the heat flow curve. The wax appearance temperature (T_{wa}) of crude oils was determined as the onset of an exothermic peak on the heat flow curves under cooling (Table 1). The experimental uncertainties in the measurements of temperatures and enthalpies were ± 0.5 °C and ± 0.01 J/g, respectively.

3 Results and discussion

3.1 High molecular weight paraffin hydrocarbons in oils

It was found that the n-alkanes with a wide range of carbon numbers were present in oil asphaltenes, which were determined by GC (Fig. 1). In asphaltenes 1–3, n-alkanes were present with a carbon number of 10–60; their MMDs were polymodal with clearly marked predominance of one of the modes at C_{16}–C_{18}, C_{12}–C_{14}, and C_{54}–C_{56}, respectively. In asphaltenes 4, the n-alkanes had a carbon number of 20–54. MMD of these n-alkanes was bimodal with the dominating maximum at carbon numbers of 40–42. The asphaltenes 5 contained n-alkanes C_{20}–C_{60}, and their MMD was unimodal with the maxima at a carbon number of 52–54. It should be noted that all asphaltenes were black and shiny and nothing indicated that they contain paraffin hydrocarbons.

The presence of high molecular weight mode in MMD of n-alkanes in asphaltenes suggests that these coprecipitated paraffinic hydrocarbons can form a crystalline phase. Indeed, on the heat flow curves of the asphaltenes 3–5 at a temperature interval of 80–90 °C, a well-resolved narrow (10–15 °C width) endothermic peak indicating the melting of the crystalline phase of waxes (Fig. 2) was observed. The heat flow curve of the asphaltenes 5 is not shown on Fig. 2 because the content of the crystalline phase of paraffin hydrocarbons in them was low (enthalpy of melting of the crystalline phase is only 2.5 J/g) and at the scale of figure, the effect of melting was not visible. Values of the enthalpy of melting of the crystalline phase in asphaltenes 3 and 4 were 119.6 and 108.4 J/g, respectively. In asphaltenes 1 and 2, in which the low molecular weight n-alkanes dominated, the crystalline phase was not observed.

A comparison was made between compositions of waxes coprecipitated with asphaltenes and waxes which were isolated from maltenes using acetone. In most cases, the isolation of waxes from crude oil was carried out using acetone by various modifications of the Burger method (Burger et al. 1981), when the waxes were isolated from maltenes (Branthaver et al. 1983; Elsharkawy et al. 2000; Fazeelat 2006; Alcazar-Vara and Buenrostro-Gonzalez 2011) or saturates (Yang and Kilpatrick 2005; Lu et al. 2008).

According to the GC data on waxes isolated from crude oils by the standard method, the low molecular weight n-alkanes with carbon atoms <16 and high molecular weight homologs with a carbon number >50 were absent (Fig. 3). The waxes from investigated oils contained n-alkanes C_{18}–C_{48} and exhibited unimodal MMDs with maxima at C_{26}–C_{28}. It is evident that the waxes isolated from different oils by the standard method were almost the same, did not contain high molecular weight hydrocarbons, and did not characterize the peculiarities of hydrocarbon composition of oils studied. The MMDs of n-alkanes of waxes in asphaltenes and waxes isolated from crude oils by the standard method demonstrated (Figs. 1, 3) that in asphaltenes, there were n-alkanes with a wider range of

carbon numbers of 12–60+, their MMD was polymodal, and one of the maxima of MMD was shifted to higher molecular mass.

On the example of the oils from oil reservoirs of different ages in the Samara region, Russia (oils 1–3), it is shown that the MMDs of n-alkanes in the asphaltenes were quite different depending on the age of the oil reservoir from which the oils were produced (Fig. 1). So, in the asphaltenes of crude oil 1 taken from the Bashkirian stage (Carboniferous period), the main wide maximum of MMD was C_{16}–C_{28}. With an increase in the age of oil reservoirs (Tournaisian stage, Carboniferous period), the main maximum of MMD of n-alkanes in asphaltenes was shifted to low molecular weight n-alkanes C_{12}–C_{13} (asphaltenes of oil 2) when the occurrence depth increased. In asphaltenes of oil 3 taken from greater depth (Devonian period), low weight molecular n-alkanes were practically absent and the main maximum of MMD was at carbon numbers of 54–56.

Most possibly, the presence of low molecular n-alkanes in the asphaltenes is caused by their being trapped by porous structure of asphaltenes, while the presence of high molecular weight homologs in asphaltenes is a result from coprecipitating when the standard procedure of deasphalting by low molecular weight n-alkanes is used. The decrease in the share of low molecular weight n-alkanes in asphaltenes with the increase of the age of oil reservoir may result from: (1) decrease in the share of low molecular weight n-alkanes in greater depth crude oils and/or (2) compression of asphaltene aggregates: when the occurrence depth of oil reservoirs increases, the asphaltene aggregates in oil become denser and their ability to occlude low molecular weight oil hydrocarbons reduces. The increase in the share of high molecular weight n-alkanes in asphaltene with increasing age of oil reservoir is explained by the increase in their share in the oils, from which asphaltenes are precipitated (Ganeeva et al. 2010).

Thus, asphaltenes contain paraffin hydrocarbons with a wide range of carbon numbers from 10 to 60 (depending on the composition of oil), and their MMD can depend on the structure of asphaltenes and hydrocarbon composition of oil. High molecular weight paraffin hydrocarbons in asphaltenes can form crystalline phases. The content and composition of waxes isolated from deasphalting resin-free oil do not represent the total content, and compositional features of all waxes present in crude oil. For a more complete study of the composition and properties of oil waxes, especially as regard to their most high molecular weight paraffins, an accurate study of the asphaltenes is required.

The phenomenon of coprecipitation of the most high molecular weight paraffin hydrocarbons with asphaltenes at the standard procedure of deasphalting was used by us for studying the redistribution of high molecular weight

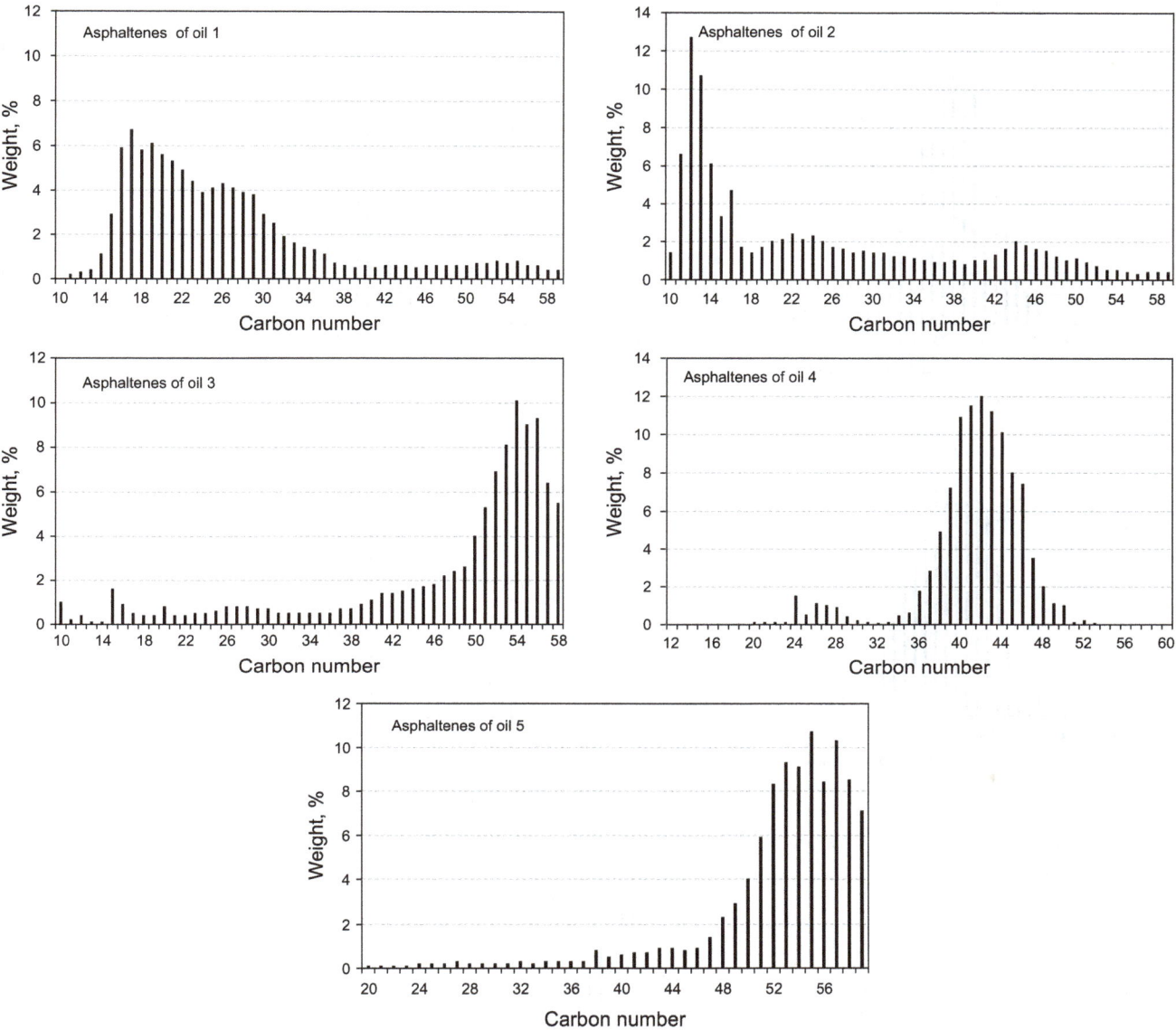

Fig. 1 MMD of *n*-alkanes in oil asphaltenes

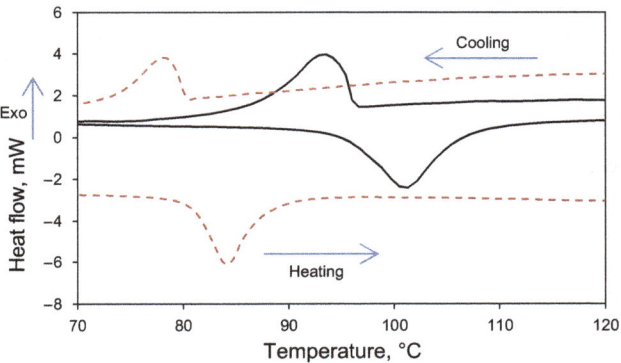

Fig. 2 Heat flow curves of asphaltenes from oils 3 (*solid lines*) and 4 (*dashed lines*)

paraffin hydrocarbons between oils and solid deposits formed in well equipment during oil production. For this purpose, we studied the composition of paraffin hydrocarbons in corresponding solid deposits focusing on the study of waxes in asphaltenes from solid deposits.

3.2 The composition and properties of waxes in wax deposits

The solid deposit formed in producing wells, in borehole equipment, and pipelines is known to represent a complex mixture of waxes, asphaltenes, and resins along with trapped oil, water, and inorganic material. The amount and composition of the solid deposits are a function of

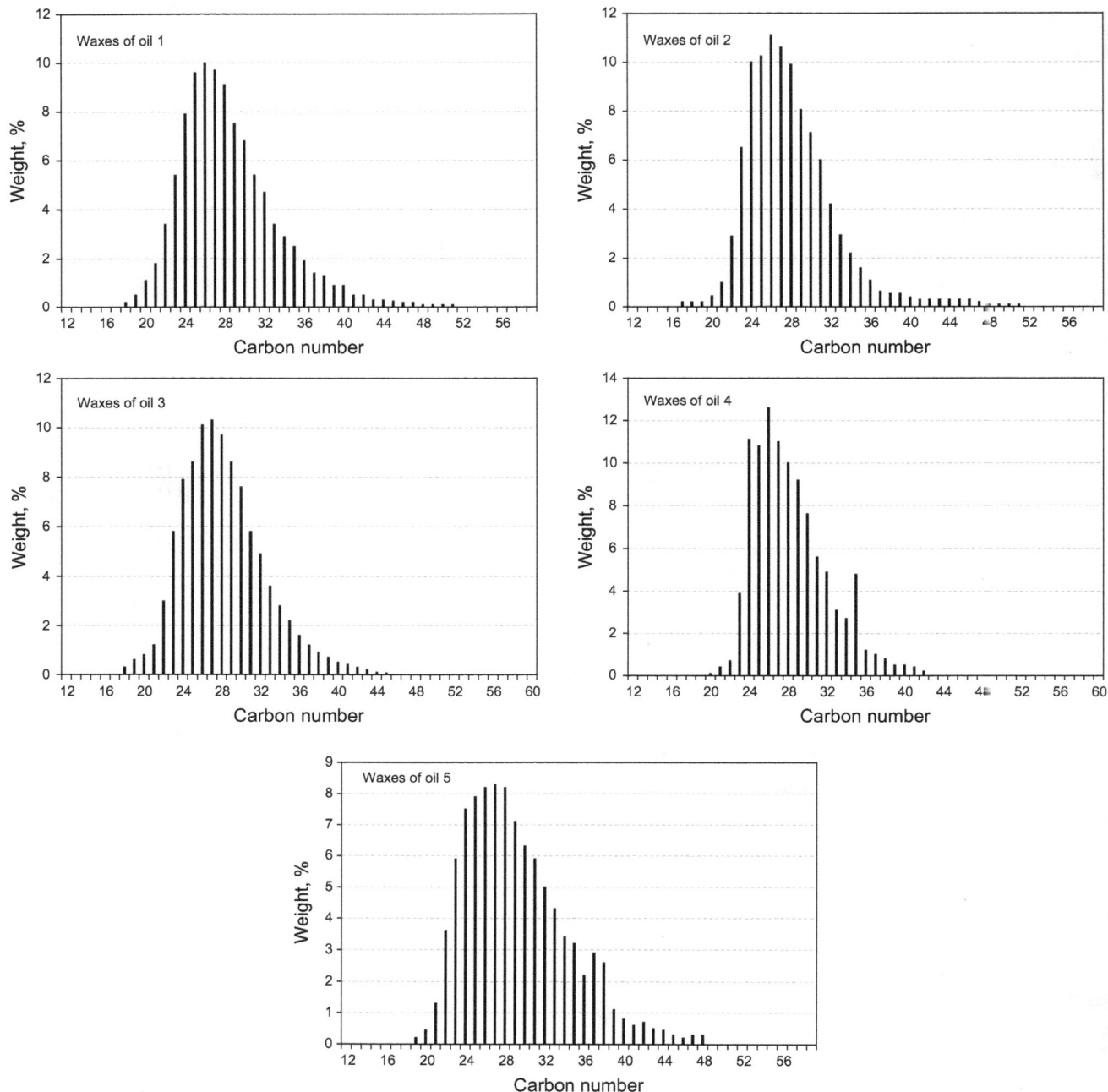

Fig. 3 MMD of *n*-alkanes in waxes of oils 1–5

changing temperature, pressure, and oil composition during oil production (Leontaritis 1996; Chouparova et al. 2004). It is known that solid deposits produced from different depths of the same well vary in amount and composition: the closer to the surface the bigger the amount of solid deposits and the bigger the share of low molecular weight *n*-paraffin hydrocarbon (macrocrystalline waxes) in them. At greater depths, little solid deposit is formed, but it contains a high proportion of the most high molecular weight paraffin hydrocarbons (microcrystalline waxes) (Chouparova et al. 2004). Thus, it is believed that most unstable macro- and microcrystalline waxes may precipitate from oil and form the solid deposits at different depths in the well, in borehole equipment, and pipelines, and therefore high molecular weight hydrocarbons are practically absent in produced oils. This fact has been demonstrated repeatedly by comparative analyses of MMD of paraffin hydrocarbons in oils and in the corresponding solid deposits (Philp et al. 1995; Garcia 2001; Paso and Fogler 2004; Yang and Kilpatrick 2005). Since we have shown that the most high molecular weight paraffin hydrocarbons are concentrated in asphaltenes; then in order to study the peculiarities of redistribution of the most high molecular weight paraffin hydrocarbons between oil and solid deposit, we studied the composition of *n*-alkanes in asphaltenes of solid deposits and compared it with that of the corresponding oils.

The solid deposits studied were dark brown semi-solid materials, characterized by high melting point, and for this reason, they were classified as wax deposits (Table 1). It should be noted that asphaltenes isolated from the wax deposits were light brown and had a greasy consistency. This testified that asphaltenes were polluted by waxes.

According to calorimetry data on these asphaltenes, there was a crystalline phase of paraffin hydrocarbons with enough high melting temperatures from 80 to 90 °C and high enthalpies of melting 150–170 J/g (Fig. 4).

Asphaltenes were investigated with GC. Figure 5 shows the MMDs of *n*-alkanes in asphaltenes of the wax deposits 3, 4, and 5. As can be seen, asphaltenes of wax deposits contained *n*-alkanes with a carbon number of 16–60 as the oil asphaltenes. The maxima of MMD of the *n*-alkanes in asphaltenes of wax deposit 3 was at 41–44, in asphaltenes of wax deposit 5 at 49–51. The MMD of *n*-alkanes in asphaltenes of wax deposit 4 was bimodal with the maxima at 27–29 and 40–42.

A comparative analysis of the composition and thermal properties of paraffin hydrocarbons in asphaltenes of crude oils and the corresponding wax deposits shows that the asphaltenes of crude oils contain higher molecular weight paraffin hydrocarbons (Figs. 1, 5). Indeed, the maxima of MMD of *n*-alkanes in oil asphaltenes 3, 4, and 5 was at 54–56, 40–44, and 54–56 (Fig. 1), whereas the maxima of MMD of *n*-alkanes in corresponding wax deposit asphaltenes showed *n*-alkanes with a smaller number of carbon atoms with maxima at 41–43, 40–42, and 49–51,

respectively (Fig. 5). The temperatures of melting of waxes in oil asphaltenes 3–5 were 101.2, 85.2, and 89.5 °C while those in the corresponding asphaltenes of wax deposits were 88.1, 81.4, and 81.2 °C. The higher temperatures of melting of waxes in oil asphaltenes confirm the higher molecular weight composition of waxes in them.

Thus, it is found that in produced oil there are higher molecular weight paraffin hydrocarbons that are absent in the wax deposits formed in the borehole equipment during production of this crude oil. In other words, the highest

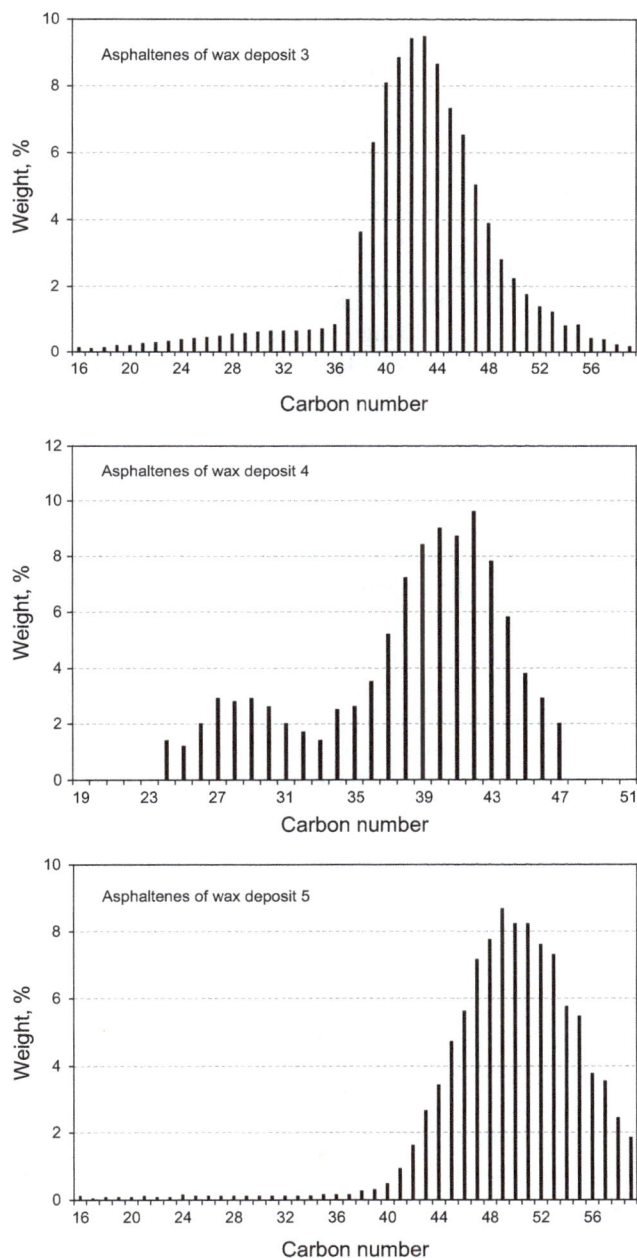

Fig. 5 MMD of *n*-alkanes in asphaltenes of wax deposits 3–5

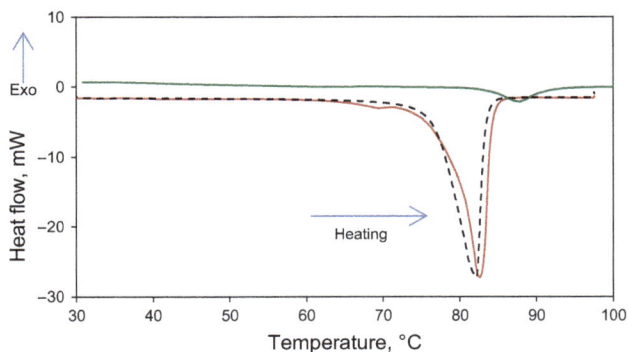

Fig. 4 Heat flow curves of asphaltenes from wax deposits 3 (*green line*), 4 (*red solid line*), and 5 (*dashed line*)

molecular weight n-alkanes do not precipitate out and remain dispersed in the oil. It should be noted that this fact is not surprising for mixture of individual n-alkanes in solvents and is predictable for crude oil. So, previously the polydispersity and cocrystallization of individual n-alkanes in solvents reduce the mass and high molecular weight homolog fraction in a deposit formed using a coldfinger apparatus (Senra et al. 2008). As to crude oils, the share of high molecular weight hydrocarbons in them is very low [a recurrence ratio of C_{n+1}/C_n is approximately 0.7–0.8 (Senra 2009)]; therefore, the n-alkanes that will crystallize out at the cloud point should not necessarily be the highest molecular weight n-alkanes present in the crude oil. However, it is known from papers (Philp et al. 1995; Garcia and Carbognani 2001; Paso and Fogler 2004; Yang and Kilpatrick 2005) that paraffin hydrocarbons in the wax deposits are higher molecular weight in comparison with those in the oil. The investigation of asphaltenes in which the highest molecular weight paraffin hydrocarbons are concentrated shows that indeed the wax deposit consists not the most high molecular weight hydrocarbons of crude oils. Evidently, more high molecular weight n-alkanes initiate the crystallizing of the more low molecular homologs causing their deposition from crude oil in a greater degree.

4 Conclusions

The research found that asphaltenes contain paraffin hydrocarbons with a wide MMD from 12 to 60. On the example of the oils of different ages from the Samara region, Russia, it is shown that the composition of paraffin hydrocarbons in asphaltenes depends on the age of the oil reservoir from which oil is produced. The higher the age of oil reservoir, the smaller the share of low molecular weight n-alkanes in asphaltenes. That may be due to the decrease in the share of low molecular weight n-alkanes in oil and the fact that with increasing depth of occurrence of the oil reservoir, the asphaltenes become denser and their ability to occlude low molecular weight hydrocarbons reduces. The increase in the share of high molecular weight n-alkanes in asphaltenes with an increase in the age of oil reservoir is explained by the increase in their share in the oil, from which asphaltenes are precipitated.

The higher molecular weight waxes may coprecipitate with asphaltenes and in some cases can form a crystalline phase in them with melting temperatures of 80–90 °C. Therefore, the commonly used methods of determination of the content of waxes in oils (by acetone from maltenes or saturates) do not allow the investigation of the total content and composition of waxes of crude oil, especially their high molecular weight constituents.

The study of the composition of paraffin hydrocarbons in oil asphaltenes allowed identifying the peculiarities of redistribution of high molecular weight paraffin hydrocarbons between oil and wax deposits. In asphaltenes of oils higher molecular weight, n-alkanes are found in comparison with the corresponding wax deposits. This indicates that in the wax deposits not all high molecular weight paraffin hydrocarbons are concentrated. A part of higher molecular n-alkanes remains in oil and initiates crystallization of more low molecular weight homologs, causing their deposition from crude oil to a greater degree.

Thus, the performed study shows the need for a careful analysis of asphaltenes to determine the exact composition of paraffin hydrocarbons of oils, especially their high molecular weight constituents, and to detect the peculiarities of redistribution of high molecular weight paraffin hydrocarbons in crude oils.

Acknowledgments The work is performed according to the Russian Government Program of Competitive Growth of Kazan Federal University.

References

Acevedo S, Castro A, Negrin JG, Fernandez A, Escobar G, Piscitelli V. Relations between asphaltene structures and their physical and chemical properties: the rosary-type structure. Energy Fuels. 2007;21(4):2165–75. doi:10.1021/ef070089v.

Acevedo S, Cordero JMT, Carrier H, Bouyssiere B, Lobinski R. Trapping of paraffin and other compounds by asphaltenes detected by laser desorption ionization-time of flight mass spectrometry (LDI-TOF MS): role of A1 and A2 asphaltene fractions in this trapping. Energy Fuels. 2009;23 842–8. doi:10.1021/ef8007745.

Alcazar-Vara LA, Buenrostro-Gonzalez E. Characterization of the wax precipitation in Mexican crude oils. Fuel Process Technol. 2011;92:2366–74. doi:10.1016/j.fuproc.2011.08.012.

Branthaver JF, Thomas KR, Dorrence SM, Heppner RA, Ryan MJ. An investigation of waxes isolated from heavy oils produced from northwest asphalt ridge tar sands. Liq Fuels Technol. 1983;1:127–46.

Burger ED, Perkins TK, Striegler JH. Studies of wax deposition in the Trans-Alaska pipeline. J Pet Technol. 1981;33:1075–86. doi:10.2118/8788-PA.

Chouparova E, Philp RP. Geochemical monitoring of waxes and asphaltenes in oils produced during the transition from primary to secondary water flood recovery. Org Geochem. 1998;29:449–61. doi:10.1016/S0146-6380(98)00056-7.

Chouparova E, Lanzirotti A, Feng H, Jones KV, Marinkovic N, Whitson C, Philp P. Characterization of petroleum deposits formed in a producing well by synchrotron radiation-based microanalysis. Energy Fuels. 2004;18:1199–212. doi:10.1021/ef030108a.

Coto B, Martos C, Espada JJ, Robustillo MD, Merino-Garcia D, Pena JL. Study of new methods to obtain the n-paraffin distribution of crude oils and its application to the flow assurance. Energy Fuels. 2011;25:487–92. doi:10.1021/ef100987v.

Elsharkawy AM, Al-Sahhaf TA, Fanhim MA. Wax deposition from Middle East crudes. Fuel. 2000;79:1047–55. doi:10.1016/S0016-2361(99)00235-5.

Fazeelat TJ. Chemical composition and geochemical applications of waxes isolated from Pakistani crude oil. J Chem Soc Pak. 2006;26:187–90.

Fazeelat TJ, Saleem A. GC-FID analysis of wax paraffin from Khaskheli crude oil. J Chem Soc Pak. 2007;29:492–9.

Ganeeva YM, Foss TR, Yusupova TN, Romanov GV. Distribution of high-molecular-weight n-alkanes in paraffinic crude oils and asphaltene-resin-paraffin deposits. Pet Chem. 2010;50:17–22. doi:10.1134/S0965544110010020.

Ganeeva YM, Yusupova TN, Romanov GV, Bashkirtseva NY. Phase composition of asphaltenes. J Therm Anal Calorim. 2014;115:1593–600. doi:10.1007/s10973-013-3442-3.

Garcia MC. Paraffin deposition in oil production. In: SPE international symposium on oilfield chemistry, 13–16 February, Houston, TX, 2001. doi:10.2118/64992-MS.

Garcia MC, Carbognani L. Asphaltene-paraffin structural interactions: effect on crude oil stability. Energy Fuels. 2001;15:1021–7. doi:10.1021/ef0100303.

Gray MR, Tykwinski RR, Stryker JM, Tan X. Supramolecular assembly model for aggregation of petroleum asphaltenes. Energy Fuels. 2011;25:3125–34. doi:10.1021/ef200654p.

Groenzin H, Mullins OC. Molecular size and structure of asphaltenes from various sources. Energy Fuels. 2000;14:677–84. doi:10.1021/ef990225z.

Hawthorne SB, Miller DJ. Analysis of commercial waxes using capillary supercritical fluid chromatography–mass spectrometry. J Chromatogr. 1987;388:397–409. doi:10.1016/S0021-9673(01)94500-5.

Jowett F. Petroleum waxes. In: Hobson GD, editor. Petroleum technology. New York: Wiley; 1984. p. 1021–42.

Leontaritis KJ. The asphaltene and wax deposition envelopes. Fuel Sci Technol Int. 1996;14:13–39.

Liao Z, Zhou H, Graciaa A, Chrostowka A, Creux P, Geng A. Adsorption/occlusion characteristics of asphaltenes: some implication for asphaltene structural features. Energy Fuels. 2005;19:180–6. doi:10.1021/ef049868r.

Liao Z, Geng A, Graciaa A, Creux P, Chrostowka A, Zhang Y. Different adsorption/occlusion properties of asphaltenes associated with their secondary evolution processes in oil reservoirs. Energy Fuels. 2006;20:1131–6. doi:10.1021/ef050355+.

Lipsky SR, Duffy ML. High temperature gas chromatography: the development of new aluminum clad flexible fused silica glass capillary columns coated with thermostable nonpolar phases. Part 2. J High Resolut Chrom Chrom Commun. 1986;9:725–30. doi:10.1002/jhrc.1240090702.

Lu X, Kalman B, Redelius P. A new test method for determination of wax content in crude oils, residues and bitumens. Fuel. 2008;87:1543–51. doi:10.1016/j.fuel.2007.08.019.

Mullins OC, Sabbah H, Eyssautier J, Pomerantz AE, Barre L, Andrews AB, et al. Advances in asphaltene science and the Yen-Mullins model. Energy Fuels. 2012;26:3986–4003. doi:10.1021/ef300185p.

Murgich J. Intermolecular forces in aggregates of asphaltenes and resins. Pet Sci Technol. 2002;20:983–97.

Musser BJ, Kilpatrick PK. Molecular characterization of waxes isolated from a variety of crude oils. Energy Fuels. 1998;12:715–25. doi:10.1021/ef970206u.

Paso KG, Fogler HS. Bulk stabilization in wax deposition systems. Energy Fuels. 2004;18:1005–13. doi:10.1021/ef034105+.

Philp RP. High molecular weight paraffins ($>C_{40}$) in crude oils and source rocks. In: AAPG international conference, Mexico; 2004. p. 24–27.

Philp RP, Bishop A, del Rio J-C, Allen J. Characterization of high molecular weight hydrocarbons ($>C_{40}$) in oils and reservoir rocks. In: Cubitt JM, England WA, editors. Geochemistry of reservoirs. London: Geological Society Special Publication; 1995;86:71–85.

Schabron JF, Rovani JF, Sanderson MM, Loveridge JL, Nyadong L, McKenna AM, Marshall AG. Waxphaltene determinator method for automated precipitation and redissolution of wax and asphaltene components. Energy Fuels. 2012;26:2256–68. doi:10.1021/ef300184s.

Senra MJ. Assessing the role of polydispersity and cocrystallization on crystallizing n-alkanes in n-alkane solutions. Ph.D, Dissertation. University of Michigan; 2009.

Senra MJ, Panacharoensawad E, Kraiwattanawong K, Singh P, Fogler HS. Role of n-alkane polydispersity on the crystallization of n-alkanes from solution. Energy Fuels. 2008;22:545–55. doi:10.1021/ef700490k.

Stachowiak C, Viguie J-R, Grolier J-PE, Rogalski M. Effect of n-alkane on asphaltene structuring in petroleum oils. Langmuir. 2005;21:4824–9. doi:10.1021/la047126k.

Strausz OP, Safarik I, Lown EM, Morales-Izquierdo AA. Critique of asphaltene fluorescence decay and depolarization-based claims about molecular weight and molecular architecture. Energy Fuels. 2008;22:1156–66. doi:10.1021/ef700320p.

Thanh NX, Hsieh M, Philp RP. Waxes and asphaltenes in crude oils. Org Geochem. 1999;30:119–32. doi:10.1016/S0146-6380(98)00208-3.

Tissot BP, Welte DH. Petroleum Formation and occurrence: a new approach to oil and gas exploration. Berlin: Springer; 1978.

Yang X, Kilpatrick P. Asphaltenes and waxes do not interact synergistically and coprecipitate in solid organic deposits. Energy Fuels. 2005;19:1360–75. doi:10.1021/ef050022c.

17

Enhanced oil recovery by nonionic surfactants considering micellization, surface, and foaming properties

Achinta Bera[1] · Ajay Mandal[2] · Hadi Belhaj[1] · Tarkeswar Kumar[2]

Abstract Surfactants for enhanced oil recovery are important to study due to their special characteristics like foam generation, lowering interfacial tension between oleic and aqueous phases, and wettability alteration of reservoir rock surfaces. Foam is a good mobility control agent in enhanced oil recovery for improving the mobility ratio. In the present work, the foaming behavior of three nonionic ethoxylated surfactants, namely Tergitol 15-S-7, Tergitol 15-S-9, and Tergitol 15-S-12, was studied experimentally. Among the surfactants, Tergitol 15-S-12 shows the highest foamability. The effect of NaCl concentration and synthetic seawater on foaming behavior of the surfactants was investigated by the test-tube shaking method. The critical micelle concentrations of aqueous solutions of the different nonionic surfactants were measured at 300 K. It was found that the critical micelle concentrations of all surfactants also increased with increasing ethylene oxide number. Dynamic light scattering experiments were performed to investigate the micelle sizes of the surfactants at their respective critical micelle concentrations. Core flooding experiments were carried out in sand packs using the surfactant solutions. It was found that 22% additional oil was recovered in the case of all the surfactants over secondary water flooding. Tergitol 15-S-12 exhibited the maximum additional oil recovery which is more than 26% after water injection.

✉ Achinta Bera
  achintachm@gmail.com

[1] Petroleum Engineering Department, The Petroleum Institute, P.O. Box 2533, Abu Dhabi, United Arab Emirates

[2] Department of Petroleum Engineering, Indian Institute of Technology (Indian School of Mines), Dhanbad, Jharkhand 826004, India

Edited by Yan-Hua Sun

Keywords Enhanced oil recovery · Ethylene oxide number · Foaming properties · Surfactant flooding · Micellization

1 Introduction

Surfactants play a vital role in chemical flooding for their abilities to reduce interfacial tension (IFT) and to alter the wettability of reservoir rock surfaces to facilitate mobilization of trapped oil from natural reservoirs (Bera et al. 2014b; Jiang et al. 2014; Ko et al. 2014). In recent years, surfactant flooding has become one of the most useful tools in enhanced oil recovery (EOR) methods (Elraies et al. 2010; Flaaten et al. 2010; Kumar and Mohanty 2010; Santanna et al. 2009; Southwick et al. 2010) Selection of proper surfactants for EOR is an important issue in surfactant flooding for better economic recovery. Therefore, laboratory characterization of surfactants is one of the major steps before implementing EOR techniques. The main aim of the EOR process is to increase the capillary number by reducing the IFT between water and oil (Babadagli and Boluk 2005). As an effective candidate for wettability alteration, surfactants also help to contribute significantly to the production characteristics of oil during chemical flooding (Zhang et al. 2006). Due to the interrelationship between IFT and capillary number, interfacial phenomena of surfactants are studied in laboratory to screen the surfactants with respect to their activities. It has been reported the surface activities of several surface active agents and their mixtures play an important role in EOR (Babadagli 2005; Babadagli and Boluk 2005; El-Batanoney et al. 1999; Gong et al. 2005; Zhang et al. 2006).

In general, foam is defined as complex, highly nonequilibrium dispersions of gas bubbles in a relatively small amount of liquid generally containing surfactants (Bera et al. 2013). The main mechanism of foam stability and foamability is the absorption of surfactants at the liquid–gas interface. As a result, the intrinsic resistance of the lamella and interfacial area are directly responsible for the foam stability in the sense of thermodynamics (Huang et al. 1986). Among the various applications of foams, their importance in EOR process is very widespread (Aveyard et al. 1994a, b; Exerowa and Kruglyakov 1998; Sadoc and Rivier 1999; Shirtcliffe et al. 2003; Zochhi 1999). For surfactant flooding, foamability and foam stability of surfactant solutions are essential. During the last few years, the effect of solid particles on foam formation and foam stability has been studied intensively. It is just noted here that the effects on foamability and foam stability of the introduction of nanoparticles are becoming of considerable interest in current research for application in oil fields. The common procedures of foam preparation include shaking (Aronson 1986; Alargova et al. 2004; Binks and Tommy 2005; Dickinson et al. 2004; Dippenaar et al. 1978; Dippenaar 1982a, b; Frye and Berg 1989; Garrett 1979; Garrett et al. 2006), bubbling (Frye and Berg 1989; Kulkarni et al. 1977; Johansson and Pugh 1992; Pugh 2005; Vijayaraghavan et al. 2006), bubbling and shaking (Frye and Berg 1989), bubbling and stirring (Aktas et al. 2008; Johansson and Pugh 1992; Schwarz and Grano 2005), and sudden drop in pressure (Dickinson et al. 2004; Kostakis et al. 2006). It was established that the particle hydrophobicity (Aktas et al. 2008; Du et al. 2003; Horozov 2008; Hunter et al. 2008), size (Aktas et al. 2008; Ata 2008; Dippenaar 1982a, b; Frye and Berg 1989; Binks and Tommy 2005), and concentration (Dippenaar 1982a, b; Gonzenbach et al. 2006; Zhang et al. 2008) affect the foam stability. For foam flooding, the foamability and foam stability tests are one of the major laboratory steps for the EOR method.

In EOR, different types of surfactants are used. They have different interfacial properties for improving oil recovery. All the components of the surfactant slug are based on trial and error methods (Rosen 1989). For EOR technique, it is necessary to choose the proper surfactant with the best surface activities, i.e., ability to reduce surface tension or IFT. There has been a considerable focus on surfactant design in EOR methods. In most of the cases, anionic surfactants are used because these surfactants have several applications like emulsifiers, foam generating agents, detergents, and effective wetting agents. Depending on these adequate properties as well as low cost of anionic surfactants, they are considered as potential EOR candidates in actual reservoir cases.

In this work, foamability and foam stability of all the nonionic surfactants have been studied by the standard shaking method to understand their efficiencies as EOR candidates in different brine solutions and synthetic seawater (SSW). The critical micelle concentrations (CMCs)

of the nonionic surfactants were measured at a temperature of 300 K, and a relationship between ethylene oxide numbers (EONs) and CMCs of the surfactants used has been established from the results. Dynamic light scattering (DLS) experiments have also been performed with the surfactant solutions at their corresponding CMCs to study the sizes of the micelles of the surfactants. Core flooding experiments have been performed with the surfactant solutions. A comparison was made of the efficiencies of the surfactants to recover additional oil.

2 Experimental

2.1 Materials used

In this work, the surfactants Tergitol 15-S-7, Tergitol 15-S-9, and Tergitol 15-S-12 were purchased from Sigma-Aldrich, Germany, and the chemical name of these surfactants is secondary alcohol ethoxylate. Their general structural formula is: $C_{12-14}H_{25-29}O[CH_2CH_2O]_xH$. Table 1 shows the properties of all surfactants. The purities of the surfactants are 99.9%. The total acid number, gravity, and viscosity of the oil were found to be 0.038 mg KOH/g, 38.86°API, and 5.12 Pa s at 45 °C, respectively. The SSW was prepared by mixing different salts (NaCl, 23.54 g/L; KCl, 0.675 g/L; CaCl$_2$, 0.115 g/L; MgCl$_2$, 5.84 g/L; Na$_2$SO$_4$, 3.84 g/L; SrCl$_2$, 0.024 g/L; KBr, 0.110 g/L; NaF, 0.090 g/L; NaHCO$_3$, 0.200 g/L; H$_3$BO$_3$, 0.030 g/L) in distilled water. All these chemicals were procured from Merck Specialties Pvt. Ltd., Mumbai, India, and all of the chemicals are more than 98% pure. All the solutions and different concentrated brines were prepared by using reverse osmosis water from a Millipore water system (Millipore SA, 67120 Molsheim, France).

2.2 Measurement of surface tension and CMC

Surface tensions of surfactant solutions were measured with a programmable tensiometer (Kruss GmbH, Germany, model: K20 EasyDyne) using the Du Noüy ring method at 300 K. Special attention was paid to the cleaning of the platinum ring. The ring was cleaned with acetone and then flame dried for each measurement. The standard deviation was ±0.1 mN/m. For determination of CMCs of the surfactants, the concentration versus surface tension graph was plotted and the concentrations at the inflexion points of the curves are considered as CMCs of the surfactants.

2.3 DLS study of the surfactants at their CMCs

The surfactant solutions were prepared at their corresponding CMC values. The sizes of micelles were

Table 1 Physicochemical properties of the surfactants employed in the work

Trade name	Linear formula	EON	Molecular weight	Hydrophilic-lipophilic balance (HLB) value	Surface excess $\Gamma \times 10^{10}$, mol/cm^2	Molecular cross-sectional area A, Å^2
Tergitol 15-S-7	$C_{12-14}H_{25-29}O[CH_2CH_2O]_7H$	7	515	12.1	4.55	36.5
Tergitol 15-S-9	$C_{12-14}H_{25-29}O[CH_2CH_2O]_9H$	9	584	13.3	4.40	37.7
Tergitol 15-S-12	$C_{12-14}H_{25-29}O[CH_2CH_2O]_{12}H$	12	738	14.7	3.74	44.4

measured by a laser diffraction method using a Zetasizer version 6.00 (Malvern Instruments Ltd., Worcestershire, UK) at 300 K. The size distribution of micelles was obtained by the inbuilt software of the instrument. The software uses a reflective index (RI) of 1.465 (SBO) and a dispersant RI of 1.33 (water) during the measurement. Drops of the surfactant solution were introduced into the sample-containing cuvette, and the optimum volume was indicated by the instrument.

2.4 Foamability and foam stability tests

Foamability and foam stability experiments were conducted in a graduated measuring cylinder with 0.5wt% surfactant solution. We used bottle shaking tests (ASTM D-3601) to evaluate the foaming capacity of different surfactants in the presence of salts, where the volume of gas (air) is fixed in the container (centrifuge tube). Bottle shaking test (ASTM D-3601) for foam generation study is a standard method, which is reported by many authors (Schramm and Wassmuth 1994; Tamura and Kaneko 2004; Nadkarni 2007; Moayedi et al. 2014). For foam study, a constant volume of the aqueous sample in a 10-mL graduated centrifuge tube was shaken manually at a fixed frequency for fixed time (15 min for each case) and then left untouched on a flat surface (make necessary corrections for volume). The foam height and liquid holdup of the generated foam for each respective sample were recorded over time by visual observation. A plot of time versus foam volume indicates the foam stability. Foam stability was measured assuming that mechanical vibrations are absent. The foamability of the surfactant solution was measured by taking the initially produced foam volume after constant time shaking with different brine concentrations and SSW at 300 K.

2.5 Apparatus and methods for surfactant flooding for oil recovery

A schematic of the experimental setup for surfactant flooding is shown in Fig. 1. The whole experimental setup contains a core holder for the sand pack, different cylinders for surfactant solution and crude oil, a pump from ISCO,

and an effluent collecting cylinder. The core holder was fully filled with 60–70 mesh sand, and while filling the core holder brine was used to saturate the sand pack for measuring its porosity. Permeability was measured by brine flooding through the sand pack. After that the sand pack was flooded with the crude oil until water production reaches about 1% at 400 psig. The initial water saturation of the core was determined by mass balance. After water flooding, a ~ 0.6 pore volume (PV) surfactant slug was injected followed by ~ 1.25 PV water injection as chase water flooding. The above-mentioned method has also been described in our previous work in details (Bera et al. 2014a).

3 Results and discussion

3.1 CMCs and micelle sizes of the surfactants

CMC measurement of surfactant is very important for foamability and foam stability studies as well as their applications in further preliminary screening of surfactants. Before selecting a surfactant for application in oil fields, it is necessary to characterize the surfactant initially. It is well known that surfactants start to undergo micelle formation at CMCs (Hoff et al. 2001). Figure 2 shows the CMC values of the surfactants studied in this work. The present work shows that CMCs of the surfactants (Tergitol 15-S-7, Tergitol 15-S-9, and Tergitol 15-S-12) increase with increases in EON of the surfactants which shows a strong similarity to another study (Wu et al. 2006). It is found that the CMC values are 0.0031wt%, 0.0042wt%, and 0.0051wt% for the surfactants of EON of 7, 9, and 12, respectively. The data clearly illustrate the relationship between EON and CMC values of the surfactants used. For ethoxylated nonionic surfactants, the steric hindrances between head groups of the surfactants can be expected to increase with increasing EON. Subsequently with an increase in EON, the head group parameter also increases within the head group. Therefore, low EON may reduce the head group parameter as well as the area per molecule; therefore, the packing parameter is large which results in bilayer aggregates (lamella). For high EON, the head group

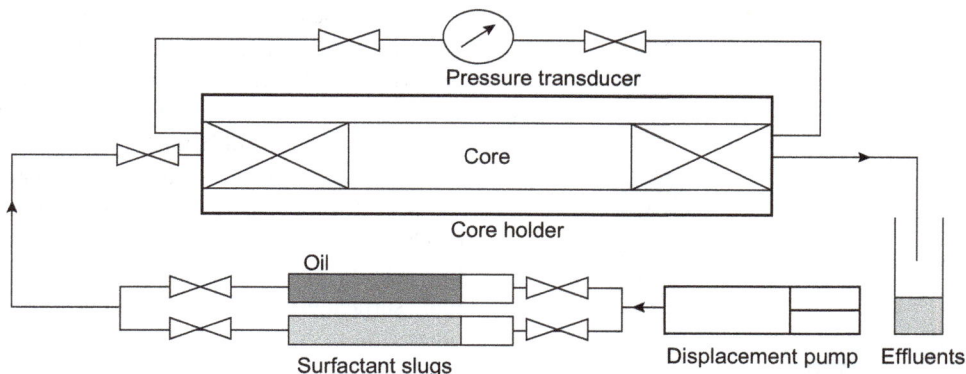

Fig. 1 Schematic diagram of experimental setup for surfactant flooding in a sand pack

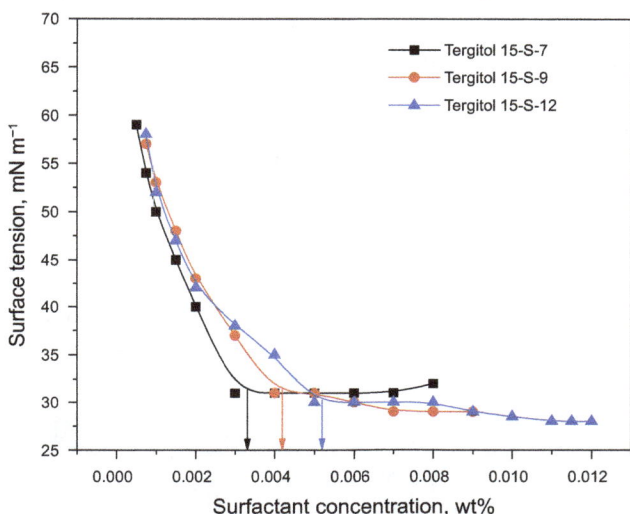

Fig. 2 Surface tensions of the nonionic surfactants used at different concentrations and CMC determination at 300 K

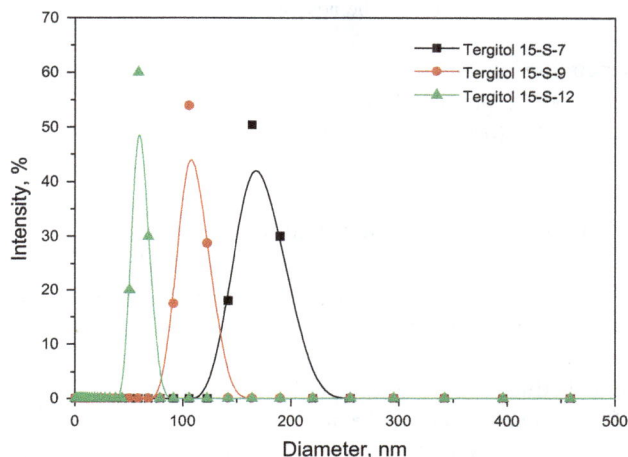

Fig. 3 Size distribution of the micelles of the surfactants at their CMCs

parameter and the area per molecule increase, but the packing parameter decreases; thus, it is possible to form cylindrical micelles. With increasing EON, the packing parameter decreases and spherical micelles may be formed by decreasing the aggregation number. Therefore, the increases in head group parameter and area per molecule for nonionic surfactants with increasing EON give rise to an increase in CMC as the head group size increases. Figure 2 clearly shows that the surfactants are very active in reducing the surface tension of the liquid–air system. Among all the nonionic surfactants, Tergitol 15-S-12 shows the lowest surface tension value at its CMC. The surface tension values of the surfactant solutions at their CMCs are 31, 29, and 28 mN/m for Tergitol 15-S-7, Tergitol 15-S-9, and Tergitol 15-S-12, respectively.

It is important to consider the micelle sizes of surfactants in oil recovery processes. Figure 3 shows the size distributions of the micelles of the prepared surfactant solutions at their CMC values. The z-average diameter can be calculated in dynamic light scattering as follows (Zheng et al. 2016; Bera et al. 2012b):

$$D_z = \sum S_i \bigg/ \sum (S_i/D_i) \tag{1}$$

where S_i is the scattered intensity from particle i and D_i is the diameter of particle i.

The sizes of the micelles are shown in Fig. 4. In the present study, the typical micelle sizes of the surfactants range from 0.5 μm to 10 μm which are also desired for core flooding experiments when compared to the pore size distribution of the sand grains in the sand pack model. Figure 4 shows that the micelle size of Tergitol 15-S-12 is higher than that of the other two surfactants. This is because the ethylene oxide chain length of Tergitol 15-S-12 is greater than the other two surfactants. From this study, it can be also possible to establish a relationship between the micelle sizes and CMCs of the surfactants. The results show that with an increase in CMC of the surfactant the micelle size also increases. Therefore,

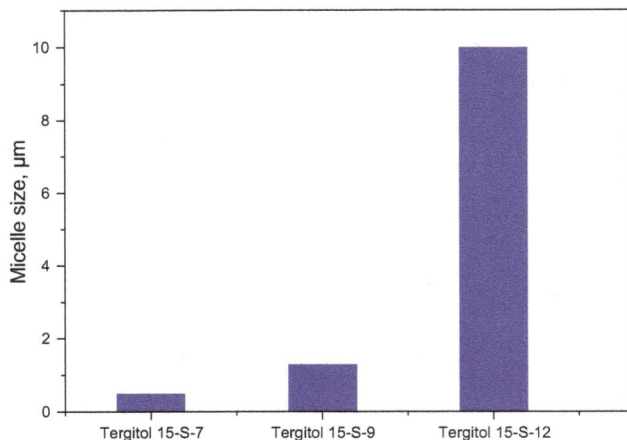

Fig. 4 Sizes of micelles of the nonionic surfactants

depending on the CMCs of the surfactants their micelle sizes can be predicted and it is possible to screen the surfactants for further investigation for implementing in EOR methods.

3.2 Foamability and foam stability

The foamability test is a special test of surfactants for their selection in EOR by foam injection. In the present work, foaming properties of nonionic surfactants were tested in pure distilled water, NaCl solutions (2wt% and 4wt%), and SSW. Figure 5 shows the foamability at 0.5 wt% of different surfactants. Results indicate that Tergitol 15-S-12 generated the highest amount of foam, i.e., higher foamability than the other nonionic surfactants used. The higher foaming properties of Tergitol 15-S-12 may be explained on the basis of higher EON. Nonionic surfactants like Tergitol 15-S-12, Tergitol 15-S-9, and Tergitol 15-S-7 can form

good foam below their cloud points. Nonionic surfactants can only produce foam in good amounts when they are able to form a well-packed adsorption monolayer at the air–water interface. In the present case, the foaming is not high enough for all the surfactants and the reason is the instability of the foam films. Again, the higher foamability of Tergitol 15-S-12 compared to other nonionic surfactants can be explained on the basis of stability of monolayers. As the EON of Tergitol 15-S-12 is higher than the other nonionic surfactants, a more stable monolayer is formed and for other cases a bilayer might be favorable. As a result, Tergitol 15-S-12 shows higher foamability than the others (Patrick et al. 1997; Mittal and Shah 2002). Figure 6 shows the foaming of the Tergitol surfactants at their corresponding CMC values after several hours of equilibrium.

Adsorption of surfactants at the air–water interface plays an important role in the formation of foam and its stability. The Gibbs surface adsorption equation was used to calculate the surface excess for all the surfactants as follows (Amaral et al. 2008; Azira et al. 2008; Tan et al. 2005; Wang and Chen 2006):

$$\Gamma = -\frac{1}{RT}\left(\frac{d\gamma}{d\ln C}\right) \tag{2}$$

where Γ is the surface excess, mmol/cm^2; R is the universal gas constant, 8.314 J mol^{-1} K^{-1}; γ is the surface

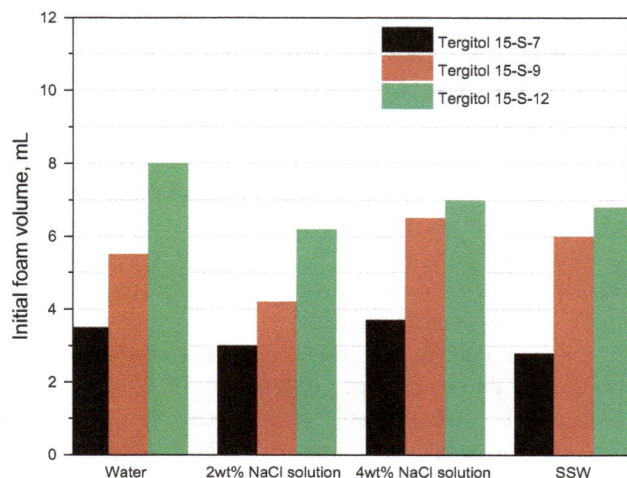

Fig. 5 Initial amount of produced foam for different surfactant solutions in distilled water, 2wt% NaCl, 4wt% NaCl, and SSW

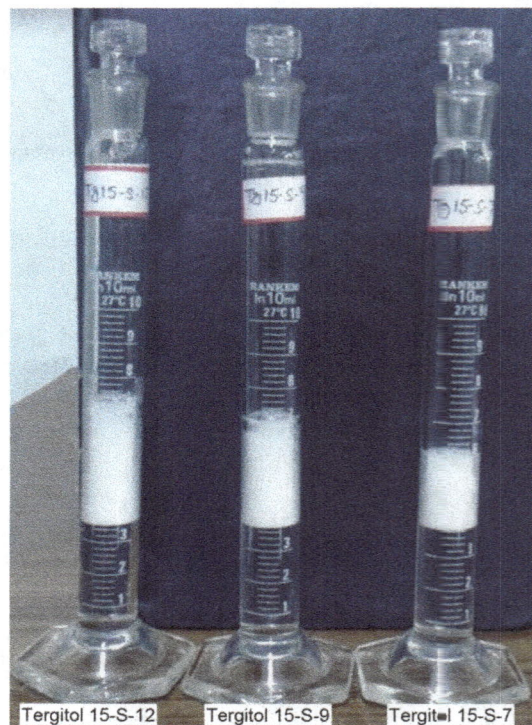

Fig. 6 Photograph of the foaming of the Tergitol surfactants in distilled water after several hours of equilibrium (from left to right: Tergitol 15-S-12, Tergitol 15-S-9, and Tergitol 15-S-7)

tension, mN m^{-1}; T is the thermodynamic temperature, K; and C represents the concentration of surfactant, mmol L^{-1} at corresponding CMC value.

The slope of the plot of logarithmic concentration of surfactant versus surface tension gives the value of $\left(\dfrac{d\gamma}{d\ln C}\right)$. On the other hand, the molecular cross-sectional area of the polar head group (A) was calculated from the following equation:

$$A = \frac{1}{\Gamma N_A} \quad (3)$$

where N_A indicates Avogadro's number, 6.023×10^{23} mol^{-1}; A is the molecular cross-sectional area of the polar head group, Å^2.

Table 1 shows the calculated values of the surface excess (Γ) and molecular cross-sectional area for all the nonionic surfactants from Eqs. (2) and (3). In our previous work, we also determined the surface excess and molecular cross-sectional area for several cationic, nonionic, and anionic surfactants (Bera et al. 2013). A significant result was found in the case of nonionic surfactant systems that as the EON of the surfactant increases the adsorption also increases. In the case of Tergitol 15-S-12, the surfactant has higher EON value and its Γ value is greater than that of the other surfactants (Tergitol 15-S-7 and Tergitol 15-S-9). It is also found from Table 1 that as Γ values increase the values of the molecular cross section of the polar head group (A) of the nonionic surfactants decrease accordingly. Surface excess is the number of moles of surfactant per unit area at the liquid–air interface. In the solution phase, most of the surfactants form a vertically monolayer just below the CMC values. The area/molecule is usually determined by the cross-sectional area of the head group. For nonionic surfactants with ethylene oxide, the area per molecule is generally much larger than other types of surfactants. Therefore, with an increase in the ethylene oxide number in the head group the area per molecule also increases. As Tergitol 15-S-12 has the highest ethylene oxide number; therefore, it has the lowest surface excess and the highest area per molecule.

The foam stability of a surfactant solution can be defined as the change in foam volume, i.e., the volume of liquid drained from the foam, per unit time (Bera et al. 2013). The foam stability of the surfactants in distilled water, 2wt% NaCl solution, 4wt% NaCl solution, and SSW is depicted in Figs. 7, 8, 9, and 10, respectively. The common way to determine the foam stability is to measure the foam volume after production of certain amount of foam with different time intervals. The foam structure is related to time which quantifies the foam stability of a certain surfactant (Kroschwitz 1994). Several factors like drainage, disproportionation, and coalescence influence the

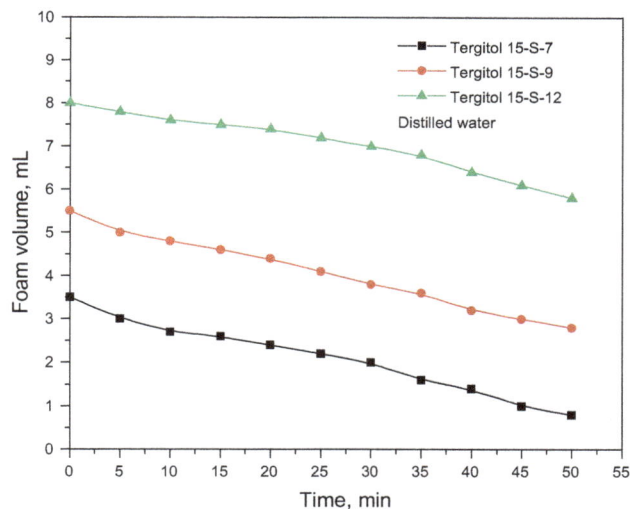

Fig. 7 Foam volume versus time for different surfactants in distilled water

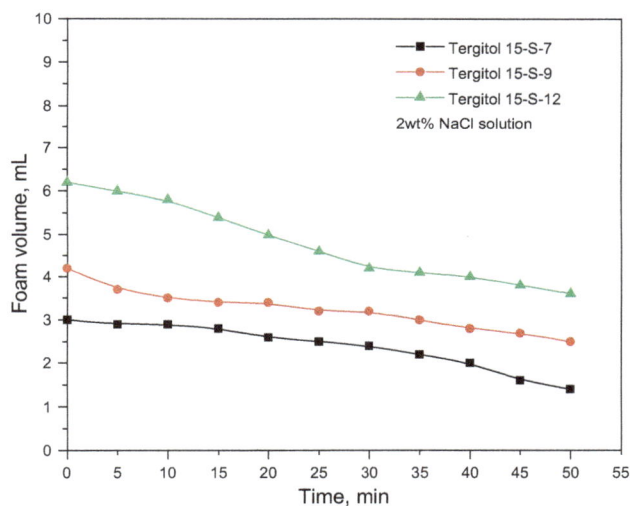

Fig. 8 Foam volume versus time for different surfactants in 2wt% NaCl solution

foam stability. Foam stability depends on dispersed particles of colloid in the continuous phase. These dispersed particles control the liquid drainage from the foam which significantly accounts for the high stability of a foam. Mainly these particles reduce the liquid drainage rate and increase the surface viscosity of the continuous phase. These colloid dispersions greatly affect the foam stability (Kaptay 2004; Sethumadhavan et al. 2001).

In all solutions, Tergitol 15-S-12 shows the highest foam stability. Foam stabilization is mainly caused by van der Waals forces between the molecules in the foam, electrical double layers created by dipolar surfactants (Zwitterionic surfactant) and the Marangoni effect, which acts as a restoring force to the lamellae. Other important factors that

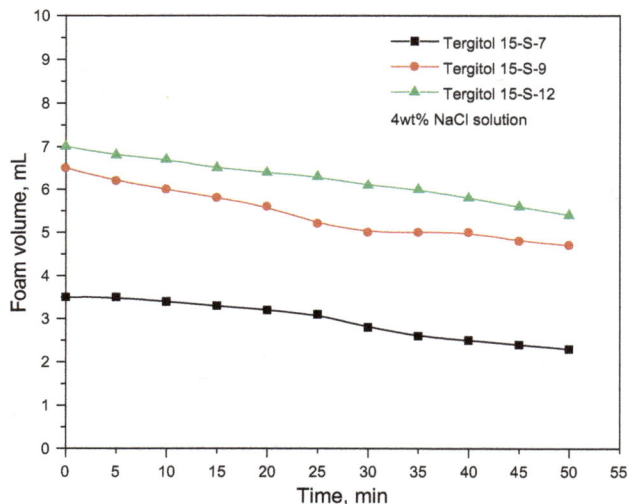

Fig. 9 Foam volume versus time for different surfactants in 4wt% NaCl solution

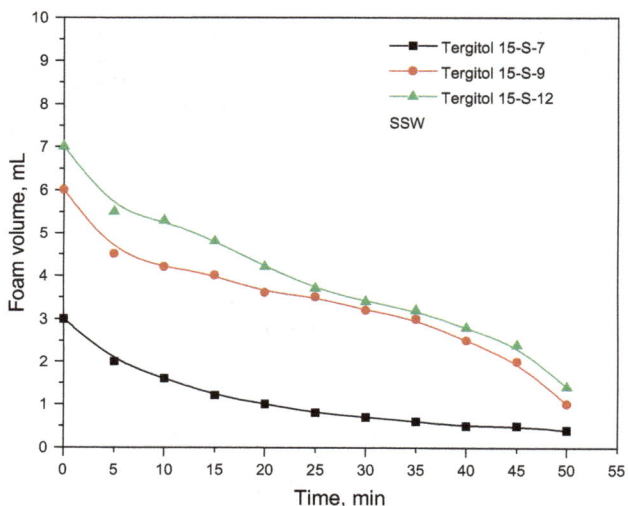

Fig. 10 Foam volume versus time for different surfactants in synthetic seawater (SSW)

control the foam stability are surface viscosity and film elasticity. Of all the surfactants used in this study, irrespective of other influencing factors, the main controlling parameter is the surface viscosity. The surface viscosity of Tergitol 15-S-12 is high due to the presence of the high number of ethylene oxide units in the head group. The viscous nature of the surfactant solution results in a slow drainage of liquid through the bubble interfaces. As a result, the foam produced by the surfactant Tergitol 15-S-12 is also stable and shows higher stability than the other surfactants. Different salts in SSW affect the foam stability; hence, foam stability is low in the case of SSW. Some researchers suggested two different regimes of foam decay.

One is during the initial stage immediately after foam formation, and the other is the comparatively slow drainage (Lunkenheimer and Malysa 2003; Carey and Stubenrauch 2009). The produced thin films in the air–water interface indicate the foam stability. Foam stability and quality of foam are very important in oil recovery to increase the sweep efficiency. Therefore, laboratory study of foam stability is a crucial step for foam flooding for EOR.

3.3 Surfactant flooding and oil recovery

Surfactants are considered to be important chemicals for tertiary recovery by reducing IFT and changing wettability of rock surfaces. In the present work, surfactant solutions were injected into the sand pack after water flooding. Due to the high porosity of the sand pack ($\sim 37\%$), water flooding is able to produce high recovery ($\sim 52\%$). Figure 11 shows the performances of surfactants in oil recovery after injection of different pore volumes into the sand pack and the variations of oil and water cuts. Figure 11 shows that after injection of 1.15 PV water the oil cut decreases to 5% and the water cut goes to 95%. At this moment, the surfactant injection is started and it is found that the water cut declines gradually and the oil cut again increases to produce the highest recovery. The enhanced recovery can be explained on the basis of IFT reduction or increase in capillary number and consequent mobilization of the oil trapped inside pore throats. As a result, the oil saturation increases due to coalescence of oil drops and retrapping of oil drops mobilized from the oil bank by the surfactant slug. Figure 11 also shows that the additional of recovery by Tergitol 15-S-12 is higher than others due to significant interfacial surface active properties of the surfactant. On the other hand, due to the presence of 12

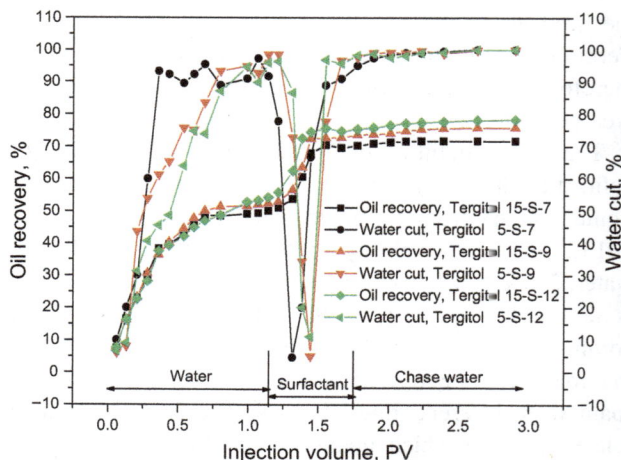

Fig. 11 Oil recoveries by Tergitol surfactants in sand pack flooding experiments

Table 2 Recovery of oil by surfactant flooding of sand packs for three different systems

Surfactant	Porosity, %	Permeability k, Darcy		Design of surfactant slug for flooding	Recovery of oil after water flooding at 95% water cut, %OOIP	Additional recovery, %OOIP	Saturation, %		
		k_w ($S_w = 1$)	k_o (S_{wi})				S_{wi}	S_{oi}	S_{or}
Tergitol 15-S-7	37.3	0.322	0.002	0.6 PV surfactant slug + chase water	51.6	22.1	18.5	81.5	23.0
Tergitol 15-S-9	37.3	0.303	0.002	0.6 PV surfactant slug + chase water	51.5	24.6	17.3	82.7	23.8
Tergitol 15-S-12	37.7	0.296	0.002	0.6 PV surfactant slug + chase water	51.8	26.9	19.5	80.5	21.0

ethylene oxide groups in Tergitol 15-S-12, the micelle size is larger than the other two surfactants. Therefore, the micelles formed by surfactant Tergitol 15-S-12 are more stable than those of the other two favored by the entropy and enthalpy of micellization. Therefore, during surfactant flooding for Tergitol 15-S-9 and Tergitol 15-S-7 it is difficult to form micelles in the system and they cannot reduce the IFT as required. The oil trapped in the porous matrix cannot be recovered easily, and residual oil saturation remains high. As Tergitol 15-S-12 can produce a microemulsion easily, it is also able to produce ultra-low IFT and shows higher additional oil recovery as described in the oil recovery section. It is also worth mentioning at this point that the foaming property of Tergitol 15-S-12 is better than other surfactants and as a result Tergitol 15-S-12 shows higher oil recovery than the other surfactants. Therefore, a bridge can be established from all the studies of surfactant solutions and their influences on oil recovery to provide an overview of the importance and significance of this work. The additional recovery was 22.3, 24.6, and 26.9% for Tergitol 15-S-7, Tergitol 15-S-9, and Tergitol 15-S-12, respectively. The detail of the surfactant flooding results with properties of the sand pack is given in Table 2.

Another important mechanism that plays a vital role in oil recovery is wettability alteration by surfactants (Bera et al. 2012a, 2015). Oil–water relative permeabilities are highly influenced by wettability alteration during surfactant flooding through an oil-saturated sand pack. Surfactant flooding enhances the relative permeability of oil as the oil-wet state of the sand pack is altered to a water-wet state. As a result, flows of oil and water through the sand pack are changed accordingly. Water finds the path to move forward along the pore wall with water-wet state and reaches the center of the pores in the oil-wet state during the surfactant flooding. Therefore, the capillary forces act along the direction of water flooding for a water-wet surface and the opposite for an oil-wet surface. This phenomenon also reflects the increase in displacement efficiency by surfactant flooding which enhances the oil recovery after water flooding. As the sand pack is water-wet in general, injection of the surfactant solution can recover more oil from the oil-saturated core during surfactant flooding by alteration of wettability toward a water-wet state. So surfactant flooding can reduce the residual oil saturation. Therefore, after water flooding in an oil field surfactant flooding can significantly recover the residual oil.

Apart from the wettability alteration, in situ microemulsion formation during surfactant flooding causes increased oil recovery (Glinsmann 1979; Jeirani et al. 2014). During surfactant flooding, a multiphase microemulsion system is formed by mixing the surfactant slug with the remaining oil in the reservoir after water flooding. As the microemulsion is formed, IFT between the oil and surfactant slug reduces to ultra-low one which helps to increase the capillary number followed by enhanced oil recovery. The in situ generated microemulsion is assumed to disseminate toward the producing well by sweeping a significant amount of oil in the reservoir. The process depends on the compositions present in the surfactant solution and overall compositions of the formed microemulsion system.

4 Conclusions

Present studies of micellization and sizes of micelles, foamability and foam stability of different nonionic surfactants provide a sound background for selection and application of surfactants for EOR. Based on the results of the work, the following conclusions can be drawn:

1. Tergitol 15-S-12 shows maximum foamability in distilled water, NaCl solutions of different concentrations, and SSW.
2. The present study provides useful information for selection of surfactant systems for EOR. When the molecular cross section of the polar head group (A) decreases, the surface excess (Γ) value increases for all the nonionic surfactants.
3. DLS study shows that with an increase in EON of the surfactants the micelle sizes of the surfactants also increase at their corresponding CMCs.

4. The additional recovery is more than 22%. The surfactant with the highest EON value, Tergitol 15-S-12, shows the greatest efficiency to recover more than 26% additional oil compared to straight water flooding.

Acknowledgements The first author (AB) thanks the Petroleum Institute, Abu Dhabi, for providing the fellowship for his postdoctoral research. The second author (AM) gratefully acknowledges the financial support provided by Council for Scientific and Industrial Research [22(0649)/13/EMR-II], New Delhi, to the Department of Petroleum Engineering, Indian Institute of Technology (Indian School of Mines), Dhanbad, India.

References

Aktas Z, Cilliers JJ, Banford AW. Dynamic froth stability: particle size, airflow rate and conditioning time effects. Int J Miner Process. 2008;87(1–2):65–71. doi:10.1016/j.minpro.2008.02.001.

Alargova RG, Warhadpande DS, Paunov VN, et al. Foam superstabilization by polymer microrods. Langmuir. 2004;20(24):10371–4. doi:10.1021/la048647a.

Amaral MH, Neves JD, Oliveria AZ, et al. Foamability of detergent solutions prepared with different types of surfactants and waters. J Surfactants Deterg. 2008;11:275–8. doi:10.1007/s11743-008-1088-0.

Aronson MP. Influence of hydrophobic particles on the foaming of aqueous surfactant solutions. Langmuir. 1986;2(5):653–9. doi:10.1021/la00071a023.

Ata S. Coalescence of bubbles covered by particles. Langmuir. 2008;24(12):6085–91. doi:10.1021/la800466x.

Aveyard R, Binks BP, Fletcher PDI, et al. Aspects of aqueous foam stability in the presence of hydrocarbon oils and solid particles. Adv Colloid Interface Sci. 1994a;48:93–120. doi:10.1016/0001-8686(94)80005-7.

Aveyard R, Binks BP, Fletcher PDI, et al. Contact angles in relation to the effects of solids on film and foam stability. J Dispers Sci Technol. 1994b;15(3):251–71. doi:10.1080/01932699408943557.

Azira H, Tazerouti A, Canselier JP. Study of foaming properties and effect of the isomeric distribution of some anionic surfactants. J Surfactants Deterg. 2008;11(4):279–86. doi:10.1007/s11743-008-1093-3.

Babadagli T. Analysis of oil recovery by spontaneous imbibitions of surfactant solution. Oil Gas Sci Technol Rev. 2005;60(4):697–710. doi:10.2516/ogst:2005049.

Babadagli T, Boluk Y. Oil recovery performances of surfactant solutions by capillary imbibitions. J Colloids Interface Sci. 2005;282(1):162–75. doi:10.1016/j.jcis.2004.08.149.

Bera A, Kissmathulla S, Ojha K, et al. Mechanistic study of wettability alteration of quartz surface induced by nonionic surfactants and interaction between crude oil and quartz in presence of sodium chloride salt. Energy Fuels. 2012a;26(6):3634–43. doi:10.1021/ef300472k.

Bera A, Kumar S, Mandal A. Temperature-dependent phase behavior, particle size, and conductivity of middle-phase microemulsions stabilized by ethoxylated nonionic surfactants. J Chem Eng Data. 2012b;57(12):3617–23. doi:10.1021/je300845q.

Bera A, Kumar T, Ojha K, et al. Screening of microemulsion properties for application in enhanced oil recovery. Fuel. 2014a;121:198–207. doi:10.1016/j.fuel.2013.12.051.

Bera A, Mandal A, Guha BB. Effect of synergism of surfactant and salt mixture on interfacial tension reduction between crude oil and water in enhanced oil recovery. J Chem Eng Data. 2014b;59(1):89–96. doi:10.1021/je400850c.

Bera A, Mandal A, Kumar T. Effect of rock-crude oil-fluid interactions on wettability alteration of oil-wet sandstone in presence of surfactants. Pet Sci Technol. 2015;33(5):542–9. doi:10.1080/10916466.2014.998768.

Bera A, Ojha K, Mandal A. Synergistic effect of mixed surfactant systems on foam behavior and surface tension. J Surfactants Deterg. 2013;16(4):621–30. doi:10.1007/s11743-012-1422-4.

Binks BP, Tommy SH. Aqueous foams stabilised solely by silica nanoparticles. Angew Chem Int Ed Engl. 2005;44(24):3722–5. doi:10.1002/anie.200462470.

Carey E, Stubenrauch C. Properties of aqueous foams stabilized by dodecyl trimethyl ammonium bromide. J Colloid Interface Sci. 2009;333(2):619–27. doi:10.1016/j.jcis.2009.02.038.

Dickinson E, Ettelaie R, Kostakis T, et al. Factors controlling the formation and stability of air bubbles stabilized by partially hydrophobic silica nanoparticles. Langmuir. 2004;20(20):8517–25. doi:10.1021/la048913k.

Dippenaar A. The destabilization of froth by solids. I. The mechanism of film rupture. Int J Miner Process. 1982a;9(1):1–14. doi:10.1016/0301-7516(82)90002-3.

Dippenaar A. The destabilization of froth by solids. II. The rate-determining step. Int J Miner Process. 1982b;9(1) 15–22. doi:10.1016/0301-7516(82)90003-5.

Dippenaar A, Harris PJ, Nicol MJ. The effect of particles on the stability of flotation froths. Rep. Natl. Inst. Metall South Africa, 34 pp, 1978.

Du Z, Bilbao-Montoy MP, Binks BP, et al. Outstanding stability of particle-stabilized bubbles. Langmuir. 2003;19(8):3106–8. doi:10.1021/la034042n.

El-Batanoney M, Abdel-Monghny T, Ramazi M. The effect of mixed surfactants on enhancing oil recovery. J Surfactants Deterg. 1999;2(2):201–5. doi:10.1007/s11743-999-0074-7.

Elraies KA, Tan IM, Awang M, et al. A new approach to low-cost, high performance chemical flooding system. In: SPE production and operation conference and exhibition, June 8–10, Tunis, Tunisia; 2010. doi:10.2118/133004-MS.

Exerowa DR, Kruglyakov PM. Foams and foam films Amsterdam: Elsevier; 1998.

Flaaten AK, Nguyen QP, Zhang J, et al. Alkaline/surfactant/polymer chemical flooding without the need for soft water. SPE J. 2010;15(1):184–96. doi:10.2118/116754-PA.

Frye GC, Berg JC. Antifoam action by solid particles. J Colloid Interface Sci. 1989;127(1):222–38. doi:10.1016/0021-9797(89)90023-4.

Garrett PR. The effect of polytetrafluoroethylene particles on the foamability of aqueous surfactant solutions. J Colloid Interface Sci. 1979;69(1):107–21. doi:10.1016/0021-9797(79)90085-7.

Garrett PR, Wicks SP, Fowles E. The effect of high volume fractions of latex particles on foaming and antifoam action in surfactant solutions. Colloids Surf A. 2006;282–283:307–28. doi:10.1016/j.colsurfa.2006.01.054.

Glinsmann GR. Surfactant flooding with microemulsions formed in situ—effect of oil characteristics. In: SPE annual technical conference and exhibition, September 23-26, Las Vegas, Nevada, 1979. doi:10.2118/8326-MS.

Gong Y, Li Z, An J. The properties of sodium naphthalene sulfonate in lowering interfacial tension and its possibility of application in EOR. J Dispersion Sci Technol. 2005;26(4):503–7. doi:10.1081/DIS-200054612.

Gonzenbach UT, Studart AR, Tervoort E, et al. Ultrastable particle-stabilized foams. Angew Chem Int Ed. 2006;45(2):3526–30. doi:10.1002/anie.200503676.

Hoff E, Nystrom B, Lindman B. Polymer–surfactant interactions in dilute mixtures of a nonionic cellulose derivative and an anionic surfactant. Langmuir. 2001;17(1):28–34. doi:10.1021/la001175p.

Horozov TS. Foams and foam films stabilized by solid particles. Curr Opin Colloid Interface Sci. 2008;13(3):134–40. doi:10.1016/j.cocis.2007.11.009.

Huang D, Nimolvo A, Wasan D. Foams: basic properties with applications to porous media. Langmuir. 1986;2(5):672–7. doi:10.1021/la00071a027.

Hunter TN, Pugh RJ, Franks GV, et al. The role of particles in stabilizing foams and emulsions. Adv Colloid Interface Sci. 2008;137(2):57–81. doi:10.1016/j.cis.2007.07.007.

Jeirani Z, Mohamed Jan B, Si Ali B, et al. In-situ prepared microemulsion-polymer flooding in enhanced oil recovery—a review. J Pet Sci Technol. 2014;32(2):240–51. doi:10.1080/10916466.2011.588644.

Jiang P, Li N, Ge J, et al. Efficiency of a sulfobetaine-type surfactant on lowering IFT at crude oil-formation water interface. Colloids Surf A. 2014;443:141–8. doi:10.1016/j.colsurfa.2013.10.061.

Johansson G, Pugh RJ. The influence of particle size and hydrophobicity on the stability of mineralized froths. Int J Miner Process. 1992;34(1–2):1–21. doi:10.1016/0301-7516(92)90012-L.

Kaptay G. Interfacial criteria for stabilization of liquid foams by solid particles. Colloids Surf A. 2004;230(1–3):67–80. doi:10.1016/j.colsurfa.2003.09.016.

Ko KM, Chon BH, Jang SB, et al. Surfactant flooding characteristics of dodecyl alkyl sulfate for enhanced oil recovery. J Ind Eng Chem. 2014;20(1):228–33. doi:10.1016/j.jiec.2013.03.043.

Kostakis T, Ettelaie R, Murray BS. The effect of high salt concentrations on the stabilization of bubbles by silica particles. Langmuir. 2006;22(3):1273–80. doi:10.1021/la052193f.

Kroschwitz JI. Kirk-Othmer encyclopedia of chemical technology. 4th ed. New York: Wiley; 1994.

Kulkarni RD, Goddard ED, Kanner B. Mechanism of antifoaming: role of filler particle. Ind Eng Chem Fundam. 1977;16(4):472–4. doi:10.1021/i160064a014.

Kumar R, Mohanty KK. ASP flooding of viscous oils. In: SPE annual technical conference and exhibition, September 19–22, Florence, Italy; 2010. doi:10.2118/-135265-MS.

Lunkenheimer K, Malysa K. A simple automated method of quantitative characterization of foam behaviour. Polymer Int. 2003;52(4):536–54. doi:10.1002/pi.1105.

Mittal KL, Shah DO, editors. Adsorption and aggregation of surfactants in solution. Basel: CRC Press; 2002.

Moayedi M, James LA, Mahmoodi M. An experimental study on optimization of SAG process utilizing nonionic surfactants and sodium lignosulfonate. In: International symposium of the society of core analysts, 8–11 September Avignon, France, 2014, SCA2014-087.

Nadkarni RAK. Guide to ASTM test methods for the analysis of petroleum products and lubricants. ASTM manual series; no. mnl44-2nd, West Conshohocken, PA, p. 111, 2007.

Patrick HN, Warr GG, Manne S, Aksay IA. Self-assembly structures of nonionic surfactants at graphite/solution interfaces. Langmuir. 1997;13(16):4349–56. doi:10.1021/la9702547.

Pugh RJ. Experimental techniques for studying the structure of foams and froths. Adv Colloid Interface Sci. 2005;114–115:239–51. doi:10.1016/j.cis.2004.08.005.

Rosen MJ. Selection of surfactant pairs for optimization of interfacial properties. J Am Oil Chem Soc. 1989;66(12):1840–3. doi:10.1007/BF02660759

Sadoc JF, Rivier N, editors. Foams and emulsions NATO ASI series E, vol. 354. Dordrecht: Kluwer; 1999.

Santanna VC, Curbelo FDS, Castro Dantas TN, et al. Microemulsion flooding for enhanced oil recovery. J Pet Sci Eng. 2009;66(3–4):117–20. doi:10.1016/j.petrol.2009.01.009.

Schramm LL, Wassmuth F. Foams: basic principles. foams: fundamentals and applications in the petroleum industry. In: Schramm LL, editor. Advances in chemistry series 242. Washington, DC: American Chemical Society; 1994. p. 3–45.

Schwarz S, Grano S. Effect of particle hydrophobicity on particle and water transport across a flotation froth. Colloids Surf A. 2005;256(2–3):157–64. doi:10.1016/j.colsurfa.2005.01.010.

Sethumadhavan GN, Nikolov AD, Wasan DT. Stability of liquid films containing monodisperse colloidal particles. J Colloid Interface Sci. 2001;240(1):105–12. doi:10.1006/jcis.2001.7628.

Shirtcliffe NJ, McHale G, Newton MI, et al. Intrinsically superhydrophobic organosilica sol–gel foams. Langmuir. 2003;19(13):5626–31. doi:10.1021/la034204f.

Southwick JG, Svec Y, Chilek G, et al. The effect of live crude on alkaline–surfactant–polymer formulations: implications for final formulation design. In: SPE annual technical conference and exhibition, September 19-22, Florence, Italy; 2010. doi:10.2118/135357-MS.

Tamura T, Kaneko Y. Foam film stability in aqueous systems. In: Hartland S, editors. Surface and interfacial tension: measurement, theory, and applications. Surfactant Science Series 119, CRC Press, New York, p. 90, 2004.

Tan SN, Fornasiero D, Sedev R, et al. The role of surfactant structure on foam behavior. Colloids Surf A. 2005;263(1–3):233–8. doi:10.1016/j.colsurfa.2004.12.060.

Vijayaraghavan K, Nikolov A, Wasan D. Foam formation and mitigation in a three-phase gas-liquid-particulate system. Adv Colloid Interface Sci. 2006;123–126:49–61. doi:10.1016/j.cis.2006.07.006.

Wang HR, Chen KM. Preparation and surface active properties of biodegradable dextrin derivative surfactants. Colloids Surf A. 2006;281(1–3):190–3. doi:10.1016/j.colsurfa.2006.02.039.

Wu Y, Shuler PJ, Blanco M, et al. An experimental study of wetting behavior and surfactant EOR in carbonates with model components. In: SPE/DOE symposium on improved oil recovery, April 22–26, Tulsa, Oklahoma, 2006. doi: 10.2118/99612-PA.

Zhang DL, Shunhua L, Puerto M, et al. Wettability alteration and spontaneous imbibitions in oil-wet carbonate formations. J Pet Sci Eng. 2006;52(1–4):213–26. doi:10.1016/j.petrol.2006.03.009.

Zhang S, Sun D, Dong X, et al. Aqueous foams stabilized with particles and nonionic surfactants. Colloids Surf A. 2008;324(1–3):1–8. doi:10.1016/j.colsurfa.2008.03.020.

Zheng T, Bott S, Huo Q. Techniques for accurate sizing of gold nanoparticles using dynamic light scattering with particular application to chemical and biological sensing based on aggregate formation. ACS Appl Mater Interfaces. 2016;8(33):21585–94. doi:10.1021/acsami.6b06903.

Zochhi G. In: Broze G, editors. Handbook of detergents, part A: properties. Surfactant science series. vol. 59, Marcel Dekker, New York, 1999.

18

Improvements to the fuzzy mathematics comprehensive quantitative method for evaluating fault sealing

Da-Wei Dong[1] · Ji-Yan Li[2,3] · Yong-Hong Yang[2] · Xiao-Lei Wang[4] ·
Jian Liu[2]

Abstract Fuzzy mathematics is an important means to quantitatively evaluate the properties of fault sealing in petroleum reservoirs. To accurately study fault sealing, the comprehensive quantitative evaluation method of fuzzy mathematics is improved based on a previous study. First, the single-factor membership degree is determined using the dynamic clustering method, then a single-factor evaluation matrix is constructed using a continuous grading function, and finally, the probability distribution of the evaluation grade in a fuzzy evaluation matrix is analyzed. In this study, taking the F1 fault located in the northeastern Chepaizi Bulge as an example, the sealing properties of faults in different strata are quantitatively evaluated using both an improved and an un-improved comprehensive fuzzy mathematics quantitative evaluation method. Based on current oil and gas distribution, it is found that our evaluation results before and after improvement are significantly different. For faults in "best" and "poorest" intervals, our evaluation results are consistent with oil and gas distribution. However, for the faults in "good" or "poor" intervals, our evaluation is not completely consistent with oil and gas distribution. The improved evaluation results reflect the overall and local sealing properties of target zones and embody the nonuniformity of fault sealing, indicating the improved method is more suitable for evaluating fault sealing under complicated conditions

Keywords Fault sealing property · Fuzzy mathematics · Dynamic clustering method · Quantitative study

1 Introduction

Throughout geological history, faults have played a very important role in hydrocarbon migration and accumulation. Large faults, which form the boundaries of oil and gas fields, can generally control hydrocarbon accumulation (Agosta et al. 2012; Allan 1989; Balsamo et al. 2010; Bouvier et al. 1989; Braathen et al. 2009; Brogi and Novellino 2015; Choi et al. 2015; Collettini et al. 2014; Fisher and Jolley 2007). However, small faults generally separate oil and gas, resulting in an increase in well spacing that poses challenges in oil and gas exploration. Thereby, studies of fault sealing have become a very important focus for oil and gas exploration and development (Ciftci et al. 2013; Collettini et al. 2014; Davatzes and Aydin 2005; Fachri et al. 2013a, b). Studies of fault sealing have evolved from trap theory, which is used to identify hydrocarbon sealing mechanism(s) and physical parameters of faults (Hubbert 1953; Smith 1980). Based on different mathematical principles and modes, the sealing properties of one or more given faults are quantitatively identified and predicted by considering various sealing mechanisms and

✉ Da-Wei Dong
327344321@qq.com

[1] Shengli College of China University of Petroleum, Dongying 257061, Shandong, China

[2] Research Institute of Petroleum Exploration and Development, Shengli Oilfield Company, Dongying 257000, Shandong, China

[3] Working Stations for Post Doctors, Dongying 257000, Shandong, China

[4] Oil Production Plant of Dongxing, Shengli Oilfield Company, Sinopec, Dongying 257061, Shandong, China

Edited by Jie Hao

factors. (Childs et al. 2009; Egholm et al. 2008; Fachri et al. 2011; Faulkner et al. 2003; Fisher and Knipe 1998; Wu et al. 2010; Knipe et al. 1997).

With constant improvements in petroleum geology, scholars in China and abroad have gradually realized that fault sealing is jointly controlled by multiple factors, rather than just one or two single factors (Lü et al. 2007; Knott 1993; Knipe et al. 1997; Faulkner et al. 2010; Gibson 1994; Zhang et al. 2013; Pei et al. 2015). With a deepening of quantitative single-factor studies of fault sealing, many scholars tried to comprehensively evaluate fault sealing using mathematical theory methods, including nonlinear mapping methods, gray relational, logical information, fault connectivity probabilistic methods and fuzzy comprehensive evaluation methods (Fu et al. 2005, 2008, 2012; Lü et al. 1995; Lü and Fu 2002; Zhang et al. 2007; Jiang et al. 2008; Li 2009; Zhang et al. 2015). However, for all these methods, some key parameters are difficult to obtain and have regional limitations, so that human factors are generally introduced for value assignment or adjustment. Thus, scholars manage to optimize these parameters using a variety of mathematical approaches. Among them, the fuzzy comprehensive evaluation method, which is strongly systematic and provides definite results, is accepted by many scholars and has been gradually improved over the years.

In this paper, based on the principles of fuzzy comprehensive evaluation, the single-factor membership degree was established using the dynamic clustering method. Then, a single-factor evaluation matrix was constructed using the continuous grading function in order to optimize the probability distribution of the evaluation grade in the fuzzy comprehensive evaluation method. Next, we considered a fault in the Chepaizi Bulge in the northwestern margin of the Junggar Basin as an example, and the sealing properties in the vertical and strike directions were evaluated. Finally, our results are compared with oil and gas exploration results.

2 Un-improved fuzzy mathematics evaluation of fault sealing

The physical principles that affect fault sealing include principal stress sealing, lithological allocation sealing, shale smear sealing, time allocation sealing and occurrence allocation sealing (Liu 1998). Single factors include fault properties, fault plane pressure, lithological allocation, fault dip angle, shale smear and fault active time. Thus, being affected by numerous factors, fault sealing shows very strong complexity and randomness. Fortunately, the fuzzy evaluation method can take various types of single factors that affect fault sealing into consideration, and

hence realize a comprehensive evaluation using the principle of fuzzy transformation and maximum membership degree. Using this mathematical method, fault sealing properties can be effectively identified. The process is as follows: First, based on regional geology and fault development, the single factors that affect fault sealing are selected and quantified in order to obtain a single-factor quantization matrix $U_{n \times 1}$ (n is the number of single factors). Then, based on the general division of fault sealing and regional requirements, the number of evaluation grades is divided (m). Finally, a single-factor membership degree matrix is constructed based on the number of evaluation grades and the oil and gas exploration results of the work area, $V_{1 \times m}$, based on which, $U_{n \times 1}$ is evaluated in order to construct a fuzzy evaluation matrix $R_{n \times m} = U_{n \times 1} \cdot V_{1 \times m}$. Since each single factor has different contributions to fault sealing, once the weighted matrix of each single factor is provided, $W_{1 \times n}$, the fuzzy evaluation matrix $B_{1 \times m} = W_{1 \times n} \cdot R_{n \times m}$ can be obtained. Finally, the maximum value of the fuzzy evaluation matrix $B_{1 \times m}$ is taken as the corresponding evaluation grades.

However, there are various methods that can be used to acquire the parameters mentioned above, both qualitatively and quantitatively, and which are constantly optimized (Lü and Fu 2002; Russell et al. 2003; Fu et al. 2012; Li et al. 2009). There are many single factors that affect fault sealing, generally including fault properties, fault plane pressure, fault lithological allocation and shale smear. Evaluation grades can be divided into five ranks: good, better, moderate, fairly poor and poor. The single-factor membership degree can be obtained via a discrete function method and/or a continuous function method. The former applies to qualitative single-factor membership degree, such as fault properties, while the latter applies to quantitative single-factor membership degree, such as fault plane pressure. Weighting coefficients can be obtained via the expert survey method, analytic hierarchy process (AHP), Delphi method and/or the weight matrix method (Ding and Jin 2012; Sun and Wu 1995; Qiu et al. 2007). The larger the weighting coefficient of a single factor, the greater the impact of the single factor on fault sealing. The mathematical model of fuzzy comprehensive evaluation includes the weighted-average type, the main factor highlight type and the main factor determination type (Sun et al. 2010). The weighted-average type can take a variety of single factors into consideration in order to avoid information loss. The main factor highlight type and the main factor determination type emphasize the main controlling factors and prevent interference factors. In the evaluation process of fault sealing, many scholars use the weighted-average type. However, determining the above parameters and effectively predicting fault sealing must be based on the actual petroleum geology of the study area and from suggestions made by local experts in order to reduce exploration risks.

3 Improved fuzzy mathematics of fault sealing

3.1 The establishment of the single-factor membership degree using the dynamic clustering method

In this study, the membership degree of each single factor is established by applying the dynamic clustering method. Based on a large sample dataset, pre-classification is roughly performed. Then, gradual adjustment is made until a reasonable classification is obtained. The maximum and minimum values of each class constitute the range of membership degrees. For fault sealing, the single-factor membership degree is established as follows: Firstly, a large number of samples of one single factor of the work area and their corresponding oil–gas show are constructed. Then, taking best, good, poor and poorest oil and gas show as the standard, the range of the membership degrees of the fault gouge ratio is determined. This means that the fault gouge ratio of a given area is clustered. In the process of clustering, the maximum and minimum values are the range limits of membership degree. The so-called dynamic clustering means the gradual adjustment of a value to a reasonable range using the dynamic clustering principle. This method has the advantages of incorporating the oil and gas geology of the study area, so that it has regional eigenvalues. It is closely related to the fault sealing properties of the study area and hence has a certain degree of predictability. The advantage of this method is that the regional petroleum geology has a regional characteristic value which is related to fault sealing, and is predictive.

3.2 The establishment of a single-factor evaluation matrix using the continuous grading function

After the single-factor membership degree is determined via the dynamic clustering method, fuzzy evaluation should be carried out for all single factors in order to establish a single-factor evaluation matrix. The division of the single-factor membership degree $S(i)$ only gives evaluation results of good and poor intervals, and it does not provide probabilistic evaluation results of each interval. For example, the range of "good" and "poor" is (1–0.5) and (0.5–0). If the single factor value is 0.7, it cannot be regarded simply as "good," thus implying that a further probabilistic evaluation is needed. Hence, the definition "good" accounts for 70% good, and "poor" accounts for 30%. $e(i)$ is the boundary value of the re-probabilistic evaluation. The precise method followed is as follows: First, a range of values for fault sealing is determined using single factors, $S(i)$. Then, the threshold value of each adjacent class is determined, namely the classification representative value $e(i)$, which is determined as per the following principles (Zhao 2001):

$$\left.\begin{array}{l} e(1) = S(1) \\ e(2) = [S(1) + S(2)]/2 \\ e(3) = [S(2) + S(3)]/2 \\ e(4) = S(3) \end{array}\right\}$$

Fault properties, fault gouge ratios and fault sealing are not always linearly related with each other. However, they have a piecewise function relation based on single-factor membership degree. According to Newton's iteration principle, it can be assumed that the single-factor evaluation criteria $r(\chi)$ are a linear piecewise function, which can be solved by successive approximation. Then, fuzzy subsets of the fault evaluation criteria can be determined using the following methods:

$$r_1(\chi) = \begin{cases} 1 & \chi \leq e(1) \\ \dfrac{e(2) - \chi}{e(2) - e(1)} & e(1) \leq \chi \leq e(2) \\ 0 & \chi \geq e(2) \end{cases} \qquad r_2(\chi) = \begin{cases} 1 - r_1(\chi) & e(1) \leq \chi \leq e(2) \\ \dfrac{e(3) - \chi}{e(3) - e(2)} & e(1) \leq \chi \leq e(2) \\ 0 & \chi \leq e(1), \chi \geq e(3) \end{cases}$$

$$r_3(\chi) = \begin{cases} 1 - r_2(\chi) & e(2) \leq \chi \leq e(3) \\ \dfrac{e(4) - \chi}{e(4) - e(3)} & e(3) \leq \chi \leq e(4) \\ 0 & \chi \leq e(2), \chi \geq e(4) \end{cases} \qquad r_4(\chi) = \begin{cases} 0 & \chi \leq e(3) \\ 1 - r_3(\chi) & e(3) \leq \chi \leq e(4) \\ 1 & \chi \geq e(4) \end{cases}$$

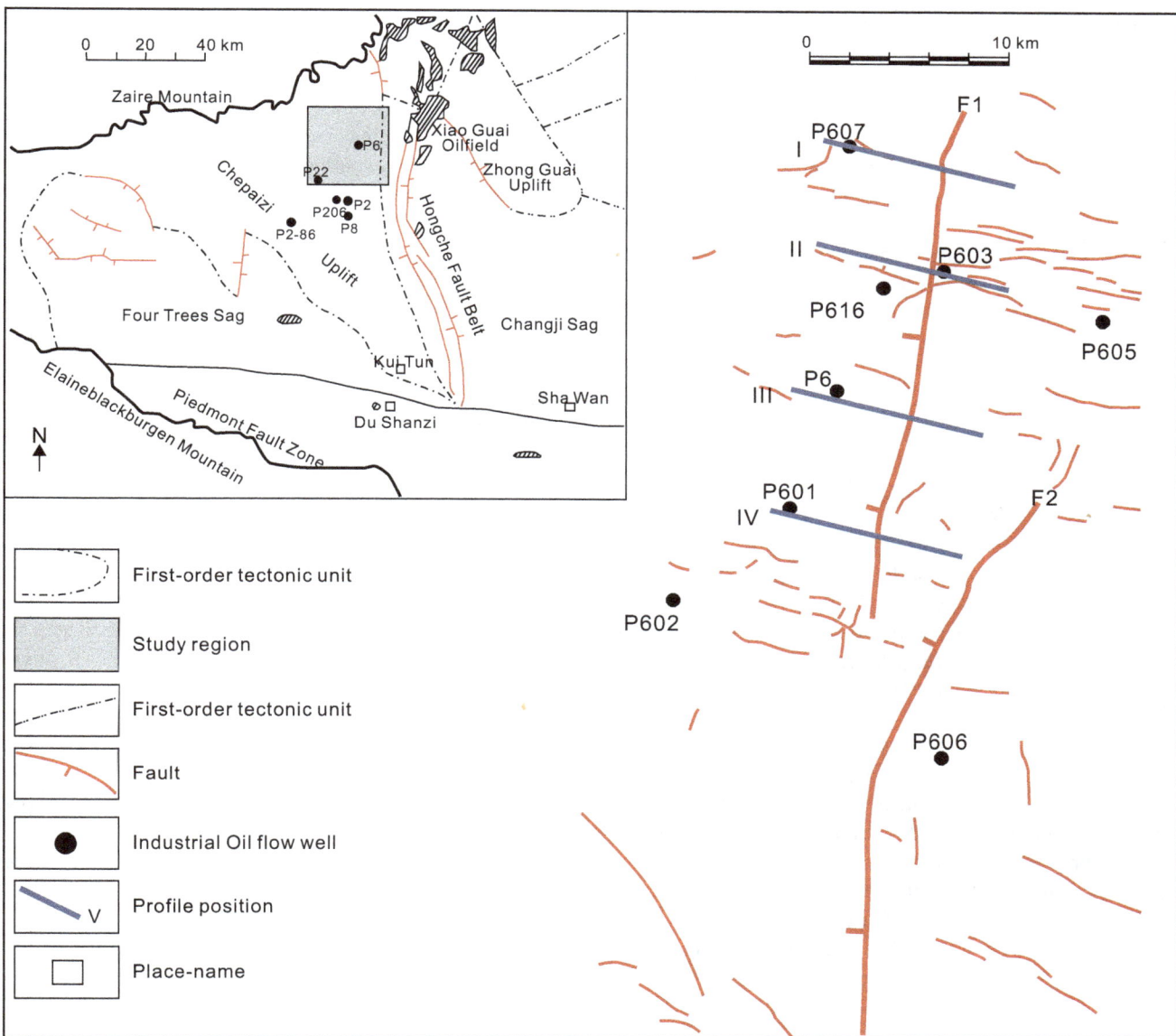

Fig. 1 The position of the seismic profile study of fault sealing

The fuzzy subset obtained based on N single-factor evaluation indices constitutes fuzzy sets:

$$R = \begin{bmatrix} r_{11} & r_{12} & r_{13} & r_{14} \\ \bullet & \bullet & \bullet & \bullet \\ \bullet & \bullet & \bullet & \bullet \\ r_{n1} & r_{n2} & r_{n3} & r_{n4} \end{bmatrix}_{n \times 4}$$

4 Case studies

Taking the fault F1 located in the northeastern Chepaizi Bulge as an example (Fig. 1), the fault sealing properties of the main target zones were evaluated using both the improved and un-improved methods. This approach was used to determine whether the improved comprehensive

evaluation method for fault sealing is more reasonable. The Chepaizi Bulge is located at the southern end of the northwestern margin of the Junggar Basin. It is a secondary structural unit in the western uplift of the Junggar Basin, adjacent to the Changji Sag and the Zhongguai bulge to the east, Sikeshu Sag to the south and Zaire Mountain to the northwest (Fig. 1). Conditions in the Chepaizi Bulge make this a good place for hydrocarbon accumulation. The main reservoirs are Cretaceous (K) and Neogene Sha1 Member (N_1s_1). The F1 fault in the northeastern Chepaizi Bulge is a very important oil control fault, and its sealing properties directly affect hydrocarbon accumulation in this area. The fault has NNE-trending, EW-trending compression in the Yanshanian Period and NWW-trending extension in the

Himalayan Period. Vertically, there are reverse faults in the Cretaceous strata and basement and normal faults in Cenozoic strata, which represent a negative inverted structure as a whole.

4.1 Single-factor selection and quantitative calculation

Since the study area has undergone multiple phases of tectonic evolution, various types of faults with different occurrences have developed. There are many single factors that affect fault sealing in response to tectonic evolution, so that the weighting coefficients of the single factors are different. However, in this paper, and taking fault F1 as an example, the sealing properties in the vertical and strike directions were analyzed. Four single factors with comparable significance were selected, including fault surface normal stress (σ), fault properties, sand/formation ratio (N/G) and the shale content of fault zone fillings. In addition, four seismic profiles vertical to fault F1 were selected (the location is shown in Fig. 1), which were converted into four geological sections via a time-depth conversion. Then, lithological sections were recovered using well logs in order to obtain the quantitative and qualitative values of each single factor (Table 1).

Fault surface normal stress, which is an important parameter used to characterize the fault opening degree, is crucial to fault sealing (Fu et al. 2005). Typically, this value is the vector sum of the gravitational force imparted by the overlying strata, the regional principal compressive stress, the regional principal stress as well as the fault dip angle. Fault F1 is a shovel fault, where the fault plane normal stress increases from the bottom to the top, where the fault sealing properties become better. However, the Chepaizi Bulge is located in a slope belt of a foreland basin, where the buried depth is shallow, so that the fault surface pressure is small.

In general, the sealing properties of compressional, shear and compressional-shear faults are good, while the sealing properties of extensional and extensional-shear faults are poor. Fault F1 is a negative inverted fault, and it is an extensional fault that formed during the Neogene. However, the fault occurrence changes significantly, which is steep in the upper part and gentle in the lower part. That is, regional tensile stress can only cause the Neogene "steep" fault section to extend, but cannot completely make the Cretaceous "gentle" fault section extend. In addition, the pressure experienced by the Cretaceous fault plane is significantly larger than that of the Neogene fault plane, which can also prove this change. After undergoing compressional deformation, the Cretaceous fault section has better sealing properties than the Neogene extensional fault section due to its fault structure and mudstone smearing. Therefore, the Neogene fault section has been assigned as "extensional", while the Cretaceous fault section is assigned as "compressional".

Sand shale contraposition is an important process for oil and gas lateral sealing. In fault dislocation, if sand shale contraposition occurs in one interval, the fault in this interval is closed, while if multiple sand layers are connected in one interval, the fault in this interval is open. The parameter characterizing possible sand shale contraposition is the sand/formation ratio, which is the ratio between the sandstone and layer formation thicknesses, N/G. When N/G is large, the possibility of a sand–sand connection is large, and hence the fault sealing ability is poor. When the N/G ratio is small, the fault sealing ability is in fact good. However, this single factor does not take the fault throw generated by fault slipping into consideration. Therefore, the following single factor is added.

Faults will form fault zones during displacement and dislocation processes. When a given fault zone is filled with shale, it can seal oil and gas laterally due to the plastic flow and compaction of shale. When a fault zone is filled with sandstone, it can act as a hydrocarbon migration pathway since sandstone compacts poorly and has good porosity and permeability. Accordingly, Fu et al. (2012) improved the fault gouge ratio (SGR) proposed by Yielding et al. (1997) and proposed to characterize the sealing property of faults using the ratio between the mudstone thickness of a fault belt and the sum of the fault throw and

Table 1 Evaluation parameter list of single factors on the sealing ability of the F1 fault

Section line no.	Formation	σ, MPa	Fault surface	N/G	Rm	Section line no.	Formation	σ, MPa	Fault surface	N/G	Rm
I	N_1s_3	0.46	Tensile	0.95	0.04	III	N_1s_3	0.66	Tensile	1	0
	N_1s_2	0.52	Tensile	0.02	0.74		N_1s_2	0.77	Tensile	0.12	0.67
	N_1s_1	0.58	Tensile	0.56	0.32		N_1s_1	0.86	Tensile	0.33	0.5
	K	0.82	Compressive	0.24	0.7		K	1.29	Compressive	0.26	0.61
II	N_1s_3	0.25	Tensile	0.99	0.01	IV	N_1s_3	0.09	Tensile	0.99	0.02
	N_1s_2	0.28	Tensile	0	0.79		N_1s_2	0.10	Tensile	0.30	0.70
	N_1s_1	0.31	Tensile	0.46	0.40		N_1s_1	0.11	Tensile	0.28	0.49
	K	0.48	Compressive	0.17	0.79		K	0.19	Compressive	0.28	0.70

faulted formation thicknesses, namely the shale content of the fillings in a given fault belt. Due to the negative inversion of fault F1, Cretaceous shale smeared the fault surface repeatedly. Therefore, the shale in the Cretaceous fault zone fillings was formed in compressional and extensional stages. Hence, the faults that formed in compressional and extensional stages were obtained from structural section restoration.

4.2 Single-factor weighting coefficients and membership degree

Single-factor weighting coefficients are significantly different in different areas. The numerous methods that are used to determine these coefficients are all based on the expert investigation method. Therefore, in this paper, the weighting coefficients of four single factors, including the normal stress of the fault plane (w_1), fault properties (w_2), sand/formation ratio (w_3) and the shale content of the fillings in the fault belt (w_4), were determined using the expert investigation method (Table 2). It is worth mentioning that the formation pressure of the oil and gas discovery wells in this area is abnormal, while the formation pressure of dry wells and water wells is normal, meaning pressure is sensitive to oil and gas accumulation so that the weighting coefficient of fault plane normal stress is the largest.

The statistics used in this study were applied to the well logging data of 85 wells around fault F1 and the corresponding oil and gas shows. Moreover, the values of four single factors were calculated, which were classified into four categories using the dynamic clustering method: best, good, poor and poorest. The maximum and minimum values of each category were taken as the boundary values of each membership degree. Based on the advice of oilfield experts, membership degree was then slightly adjusted. The final results are shown in Table 3.

4.3 Optimized fuzzy evaluation matrix

The advantage of the dynamic clustering method in comprehensive fault sealing property evaluation is to establish a single-factor evaluation matrix. In order to illustrate the establishment process of the un-improved and improved single-factor evaluation matrices, the establishment of the

Table 2 Single-factor weight coefficients of fault sealing in the northeast of Chepaizi Uplift

Influence factor	w_1	w_2	w_3	w_4
Weight coefficient	0.3	0.2	0.25	0.25

Table 3 The single-factor membership list in the study area

Standard	Best	Good	Poor	Poorest
N/G	≤0.3	0.3–0.55	0.55–0.8	≥0.8
Rm	≥0.70	0.55–0.70	0.40–0.55	≤0.40
Fault properties	≥0.9	0.75–0.9	0.5–0.75	≤0.5
σ, MPa	≥1	0.5–1	0.3–0.5	≤0.3

single-factor matrix in the first member of the Shawan Formation in profile I was analyzed.

Before improvement, a single value was directly assigned to a single-factor membership degree evaluation matrix using the discrete function and continuous function methods. When the N/G of the first member of the Shawan Formation was 0.56, (i.e., the grade of the maximum membership degree is "poor"), this grade was directly assigned a value of 0.5 when using the discrete function method and 0.3 when using the continuous function method. The single-factor membership degree evaluation matrix is $R_{N/G} = (0.5)$ or (0.3). However, although a value of 0.56 implies a grade of "poor," which is close to a grade of "good," a value of 0.56 should have some probability evaluation in the grade "good." However, the un-improved single-factor membership degree evaluation matrix does not embody the probability in this transitional type.

Instead, a single-factor evaluation matrix that is constructed by the continuous grading membership function can properly address the problem mentioned above. Its calculation process after improvement is as follows: First, $S(i)$ is determined based on N/G, namely $S(1) = 0.3$, $S(2) = 0.55$, $S(3) = 0.8$. Then, the threshold value of each adjacent evaluation level is calculated, i.e., grading representative value $e(i)$:

$$\begin{cases} e(1) = S(1) = 0.3 \\ e(2) = [S(1) + S(2)]/2 = 0.425 \\ e(3) = [S(2) + S(3)]/2 = 0.675 \\ e(4) = S(3) = 0.8 \end{cases}$$

Finally, when the N/G of the first member of the Shawan Formation is 0.56, the single-factor evaluation matrix is calculated: $e(2) < 0.56 < e(3)$,

$$\begin{cases} r_1 = 0 \\ r_2 = [e(3) - 0.56]/[e(3) - e(2)] = 0.46 \\ r_3 = 1 - r_2 = 0.54 \\ r_4 = 0 \end{cases}$$

After improvement, the evaluation matrix of the sand/formation ratio of the first member of the Shawan Formation is $R_{N/G} = (0 \quad 0.46 \quad 0.54 \quad 0)$.

The comprehensive evaluation matrix of the four factors of the first member of the Shawan Formation is:

$$R = \begin{bmatrix} 0.00 & 0.51 & 0.49 & 0.00 \\ 0.00 & 0.00 & 0.80 & 0.20 \\ 0.00 & 0.46 & 0.54 & 0.00 \\ 0.00 & 0.00 & 0.00 & 1.00 \end{bmatrix}$$

The fuzzy evaluation matrix of the first member of the Shawan Formation is $B = W \cdot R = (0 \quad 0.27 \quad 0.44 \quad 0.29)$. According to the principle of the maximum membership degree, the evaluation result of the first member of the Shawan Formation is poor. However, the probability distribution of this evaluation grade is scattered, indicating the fuzziness of the fault sealing property is large. Conversely, if the probability distribution of the evaluation grade is concentrated, the evaluation result of fault sealing property is obvious.

4.4 Fuzzy evaluation results and analysis

Using the improved fuzzy evaluation method described above, the sealing properties of fault F1 in the vertical and strike directions were evaluated. The results are shown in Table 4. Cross sections I, II, III and IV are well-tied cross sections that span from south to north along the fault strike. Cretaceous (K), the first member of Shawan Formation (N_1s_1), the second member of Shawan Formation (N_1s_2) and the third member of Shawan Formation (N_1s_3)

constitute intervals from top to bottom in the vertical direction. The evaluation results show that the valuation result of Cretaceous faults is "good," but the maximum probability is only 0.58, indicating oil and gas are sealed when they migrate to the footwall, but it may be oil and gas bearing in the hanging wall. The valuation result of the first member of the Shawan Formation is "poor" or "poorest." The probability distribution of the evaluation grade is dispersed, showing the ambiguity of footwalls and hanging walls on its oil-bearing properties. The evaluation result of the second member of the Shawan Formation is "good." A regional survey showed that this interval is a stably distributed shale bed, namely regional caprock. The evaluation result of the third member of the Shawan Formation is "poor," but mud logging results show this interval is conglomerate. There were no oil and gas shows discovered in this interval in the work area.

Next, we compared the variations between the improved and un-improved fuzzy evaluations and their impact on the evaluation results. In this paper, a single-factor membership degree evaluation matrix was directly assigned a value using the discrete function method. The assigned values of evaluation grades $R = (best, good, poor, poorest) = (1, 0.66, 0.33, 0)$ and the fuzzy evaluation results are shown in Table 4. Upon inspection, the fuzzy evaluation values of the first and second members of the Shawan Formation change significantly, but they do have consistent results. However, the fuzzy evaluation values and evaluation

Table 4 Evaluation form of fault sealing

Section line no.	Formation	Improved evaluation value					Unimproved evaluation value		Oil and gas display	
		Best	Good	Poor	Poorest	Value	Result	Value	Hanging wall	Footwall
I	N_1s_3	0.00	0.05	0.41	0.54	Poorest	0.10	Poorest	Water	Water
	N_1s_2	0.50	0.10	0.36	0.04	Best	0.70	Best	–	–
	N_1s_1	0	0.27	0.44	0.29	Good	0.36	Poor	Oil	Oil
	K	0.58	0.39	0.03	0.00	Best	1.00	Best	Water	Oil
II	N_1s_3	0.00	0.00	0.16	0.84	Poorest	0.00	Poorest	Water	Water
	N_1s_2	0.50	0.00	0.16	0.34	Best	0.50	Good	–	–
	N_1s_1	0.00	0.22	0.22	0.56	Good	0.26	Poorest	Oil–Water	Oil
	K	0.50	0.24	0.26	0	Best	0.80	Best	Water	Water
III	N_1s_3	0.00	0.22	0.24	0.54	Poorest	0.20	Poorest	Water	Water
	N_1s_2	0.42	0.38	0.16	0.04	Best	0.61	Poor	Empty	Empty
	N_1s_1	0.32	0.27	0.37	0.04	Best	0.45	Poor	Oil	Oil
	K	0.55	0.40	0.05	0.00	Best	0.92	Best	Oil	–
IV	N_1s_3	0.00	0.00	0.16	0.84	Poorest	0.00	Poorest	Water	Water
	N_1s_2	0.50	0.00	0.16	0.34	Best	0.5	Good	–	–
	N_1s_1	0.25	0.03	0.38	0.34	Good	0.33	Poor	Oil	Water
	K	0.50	0.18	0.02	0.30	Best	0.7	Best	Oil	–

"–" in the table means there are no drilling test data or wells are not drilled in this horizon

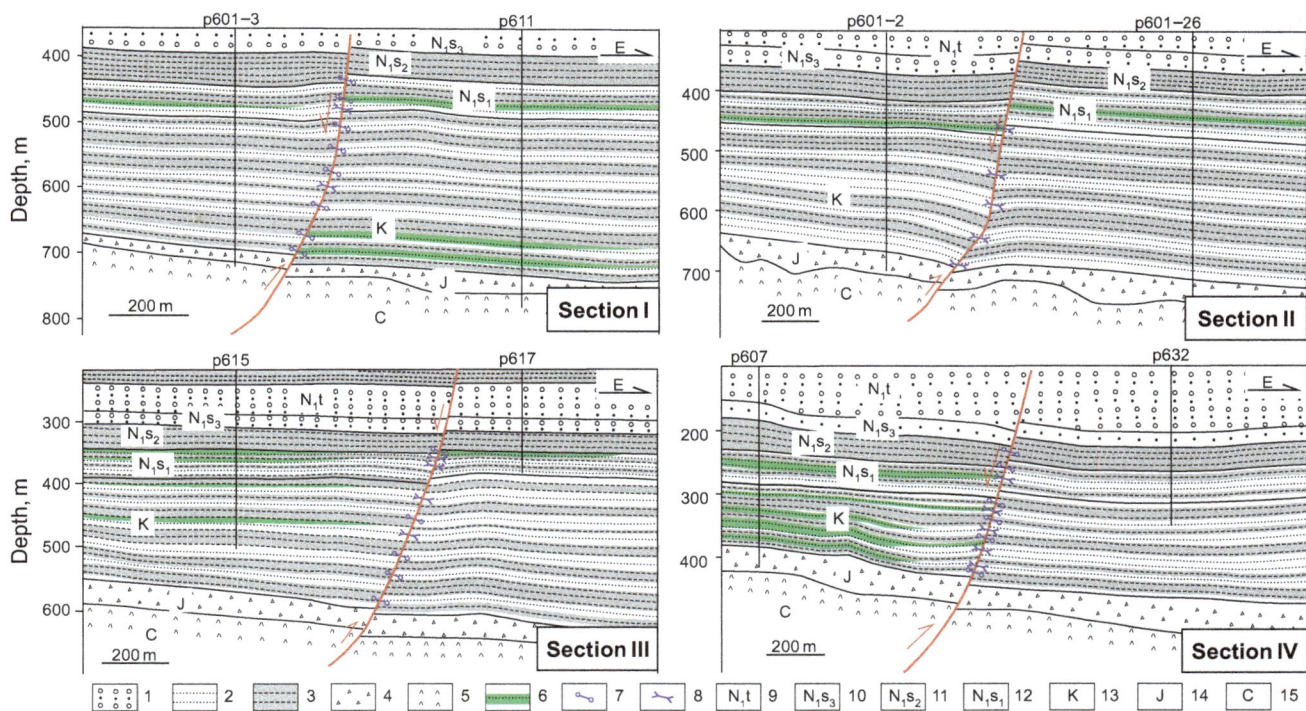

Fig. 2 Analysis of fault sealing and its oil reservoir in sections **I–IV** *1* Conglomerate, *2* Sandstone, *3* Mudstone, *4* Breccia, *5* Igneous rocks, *6* Hydrocarbon migration direction, *7* Faults open, *8* Faults closed

results of the first member of the Shawan Formation change significantly before and after improvement, and the evaluation results before improvement are not well matched with the oil and gas shows in footwall and hanging wall. Taking cross section III as an example, the evaluation results before improvement are poor, but the footwall and hanging walls are both oil bearing. After improvement, the probability distribution of the evaluation grades is scattered, indicating oil and gas may be trapped locally after having continuously migrated from the footwall to the hanging wall in the local open interval (Fig. 2). The above analysis shows that the improved method is more advantageous for regions with larger degrees of vagueness of their fault sealing properties, and hence higher exploration risks.

5 Conclusions

The evaluation of fault sealing properties is an important part of oil and gas exploration and development. Fuzzy comprehensive evaluation is a systematic evaluation method for fault sealing. Since traditional methods are greatly affected by human factors during the establishment of the single-factor membership degree, the assignment method results in fuzzy evaluation values, so that the result only reflects the target interval. In this paper, the dynamic clustering method was introduced to determine the single-factor membership degree. Then, a single-factor evaluation matrix was constructed using the continuous grading function in order to determine the optimum comprehensive evaluation matrix and make a fuzzy evaluation of the fault sealing properties. A comparison of the fuzzy evaluation and its result before and after improvement, combined with current oil and gas distribution regularity, showed that the evaluation results before and after improvement are significantly different. For faults designated as "best" and "poorest," the evaluation results are consistent with oil and gas distribution. However, for faults designated as "good" or "poor," the evaluation results of sealing property are not completely consistent with the oil and gas distributions. The improved results reflect the overall and local sealing properties of target zones and embody the fuzziness of fault sealing, indicating the improved method is more precise for evaluating fault sealing properties under complicated conditions.

However, the improved method still has its limitations. First of all, this method still cannot solve the multi-scale fault sealing problem. As mentioned above, the N_1s_1 section in profile III displays strong heterogeneity, and although the evaluation results of large scale are "poor," the fault sealing properties exist at small scales. Secondly, this method can only explain the sealing properties of the faults in profile. When oil and gas migrate in multiple directions, the evaluation results will not be consistent with drilling results. For example, the evaluation results of section K in the III and IV profiles are good, but the fault is oil bearing in the hanging wall. This phenomenon is related to a two-way oil supply. Oil source comparison shows that the oil and gas migration in the study area is from east to west in the Changji Sag and from south to north in the Sikeshu Sag (Song et al. 2007). Therefore, this method is only applicable for studies of single-scale fault sealing properties. In addition, the direction of oil and gas migration should be vertical to the fault plane.

Acknowledgements This study is supported by the Science and Technology Project of Universities and Colleges in Shandong Province "Investigation on diagenetic environment and transformation pattern of red-bed reservoirs in the rift basins" (No. J16LH52).

References

Agosta F, Ruano P, Rustichelli A, et al. Inner structure and deformation mechanisms of normal faults in conglomerates and carbonate grainstones (Granada Basin, Betic Cordillera, Spain): inferences on fault permeability. Struct Geol. 2012;45:4–20. doi:10.1016/j.jsg.2012.04.003.

Allan US. Model for hydrocarbon migration and entrapment within faulted structures. AAPG Bull. 1989;70(7):803–11. doi:10.1306/94885962-1704-11D7-8645000102C1865D.

Balsamo F, Storti F, Salvini F, et al. Structural and petrophysical evolution of extensional fault zones in low-porosity, poorly lithified sandstones of the Barreiras Formation, NE Brazil. Struct Geol. 2010;32(11):1806–26. doi:10.1016/j.jsg.2009.10.010.

Bouvier JD, Kaars-Sijpesteijn CH, Kluesner DF, et al. Three-dimensional seismic interpretation and fault sealing investigations, Nun River Field, Nigeria. AAPG Bull. 1989;73(11):1397–414.

Braathen A, Tveranger J, Fossen H, et al. Fault facies and its application to sandstone reservoirs. AAPG Bull. 2009;93(7):891–917. doi:10.1306/03230908116.

Brogi A, Novellino R. Low angle normal fault (LANF)-zone architecture and permeability features in bedded carbonate from inner Northern Apennines (RapolanoTerme, Central Italy). Tectonophysics. 2015;638(638):126–46. doi:10.1016/j.tecto.2014.11.005.

Childs C, Manzocchi T, Walsh JJ, et al. A geometric model of fault zone and fault rock thickness variations. Struct Geol. 2009;31(2):117–27. doi:10.1016/j.jsg.2008.08.009.

Choi JH, Yang SJ, Han SR, et al. Fault zone evolution during Cenozoic tectonic inversion in SE Korea. Asian Earth Sci. 2015;98:167–77. doi:10.1016/j.jseaes.2014.11.009.

Ciftci NB, Giger SB, Clennell MB. Three-dimensional structure of experimentally produced clay smears: implications for fault seal analysis. AAPG Bull. 2013;97(5):733–57. doi:10.1306/10161211192.

Collettini C, Carpenter BM, Viti C, et al. Fault structure and slip localization in carbonate-bearing normal faults: an example from the Northern Apennines of Italy. Struct Geol. 2014;67:154–66. doi:10.1016/j.jsg.2014.07.017.

Davatzes N, Aydin A. Distribution and nature of fault architecture in a layered sandstone and shale sequence: an example from the Moab fault, Utah. AAPG Memoir. 2005;85:153–80. doi:10.1306/1033722M853134.

Ding WL, Jin WZ. Research and application of the comprehensive evaluation system of multi information to fault sealing. Beijing: Geological Publishing House; 2012. p. 70–104 (in Chinese).

Egholm DL, Clausen OR, Sandiford M, et al. The mechanics of clay smearing along faults. Geology. 2008;36(10):787–90. doi:10.1130/G24975A.1.

Fachri M, Rotevatn A, Tveranger J. Fluid flow in relay zones revisited: towards an improved representation of small-scale structural heterogeneities in flow models. Mar Pet Geol. 2013a;46(3):144–64. doi:10.1016/j.marpetgeo.2013.05.016.

Fachri M, Tveranger J, Braathen A, et al. Sensitivity of fluid flow to deformation-band damage zone heterogeneities a study using fault facies and truncated Gaussian simulation. Struct Geol. 2013b;52(1):60–79. doi:10.1016/j.jsg.2013.04.005.

Fachri M, Tveranger J, Cardozo N, et al. The impact of fault envelope structure on fluid flow: a screening study using fault facies. AAPG Bull. 2011;95(4):619–48. doi:10.1306/09131009132.

Faulkner DR, Jackson CAL, Lunn RJ, et al. A review of recent developments concerning the structure, mechanics and fluid flow properties of fault zones. Struct Geol. 2010;32(11):1557–75. doi:10.1016/j.jsg.2010.06.009.

Faulkner DR, Lewis AC, Rutter EH. On the internal structure and mechanics of large strike-slip fault zones: field observations of the Carboneras fault in southeastern Spain. Tectonophysics. 2003;367(3):235–51. doi:10.1016/S0040-1951(03)00134-3.

Fisher QJ, Jolley SJ. Treatment of faults in production simulation models. Geol Soc Lond Spec Publ. 2007;292(1):219–33. doi:10.1144/SP292.13.

Fisher QJ, Knipe RJ. Fault sealing processes in siliciclastic sediments. Geol Soc Lond Spec Publ. 1998;147(1):117–34. doi:10.1144/GSL.SP.1998.147.01.08.

Fu G, Liu HX, Du HF. Seal mechanisms of different transporting passways of fault and their research methods. Pet Geol Exp. 2005;27(4):404–8. doi:10.3969/j.issn.1001-6112.2005.04.016 (in Chinese).

Fu G, Shi JJ, Lü YF. An improvement in quantitatively studying lateral seal of faults. Acta Pet Sin. 2012;33(3):414–8 (in Chinese).

Fu G, Yin Q, Du Y. A method studying vertical seal of fault with different filling forms and its application. Pet Geol Oilfield Dev Daqing. 2008;27(1):1–5. doi:10.3969/j.issn.1000-3754.2008.01.001 (in Chinese).

Gibson RG. Fault-zone seals in siliciclastic strata of Columbus Basin, Offshore Trinidad. AAPG Bull. 1994;78(9):1372–85. doi:10.1306/A25FECA7-171B-11D7-8645000102C1865D.

Hubbert MK. Entrapment of petroleum under hydrodynamic conditions. AAPG Bull. 1953;37(8):1954–2026. doi:10.1306/5CEADD61-16BB-11D7-8645000102C1865D.

Jiang Z, Dong Y, Li H, et al. Limitation of fault-sealing and its control on hydrocarbon accumulation—an example from the Laoyemiao Oilfield of the Nanpu Sag. Pet Sci. 2008;5:295–301. doi:10.1007/s12182-008-0049-6.

Knipe RJ, Fisher RJ, Jones G, et al. Fault seal analysis: successful methodologies, application and future directions. Nor Pet Soc Spec Publ. 1997;7(97):15–38. doi:10.1016/S0928-8937(97)80004-5.

Knott SD. Fault seal analysis in the North Sea. AAPG Bull. 1993;77(5):778–92. doi:10.1306/BDFF8D58-1718-11D7-8645000102C1865D.

Li CH, Luo LY, Chen PY, et al. Analysis and application of weight coefficient and attribution function during fuzzy evaluation of fault sealing. Complex Hydrocarb Reserv. 2009;2(1):5–13. doi:10.16181/j.cnki.fzyqc.2009.01.018 (in Chinese).

Li Y. Shale smearing and its quantitative characterization in the perspective of fault rocks—a case study of the Ying 32 fault in the Dongxin field of Jiyang depression. Acta Geol Sin. 2009;83(3):426–34. doi:10.3321/j.issn:0001-5717.2009.03.010 (in Chinese).

Liu ZR. The mechanism and structural model of fault-block reservoir. Beijing: Petroleum Industry Press; 1998 (in Chinese).

Lü YF, Chen ZM, Chen FJ. Evaluation of sealing ability of faults using nonlinear mapping analysis. Acta Pet Sin. 1995;16(2):36–41 **(in Chinese)**.

Lü YF, Fu G. Study of fault sealing. Beijing: Petroleum Industry Press; 2002 **(in Chinese)**.

Lü YF, Sha ZX, Fu XF, et al. Quantitative evaluation method for fault vertical sealing ability and its application. Acta Pet Sin. 2007; 28(5):34–8 **(in Chinese)**.

Pei YW, Paton DA, Knipe RJ, et al. A review of fault sealing behaviour and its evaluation in siliciclastic rocks. Earth Sci Rev. 2015;150:121–38. doi:10.1016/j.earscirev.2015.07.011.

Qiu YB, Zha M, Qu JX. Fuzzy comprehensive evaluation of the sealability of the faults in Chenbao Oilfield. J Xi'an Shiyou Univ (Nat Sci Ed). 2007;22(3):28–32. doi:10.3969/j.issn.1673-064X. 2007.03.006 **(in Chinese)**.

Russell KD, An LJ, Paul J, et al. Fault-seal analysis south Marsh Island 36 field, Gulf of Mexico. AAPG Bull. 2003;87:479–91. doi:10.1306/08010201133.

Smith DA. Sealing and nonsealing faults in Louisiana Gulf coast salt basin. AAPG Bull. 1980;64(2):145–72. doi:10.1306/2F918946-16CE-11D7-8645000102C1865D.

Song CC, He LJ, Ma LQ, et al. Characteristic of hydrocarbon accumulation of Chepaizi swell in Junggar Basin. Xinjiang Pet Geol. 2007;28(2):136–8 **(in Chinese)**.

Sun H, Xu TW, Fan SW, et al. Synthetic fuzzy judgment of fault sealing in MQ area. J Oil Gas Technol. 2010;32(2):200–3 **(in Chinese)**.

Sun SX, Wu Z. Application of fuzzy synthesized evaluation to the hydrocarbon generating conditions. J Chengdu Univ Technol. 1995;22(3):6–10 **(in Chinese)**.

Wu ZP, Wei C, Xue Y. Structural characteristics of faulting zone and its ability in transporting and sealing oil and gas. Acta Geol Sin. 2010;84(4):570–8 **(in Chinese)**.

Yielding G, Freeman B, Needham DT. Quantitative fault seal prediction. AAPG Bull. 1997;81(6):897–917. doi:10.1016/S09 28-8937(97)80010-0.

Zhang LK, Luo XR, Liao QJ, et al. Quantitative evaluation of fault sealing property with fault connectivity probabilistic method. Oil Gas Geol. 2007;28(2):181–90 **(in Chinese)**.

Zhang LK, Luo XR, Song GQ, et al. Quantitative evaluation of parameters to characterize fault opening and sealing during hydrocarbon migration. Acta Pet Sin. 2013;34(1):92–100. doi:10.7623/syxb201301010 **(in Chinese)**.

Zhang J, Wu ZP, Li W, et al. The fault system of the north section of Liaodong Salient, constraints from seismic data and physical modeling experiment. Acta Geol Sin. 2015;89(s1):217–27. doi:10.1111/1755-6724.12303_28 **(in Chinese)**.

Zhao HP. The comparison of four operators in synthetic fuzzy judgment of environment. Environ Technol Guizhou. 2001;7(3): 28–35 **(in Chinese)**.

PERMISSIONS

LIST OF CONTRIBUTORS

Shi-Qun Li, Bao-Sheng Zhang and Xu Tang
School of Business Administration, China University of Petroleum, Beijing 102249, China

Yongseok Lee, Changjun Ko, Hodong Lee, Kyeongwoo Jeon, Seolin Shin and Chonghun Han
School of Chemical and Biological Engineering, Seoul National University, Seoul, South Korea

Chun-Yu Di, Xiao-Feng Li, Ping Wang, Zhi-Hong Li and Bin-Bin Fan
College of Chemistry and Chemical Engineering, Taiyuan University of Technology, Taiyuan 030024, Shanxi, China

Tao Dou
CNPC Key Laboratory of Catalysis, College of Chemical Engineering, China University of Petroleum, Beijing 102249, China

Ezzat Rafiee
Faculty of Chemistry, Razi University, Kermanshah 67149, Iran
Institute of Nano Science and Nano Technology, Razi University, Kermanshah 67149, Iran

Sadegh Sahraei and Gholam Reza Moradi
Department of Chemical Engineering, Faculty of Engineering, Razi University, Kermanshah 67149, Iran

Chao-Dong Wu
School of Earth and Space Sciences, Peking University, Beijing 100087, China

Yu-Wen Cai
School of Earth and Space Sciences, Peking University, Beijing 100087, China
Research Institute of Petroleum Exploration and Development (RIPED), Beijing 100083, China
State Key Laboratory of Enhanced Oil Recovery, Beijing 100083, China

Shui-Chang Zhang, Kun He, Jing-Kui Mi, Wen-Long Zhang, Xiao-Mei Wang and Hua-Jian Wang
Research Institute of Petroleum Exploration and Development (RIPED), Beijing 100083, China
State Key Laboratory of Enhanced Oil Recovery, Beijing 100083, China

Yan-Bin Cao
State Key Laboratory of Heavy Oil Processing, China University of Petroleum, Qingdao 266580, Shandong, China
Research Institute of Petroleum Engineering and Technology, Shengli Oilfield Company, Sinopec, Dongying 257000, Shandong, China

Long-Li Zhang and Dao-Hong Xia
State Key Laboratory of Heavy Oil Processing, China University of Petroleum, Qingdao 266580, Shandong, China

Yan Li, Jian Shuai, Ya-Tong Zhao and Kui Xu
Faculty of Mechanical and Transportation Engineering, China University of Petroleum, Beijing 102249, China

Zhong-Li Jin
China ENFI Engineering Corporation, Beijing 100038, China

Mehdi Habibpour and Peter E. Clark
School of Chemical Engineering, Oklahoma State University, Stillwater, OK 74078, USA

Jing Zhao
Key Laboratory of Marine Mineral Resources, Guangzhou Marine Geological Survey, Ministry of Land and Resources, Guangzhou 510760, Guangdong, China

Qian-Yong Liang
Key Laboratory of Marine Mineral Resources, Guangzhou Marine Geological Survey, Ministry of Land and Resources, Guangzhou 510760, Guangdong, China
State Key Laboratory of Organic Geochemistry, Guangzhou Institute of Geochemistry, Chinese Academy of Sciences, Guangzhou 510640, Guangdong, China

Yong-Qiang Xiong and Yun Li
State Key Laboratory of Organic Geochemistry, Guangzhou Institute of Geochemistry, Chinese Academy of Sciences, Guangzhou 510640, Guangdong, China

Chen-Chen Fang
PetroChina Research Institute of Petroleum Exploration and Development, Beijing 100083, China

Hu Jia, Peng-Gang Liu, Wan-Fen Pu and Lu Gan
State Key Laboratory of Oil and Gas Reservoir Geology and Exploitation, Southwest Petroleum University, Chengdu 610500, Sichuan, China

Xian-Ping Ma and Jie Zhang
Oil Production Technology Institute of Dagang Oilfield, CNPC, Tianjin 300280, China

Yong-Le Hu and Yong Li
1 Research Institute of Petroleum Exploration & Development, PetroChina, Beijing 100083, China

Dai-Gang Wang
Research Institute of Petroleum Exploration & Development, PetroChina, Beijing 100083, China
School of Earth and Space Sciences, Peking University, Beijing 100871, China

Jing-Jing Sun
College of Petroleum Engineering, China University of Petroleum, Qingdao, Shandong 266580, China

Xia Zhang and Chun-Ming Lin
State Key Laboratory for Mineral Deposits Research, School of Earth Sciences and Engineering, Nanjing University, Nanjing 210023, Jiangsu, China

Xiao-Fei Fu, Ling-Dong Meng, Hai-Xue Wang and Zong-Bao Liu
Laboratory of CNPC Fault Controlling Reservoir, Northeast Petroleum University, Daqing 163318, Heilongjiang, China
Science and Technology Innovation Team in Heilongjiang Province "Fault Deformation, Sealing and Fluid Migration", Daqing 163318, Heilongjiang, China
The State Key Laboratory Base of Unconventional Oil and Gas Accumulation and Exploitation, Earth Science College, Northeast Petroleum University, Daqing 163318, Heilongjiang, China
Earth Science College, Northeast Petroleum University, Daqing 163318, Heilongjiang, China

Xiao Lan
Laboratory of CNPC Fault Controlling Reservoir, Northeast Petroleum University, Daqing 163318, Heilongjiang, China

Earth Science College, Northeast Petroleum University, Daqing 163318, Heilongjiang, China

Zhi-Qiang Guo and Zai-He Chen
NO. 5 Production Plant, Huabei Oilfield Co., PetroChina, Xinji 052360, Hebei, China

Lei Gu
College of Earth and Environmental Sciences, Lanzhou University, Lanzhou 730000, China

Yi-Xin Zhang
School of Mathematics and Statistics, Lanzhou University, Lanzhou 730000, China

Jian-Zhou Wang
School of Statistics, Dongbei University of Finance and Economics, Dalian 116025, China

Gina Chen
School of Law, Northwestern University, Chicago, IL 60611, USA

Hugh Battye
School of History and Culture, Lanzhou University, Lanzhou 730000, China
Department of Economics, The London School of Economics and Political Science, London WC2A 2AE, UK

Jin-Jiang Wang, Ying-Hao Zheng, Lai-Bin Zhang and Li-Xiang Duan
School of Mechanical and Transportation Engineering, China University of Petroleum, Beijing 102249, China

Rui Zhao
School of Electrical and Electronic Engineering, Nanyang Technological University, Singapore 639798, Singapore

Yulia M. Ganeeva, Tatiana N. Yusupova and Gennady V. Romanov
A.E. Arbuzov Institute of Organic and Physical Chemistry, Kazan Scientific Center, Russian Academy of Sciences, 8 Arbuzov St., Kazan, Russian Federation 420088
Institute of Geology and Petroleum Technologies, Kazan (Volga region) Federal University, 18 Kremlyovskaya St., Kazan, Russian Federation 420008

Achinta Bera and Hadi Belhaj
Petroleum Engineering Department, The Petroleum Institute, P.O. Box 2533, Abu Dhabi, United Arab Emirates

Ajay Mandal and Tarkeswar Kumar
Department of Petroleum Engineering, Indian Institute of Technology (Indian School of Mines), Dhanbad, Jharkhand 826004, India

Da-Wei Dong
Shengli College of China University of Petroleum, Dongying 257061, Shandong, China

Ji-Yan Li
Research Institute of Petroleum Exploration and Development, Shengli Oilfield Company, Dongying 257000, Shandong, China
Working Stations for Post Doctors, Dongying 257000, Shandong, China

Yong-Hong Yang and Jian Liu
Research Institute of Petroleum Exploration and Development, Shengli Oilfield Company, Dongying 257000, Shandong, China

Xiao-Lei Wang
Oil Production Plant of Dongxing, Shengli Oilfield Company, Sinopec, Dongying 257061, Shandong, China

Index

www.ingramcontent.com/pod-product-compliance
Lightning Source LLC
Chambersburg PA
CBHW080245230326
41458CB00097B/3537